THE LOS ALAMOS SYMPOSIUM—1989

HIGH TEMPERATURE SUPERCONDUCTIVITY

PROCEEDINGS

December 6–9, 1989
Los Alamos, New Mexico

THE LOS ALAMOS SYMPOSIUM—1989

HIGH TEMPERATURE SUPERCONDUCTIVITY

P R O C E E D I N G S

Editors

K.S. BEDELL
Los Alamos National Laboratory

D. COFFEY
Los Alamos National Laboratory

D.E. MELTZER
University of Florida, Gainesville

D. PINES
University of Illinois, Urbana

J.R. SCHRIEFFER
University of California, Santa Barbara

ADDISON-WESLEY PUBLISHING COMPANY
The Advanced Book Program
Redwood City, California • Menlo Park, California • Reading, Massachusetts
New York • Don Mills, Ontario • Wokingham, United Kingdom • Amsterdam
Bonn • Sydney • Singapore • Tokyo • Madrid • San Juan

Publisher: *Allan M. Wylde*
Production Manager: *Jan V. Benes*
Promotion Manager: *Laura Likely*
Electronics Production Consultant: *Laurie Petrycki*
Cover Design: *Iva Frank*

Library of Congress Cataloging in Publication Data

High temperature superconductivity: proceedings/edited by Kevin
   Bedell...[et al.].
     p. cm.
   Proceedings of a symposium held at the Los Alamos National
Laboratory in Nov. 1989.
    1. High temperature superconductivity—Congress. 2. High
temperature superconductors—Congresses. I. Bedell, Kevin.
QC611.98.H54H54345   1990   537.6'23—dc20   90-32376
ISBN 0-201-51249-1

This book was typeset by Mary Louise Garcia on a Macintosh II using TEX programming language. Camera-ready copy was produced on Apple LaserWriter II.

Copyright © 1990 by Addison-Wesley Publishing Company, Advanced Book Program, 350 Bridge Parkway, Redwood City, CA 94065

All rights reserved. No part of this publication may be reproduced, stored in a retrieval system, or transmitted in any form or by any means, electronic, mechanical, photocopying, recording, or otherwise, without the prior permission of Addison-Wesley Publishing Company.

ABCDEFGHIJ-MA-943210

FOREWORD

The idea for the Symposium came in discussions among the organizers on ways in which the Advanced Studies Program might be able to improve communication among researchers on high temperature superconductivity. The problem is not that there are too few international conferences, it is that there are so many with a plethora of invited and contributed papers, and too little time for informal discussion. We therefore were led to look back to a different time in physics, when it was possible to gather, in one not-too-large room, a group of leaders in a sub-field under circumstances which encourage both reflection and extended informal discussion.

The Solvay Congresses of the period 1927-1955 provide a paradigm for the sort of conference we hoped might be useful: some 30-40 participants in a meeting organized around a program of a small number of comparatively long (1 1/2 hour) talks by invited rapporteurs, followed in each case, by a long period devoted to informal discussion and a few brief research reports. We believed that such a symposium would be highly beneficial to the participants, but that for it to have impact on the wider scientific community it would be necessary to publish proceedings, including a detailed account of the informal discussion following each report, and to publish these promptly.

Don M. Parkin, the Director of the Center for Materials Science at Los Alamos, then agreed to underwrite the expense of the Symposium, and Addison-Wesley agreed to publish the proceedings of the symposium in near-record time (the final manuscripts reached the publisher on February 2, 1990, and bound volumes were scheduled to be available on March 12, 1990), at a comparatively modest cost.

The Symposium was structured around six half-day sessions: a keynote address plus a summary of the experimental facts; normal state properties; spin fluctuations; the quantum spin liquid; numerical studies; and phenomenology. Each session began with a paper by a rapporteur (or rapporteurs) who reviewed critically the topic in question, and continued with extensive discussion of that paper and brief research reports. As these proceedings indicate, this format worked well, in that there was time enough for participants to explore with each rapporteur a number of open questions, and, where the occasion seemed to warrant it, present a different perspective either at that, or a subsequent session. Because the informal discussions were such a key part of the symposium, we have made every effort to reproduce both their content and their flavor, as a way of enabling those who could not be present to share in the dialogue, debate, and, at times, dissent.

The logistic problems of reproducing the character and content of the informal discussions proved formidable, so we made every effort to enable speakers and discussants to review the transcript of their spoken remarks, with the proviso that they not attempt to take advantage of this opportunity to rewrite history. In this effort, we have received considerable assistance from our fellow Editors, Dr. David Meltzer, who prepared a draft transcript from the tapes of the discussions, Dr. Dermot Coffey, who took primary responsibility for organizing the typed discussion remarks. We would also like to thank Mary Louise Garcia, who has done an outstanding job of meeting all deadlines and preparing for publication the author-submitted manuscripts and discussion remarks. It gives us great pleasure to thank them and our fellow participants, for their remarkable efforts on behalf of the readers of this volume.

Kevin S. Bedell
David Pines
Robert Schrieffer

Los Alamos, NM
February 1, 1990

HTSC SYMPOSIUM

PREFACE

High Temperature Superconductivity, The Los Alamos Symposium: 1989 was sponsored by the Los Alamos National Laboratory, Center for Materials Science and the Advanced Studies Program on High Temperature Superconductivity Theory. The Advanced Studies Program (ASP) is a unique melding of the talents of outstanding academic theoretical physicists with the theoretical and experimental resources of Los Alamos.

The ASP was conceived as a unique partnership between a DOE National Laboratory and a public utility, the Public Service Company of New Mexico. In recognition of the potential impact of high temperature superconductivity on energy consumption, Gerry Geist and David Rusk of PNM approached Los Alamos with a desire to contribute to the advance of this field in a general way. Los Alamos committed resources and manpower in addition to the resources provided by PNM for ASP. In forming the program, both parties committed themselves to a program focused on basic science with a principal objective of advancing the field on a broad front with no specific point of view by bringing a team of leading academic theorists into a joint effort with the theoretical and experimental scientists of a major national laboratory. In April, 1989 Los Alamos assumed full responsibility for the program and designated Kevin S. Bedell as Deputy ASP Director. The Laboratory is fully committed to advancing high temperature superconductivity research using this model.

The ASP fellows lead by senior scientist Robert Schrieffer (UCSB) and joined by David Pines (Univ. of Ill.), Elihu Abrahams (Rutgers), Sebastian Doniach (Stanford) and Maurice Rice (ETH, Zurich) spend considerable time at Los Alamos interacting with the local experimentalists and theorists and supervising two postdoctorals, Arno Kampf and Dermot Coffey. This combined community of academic and laboratory scientists has significant impact on the scientific content of the Los Alamos program in high temperature superconductivity as well as enriching the reseach activities of the academics. This synergism has developed interactions that have brought a level of scientific excitement and progress that would not be possible otherwise. This publication is but one example of how the ASP is advancing the field of high temperature superconductivity on a broad front.

Don M. Parkin
Director, Center for Materials Science

Contents

Foreword   vii

Preface   ix

1   **Keynote Address**
       Chair — J. R. Schrieffer

     The Normal State of High T_c Superconductors: A New Quantum Liquid
       P. W. Anderson   3

2   **Experimental Facts**
       Chair — H. Ott

     Selected Experiments on High T_c Cuprates
       B. Batlogg   37

3   **Normal State Properties**
       Chair — S. Doniach

     Normal State Properties of the Oxide Superconductors: A Review
       P. A. Lee   96

     Discussions and Contributed Talks:

     Large Doping Approach to High T_c
       J. R. Schrieffer   123

Recent Angle Resolved Photoemission Results from
$Bi_2Sr_2CaCu_2O_8$ Single Crystals
    R. S. List   129

4 Spin Fluctuations in the Insulating and Metallic Phases
    Chair — D. Pines   1

(a) Insulating Phase:

Magnetic Properties of La_2CuO_4
    S. Chakravarty   136

Discussions and Contributed Talk:

Anistropic Antiferromagnets from the Hubbard Model
    Z. Tesanovic   193

(b) Metallic Phase:

Spin Dynamics of Superconducting Cuprates
    A. J. Millis   198

Discussions and Contributed Talk:

Antiferromagnetic Fluctuation Dynamics in the YBCO Superconductors
    R. Walstedt   229

Copper and Oxygen Nuclear Magnetic Resonance in $YBa_2Cu_3O_{6.63}$ ($T_c = 62K$)
    M. Takigawa   236

Spin Correlations and Pairing States in Superconducting $YBa_2Cu_3O_7$
    D. Pines   240

5 Quantum Spin Liquids
    Chair — E. Abrahams

Semionics: a theory of high temperature superconductivity
    A. Zee   248

Discussions and Contributed Talks:

Does the "background" in ARPES belong to $G_1(\mathbf{K},\omega)$?
    G. A. Sawatzky   297

Orbital Magnetic Moments in Anyon and Flux Models
    T. M. Rice   302

Comment on μSR Experiments to Search for Magnetic Fields from Flux Phases in High Temperature Superconductors
    R. H. Heffner   307

6   Numerical Studies
    Chair — T. M. Rice

Results from Numerical Simulations of the 2D Hubbard Model
    D. J. Scalapino   314

Discussions and Contributed Talks:

Spectral Functions from Quantum Monte Carlo
    R. N. Silver   373

From Quantum-Chemical to Strong Coupling Models: Support for a Single-Band Model
    E. B. Stechel   377

Comments on the Nature of the Normal Metallic Phase of the CuO Planar Materials
    R. M. Martin   383

Phase Separation of Holes in Antiferromagnets
    V. J. Emery   388

Antiferromagnetic Paramagnons in the Cuprate Oxides Superconductors: A Novel Fermi Liquid
    D. Pines   392

Neutron Scattering Studies of Spin Fluctuations in Metallic $YBa_2Cu_3O_{6+x}$
J. Tranquada — 397

Some Comments on Nuclear Spin Relaxation Below T_c
D. J. Scalapino — 400

7 Phenomenology
Chair — A. P. Malozemoff

The Phenomenology of Flux Motion in High Temperature Superconductors
S. Doniach — 406

Discussions and Contributed Talks:

Dissipation in very Anisotropic Superconductors
P. H. Kes — 436

Chiral spin state and guage field in the extended t-J model
H. Fukuyama — 440

Comment on the Transport Properties of YBaCuO Superconductors
A. P. Malozemoff — 443

8 Summary
Chair — J. L. Smith

Experimental Summary
Z. Fisk — 448

Chair — A. R. Bishop

Summary Remarks on the Theory of High-T_c Superconductivity
E. Abrahams — 464

List of Attendees — 471

1. KEYNOTE ADDRESS

Chair — J. R. Schrieffer

P. W. Anderson and Y. Ren
Joseph Henry Laboratories of Physics
Jadwin Hall, Princeton University Princeton, NJ 08544

The Normal State of High T_c Superconductors: A New Quantum Liquid

A renormalization group analysis of interacting Fermion systems shows that there are two fundamentally different fixed points, Landau Fermi liquid theory and what has been called "Luttinger Liquid Theory" by Haldane, a type of state in which fluctuations of different quantum numbers—*e.g.*, charge and spin—acquire distinct spectra, and Fermi surface correlations have unusual exponents. The Luttinger liquids comprise most interacting 1-dimensional systems, and some higher-dimensional systems in which, as in the Hubbard model, the band spectrum is bounded above. All higher-dimensional systems with true Mott-Hubbard gaps and an upper Hubbard band, including most 2d Hubbard models, are Luttinger liquids. We summarize the experimental evidence in favor of the 2d Hubbard model for high T_c superconductors and the constraints pointing towards its non-Fermi liquid nature. We describe progress towards explaining normal state and superconducting properties and elucidating the high T_c mechanism.

This work was supported by the NSF, Grant # DMR-8518163 and AFOSR # 87-0392.

I. INTRODUCTION

We find that the "normal" metal from which high T_c arises in the cuprate larger compounds is in a new metallic state for which we borrow Haldane's phase "Luttinger Liquid", which is appropriate both because it has a pseudo-Fermi surface satisfying Luttinger's theorem—*i.e.*, it is a "$Z = 0$" generalization of a Landau Fermi liquid—and because it is a higher-dimensional generalization of the soluble one-dimensional interacting Fermi systems described in this way by Haldane. It is also a generalization of the Bethe-ansatz soluble systems.

We will first describe the experimental facts which require us to consider the two-dimensional Hubbard model as basic to the physics, and show that the state of this system is not a Fermi liquid. Next we show that the 2D Hubbard Model, as well as any higher-dimensional system which has an "upper Hubbard band", can not be a normal Fermi liquid, because the wave-function renormalization parameter $Z = 0$. This central proof forces us to examine the only previously known case of $Z = 0$, the 1D "Luttinger Liquid" and in particular, the 1D Hubbard model, its exact solution and the general Haldane formalism. We then show how to generalize this formalism to higher dimensions, retaining the separation of charge and spin and the Luttinger liquid formalism. Finally, we relate this state to the experimental response functions of the normal cuprate systems, in particular, proposing a superconductivity mechanism. A final section lists experimental phenomena, and assesses the state of agreement with the theory, in quantitative or qualitative terms.

II. EXPERIMENTAL CONSTRAINTS ON THE NORMAL STATE OF THE CUPRATES

Two core questions must be decided before a serious approach to the high T_c problem can be mounted: "Is the normal state a Fermi liquid?," and "What is the appropriate model?." Although there is no consensus in the community on either, the evidence on both is quite overwhelming if one examines the situation as a whole and does not narrow one's focus to one individual measurement to the exclusion of others. Any single fact can always be explained away, but the overall picture has only one plausible meaning.

I shall only mention the broad range of computational, structural, and chemical evidence which force us to the one-band Hubbard model, with dominantly repulsive interactions, as an underlying model; much more detailed discussions are in my three sets of lectures on this subject.[1-3] Most conclusive are the cluster calculations of Stechel, Martin and Schluter,[4] which parametrize perfectly to such a model, but the overall structural evidence—such as the structural integrity of the CuO_2 planes, the magnetic nature of all undoped phases, the cleavage of the BSCCO compounds, the wide optical and photoelectric gaps even after doping, and many other indications, leave one with no alternative to the following conclusions: that the strongest bonds are the $d_{x^2-y^2} - Op_\sigma$ hybrids which hold the planes together, and push the corresponding antibonding state up to the neighborhood of the chemical potential. The only other bond which plays an electronic role is the d_z^2 antibonding hybrid, which must be beginning to empty in overdoped $(La - Sr)_2CuO_4$. (And which must also be split off from the $d_{x^2-y^2}$ hybrid which it would touch in the absence of second-neighbor $O - O$ hopping integrals). Of course, by dint of hybridizing into the active bands, many other electronic states play a role, but one must reject the many band theorists' suggestions of "pockets" of carriers in Bi or other orbitals, in that these are totally incompatible with the very high c-axis resistivity, far above the Mott limit, which is shown by all compounds, and with the excellent infrared data which are compatible with only one set of carriers (and confirm zero c-axis metallic conduction).

The chains in YBCO are a special problem. Clearly in the lower O members of the series their oxidation state moves from Cu^+ to Cu^{++}, indicating that at $\sim O_{6.5}$ Cu^+ has just moved above μ. It is implausible therefore, since on the planes Cu^{+++} is now at the chemical potential, that many carriers ever flow on the chains, which, I would propose, are simple Heisenberg chains throughout the superconducting region (with, of course, 1d spinon excitations mimicking the magnetic properties of carriers, as well as, very likely, chain-end magnetic centers confusing the magnetic properties except for $O_{6.9-7}$). Many lines of evidence, not least the NMR and Raman data described at Los Alamos by Millis, make it clear that electronic parameters do not shift with doping, which makes almost inescapable the dominance of repulsive Hubbard interactions in the planes. (These "Hubbard" interactions are, of course, not to be taken literally as "U"'s but result predominantly from $Cu - O$ bonding integrals and Hilbert space restrictions)

Next let us take up the other question. "Is it a Fermi liquid?" Many of the relevant facts were adduced in a recent publication[5] but a more complete summary is in my own lecture notes of June this year. I will list a sequence of types of measurement with "pros" and "cons" on the Fermi liquid question. Let me make it clear that the question is about the normal state; superconductivity I only touch on.

(1) TUNNELING DENSITY OF STATES

Here the relatively few measurements on the normal state are almost the whole of the worthwhile story, since they are immune to the ubiquitous surface problems (themselves a strong piece of indirect evidence that superconductivity is an interlayer phenomenon and not intrinsic to the individual layers). Tunneling into the normal state invariably produces the very reproducible "V" shaped curves, with $\sigma \simeq A + B|V|$. (The few studies which conflict were point contact measurements primarily concerned with superconducting results). Pro: there is a finite zero-bias conductance, indicating a finite angle-averaged density of states at the Fermi level. Con: but the $|V|$ behavior is unique (except for the equally mysterious case of

$BaBiO_3$-based materials.) Really the only reasonable explanation is that of Zou-Anderson, that electrons are composite, not elementary, in the normal layers; and no other reasonable suggestion has been made. **First constraint:** *particles are composite.*

(2) RESISTIVITY AND TRANSPORT

Pro: the resistivity in the ab plane is of metallic magnitude, coming down to $\sim 50\mu\Omega$ cm at $\sim 100°$K. Con: almost all other facts about transport are unusual. Strangest of all is that the non-stoichiometric cases are as good conductors as pure ones, and that no residual resistance occurs except in the cases where doping is in the planes. The potential scattering resistance calculated from a reasonable estimate for most of these substances should be of least 10 times larger and should appear as a residual resistance. I calculated $\hbar/\tau \sim 1/4 ev$ for BSCCO, about equal to the mean free time extrapolated to $2000°$ K. Even stranger is that ρ_c is 300-10^4 times larger (well above the Mott limit) and usually has inverse T-dependence. Localization theory for a Fermi liquid does not permit anisotropic localization. The only viable suggestion is that of Zou and Anderson. **Second constraint:** *the real excitations are confined* to 2d planes.

The sign and behavior of the Hall resistance are anomalous. It is nearly certain that the Fermi surface is a continuous ring about the origin, *i.e.*, electron-like; yet R_H becomes increasingly positive at low T. This is among the key facts which should essentially be explained by any theory. Clearly, however, we have a **third constraint**. The carriers obey lower (or possibly upper) Hubbard band counting, not conventional. What little is known about the thermoelectric power concurs. An interesting fact about thermoelectric power is that, in general, the entropy carried per carrier does not extrapolate to zero at $T = 0$.

Heat conductivity seems to obey a Wiedemann-Franz law with the constant > 1, above T_c; there is an anomalous absence of phonon transport for the ab plane which suddenly reappears at T_c. In the c-direction, confirming ρ_c, there appears to be no electronic heat conduction.

(3) SPECTROSCOPY

In the infrared for the normal state, this mimics conductivity if we allow $\omega \leftrightarrow T$—*e.g.*, σ_{ir} extrapolates to σ_{dc} at $\omega = 0$. This means that the resistivity is a good measure of the inelastic electron scattering, and hence that $\frac{\hbar}{\tau} \simeq E_k$, or that $Im\Sigma(\omega) \simeq \omega$. This, in turn, implies that $Re\Sigma \sim \omega \ln \omega$, $Z \equiv 0$, *i.e.*, that the renormalization constant of any possible Fermi liquid theory is zero and no quasiparticles exist in the usual sense (**4th constraint**). Incidentally, the superconducting state is revealed by the IR to be conventional in the sense that the sum-rule agrees with the penetration depth λ. (Also, IR reveals a reasonably good gap, of order $2\Delta \gtrsim 6kT_c$. It also reveals no sign of an infrared excitation strong enough to cause high T_c.)[6] A strong Raman background, very nearly frequency-independent and obviously a bulk electronic property, is observed. This is a second strong evidence for $Im \sum \sim \omega$, which, as far as we can see, implies composite quasiparticles.

(4) ANGLE-RESOLVED PHOTOEMISSION

Here we have one pro: a sharp Fermi surface is revealed. But again, it is far *too* sharp: its breadth Δk seems to be less than .1 of the Brillouin zone, yet the trivial calculations on inelastic scattering carried out above show that $l \sim 1/k_F$ so that $\Delta k \sim k_F$. *The entities which form the Fermi surface cannot be scattered by charged impurities!* Hence, as above, $Z \equiv 0$ and $F_0^s \to \infty$: the Fermi surface must be formed by particles with spin only. Important **constraint 5:** *There is a Luttinger-Theorem Fermi surface,* formed by spin-only entities.

When we look at the detailed spectra (see e.g., Fig. 1), we see a remarkably sharp cusp dispersing away from the Fermi surface at a reasonable clip, a broad background extending to high energies both above the cusp and even above the Fermi surface, and a linear decrease to a sharp second cusp at the Fermi energy. We will see that this spectrum is easily explained by a composite picture, while I do not feel that the conventional interpretation attempted by the experimentalists can be made consistent with all the facts. Photoemission in the superconducting state roughly confirms the large IR gap.

Angle-averaged photoemission gives us no new information on density of states which is not available from tunneling into the normal state. It confirms sensible estimates of the hybridization of the carrier bands, and that the carriers are in a smaller band above the Op non-bonding mass of bands.

(5) NMR

NMR relaxation and Knight shifts in the normal state reveal the **sixth constraint:** *there are two different suceptibilities.* A conventional spin susceptibility seems to control the relaxations on O^{17} in the plane and on Y, while a second, very unconventional and large, relaxation process occurs *in addition* (not alone) on the Cu II sites in the planes. This term has no large Knight Shift or susceptibility associated: we guess it is an orbital effect, since χ is remarkably constant in the best superconducting materials, yet this term must correspond to something which fluctuates very slowly. Antiferromagnetic spin fluctuations seem utterly implausible to me: there is no independent evidence for them and they could not behave this way.

The 60° YBCO is a special case which has so many anomalies (*e.g.,* IR "knees", residual resistance, strange knight shift, etc.) that it clearly is not generic.

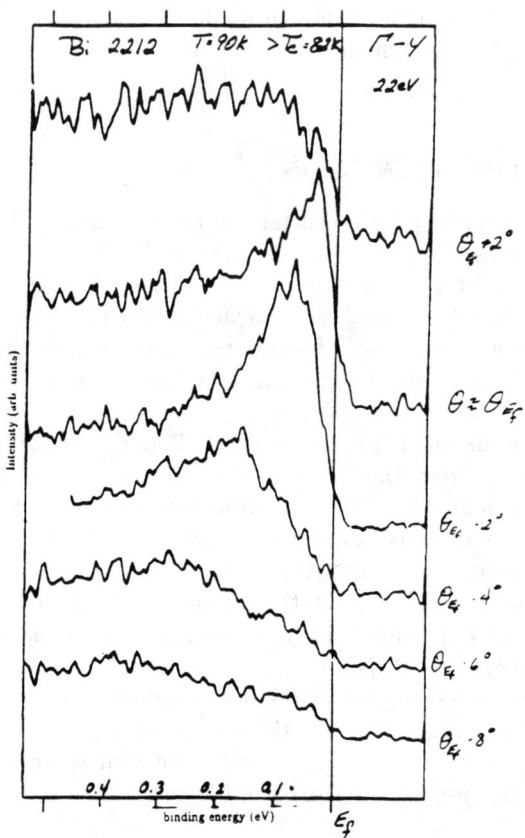

FIGURE 1a Angle-resolved photoemission data for BSCCO at $T > T_c$.

(6) HEURISTICS OF T_C

Here we make our one foray into superconducting properties with a **seventh constraint**: superconductivity occurs as a result of interlayer contact. That is, the isolated layer has no mechanism to cause superconductivity. Some hints of evidence: One-layer BSCCO material is metallic but has a T_c of 6° or (for many crystals) 0°. BSCCO T_c's vs. layer number can be fit well empirically by the empirical layer scheme.

$$\Delta_i(T) = \chi^i_{pair} \sum_j \frac{t^2_{ij}}{J} \Delta_j(T) \qquad (2-1)$$

where $\chi_{pair} \propto \frac{1}{T}$ and i,j are layer numbers. That is, χ_{pair} is only moderately large but rapidly T-dependent relative to BCS; and it does not diverge at $T > 0$. (2-1) also fits all known pressure-dependences if we add the premise that close layers with strong coupling are not much affected by pressure, only distant couplings. Tl systems fit as well, but we must assume that the Tl layer has a pretty big $T_\perp$.

Finally, epitaxial experiments and the simple observation that $\Delta \to 0$ at surfaces are *incompatible* with an intralayer mechanism.

These seven constraints alone practically direct us unequivocally to the theory we shall now discuss.

III. "LUTTINGER LIQUID" BEHAVIOR OF THE NORMAL METALLIC STATE OF THE 2D HUBBARD MODEL

Haldane[7] has characterized the behavior of a large variety of one-dimensional quantum fluids by the term "Luttinger liquid", showing that they can all be solved by common techniques based on transforming to phase and phase-shift variables for the Fermi surface excitations (a procedure often called "bosonization" even though some of the Luttinger liquids start out as bose systems). These systems are characterized by fractionation of quantum numbers — *e.g.*, in the Heisenberg spin chain the excitations are spin 1/2 Fermion-like, while in the Hubbard model they are spin 1/2 chargless spinons and $\pm e$ spinless holons with Fermion-like properties — and, often, a Fermi surface with nonclassical exponents, and unusual exponents for correlation functions.

I will here restrict the term, for my purposes, to systems based on Fermions — preferably ordinary electrons — and argue that the "Luttinger liquid" is a fixed point, or a manifold of fixed points, of the same renormalization group which, "usually," leads to the Landau-Fermi liquid as a unique fixed point. (The interaction parameters of Landau Fermi liquid theory are well-known[8] to be marginal operators around a single fixed point, the effectively free Fermi liquid.)

Some years before, Luther[9] showed that the bosonization techniques used to solve these one-dimensional models are equally applicable to d-dimensional Fermi gases, and they describe certain facts slightly more accurately than Fermi liquid theory — the existence of $2k_F$ singularities in correlation functions, for instance, for the free-particle systems. But Luther did not consider the possibility that the interacting d-dimensional problem could lead to new physics.

The first new point I want to make is that two of the reasons usually given for the unique nature of one-dimensional Fermi systems are untenable. The first is that in $1d$ one has only forward scattering, or backward scattering where the momentum of one particle is maintained, if not its spin. This is indeed the correct reason for viability of the Bethe Ansatz. But after renormalization the Landau theory has only forward or exchange scattering, and the renormalized particles indeed obey a

Bethe ansatz of the simplest form. This is the essence of Luther's argument, that the excitations can be bosonized in each direction around the Fermi surface.

Second, it is argued that particles cannot be interchanged in $1d$ without encountering phase-changing interactions, hence statistics are meaningless in $1d$: but none of Haldane's arguments seem to fail in the slightest if we introduce weak longer-range hopping integrals in any of the examples, and such hopping integrals can allow a Berry process. No one argues, in fact, that real electrons living in $3d$ space in the presence of a chain of ions, which know perfectly well that they are Fermions, will not obey the models and show the fractionization effects.

The unique effect in $1d$ is one which is also present in a class of higher-d models, specifically $2d$ repulsive Hubbard models, and in some strong coupling higher d cases. This is the presence of an *unrenormalizable Fermi surface phase shift*. Such a phase shift signals that the addition of a particle changes the Hilbert space for the entire system of particles — it requires a net motion of field amplitude through the distant boundary of the system, or a net change of wavelengths. The effects of such phase-shifts were explored thoroughly in connection with the "x-ray edge problem"[10] and are summarized in the "infrared catastrophe" theorem.[11]

$$\langle VAC(V) \mid VAC(0) \rangle \propto e^{-\frac{1}{2}(\frac{\delta}{\pi})^2 \ln N} \qquad (3-1)$$

where $|VAC(V)\rangle$ is the non-interacting Fermi sea in the presence of a potential V which causes a phase shift δ. The singularity is the result of the shifting of the entire spectrum of k values (in the presence of fixed boundary conditions) or of the displacement of wave-function nodes (for scattering boundary conditions), and is independent of the finite contribution which may ensue from local modifications of the wave-functions.

In the conventional higher d, free electron gas cases to which Landau liquid theory applies, it is implicitly assumed—and indeed, self-consistently so—that the phase-shifts caused by adding or removing a single particle can be made to vanish in favor of a renormalization of all the quasiparticle mean field energies. It is assumed that there is an effective mean field energy whose eigenstates are the precise k-states of the appropriate free particle system.[12] The formal result of this process is that the wave-function renormalization constant Z is finite: that is, the overlap integral

$$\sqrt{Z} = \langle c_{k\sigma}^+ \Psi_G(N) \mid \Psi_{k\sigma}(N+1) \rangle > 0 \qquad (3-2)$$

where Ψ_k is the exact wave function of the $N+1$ particle state with one quasiparticle added, and in particular

$$\langle c_{k_F\sigma}^+ \Psi_G(N) \mid \Psi_0(N+1) \rangle > 0 \qquad (3-3)$$

(where in a Fermi system, the $N+1$ particle system necessarily has one particle added near the Fermi surface, hence its ground state is quasi degenerate). (3) cannot be true if there is a phase-shift due to the addition of $c_{k\sigma}^+$.

In one dimension, for interacting particles, such a phase shift is unavoidable, since the effective range of interactions is necessarily (for real interactions) of the order of the wavelength ($\delta = k_F a$). Thus in all the realistic one-dimensional systems,

$Z = 0$, the Fermi liquid fixed point is excluded, and the phase-shifts due to interactions must be taken into account as relevant variables—in fact, in many cases renormalization invariants. $Z \equiv 0$ implies that the Fermi sea excitations—which may still exist—do not carry charge (but may carry spin and be spinons). I will summarize Haldane's analysis of the spectrum of the $1d$ Hubbard model shortly.

In $2d$, the scattering length for free particles and repulsive interactions diverges only as $\frac{1}{k \ln k}$ as $k \to 0$, and for higher dimensions it is $\sim \frac{1}{n} \sim k_F^{-\frac{1}{d}}$; in both cases no serious problems need ensue for shorter wavelengths. But there is one type of problem where finite phase-shifts are inevitable, namely systems with a single-particle spectrum bounded above and below in energy. In this case the introduction of an extra particle may cause a bound state to split off from the top of the spectrum (an "anti-bound state").[13] By Levinson's theorem,[14] the presence of a bound state, either above or below the band, is signalled by a difference π in phase shift in the appropriate channel between the top and bottom of the band. This corresponds to the fact that one state must be removed from the band to make up the bound state, and to Friedel's identity[15]

$$\Delta n(k) = \sum_\ell (2\ell + 1)\, \delta_\ell(k) \qquad (3-4)$$

for the change in number of states to be found below a wave vector k due to a phase shift $\delta(k)$. Continuity—or the fact that any bound state must be a superposition of all states in the band—tells us that some δ must remain finite at all energies in the band.

For any dimension, a repulsive interaction U sufficiently strong to split off an upper Hubbard band adds one state to that band for each added electron; the upper Hubbard band is the manifold of antibound states and where it is present we must have $Z \equiv 0$ in the occupied lower Hubbard band, since the Hilbert space changes when we add a carrier. In the $2d$ Hubbard model (with one band—the generalized Hubbard models recently introduced are an irrelevancy) any potential whatever will split off bound states of holes (for which the interaction potential is attractive) above the band, because of the well-known fact that in two dimensions all potentials bind. Thus, although for very low or very high occupancies the relevant singularities come in with small coefficients which are non-analytic in the interaction ($\sim e^{-t/u}$) or density ($e^{-n^2 \ln n}$), $Z = 0$ in all cases. These terms will not be picked up in series expansions.

We can identify the relevant interactions by thinking of the upper Hubbard band as a kind of "ghost" condensate in a channel of $2k$ total momentum and zero total spin reflecting the fact that each particle of down spin prevents some state of the same momentum and up spin from being occupied, so that the "condensate" represents both states being occupied. The interactions which maintain the non-occupancy are attractive in the particle-hole channel, and are of the form, in k-space

$$V_{2k_F + q}\, \rho_{2k_F + q \uparrow}\, \rho_{-2k_F - q \downarrow} \qquad (3-5)$$

which contains terms like $c^+_{k\uparrow} c_{k\downarrow} c^+_{-k+q\downarrow} c_{-k+q\uparrow}$ attractive to electron-hole pairs of approximately zero momentum, i.e., spin fluctuations.

The best representation is to pick an arbitrary origin and use angular momentum eigenfunctions, $\varphi_{k,m,\sigma}(r)$. Consider scatterings of electrons in k, m and k', m'. $m + m'$ is the total angular momentum and is a function of choice of origin; the phase-shifts at a given energy are independent of it. $m - m'$ is the relative angular momentum and the strong Hubbard interaction acts primarily in the $m = m'$ channel, $m - m' = 0$. The antibound state is the top state in this channel and the relevant phase shift occurs between ingoing and outgoing waves in this channel at the Fermi energy. (Corrections due to the lattice are minor and easy to add.)

It is worth discussing the resulting state in terms of a "renormalized Bethe ansatz" picture. If we, following Gallivotti,[7] use a "poor man's renormalization group" procedure to eliminate k states far from the Fermi surface, we will end up with a shell of low-energy excitations with momenta near the Fermi surface. Even if $Z = 0$ and even in the presence of our new interactions, for the thin shell of states near k_F every real scattering is non-diffractive in that the two k-vectors never change: charge is always scattered forward. When $Z = 0$, however, the original k-states of the Fermi liquid are not adequate to contain all the particles, and the Bethe ansatz wave-function contains a continuous spectrum of k's through the Fermi surface, exactly as in one dimension, where the Hubbard model solution for large U may be written $\sum_Q (-1)^Q \det ||e^{k_i \, x_{Q_j}}|| \times$ spin function, and the spectrum of k_i's extends continuously to $|Q|$, $Q > k_F$. We presume that the same form is valid for renormalized particles in 2d, near the Fermi surface.

The interaction terms (5), just as in one dimension, can be rewritten as a coupling between ingoing and outgoing spin fluctuations

$$V \sum_{\substack{k \sim k_F \\ k' \sim k_F \\ q}} c^+_{k\uparrow} c_{k+q\downarrow} c^+_{-k'\downarrow} c_{-(k'+q)\uparrow} \qquad (3-6)$$

and under bozonization this turns into a term proportional to $\nabla \theta_\uparrow \nabla \theta_\downarrow$ in the Hamiltonian for the phase-shift variables θ_σ, defined by $\nabla \theta_\sigma = 2\pi \rho_\sigma$. This term in the effective Boson Hamiltonian must be transformed away by a Bogoliubov transformation. But as in one dimension, when transformed back into Fermion variables the new dynamical variables, even though their equations of motion have linear dispersion relations, do not correspond to simple Fermion or Boson excitations, and have Green's functions with non-classical exponents. Charge and spin separate, the low-energy spin excitations being Fermions at the original k_F, the charge excitations centering around the spanning vectors $2k_F$.

The actual correlation functions and Green's functions in the two-dimensional case have not yet been calculated: fortunately, many experimental data can be calculated by using the photoemission data[16] to describe a semiempirical fit, and by using the one-dimensional Hubbard model as an appropriate guide to understanding.[17] The actual calculation of physical properties will be described separately.

At present, all experimental observations seem compatible with this point of view, and many puzzling ones receive almost unique explanations.

IV. SOME NEW AND OLD RESULTS ON THE 1-DIMENSIONAL HUBBARD MODEL

All of our formal structure will depend on understanding the 1-dimensional repulsive Hubbard model, since this is the model for our "Luttinger Liquid" state and since we believe the fixed point for the 2-dimensional problem maps onto this model in each separate angular momentum channel.

The repulsive Hubbard Model in one space dimension was solved, in principle, many years ago by Lieb and Wu, who calculated the ground state and the spectrum of elementary excitations using Bethe Ansatz methods. Ogata and Shiba have written the Lieb-Wu solution for large, but not infinite, U in a suggestive form:

$$\Psi_{LW} = \sum_Q (-1)^Q \; \text{Det} \; ||e^{ik_i x_{Qj}}|| \times \Psi(y_1 \ldots y_\alpha \ldots y_{n/2}) \qquad (4-1)$$

Here Q is a permutation of the N occupied sites $x_1 \ldots x_N$ in a chain of N_a atoms, and the y's are a labeling of the positions $x_1 \ldots x_{N/2}$ of the up spins (we assume $N/2$ up spins, $N/2$ down spins) by simple ordering. Ψ is a solution of the "Squeezed" Heisenberg model in which all hole sites are omitted, which depends on $N/2$ momentum-like variables Λ_α, which are labelled by a set of integers or half-odd integers J_α, $\alpha = 1 \ldots N/2$. The k_i's in turn, are related to a set of integers or half-odd integers I_j. As $U \to \infty$ the k's become evenly spaced. The energy is $\sum \cos k_i$ and the momentum is $\sum k_j = \frac{2\pi}{N_a} \sum (I_j + J_\alpha)$. The excitation spectrum for, say, holes is obtained by dropping out one I and one J, and the energy is the sum of non-interacting spinon (J) and holon (I) contributions. Both spectra are linear for low energies:

$$E_h(k) = v_h \, |k_h - 2k_F| \quad E_s(k) = v_s \, |k_s + k_F| \qquad (4-2)$$

Here $v_s \sim 2J = 2t^2/U$ and $v_h \simeq 2t \gg v_s$. The rules for shifting the I and J spectra from integer to half-integer assure that the holon and spinon have opposite velocities near k_F (There is a second zero of energy at $3k_F$, where they are parallel). Electron-like spectra are similar with "antiholon" and "antispinon" behaving as such. The spectrum is shown in Fig. 2, and a schematic of the "thermal" density of states —the number of actual one-electron-like or one-hole-like eigenstates—is shown in Fig. 3. Clearly, an angle-averaged thermal density of states would rise linearly with ω, as we postulated as an explanation for the linear tunneling DOS (which, oddly, is probably not caused by this—see Section 5.) One may assure oneself by analytical arguments that the corrresponding $n(k)$ would have only a simple linear cusp at $k = k_F$.

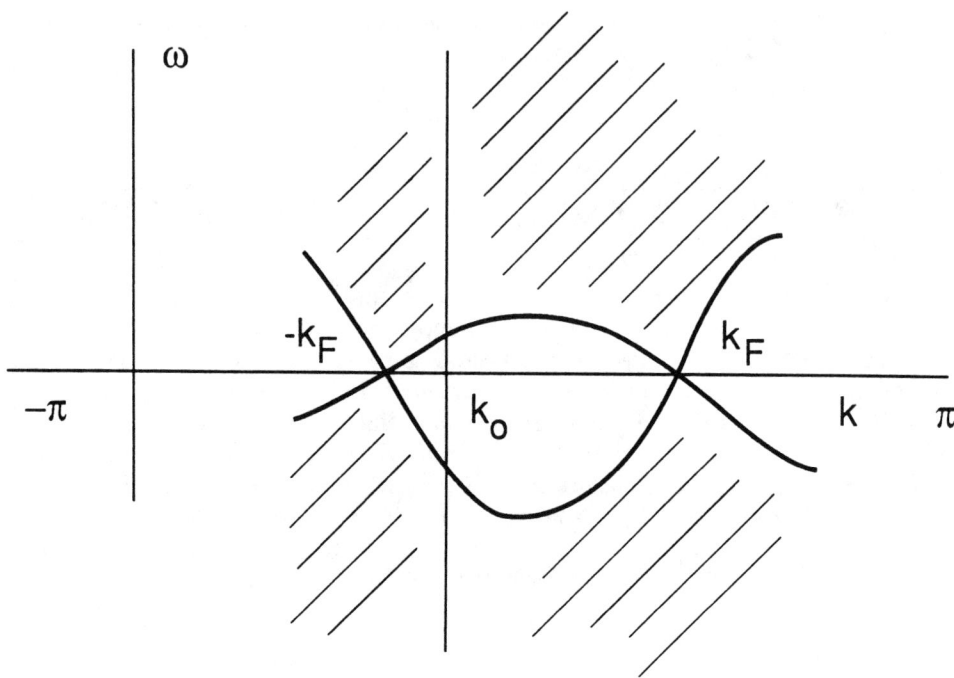

FIGURE 2 E vs. k for holons and spinons in one dimension.

The actual one-particle density of states which would be measured by PES or tunneling cannot be calculated exactly from Lieb and Wu's solutions, and has been the subject of a number of numerical calculations and formal approaches, the latter using one version or another of the "bosonization" methods used to solve the Tomonaga-Luttinger models. Calculations by Ogata and Shiba, and by Parinello *et al.* assure us that the *actual* n_k behaves like

$$n_k \simeq 1/2 - c|k - k_F|^\alpha \operatorname{sgn}|k - k_F|$$

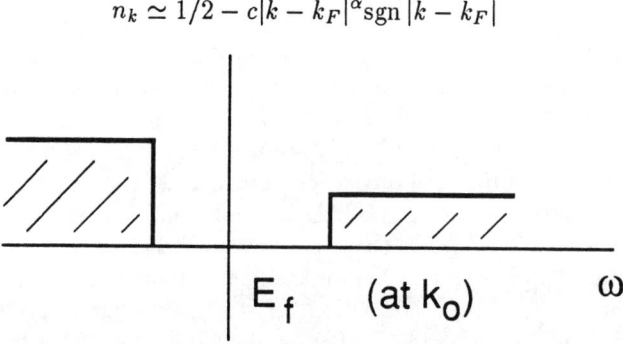

FIGURE 3 Thermal density of states for a given k in the 1d Hubbard Model.

where $\alpha \simeq .125$ for $U = \infty$, less for finite U. This requires, essentially, that there be a $k^{-1+\alpha}$ singularity in $Im\, G_1$ at k_F, since

$$n_k = \frac{1}{\pi}\int_{-\infty}^{0} Im\, G_1(k,\omega)d\omega$$

where

$$\frac{1}{\pi} Im\, G_1(k,\omega) = \rho(k,\omega) = \sum_{E_n} |\langle n|c_{k\sigma}|0\rangle|^2 \delta(\omega - E_n)$$

must vanish for $\omega < |E_k^h|$ or $|E_k^s|$ on the respective sides of k_F. Parinello et al. and Haldane claim from "Luttinger liquid" theory that $\alpha = 1/8$ for all n and $U = \infty$. We have calculated this exponent, on the basis of the wave-function (4-1) and the ΔI shift rules, also, to be $1/8$: see below. In any case, it is clear that the same exponent holds for the tunneling density of states

$$\int dk\, \rho(k,\omega) \propto |\omega|^\alpha \quad \alpha \sim 1/8 \qquad (4-4)$$

This tunneling DOS has been observed by Dynes and Gurvitch for oxide junctions with YBCO; we interpret these observations as coupling to normal planes. The "Luttinger liquid" (Haldane) or "Tomanaga-Luttinger scaling" theory (Parinello et al.) can only be sketched here. One starts by introducing two conjugate variables "ρ" and "φ" to describe a boson field

$$\Psi_{b\sigma}^+(x) = \sqrt{\rho_\sigma(x)}e^{i\varphi_\sigma(x)}$$

and a Fermion $\psi_\sigma^+(x)$ may be created from this by a Jordan-Wigner transformation. ρ_σ, in turn, may be written as the gradient of a second phase variable, $\theta_\sigma(x)$, $\rho_\sigma = 2\pi\nabla\theta_\sigma$, and θ_σ is now to be thought of as a "phase-shift" variable—i.e., we work in terms of a kind of dynamical Friedel theorem, where the number of particles to, say, the right of a given point is given by the local Fermi surface phase shift. φ and ρ are conjugate dynamical variables, so $[\varphi(x), \theta(x')] = i\theta(x - x')$. We focus on excitations which can be described in terms of density waves of constant velocity, in which the local phase and phase-shift are the position and momentum analogs. This representation is well-suited to the strong, unrenormalizable interactions in which we are interested, which can be expressed in terms of finite Fermi surface phase shifts—in fact, one can see that the Hubbard interaction behaves like $U\nabla\theta_\uparrow \nabla\theta_\downarrow$.

Details of a transcription of the large-U Hubbard model into this language, using Lieb and Wu to aid in the determination of the relevant parameters, are given in a later paper. The singular behavior is, it seems to us, unequivocally determined, but, of course, we cannot get any exact idea of multiplicative constants or behavior far from the pseudo-Fermi-surface.

The outcome is that the Green's function near k_F has the form

$$\frac{1}{\pi} Im\, G_1(k,\omega) = \frac{(\omega - qv_s)^\alpha}{(\omega + qv_n)} \qquad (4-5)$$

where $q = k - k_F$. This has singularities at the two cuts in the spectrum at $-kv_n$ and $+kv_s$ and nowhere else, and must be unique except for the apportionment of the powers $\propto$ and 1 to the two cuts.

(4-5) may be normalized by recognizing that $n = 1/2$ at k_F and introducing an upper cutoff, Λ, which must be of order several $\times J$. Then

$$\frac{1}{2} = N \int_0^\Lambda \frac{d\omega}{\omega^{1-\alpha}} = N \frac{\Lambda^\alpha}{\alpha}$$

$$N = \frac{\alpha}{2\Lambda^\alpha}$$

so that

$$\rho(k,\omega) = \frac{\alpha(\omega - qv_s)^\alpha \Delta^{-\alpha}}{2(\omega + qv_h)} \tag{4-6}$$

We will use this result repeatedly in what follows.

It is perhaps instructive to derive the exponent $\alpha = 1/8$ in the $U = \infty$ case directly from the Lieb-Wu wave function. First let us try to calculate the overlap between a state from which we have removed an electron from site i with spin σ, and the ground state with $N - 1$ particles. There are two pieces to this overlap. First, we have removed a spin from the spinon wave function $\psi(y, \ldots y_{N/2})$, which is very close, as shown by Shastry, to a Gutzwiller-projected free electron gas. Removal of an electron and Gutzwiller projection seem likely to commute so that the new state has an overlap proportional to $N^{-1/2}$ with the ground state, simply the overlap of φ_i with a φ_k. This is a simple way of seeing that the spin function renormalizes into a simple but uncharged free Fermion liquid with a conventional Fermi level.

The two ground states, however, differ not only in the removal of one plane wave, but in the fact that because of the Lieb-Wu shift rules every I_j has shifted to $I_j + 1/2$. The reason for this is that even if the extra plane wave is removed at k_F, all of the plane waves have to shift by a phase of $\pi/2$ to return to a uniform distribution—as we see in the Fig. 4 the new distribution interleaves the old one. But this encounters our old friend the orthogonality catastrophe, which leads to an additional overlap reduction by

$$\langle 0N | c_k^+ | 0N - 1 \rangle = e^{-(1/2)^2 \cdot 1/2 \ln N} = N^{-1/8} . \tag{4-7}$$

Now a thought-experiment familiar in the x-ray edge problem[18] is carried out: we realize that N and $v_F t$ are the same variable in effect if all waves move at velocity v_F, thus we find that $\langle 0|\psi^+ 0,t)\psi(0)|0\rangle \propto t^{-9/8}$ (a second $N^{-1/2}$ comes from the second $\psi^+(0)$. The velocity of the 1/8 power part is the *charge* velocity. Recognizing that t must enter either as $x - v_n t$ or $x + v_s t$, we recover (4-5).

Responses to charge perturbations or, for instance, to a pair potential must be dealt with carefully—perhaps the best way being to use variational derivatives of G_1, i.e., Ward identities, to get the correct responses; or one may calculate them directly from Luttinger Liquid Theory. Such calculations are obviously called for.

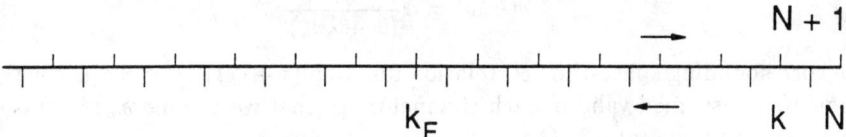

FIGURE 4 Interleaving of k values between N and $N+1$ ground states.

V. PHYSICS OF THE CUPRATES IN THE LUTTINGER LIQUID MODEL

Attempts to calculate experimental results using the above outline scheme can be based on two ideas which, one assumes, are in the end equivalent. The first and preferable one is to start from first principles using the partial wave free particle representation k,m in which the Green's function is diagonal in the low-energy limit—*i.e.*, the singular terms in the Hamiltonian are correctly handled. In this representation the correlation functions are effectively one-dimensional, in that only ingoing and outgoing waves are mixed. Many quantities—*e.g.*, pairing, NMR relaxation, tunneling, angle-averaged photoemission densities of states, and Raman background intensity may be calculated directly in this representation since they are summed over all channels. All of these are essentially the same as the one-dimensional Hubbard model would give. Transport properties such as resistivity and Hall effect, however, must be calculated in linear momentum representation and involve an extra step of partial wave resolution which we have not yet carried out.

A second technique, much less rigorous, for these transport properties is to fit to the experimental angle-resolved photoemission results and use this to estimate the Green's functions in k representation. This was the method of our original papers.[1,2] Of course the angle-resolved photoemission is then an input as far as its detailed shape is concerned, but its singularities follow from the kinematic D.O.S. which we *can* estimate; and, in principle, resistivity, etc., can be calculated this way. In so far as quantities calculated by the two schemes agree—and they do—this verifies the Luttinger Liquid Hypothesis for the angle-resolved D.O.S., which is, of course, the most fundamental measurement of the electronic state.

First let us calculate the first category of effects. In the preceding section we have conjectured the one-dimensional one-particle D.O.S.

$$\rho(k,\omega) = \alpha \frac{(\omega \frac{-kv_s}{\Lambda})^\alpha}{(\omega + kv_h)} \qquad (4-6)$$

The corresponding space-time correlation function $(x+v_h t)^{-1-\alpha} \times \theta(x - v_s t)$. This is now to be assumed valid in each channel m, so that we assume $\rho_m(k,\omega)$ is of the same form, and the total D.O.S. is

$$\rho(\omega) = \sum_m \sum_k \rho_m(k,\omega).$$

Let us start with NMR relaxation. It is important to recognize that one of our predictions is that there will be two entirely separate processes, as is actually observed. The first is mediated by the conventional nuclear spin-electron spin coupling I.S., and involves the scattering of a spinon by the nuclear spin:

$$\hbar/T_1 \propto \sum_{kk'} |A_{kk'}|^2 f_k(1-f_{k'}) \times \delta(\epsilon_k - \epsilon_{k'}) \propto K^2 T \qquad (5-1)$$

This is a pure Korringa process and should be indistinguishable from the corresponding process due to ordinary electrons. The removal of a spinon from k, m leaves behind the opposite-spin partner and does not affect the charge k_i's.

This process, when drawn in diagram form, is rather strange: it represents the scattering and reabsorption of not an electron, but a spinon acting independently of its accompanying holon. We may, apparently, for incoherent processes like this, divide the Green's function into spinon and holon parts (of course, there can be no net flow of the two because of gauge couplings!) and either can be created so long as it is reabsorbed in the same channel—that is to say, there are preexisting spinon or holon-like fluctuations which violate conventional quantum number conservation notions.

There is also the other possibility, direct coupling of orbital current to nuclear spin. In this case we couple into the orbital current and we can think of it as scattering a holon from k, m to k', m' and back, leaving the spin in a singlet state. The diagram looks like the conventional diagram,

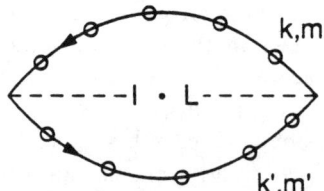

but we have to recognize that as for the spinon the scattering involves a change of m as well as k, so that there can be no coherence effects between the two channels. The holon fluctuation is sharply localized and we get an enhanced relaxation.

Our best bet for the holon Green's function is

$$\operatorname{Im} G_h = \frac{(\omega - kv_s)^\alpha}{\Lambda^\alpha \epsilon_k^h}$$

The interaction is a direct $I \cdot (L)$ with $L_{kkl}^{mm\prime}$ being an orbital moment matrix element (vanishing for $m = m\prime$). The Golden Rule gives

$$\frac{\hbar}{\tau} \propto \int d\omega \sum_{\substack{kk\prime \\ mm\prime}} |L_{kk\prime}^{mm\prime}|^2 \, Im\, G_h(k,\omega) \times$$

$$Im\, G_h(k\prime,\omega) f(\omega)(1 - f(\omega))$$

$$= N(0)^2 \int d\omega \int d\epsilon_k \int d\epsilon_{k\prime} \frac{(\omega - kv_s)^\alpha (\omega - k\prime v_s)^\alpha}{\epsilon_k \epsilon_k} f(1-f) ,$$

$$\propto T \ln^2 \frac{J}{T} \tag{5-2}$$

neglecting the k-dependence of the wavevector. Fig. 5 gives a typical fit of the sum of (5-1) and (5-2) to NMR data in the normal state.

Next we consider the pairing problem. Let us strictly confine ourselves to the interlayer tunneling mechanism of Hsu, Wheatley and Anderson, on the basis of the "experimental constraints" which seem to lead this way. The philosophy is as follows: the interlayer tunneling matrix elements $t_{kk\prime}^{ab}$ between layers a and b transport full electrons (both holon and spinon), and do not just scatter charge or spin without actual particle transport. We hypothecate a pairing self-energy $\Delta^a(k,\omega)$ in layer a, and ask whether the proximity effect induces a corresponding self-energy in layer b next door. $\Delta_k^a \tilde{\tau}$ is a self-energy insert between a k_σ electron line and a $-k_\sigma$ hole line. The corresponding anomalous part of the Green's function in layer a may be written to first order:

$$\tilde{F}_{k,-k} = \frac{\delta \tilde{G}_k^a}{\delta \Delta^a \tau} \Delta_k^a \tau \tag{5-3}$$

where $\tilde{G}^a$ is the matrix Green's function in layer a. Then the corresponding self-energy insert in layer b comes from the process of tunneling into layer a and back

$$\Delta_b \tilde{\tau}(k\prime,\omega) = \sum_{k\prime} |t_{kk\prime}^{ab}|^2 \frac{\delta \tilde{G}^a(k\prime)}{\delta \Delta^a} \Delta_a \tilde{\tau}$$

and, cancelling, T_c is determined by

$$\sum_{k\prime} |t_{kk\prime}^{ab}|^2 \frac{\delta \tilde{G}^a}{\delta \Delta^a} = 1$$

Using (4-6), we find

$$1 = t^2 N(0) \int d\epsilon_k \frac{1}{(\omega^2 - \epsilon_k^2 \pm i\delta)^{-\alpha}} \frac{\alpha}{J^\alpha}$$

$$1 = \frac{|t|^2 \, N(0)}{2T_c} \alpha \left(\frac{T_c}{J}\right)^\alpha \tag{5-4}$$

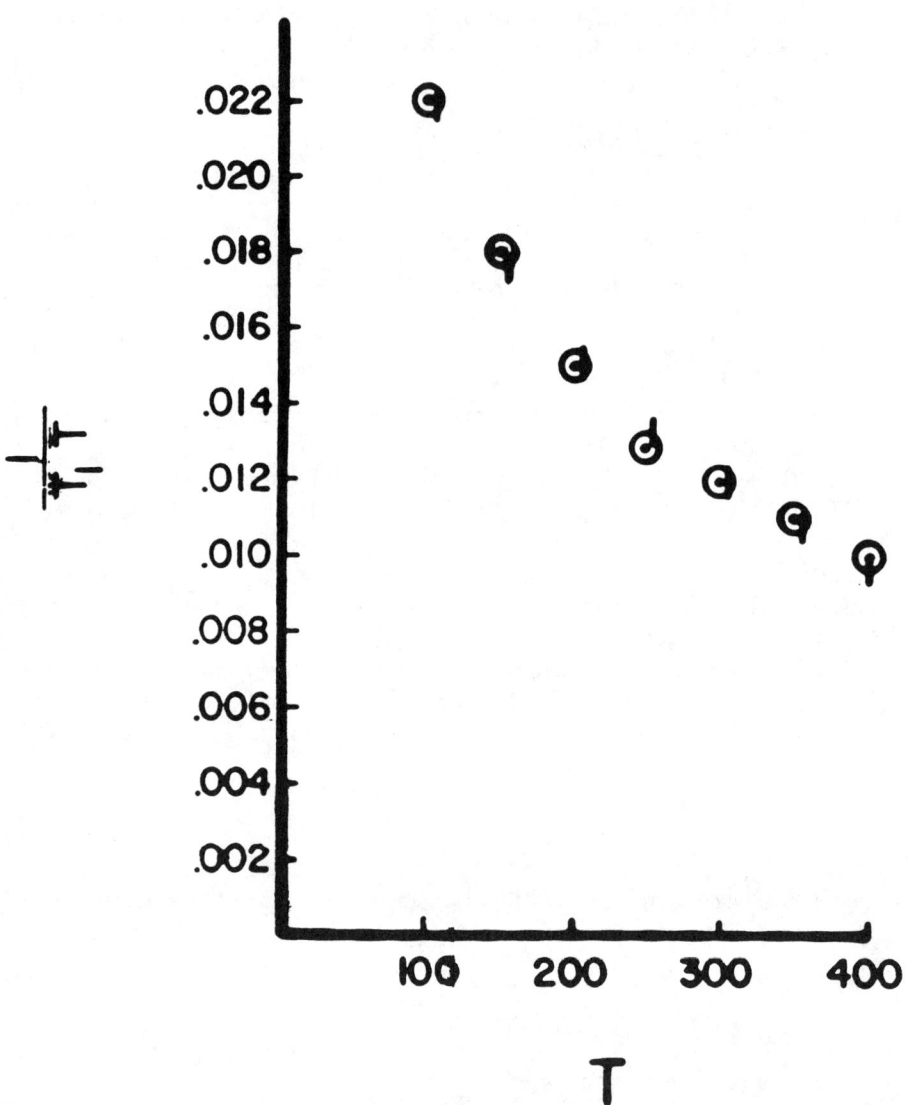

FIGURE 5 Fit to NMR relaxation data on YBCO.

Thus T_c is roughly $(\frac{m_\|}{m_\perp}) \cdot t_\perp \cdot \alpha$ If so, we are already close to the extreme limit of T_c. It is not at all clear that all normalization factors are correct in (5-4), but this seems surely the smallest possible value.

One clear implication is that the same mechanism can be operative in one-and two-dimensional organic chains, and should be tested accordingly. Is it significant that pressure brings out the superconducting state in that case?

The $\frac{1}{T}$ dependence of the effective χ_{pair} means that the steepness of the rise of $\Delta^2(T)$ and the decrease of C_{sp} below T_c will both be $\sim 3\times$ greater than in simple BCS. This prediction is in good agreement with experiment. The superconducting state below T_c is not available from this method.

Since there is a great deal of confusion about the meaning of the HWA superconductivity mechanism, we should say a word about dimensionality. Of course, two isolated layers, as occur in YBCO or BSCCO 2-layer materials, have a Kosterlitz-Thouless superconducting transition, *as would a single layer if it had an appropriate attractive interaction*. What we postulate is that the coupling between 2 *single* layers provides that attraction: above T_c, the two exist as two independent, incoherent 2-dimensional Luttinger liquids, while below T_c they are a single coherent system with no low frequency Luttinger liquid behavior. Whether charge-spin separation persists is a separate issue. The issue is not primarily dimensional but of relative sizes of interactions: this regime requires $t_\perp < J << t$.

Angle-averaged density of states is easily calculated from $Im\,G_1$. This, rather disappointingly, is quite directly simply proportional to ω^α—and this behavior has been observed by Dynes and Gurvitch. We note, however, two possibilities for the "linear $|V|$" behavior. The most interesting is that, as we pointed out, the k_F points are not the only singularities in k-space. As observed by Ogata and Shiba, there also is a singularity at $3k_F$. This is one order of ω less singular than k_F, and gives an averaged density of states $\propto (\omega/J)^{1+\alpha}$, which is of a correct form and order of magnitude to fit the $|V|$ behavior. It will be interesting to study the slight curvature in the Dynes results to see if it is compatible with this expression.

A second mechanism is simply spin-orbit tunneling, which breaks holon-spinon coherence. This will tunnel into the thermal, $|\omega|$ DOS. It seems likely that this will not have the universal magnitude and occurrence seen in the normal state tunneling results.

Many calculations call out to be done on the one-dimensional system: for instance, Raman background scattering, behavior in the presence of a pair gap, tunneling conductance (to explain ρ_c), etc.

At least two experimental phenomena require the transformation back from k, m partial waves to plane waves $e^{i(k_x x + k_y y)}$. These are angle-resolved photoemission and resistivity: in the k, m representation of $G_1(r, r')$, spinon and holon represent circular waves emanating from the origin (but, oddly, one is incoming and the other outgoing!). In the idealized case of a circular Fermi surface they have a Fermi point at k_F and $2k_F$, respectively, for each m, and in the limit of infinite time ($\omega \to 0$) they have become effectively plane waves. This picture can clearly be modified to account for only approximately circular Fermi surfaces, but it is still true that the Fermi surfaces per se are well-defined in plane wave k-space, and occur at the "Luttinger" $k_F(\theta)$ for spinons and the spanning vectors $2k_F(\theta)$ for holons. For finite frequencies or shorter times, however, the curved nature of the waves comes out, and a given electron plane wave k may be composed of holon and spinon travelling in different directions, with

$$\vec{k} = \vec{k}_s - \vec{k}_h \qquad (5-6)$$

the kinematics of the superposition of different plane-wave equivalent holons and spinons is shown in Fig. 6. It is clear that for a given k, the colinear holon and spinon represents an osculation singularity, and we expect a cusp singularity when the spinon is on the Fermi surface, at its minimum energy, and the holon has all the extra momentum $|k_F - k|$. We interpret this as the cusp singularity in Fig. 1. The kinematic density of states is inversely related to the area of the triangle formed by the three k's by a Bessel function integral formula. On the other hand, as shown, it is possible for every k to have either a hole-or an electron-like excitation at zero energy, with both holon and spinon momenta on the Fermi surface, up to $3k_F$. $3k_F$ represents a second osculation singularity which may show up in photoemission spectra (these singularities reflect in zone boundaries).

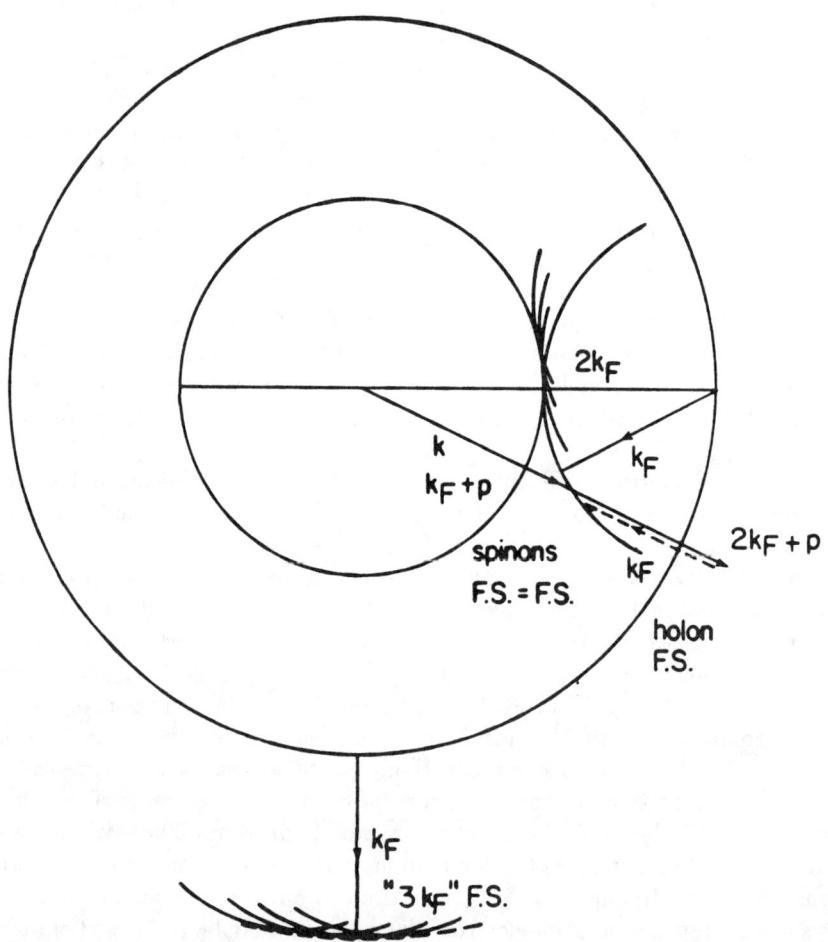

FIGURE 6 Kinematics of possible energy associated with $\vec{k} = \vec{k}_F + \vec{q}$ in 2 dimensions.

The Normal State of High T_c Superconductors: A New Quantum Liquid

We can sketch the thermal DOS spectrum but we have not calculated the detailed form of Fig. 1 yet, dynamically. Instead, we have taken the schematic fit to Fig. 1 shown in Fig. 7 as a semi-empirical starting point for the calculations of real physical properties. Fig. 7 is constrained by the following requirements.

(1) For $k = k_F$, we are probing the longest-range (in space) correlations, which must come from colinear holon and spinon; this we expect to have the same ω-dependence as the one dimensional case, $\omega^{-1+\alpha}$.

(2) $\omega = \epsilon_k^h$ must be some kind of cusp singularity, since below this point the spinon can no longer be colinear with the holon, and a decreasing volume of configuration space becomes available. In the thermal DOS, in fact, this is a vertical slope singularity.

(3) Finally, at $\omega = 0$ there is only one possible configuration of spinon and holon and the DOS vanishes: again, more sharply in the thermal DOS. We have no explanation for the observed linear slope as yet, but wonder whether the perfect lattice assumption we have so far been using is really appropriate for the actual substances.

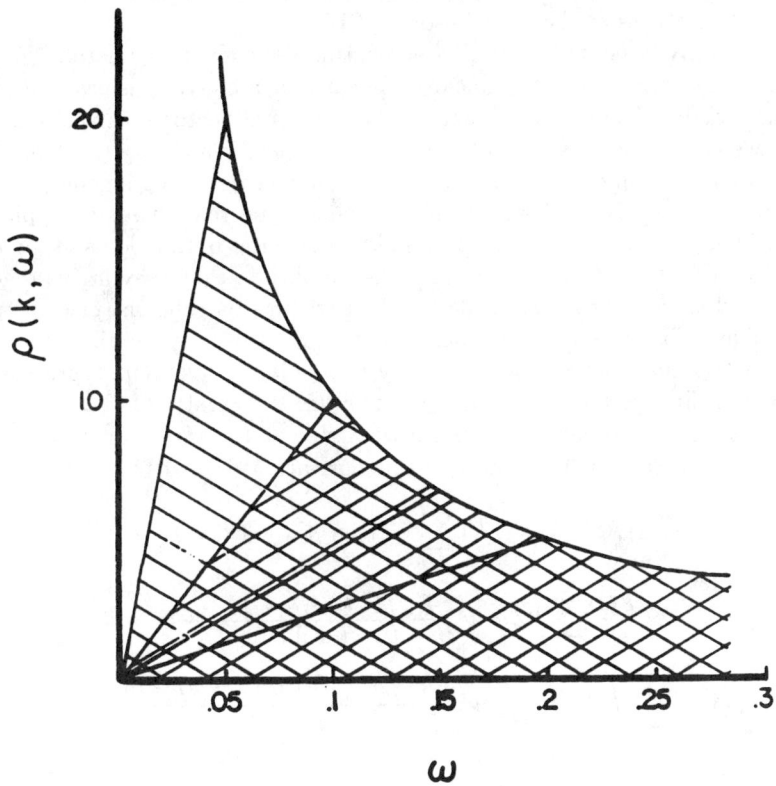

FIGURE 7 Semi-empirical fit to $\rho(k,\omega)$.

This point is worth emphasizing and discussing. The Fermi surface is established by $Z = 0$ spinons, behaving as a real spinless Fermion, and, since they refer only to the wave-function of the "squeezed Heisenberg" problem, relatively unaffected by potential scattering. The appropriate model wave function, however, is probably

$$\sum_Q (-1)^Q \det \| \varphi_n(x_{Qj}) \| \Psi_{SH}(y_1 \ldots y_{N/2})$$

where the φ_n are not plane wave but scattered plane-wave wave functions. They are roughly complete in the neighborhood of k_F, but the mean free path must be assumed very short. Under these circumstances holons away from $2k_F$ (this momentum is sharp and established by the spinon Fermi surface) will be severely scattered and the concept of "colinearity" and of m-values may rapidly become meaningless. A charge density wave can have a sharp frequency—the linear spectrum is relatively unperturbed—but scattering will probably act to connect in non-colinear holons and spinons more effectively.

Thus we have an excellent qualitative, but little quantitative, understanding of the main features of the angle-resolved PES.

Conductivity is another linear-momentum dependent calculation. What is on the face of it clear is that the linear-T, linear ω conductivity is easily understood from the one-particle Green's function. The external vector potential excites only the charge current to be sure, but implicit in the Lieb-Wu theory (though hidden in our independent spinon and holon conceptual structure) is a gauge-like interaction traceable to the constraint structure, enforcing equal real currents of spinons and holons. Thus we must assume the current vertex excites both holon and spinon parts of the electron; then $G_1(r,t)$ measures the amplitude that they arrive together at point r and time t. Any amplitude which is not in this coherent piece is instantly scattered away by the strong disorder potential. For this purpose, then, the apparent breath of G_1, due to the electron's decay into holon and spinon, represents a true electronic mean free time satisfying $\hbar/\tau \simeq E_k^h \simeq$ the greater of T, ω.

In terms of standard diagnostics we arrive at the same conclusion. I am indebted to E. Abrahams for help with this derivation. We write

$$\sigma = 1/\omega \langle j, j \rangle = 1/\omega \sum_{\omega_1, k} G(k, \omega_1 + \omega) G(k, \omega_1)$$

$$\simeq \frac{k_F^2}{\omega} \sum_k \int d\nu \, d\nu' \frac{\rho(k\,\nu)\,\rho(k,\nu')}{[\nu - (\omega_1 + \omega)](\nu' - \omega_1)}$$

$$\frac{k_F^2}{\omega} \int d\nu \sum_k \int d\nu' \frac{\rho(k\,\nu)\,\rho(k\,\nu')(th\beta(\nu-\omega) - th\beta\nu')}{\nu - \nu' - \omega}$$

and inserting the experimental fit for $\rho(k, \omega)$:

$$\frac{\rho(k,\omega)}{\alpha} \simeq \frac{\omega}{(\epsilon_k^h)^2} \quad \text{for} \quad \omega < E_k^h \qquad (5-3)$$
$$\simeq 1/\omega \quad \omega > E_k^h$$

we indeed find $\sigma \propto 1/T$. (Not $(\ln J/T)^3$ as proposed in an earlier version.) The c-axis conductivity must be thought of precisely as before,[19] as simply a part of the incoherent background which causes resistivity in the plane.

It is not trivial, really, that the obvious "$G_1 G_1$" result for σ is correct in a strongly-interacting system. It is not at all clear, as we have emphasized, *why* σ is so large, and requires the spinon mechanism. Very often $G_1 G_1$ does not give the correct answer, as in localizing systems. I feel a plausible explanation of $\rho \propto T$ is vital, but unfortunately can also be obtained for wrong reasons.

One puzzle which still remains and is of great importance is the Hall effect. It is fascinating that the Fermi line is electron-like, almost certainly, yet we see a hole-like Hall effect increasing at low T roughly as $(\ln J/T)^2$. This is, like NMR, an orbital-current phenomenon, and in some way represents a competition between spinon and holon responses; but only that qualitative remark is possible. One very important question is the charge response to phonon or other charge fluctuations. We feel this will encounter the anomalous large holon response, and low to medium frequency phonons may be very heavily damped. This damping will disappear when the gap opens up, possibly explaining the jump in thermal conductitity. Experiment needs to be refined in this area. Perhaps the biggest open question is the nature of the magnetic state below T_c. We cannot rule out either flux phases or weak Néel order.

ACKNOWLEDGEMENTS

We acknowledge vital discussions with B.S. Shastry, J. Yedidia, A. Georges, D.H. Lee, F.D.M. Haldane, and S. Girvin; also the hospitality of the Aspen Center for Physics and the IBM Yorktown Heights Laboratory.

REFERENCES

1. P. W. Anderson, Proceedings of the Enrico Fermi International School of Physics, *Frontiers and Borderlines in Many Particle Physics*, North-Holland Publ. Co., Varenna, July 1987.
2. P. W. Anderson, to be published in Proceedings of *Common Trends in Particle and Condensed Matter Physics* Workshop, Cargese, June 1988, Physics Reports Vol. 184:2–4, Elsevier Science Publ. Co.

3. P. W. Anderson, Lecture notes on *Current Trends in Condensed Matter, Particle Physics and Cosmology*, Kathmandu, June 1989.
4. M. Schluter, Proceedings of the International Conference – M²S HTSC, Materials and Mechanisms of Superconductivity Stanford University, Stanford, CA, July, 1989; also E. Stechel, discussions at the Los Alamos International Conference on the "Physics of Highly Correlated ELectron Systems," Los Alamos, December 1989.
5. C. Varma, E. Abrahams, A. Ruckenstein, P. Littlewood and S. Schmitt-Rink, Phys. Rev. Lett. **63**, 1990 (1989), which summarizes many of the ideas of Ref. 2 and 3 in a more available form.
6. Z. Schlesinger and R. Collins, preprint.
7. F. D. M. Haldane, J. Phys. C. **14**, 2585 (1981).
8. G. Benfatto and G. Gallivotti, preprint: "Perturbation Theory of the Fermi Surface..."; P.W. Anderson, to be published in Proceedings of *Common Trends in Particle and Condensed Matter Physics* Workshop, Cargese, June 1988, Physics Reports Vol. 184:2–4, Elsevier Science Publ. Co.
9. A. M Luther, Phys. Rev. B **19**, 320 (1979).
10. G. D. Mahan, Phys. Rev. **163**, 612 (1987); P. Noziéres and C. de Dominicis, Phys. Rev. **178**, 1097 (1969).
11. P. W. Anderson, Phys. Rev. **164**, 352 (1967).
12. This is the essence of the procedure of Abrikosov, Gorkov and Dzialoshinskii, (*Methods of QFT in Stat. Mechanics, Sec. 20*, Prentice Hall, 1963) which in the conventional case renormalizes the forward scattering phase shift to $\delta \sim [\ln(\Delta p)]^{-1}$ by use of the Cooper phenomenon. This renormalization procedure cannot work if there are anti-bound states present, I presume, because the assumed "non-singular" parts of the vertex are not harmless, but infinite.
13. T. Hsu and G. Baskaran, unpublished.
14. N. Levinson, Kgl. Dansk. Vid. Salsb. Mat. IF **25**, # 9 (1949).
15. J. Friedel, Adv. in Phys. **3**, 446 (1954).
16. C. G. Olson, R. Liu, and D. W. Lynch *et al*, "High-Resolution Angle-resolved Photoemission Study of the Fermi Surface and the Normal State Electronic Structure of $Bi_2Sr_2CaCu_2O_8$," to be published in Proceedings of the International Conference – M²S HTSC, Materials and Mechanisms of Superconductivity Stanford University, Stanford, CA, July, 1989; also preprint.
17. P. W. Anderson, "Theory of the Excitation Spectrum of the 'Normal' Metal in Cuprate Superconductors: Fit to Experiment and Demonstration of Enhanced Pair Susceptibility," submitted to Nature; P.W. Anderson and Y. Ren, "Theory of the Normal State of Cuprate Superconductors," to be published in Proceedings of MRS Fall Meeting-Symposium M, Boston, Nov. 17, 1989.
18. G. Yuval and P. W. Anderson, Phys. Rev. B1, 1522 (1970); P.W. Anderson, "Localized Moments" Lecture notes, Les Houches Summer School, 1968, *Probleme a N Corps Many-Body Physics*, N.Y., Gordon & Breach, pgs. 231-295, C. DeWitt & R. Balian, eds.
19. P. W. Anderson, Z. Zou, Phys. Rev. Lett. **60**, 132 (1988).

DISCUSSION

J. R. Schrieffer: Phil, to what extent can you substantiate the fact that in the two dimensional Hubbard model the holons and spinons are stable excitation or do they bind to form one composite excitation?

P. W. Anderson: I guess I can't answer first part of the last question too substantively. As to the rest of the question, yes I am sure that they are separate objects, that you certainly cannot talk in terms of a Fermi sea with funny properties. I tried to make that point very very clear indeed. If you start talking about such a Fermi sea you are driven to the conclusion that the renormalization constant for that Fermi sea is exactly zero and at that point you find that your Fermions are devoid of charge, because it's a second theorem that if $Z=0$, then $F_0^s=\infty$. Discussion of that was actually carried out between me and your colleague Walter Kohn some three years ago, although I didn't understand how important it was, and Walter has published a paper in the interim which I believe is correct. Although he seems unwilling to call his correction term F_0^s, it is an F_0^s correction and does show that the charge goes away in the limit as Z goes to zero. If the charge goes away you've got to find somewhere else to put it and the second theorem that seems to be possible to prove is that the spectrum of that second object must have a flat bottom. It must involve a mass which is, in a sense, infinite; namely, that there must be states away from the Fermi surface that have zero energy. In two-dimensions apparently that is satisfied with this peculiar $3K_F$ Fermi surface but one dimension is special.

So the answer to the question is, No, you can't get away with the only plausible single kind of excitation because just the acceptance of the fact that Z is zero immediately forces you to the conclusion that there is a second excitation somewhere. The one-dimensional example is very good for us because it shows us how that second excitation arises. That the two don't bind, if you have particle A and particle B into which particle C can decay, you can also, depending on the mass relationships, have particle C and particle B coming out of particle A and so on and the only question is which one is the cut and which ones are the poles.

Well, it is somewhat unfair to appeal to experiment but I think the experiments are good and getting better; they say that you had better make the electron the cut, and that basically is saying that the electron is only a resonance of the other two, that the other two in some sense really have an exact existence as you go to omega = 0 and that they are the particles which become eigen- excitations as the energy goes to zero. And appealing to experiment, or appealing to the one-dimensional analog, or finally appealing to this rather complex argument that you heard that if you go through the manipulations of a Fermi liquid theory in the end you find yourself doing it with a problem that has factorized itself into one-dimensional problems, you come to the same conclusion.

The second thing that is worth saying is you always have to have in mind that there are these two pieces of Tomonaga bosons or whatever you want to call them, and that they mix up, and they are the real excitations and you can't really call them fermions or bosons. You find yourself getting into unnecessary headaches if you try to identify who's what.

J. R. Schrieffer: Recently Arno Kampf and I have been studying the 2d Hubbard model in the large doping regime, where a pseudo gap exists. We find that for a range of energies within the pseudo gap the residue of the quasi particle vanishes. Thus, there is another candidate on the table.

P. W. Anderson: A candidate which has a hard time with the experiments. [And with the theorem I]

D. J. Scalapino: This may be the same question, but is it your view that if one takes a two-dimensional Hubbard model and begins to turn on U slowly from zero, we come to a critical point at which the situation changes, or is it continuous for all U?

P. W. Anderson: The situation is you're already in the mess the minute you turn on U and also the minute you turn on the density.

On the other hand these terms can be quantified, more or less, and in the case of density it's an $e^{-n^2 lnn}$ and in the case of U it's even easier, its an $e^{-t^2/U}$ and these terms are essentially invisible until you get to a point where U is really quite appreciable. So, as I said, what happens really is that you have this transformation but the transformation is so slight that you really, for experimental purposes, can think of it as the back end and the front end of a K_F and a 2 K_F excitation spectrum that are present of course in the original Fermi gas. But if you were able to look at them with a microscope you would find that they did funny things at $\omega=0$.

D. J. Scalapino: One more comment regarding the vertex correction equation. One can treat models that are finite in size (e.g., 10 × 10 Hubbard lattices) using the Monte Carlo approach to calculate the magnetic susceptibility or the pair-field susceptibility. One can run the calculations so that you average the product of the G's, which gives you basically the dressed lines interacting with all the vertex parts, so you get the full quantity χ. Alternatively, you can rig the calculations so that you get the product of the two average G's. This second calculation, I believe, is in the spirit of your calculations, where one has only the dressed G lines, either the two forward lines making up the pair-field susceptibility, or the forward and back lines giving the magnetic susceptibility, both without vertex corrections. Hence, we can look at the effect of the vertex corrections. We find, as I'll show later, that they are quite significant.

P. W. Anderson: The pair-field χ that I used is probably only the one that is appropriate if the perturbation you are considering is a pair-field made up of bare electrons. Obviously if you want to use an interior mechanism to the plane you're going to be in trouble with that. It may not even be the one appropriate for the interlayer tunneling phenomenon because that, of course, is a separate calculation too. So, although I started all of this with my enthusiam for the pair susceptibility calculation I am now worried that I don't know how to do any particular calculation and I would like to wait comments on superconductivity, except for the temperature dependence question for redoing those calculations. (Note - the interlayer calculation is in the paper as submitted.)

S. Chakravarty: Could you clarify your statements about the connection between the one-dimensional problem and the two-dimensional problem, in particular, statements about the phase-shift and orthogonality, the problem that you were referring to earlier in your talk? I understand the one-dimensional case; now how does one go about thinking about phase-shift in the two-dimensional case?

P. W. Anderson: Essentially the same way I did in the proof of the anti-bound states. You can think in terms of angular momentum channels for the relative motion, for a given total momentum, and this is a stack of states which has the characteristics of a one-dimensional stack of states. You have to think, instead of one side and the other side of the Fermi surface, you have to think of ingoing and outgoing states as interacting in that case. But once you have factorized it that way there appears to be a one-dimensional manifold.

Now why can you factorize it in that way? I dream, without having done it that rigorously but believing that it can be done rigorously, that I can confine myself, by these tricks of using phase shifts, to the neighborhood of a Fermi surface, and in the neighborhood of the Fermi surface there is only forward and backward scattering. So only the incoming and outgoing waves in those particular channels are going to strongly interact.

S. Chakravarty: One more thing about this. You talk about the cusp or the rise of the spectrum at the Fermi surface, but presumably it could also decrease. Why does it have to increase? Doesn't this depend on the sign of the phase shift that you have?

P. W. Anderson: [(First, it is not the phase shift but the effective interaction which is big.)] Of course, the spectrum at the Fermi surface could also decrease and there are response functions which go down and there are response functions which go up. And fortunately the lore of the one-dimensional Hubbard model tells you more or less which ones go up and which ones go down, although I assure you that the state of information in the one-dimensional world is not...even though it's a one-dimensional world, information doesn't travel infinitely fast, and there are people still calculating things which other people know, and people who are still calculating things wrong which other people know. I am certainly the sloppiest of all of these people, and so I don't guarantee that we've got all of them right but my feeling is that we do.

C. M. Varma: Phil has insistently stated that the clue to understanding these materials is the normal state peculiarities, and I think we owe him a great deal of gratitude in forcing us to think along those lines. So in that spirit, I would like to show these two experimental figures of $BaBi_xPb_{1-x}O_3$. This, as you know, is a 10 degree superconductor, and the conductance as a function of voltage is proportional to modulus of the voltage. In fact this led me to suggest a few months ago that all properties that barium bismuth oxide will have, all the anomalies that you see in the copper oxide compounds, and indeed...

P. W. Anderson: Chandra, do you mind if I make a positive and friendly comment on that? Some people have been saying that this phenomenon doesn't exist, and that really it's all due to low barriers.

C. M. Varma: You and I know that's nonsense??

P. W. Anderson: We know that's nonsense, but I want to point out that this graph alone has that fact in there. You can see quite clearly the curvature on this side, downward curvature, and the curvature of that side which are making it asymmetric, if you eliminate a quadratic term and for the linear term here it seems to be perfectly symmetrical. The quadratic term measures the barrier.

C. M. Varma: That's right. One of the most peculiar things of the copper oxide compounds is the Raman scattering intensity which, if you subtract out the phonons, is constant over a very constant in energy and constant in temperature over a very very wide frequency range, and this data was known before. And I have just received a preprint from Sugai in which this has been measured in barium potassium bismuth oxide, a 30 degree superconductor, in which also there is a large, significant, constant background in Raman scattering. It is a sad comment on the times that when the first experiments came out on barium potassium bismuth oxide, people were saying, "Oh, barium bismuth oxide must have the local moments." Well it does not, neither is it two-dimensional. So this is something I think not just Phil, but almost all the theorists working in this field have to face – that the normal state anomalies of this compound, are really the same as Cu-O.

H. Ott: How about the resistivity?

C. M. Varma: The normal state resistivity of these materials is usually twenty to fifty times larger than the copper oxide compound, and then you cannot see much of a temperature dependence. It's very close to the unitarity limit and if and when you make this thing good enough. I predict that it will be linear...In fact, it follows completely from an analysis of the data of tunneling and Raman. If you had one of them, you had to have the other.

The second point I want to raise is the relationship between what Phil talked about, and the paper that the five friends – friends is too high an honor, I'd be happy to be called a disciple – pointed out. Phil has this spectral weight that goes as $1/\omega$ for large ω, and is proportional to ω at small ω. We indeed pointed this out before the experiment was done. The important difference, I think, between Phil's methodology – let alone Phil's point of view – and ours is, that we believe that the central quantity in the problem is the vertex part of the two-particle susceptibilities of the problem, and given a single hypothesis about this part, all the other peculiarities follow. When Phil is calculating susceptibility, he says, assume there is something peculiar about the single-particle Green's function, and then he simply calculated this. There is a real physical difference in our thinking, and what Phil is talking about. Our peculiar Green's function is the result of the peculiar vertex.

The Normal State of High T_c Superconductors: A New Quantum Liquid

P. W. Anderson: Thank you, Chandra, and that is a very nice presentation of the barium bismuth oxide case. I knew it had the linear term but I didn't realize the data were so clean as they really are. In the first place, there is one thing common to the fundamental physics of barium lead bismuth oxide and the fundamental physics of the cuprates which is unequivocal, straightforward, and which is shared with very few other compounds –possibly only the interesting organic materials. That is that it is a system which is plausibly a Hubbard band system with an upper and lower bound on the band. One calls it a bismuth 5s band but it is primarily an oxygen 2p band with 5s symmetry, and it has an upper and lower bound. It seems very likely that it also has an upper Hubbard band and lower Hubbard band and so that it is one of the not very many truly three-dimensional systems that is going to have the same troubles. Now it has been thought that the interactions were basically attractive rather than repulsive, one has to feel that it's probably some combination of the two and the dynamics are going to unfortunately mess things up. The fundamental interactions are of course repulsive and will have the effect of kicking out an anti-bound state and one will have the same orthogonality problems. Where the two systems differ is that in the copper oxide system this strange state is confined in two dimensions and doesn't make satisfactory contact between the different layers. In the bismuth you don't have this lovely two-dimensional-to-three-dimensional transition to take advantage of, to get to a superconducting state, and some other interaction is obviously responsible for the superconductivity. I don't know what interaction it is or whether there is any enhancement or de-hancement of pairing, or whether even the fact that it has a strange normal state is relevant to its high T_c – probably it is but I don't understand that. I think we all fail to understand the dynamics and the way the phonons come in but I do feel these two have the key thing in common that I emphasize; namely the existence of an upper limit on the spectrum and the existence of a strong interaction which is kicking out an upper Hubbard band.

C. M. Varma: May I make a follow-up comment? The follow-up comment simply is that we – or at least some partial subsection of the five friends – believe that the essential physics has to do with bound states in both cases and not with anti-bound states, and the effective physics is in the attractive part of the model – effective attractive part of the model – and not in repulsive.

A. Malozemoff: Phil, I wanted to ask about some of the experiments. You neglected a group which in earlier times were very much at the focus of discussion, namely, the measurement of the doping x or concentration dependence of the number of holes. I am thinking of systems like the lanthanum strontium copper oxide. The Hall effect, μSR data, optical data, etc. have indicated a proportionality in the number of holes with x (rather than 1 + x). Do you not subscribe to those experiments now?

P. W. Anderson: Well, the short answer is Right: I do not subscribe to those experiments. The longer answer is the following: I believe that there is a phase transition, which may well be first order, that takes place between the Néel type state and these. When you dope you undoubtedly distort the Néel order...that I think is a clean, straightforward branch of theory at this point...but you still have at least on a local time-dependent basis an order parameter. The "Luttinger liquid" state clearly doesn't have magnetic order. I believe there is a sharp transition, that may be first order, between the two types of states and that any system which does not have this state in it is not superconducting. Almost all of those continuous T_c's in the early doping region, are probably mixed phases on some kind of microscopic scale. In this I agree with only half the experimentalists, but they happen to be the half I believe, so that's all right. On the other hand discontinuities in T_c elsewhere, I don't know whether they exist or not. Incidentally, one of the characteristics of the low doping limit is that almost always you find the onset of superconductivity at a higher temperature than you find any kind of midpoint of superconductivity. I don't know whether there are specific phases at higher concentrations or whether there are continuous changes.

R. M. Martin: There are two themes that come out strongly: one is lower dimension vs. higher dimension, and the other is whether there is a split-band type of solution. I would like to rephrase Doug Scalapino's question asked in the copper oxide-like systems. Namely, if you were to view the coupling between the planes as a parameter that you could turn up until you became fully three-dimensional, would there be a phase transition as a function of this parameter.

P. W. Anderson: It puzzles me. I think there might be. Yes, I guess there more or less has to be a phase transition in which you become three-dimensional, and that phase transition doesn't exist once the thing is superconducting. Once it's superconducting certainly the system is three-dimensional, it can communicate via pair tunneling, even if it isn't caused by pair tunneling. So this separation in the charge and spin probably gets destroyed, or can get destroyed, by losing some of the phase into a neighboring plane. You may ask then what's barium bismuth oxide, and you may answer on my thinking is fuzzy on this question, and I'll admit it.

R. M. Martin: Is there any quantitative way to say whether in a three-dimensional case one is in one kind of regime or the other?

P. W. Anderson: Quantitative... Yes, is there an upper Hubbard band, if you like. And that you can tell really from spectroscopy, pretty much. The other thing is, do you see Fermi surface anomalies that are not explainable any other way.

P. W. Anderson: Is that falsifiable enough for you?

J. R. Schrieffer: It's false enough. (Laughter)

D. Pines: Bob, what you meant to say is it's fanciful enough.

C. M. Varma: Don't you trust the ratio of the resistivities to give you within an order of magnitude, within two orders.

P. W. Anderson: Not for the paired layers, because of course there are much bigger hopping integrals between the pairs. Of course I do for lanthanum copper oxide, and they're reasonable. There are also band theory numbers that I've looked at, and they're very reasonable too. The order of magnitude are fine, but would like to have a better...

C. M. Varma: Bismuth-calcium, there the resistivities are parallel, the ratio is 10^{-4}.

P. W. Anderson: But they are two layer the bismuth one-layer has a very low T_c.

C. M. Varma: Maybe two-layer.

P. W. Anderson: All it needs is two layers, A and B. I don't need C. All I need is A and B. Then I've got a Kosterlitz-Thouless superconductor, and then I'm there.

T. M. Rice: Can I see that pairing susceptibility again. Is it diverging now as a power law, or...?

P. W. Anderson: Well, it is diverging as $1/T$.

T. M. Rice: As a power of logs?

P. W. Anderson: But that is this susceptibility, which is a very unusual kind of susceptibility, for this particular kind of process. So it's not the usual pair-susceptibility. That was a mistake of mine. I should have realized that it's not the usual process, and I didn't. It's just amusing that $1/T$ is about the same as (log) is. Except for the alpha.

J. R. Schrieffer: That should be $t_\perp$ in the numerator, squared?

P. W. Anderson: Yes, $t_\perp^2$.

A. Millis: Naively you would expect then that pressure in the C direction would increase $t_\perp$, and make T_c go up, whereas pressure along planes would increase $t_\parallel$ and therefore make T_c go down. But experimentally, in fact, the pressure derivative is isotropic and it's the same whichever way you do it.

P. W. Anderson: Not as far as I can see. Are there any good experiments?

A. Millis: I don't know of too many uniaxial pressure experiments on single crystals, although there are a few in the C direction, but sound velocity experiments on ceramics and single crystals suggest that T_c has an approximately isotropic pressure dependence.

P. W. Anderson: There is one very fascinating phenomenon. There are a lot of pressure measurements, and in all those pressure measurements, there are three that have very strong pressure dependences of T_c. All of those three are single-layered materials where all of the T_c has to come from the only hopping integral you have. All the others are pretty neutral, and it seems that the two t's pretty much cancel out.

C. M. Varma: I think you are predicting a very interesting effect in the sound velocity near T_c when you apply the polarization along the C axis.

P. W. Anderson: Yeah, I think I am.

C. M. Varma: You're predicting that, and I don't think that will be seen.

E. Abrahams: I have a question. Is this a theory for coupled one-dimensional Hubbard chains that you hope will work in two dimensions?

P. W. Anderson: Yes. I hope that this is also a theory of superconductivity in some of those organic superconductors. But the trouble is the one-dimensional case has this nasty Peierls instability. In the two-dimensional case we are fortunate that it doesn't seem to like to be unstable. With Peierls instabilities, and density wave instabilities, and so on, you'll only get the superconductors when the organics turn two-dimensional. That seems to agree.

C. M. Varma: Do you think any of these ideas can really be extended to two dimensions?

F. D. M. Haldane: At this time, I am rather dubious. The difference between one and two dimensions is that in one dimension, a hole can move through the background of spins without disrupting their ordering, and the system reduces to an interpenetrating Heisenberg system and a hole-gas. In two dimensions, the motion of a hole "stirs up" the spin configuration and disrupts it, so I need more convincing that a two-dimensional version of the 1D picture is possible. But we shall see.

S. Doniach: You said that the 2 K_F are pseudo-fermions, but Phil said they were bosons...

P. W. Anderson:: I don't know whether they are fermions or bosons, do you? I don't know that it matters...

F. D. M. Haldane: It doesn't matter in one dimension. The structure of the system of equations is that you can call them fermionic variables in the sense that they carry fermion number, as the electron carries fermion number, charge and spin. In this decomposition of the degrees of freedom of the Hubbard model, the electron splits into a fermion-number carrying modes (that is unrelated to charge, as empty and doubly-occupied sites have the same value of this variable) plus two bosonic degrees of freedom for spin and charge.

S. Doniach: The occupation fills it up to the chemical potential.

F. D. M. Haldane: The spinon is something like a fermion. Its spectrum is analogous to the 1D quantum Ising model in a transverse field at the critical point. There are no negative energy states, and no Fermi surface in the ground state, but if I put a magnetic field on to generate a finite spinon density, a pseudo Fermi surface develops.

P. W. Anderson:: You know, thinking of them in terms of fermions is by far the easiest way. Even though two of them make an electron.

2. EXPERIMENTAL FACTS

Chair — H. Ott

B. Batlogg
AT&T Bell Laboratories
Murray Hill, NJ 07974

Selected Experiments on High T_c Cuprates

A selection of experimental results on high temperature superconductors and their insulating parent compounds is presented. Emphasis is put on recent results and on those properties which are central to the emerging understanding of physics of these fascinating compounds.

INTRODUCTION

Three years after J. Georg Bednorz and K. Alex Müller[1] discovered cuprate superconductivity we are faced with an immense number of publications, estimated to be in excess of 10,000. An exhaustive review of all the experimental work is clearly impossible. In this chapter, an attempt is made to present those results which form a basis for our developing understanding of these fascinating compounds. The selection of topics and the emphasis put on them reflect some personal preferences, but it is mainly the result of discussions with many colleagues. May they accept my sincere thanks for their time and patience. There is little doubt which properties are distinctly characteristic of the high T_c cuprates and two recent discussions are given by P. W. Anderson[2] and by C. M. Varma.[3] Exploring how general certain "unusual" properties are will be one of the recurring themes of this chapter. Due to space and time restrictions several aspects could not be covered, such as crystallography, crystal chemistry, photo-emission and x-ray absorption spectroscopy, fluctuation effects, lattice dynamics and others.

The most significant experimental breakthrough in the experimental efforts is linked to improved awareness and understanding of the intricate material-specific problems. Those insights lead to preparation of better samples, and most importantly to growth of single crystals, which are the prerequisite to study these highly anisotropic compounds. More and more investigators have become sensitized to the subtle materials-related problems and realize, sometimes reluctantly, that a measurement can only be as good as the quality of the sample. As a clear sign of progress we find quantitative agreement between results by different research groups. Striking examples are, e.g., resistivity, optical spectroscopy, Raman scattering and NMR studies.

This chapter is not an historical account but more a selection of most recent results which, hopefully, will stand the test of time. Dealing with references is a major concern, considering the imminent potential of alienating many researchers and colleagues. It is practically impossible to be aware of all the relevant papers and to cite them all. Thus, I ask for much allowance. Several fine summary papers have appeared recently and in the preparation of this chapter extensive use was made of Refs. 4, 5, 6 and 7.

1. MAGNETISM

Here we look at a few central aspects of magnetism in the layered cuprates, including the ordered state in the insulators, its excitations, the variations of the long range ordered state due to the introduction of mobile carriers, and finally the dynamics of the metallic state. This chapter can be kept short because magnetic properties have been discussed in more detail in separate contributions at this workshop (A. Millis[8] and S. Chakravarty[9]).

A good starting point is to consider the ionic configuration of Cu^{2+} with one hole in the 3d-shell, i.e. $3d^9$ with spin 1/2. Strong antiferromagnetic super-exchange provides the dominant interaction between neighboring Cu sites. The Cu and O atoms are arranged in layers and thus La_2CuO_4 is the simplest example of a single layer compound, which one might expect to be a 2-dimensional spin 1/2 lattice Heisenberg antiferromagnet. These magnetic layers are stacked and their magnetic properties are then determined by two factors: (1) the strong antiferromagnetic interaction within the Cu-O layer $J_\parallel$ and (2) the weaker inter-layer interactions $J_\perp$ and small symmetry-lowering structural distortions which lead to long-range 3-dimensional order.

1.1 STATIC LONG RANGE ORDER IN La_2CuO_4

The magnetic properties of the parent compound La_2CuO_4 have been studied in great detail by several experimental techniques and are all well understood by now, including both static and dynamic properties. (For most Refs. see Birgeneau and Shirane,[10] also Allen, Fisk and Migliori[11]) The Néel-point of La_2CuO_4 is close to 300K and drops precipitously with the slightest deviations from ideal composition. Such deviations may involve an excess of oxygen, and/or substitution on the La or Cu sites. Most effective seem to be those substitutions which introduce extra charges, such as Sr for La. In contrast, strictly divalent substitutions for Cu are expected to depress T_N much more slowly through dilution effects. The quality of a sample is readily checked by measuring the dc susceptibility: the higher its T_N, the closer it is to the ideal composition.

Below T_N the Cu moments are antiferromagnetically ordered in three dimensions. The slight orthorhombic distortion alleviates the strict frustration of the tetragonal La_2NiF_4 structure and the observed 3-dimensional magnetic order involves anti-parallel alignment of nearest neighbor spins in adjacent layers.[12] The measured sublattice magnetization of $\sim 0.5\mu_\beta/Cu$ is significantly smaller than the $\sim 1.1\mu_\beta$ expected for a divalent Cu^{2+} moment, and most of this difference is ascribed to zero point fluctuation effects.

An additional subtlety is observed in the spin structure due to rotation of the CuO_6 octahedra, giving rise to an antiferromagnetic exchange term which in turn results in a minute net resultant moment for each pair of spins in the Cu-O layer. The equivalent net moments in neighboring layers are antiparallel aligned, but an external field of a few Tesla is enough to align them parallel.[13,14] This amounts to a field induced transition from antiferromagnetic to a ferromagnet with a small moment of $\sim 2 \times 10^{-3}\mu_B/Cu$. Numerical analysis of various experiments gives results for the coupling parameters: nearest-neighbor $J_{nn} = 116 meV$, and net inter-layer coupling $J_\perp \cong 0.002 meV$.[14] This anisotropy of almost 5 orders of magnitude is remarkable.

1.2 ANTIFERROMAGNETISM IN OTHER INSULATING ENDMEMBERS

La_2CuO_4 is by far the most intensively studied parent-compound due to the simplicity of the structure and the availability of large crystals. From the less complete information on the other compounds one might infer that their magnetic properties are essentially the same as La_2CuO_4. They order antiferromagnetically around ambient or somewhat higher temperature, as studied either by neutron scattering or μSR techniques.

The spin structure of $Ba_2YCu_3O_6$ is slightly more complicated than the one of La_2CuO_4. Each chemical unit cell contains a Cu-O double-layer (i.e. two CuO_2 layers in close proximity to each other) and one additional Cu connecting two oxygens at the apex of the CuO_5 layer-forming pyramids. This latter Cu ion is non-magnetic for O_6, but acquires a magnetic moment when oxygen starts occupying the "chain-sites" in "123" O_{6+x}. Thus, for O_6, the antiferromagnetic super-exchange within the Cu-O planes is the dominant magnetic interaction. Presumably, dipolar coupling leads to antiferromagnetic order between the two adjacent Cu-O layers, and furthermore, the double layers are also coupled antiparallel. Complications will arise when the chain-Cu acquires a moment in that it will start to frustrate the antiferromagnetic order between its neighbors which belong to different double-layers, ultimately leading to a change in the magnetic structure and a doubling of the magnetic unit cell along c.

The ordered moment is largest at O_6 and amounts to $\sim 0.66\mu_B/Cu$ for the plane-sites. This is indeed close to the expected value for the square-planar S = 1/2 Heisenberg model, and just slightly larger than in La_2CuO_4.

As in La_2CuO_4, the 3-dimensional long range order in $Ba_2YCu_3O_6$ is due to a weak interaction between Cu-O double layers with much stronger in-plane afm coupling. Therefore, it is not surprising to see a pronounced shift of T_N when this coupling is modified by external means. $NdBa_2Cu_3O_{6.35}$ is an illustrative example: T_N is $\sim 230K$ and increases under pressure by $23K/kbar$.[15] This can be seen as evidence for a rather unstable magnetic situation brought about by competing interactions trying to couple the double layers. (To start with, T_N is reduced appreciably from $\sim 500K$ in $Ba_2YCu_3O_6$).

1.3 MAGNETIC EXCITATIONS AND FLUCTUATIONS IN La_2CuO_4

The coupling between spins in the same Cu-O plane is by far the dominant magnetic interaction, and given its magnitude of $> 0.1eV$, it is not surprising to find dynamic spin fluctuations above T_N (T_N is determined by the much weaker interplane coupling.) Neutron scattering experiments show them to be 2-dimensional in character and not influenced by anisotropy in the plane. When the temperature is decreased below the Néel-point, scattering intensity due to 2-dimensional fluctuations diminishes and the intensity of the 3-dimensional magnetic Bragg peak grows. The correlation length of the 2-dimensional fluctuations increases from $\sim 40\text{\AA}$ at 500K to $\sim 400\text{\AA}$ at T_N, and the temperature dependence is quantitatively well-described by a renormalized semi-classical model.[16]

The excitation spectra of La_2CuO_4 have been studied by inelastic scattering of neutrons and photons. The main result is that conventional spin wave theory appears to provide an excellent description of the magnetic dynamics studied so far by neutrons. There is a linear relationship between excitation energy and wave vector up to at least 0.1eV, which was the limit of the neutron scattering experiment, and no novel fluctuation or lifetime effects have to be invoked to explain the dynamics for wavelengths longer than $10\text{\AA}$.[17] The excitations obey Bose-Einstein statistics. From the spin wave velocity of $0.85 \pm 0.03 eV\text{\AA}$ the exchange coupling J can be derived to be $\sim 136 meV$. An analysis based on a more recent calculation gives J = 115 meV.[18] Within the energy window probed in these studies the spin dynamics is unaffected by the 3-dimensional order at T_N, which is a reflection of the long correlation length of the strong 2-dimensional fluctuations also well above T_N. Only a modest decrease of the spin wave velocity by $\sim 9.5\%$ upon heating from 5K to 296K has been found. In this energy and wavelength regime, the system behaves as an ordered magnet.

The above values for the exchange constant J are in agreement with light scattering studies which have been performed on La_2CuO_4 and several other cuprates. Multi-magnon excitations are observed at energies up to $\sim$ 1eV and a detailed understanding of the spectral shape and selection rules has emerged.[19,20] The classical case of S = 1 K_2NiF_4 is a starting point which could be successfully explained within the framework of spin wave theory. The cuprate spectra, in comparison, are significantly broader and contain unexpected intensity with A_{1g} and B_{2g} symmetry, in addition to the usual B_{1g} symmetry. All of these novel aspects are the result of quantum fluctuations in the ground state of the S = 1/2 square planar Heisenberg antiferromagnet. In the classical Néel ground state the next-nearest-spins are parallel aligned, but quantum fluctuations for S = 1/2 induce a 30% probability of anti-parallel alignment. This leads then to new possibilities for multi-magnon scattering which are worked out theoretically and agree well with the experiments.[20] Of particular interest here are the exchange parameter values deduced from the spectra,[21] given in Table 1.

TABLE 1 Exchange parameters J (meV) for various compounds derived from inelastic light scattering, (from reference 21).

La_2CuO_4	128
Nd_2CuO_4	108
Sm_2CuO_4	110
$Ba_2YCu_3O_6$	98

Interestingly, one finds similar exchange parameters for the two structural modifications of the "214" compounds (T and T'). Although J tends to decrease slightly with increasing Cu-O distance, the data base is too narrow to establish the functional dependence further.

Light scattering estimates for the exchange parameter J are $\sim$ 25% smaller for $Ba_2YCu_3O_6$ than for La_2CuO_4 and one would expect an equivalent reduction of the spin wave velocity. Indeed, neutron scattering studies gave lower limits for the spin wave velocity of 0.5 eV Å at a composition of $O_{6.2}$.[22,23] Notable is the fact that the spins are correlated within the Cu-O double layers.[24] For most recent results on the in-plane and out-of-plane spin excitations, and their dependence upon oxygen content in $Ba_2YCu_3O_{6+x}$, see reference 25.

Inelastic scattering of neutrons and photons, therefore, provide us with reliable and consistent information about the relevant magnetic energy scale and the excitations of the undoped insulating parent compounds. While the neutron data are fully compatible with classical spin wave theory in the energy range probed so far, a picture based on quantum spin fluctuations in the ground state is necessary to explain the novel features seen in inelastic light scattering spectra.

1.4 CHARGE-CARRIER INDUCED CHANGES OF MAGNETIC PROPERTIES

This is one of the central aspects and several results will be discussed here: Néel-temperature, instantaneous spin correlation length, "spin-wave" velocity, local moment probed by μSR, and hyperfine field probed in specific heat measurements. The Néel-temperature is extremely sensitive to the presence of carriers, be they introduced by heterovalent substitution of Sr,Ba for La in La_2CuO_4 or Ce for Nd in Nd_2CuO_4, or by oxygen non-stoichiometry. T_N drops from $\sim 300K$ to 0K if only $\sim 1\%$ of Sr is present, which is to be compared to much slower depression if only the magnetic Cu sublattice were diluted. In the icostructural compounds $Pb_2Mn_{1-x}Mg_xF_4$ and $Rb_2Co_{1-x}Mg_xF_4$ long range antiferromagnetism persists up to $x = 41 \pm 1\%$.[10,26] The rapid suppression of T_N is accompanied by a reduction of the long range ordered moment as measured by neutron scattering.[27] In contrast, a local probe like μSR senses no such rapid decrease of the magnetic moment,[28-30] which indicates that the magnetic moments are disordered.

Further increase of hole concentration leads to delocalization of the carriers and superconductivity. In the superconducting compositions the average internal field seen by the muons essentially vanishes on the time scale of μs (i.e. the muon lifetime).[28-30] The same changes are also observed in $Ba_2YCu_3O_x$.[31]

Specific heat measurements provide yet another probe of static magnetic fields in that the increase of C(T) at lowest temperatures is a measure of the hyperfine magnetic field seen by the nuclei.[32] The common message of all these studies is clear: there is no evidence for static or slowly (MHz) fluctuating magnetic moments in the superconducting compounds.

The introduction of charge carriers suppresses the Néel state effectively, and reduces the instantaneous spin correlation length deduced from neutron scattering results. The correlation length is $\sim 10\text{Å}$ in the metallic range and increases towards $\sim 30-40\text{Å}$ for $La_{1.98}Sr_{.02}CuO_4$.[10] The instantaneous spin correlation length appears to be determined by the density of mobile carriers. Around x = 0.15, the spin correlation length is comparable to the superconducting coherence length.

A word of caution needs to be added here. It is often tacitly assumed that variations of oxygen content or Sr substitution lead to a homogeneous variation of the electronic, and particularly of the magnetic properties. In these compounds with short coherence lengths, it is important to consider the length scale over which homogeneity of a property is considered. For instance, NQR spectroscopy as a local probe finds two charge-differentiated Cu-sites in $(La, Sr)_2CuO_4$ which have also magnetically different behavior.[33] This crystal-chemical reality has to be kept in mind when macroscopic properties such as "T_c," susceptibility, transport properties, etc. are discussed in terms of "continuously varying carrier concentration." From this point of view, $Ba_2YCu_3O_7$ (90K T_c) with only "full Cu-O chains" and

$Ba_2YCu_3O_{6.5}$ with long-range ordered "empty and full" Cu-O chains constitute crystallographically proper phases.

1.5 SUSCEPTIBILITY $La_{2-x}Sr_xCuO_4$ AND $Ba_2YCu_3O_{6+x}$

The static susceptibility is of particular interest in the two compound series as one can follow the variations in χ when chemical modifications induce the semiconductor to superconductor transition. (Sample homogeneity on a length scale relevant for magnetism is always a concern.) Indeed, $\chi(T)$ in $Ba_2YCu_3O_{6+x}$ depends in detail on how the oxygen-deficient samples are prepared. Two representative sets of $\chi(T)$ data are shown in Fig. 1 to point out the qualitative similarities and the

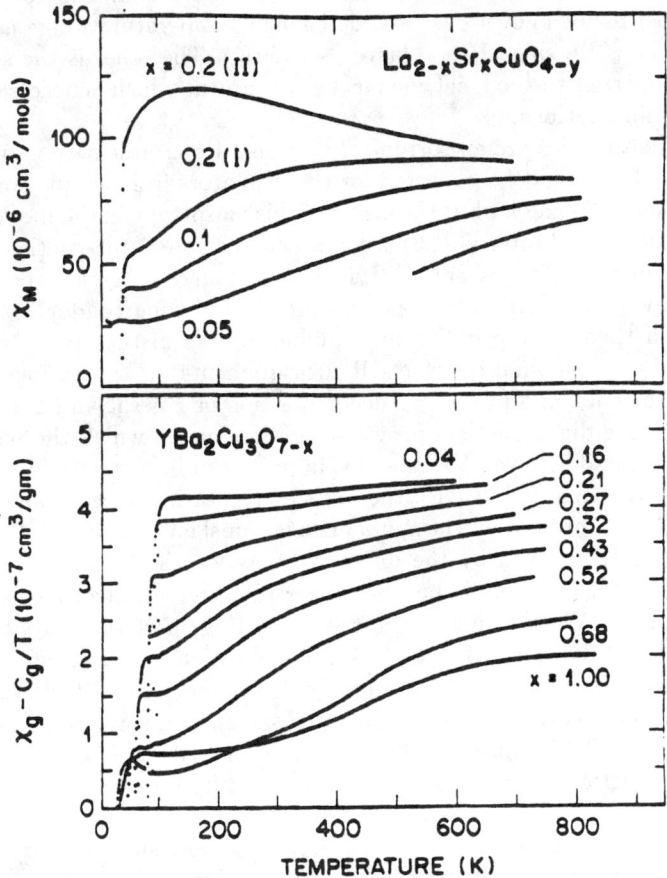

FIGURE 1 Static susceptibility of $La_{2-x}Sr_xCuO_4$ and $Ba_2YCu_3O_{7-x}$. A Curie-term has been subtracted from the $Ba_2YCu_3O_{7-x}$ data. (Johnston et al.)

quantitative differences.[34,35] Common to both series is the pronounced S-shaped $\chi(T)$ curve for the magnetic endmembers. The maxima at highest temperatures are expected for a 2-dimensional square lattice S=1/2 Heisenberg antiferromagnet. Actually, the maximum in $\chi(T)$ is expected to be at $T \cong 2.4J$, considerably higher than the measurement limit. The 3-dimensional long-range ordering near 500K is barely noticeable in $Ba_2YCu_3O_6$, but gives rise to a large peak near 240K in La_2CuO_4 (not shown in Fig. 1), which is due to the small net out-of-plane moment.

The superconducting compositions have larger susceptibilities than the semiconducting magnets with either an essentially flat $\chi(T)$ in $Ba_2YCu_3O_7$ or a slight temperature dependence for $La_{1.85}Sr_{0.15}CuO_4$. The susceptibilities for intermediate compositions interpolate smoothly between the two extrema. Although attempts have been made to map the $\chi(T)$ curves to a universal $\chi(T)$ behavior, such scaling procedures might meet with practical and conceptual difficulties.[36] One is the fact that $\chi(T)$ is not known over a wide enough temperature range as to observe the maxima in $\chi(T)$, since J is so large ($\sim 1400K$). The other is the sensitivity of electronic properties to the local chemical environment which is necessarily subject to statistical fluctuations.

The susceptibilities are anisotropic[13,37,38] and are largest perpendicular to the Cu-O planes. This might be expected for the insulators from simple considerations for a planar antiferromagnet but the origin of this anisotropy in the metals is unclear a priori. The measured total $\chi(T,x)$ is composed of a core diamagnetism, Van Vleck paramagnetism and spin susceptibilities. It is, of course, of great interest to know all these individual contributions, particularly in the superconductors, and to be able to assign them to the individual building blocks in the structure. This can be achieved by a combination of NMR measurements on the various atoms and static measurements, and in Fig. 2 a decomposition of χ is shown for $Ba_2YCu_3O_7$. In Fig. 2 the core diamagnetism provides the baseline to which the paramagnetic contributions are added. Van Vleck and spin parts can be separated in principle by comparing NMR Knight shift with the measured χ, or in superconductors through the residual NMR shifts when the spin part has vanished at $T \ll T_c$.[39-43] The spin susceptibility is then given by the difference between the $\chi_{core} + \chi_{vv}$ values and the measured total χ. Thus, it becomes clear that the anisotropy of χ_{total} is due to anisotropic Van-Vleck terms, which reflects in large part the spatial orientation of the Cu d-orbitals. According to this decomposition, the spin susceptibility is isotropic and amounts to $\sim 3 \times 10^{-4}$ emu/(mole formula unit). For illustration only, this χ_s value would correspond to a density of states with a Sommerfeld parameter γ of 22 mJ/mole K^2. The mean-field specific heat anomaly at T_c is $\sim$ 50-60 mJ/mole K^2 in zero magnetic field. For a review of heat capacity measurements see reference 44.

An interesting result of the χ analysis is that the spin susceptibility for the 90K superconductor is 3 to 4 times larger than for the semiconducting antiferromagnet (well below T_N). (2.9-3.3 vs. 0.8-1 $\times 10^{-4}$ emu/mole formula unit). The same tendency is observed also in $La_{2-x}Sr_xCuO_4$, (see Fig. 1), but the quantitative analysis

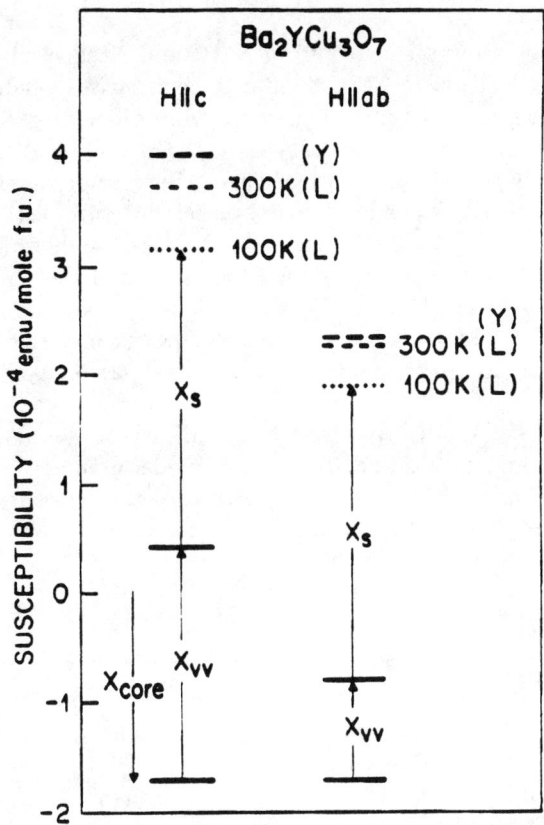

FIGURE 2 Decomposition of the static susceptibility of $Ba_2YCu_3O_7$ into contributions of the core (χ_{core}), Van Vleck (orbital) terms (χ_{vv}) and spin contributions (χ_s). The anisotropy of the total susceptibility is caused by the Van Vleck terms, while χ_s is essentially isotropic. Measurements of χ_{total} by W. C. Lee et al. (L) and Yamaguchi et al. (Y).

is not straightforward. Nevertheless, a best estimate gives a ratio of $\sim 1.5 - 1.9$ for the low temperature spin susceptibilities of $x = 0.15$ and $x = 0$. This ratio is thus only half of that in $Ba_2YCu_3O_x$.

In this context it is worthwhile to point out a peculiarity in the low temperature spin susceptibility of $Ba_2YCu_3O_6$ and La_2CuO_4 (discounting the contributions due to long range order). For both compounds, χ_s is close to 1×10^{-5} emu per one mole formula unit. Since the chain sites are magnetically inactive, this means that the "123" double-layer has the same susceptibility as the single Cu-O layer in "214."

1.6 DYNAMICS IN THE METALLIC COMPOUNDS

Much microscopic insight comes from NMR and NQR studies. In a unit cell as complex as "123" all but the Ba sites have been investigated. A wealth of information has been accumulated and the data from different groups are generally in remarkable quantitative agreement. Among the many peculiarities of the spin dynamics, only a few should be mentioned here. In the normal state, the spin relaxation rate of Cu and O depend differently on temperature, with Cu experiencing an enhanced, not Korring-type relaxation. In the superconducting state, however, the two spin systems appear to be "locked together" and the temperature dependence of the rates becomes (essentially) the same for Cu and O. No rate enhancement (Hebel-Slichter peak) is observed right below T_c, and the rates drop much faster than expected from the opening of a density-of-states gap as in traditional superconductors.

An in-depth discussion is presented by A. Millis[8] at this meeting. The basic idea is that a phenomenological one-component model with antiferromagnetic spin fluctuations can reconcile the many apparently diverse experimental results, e.g. the different relaxation rates for Cu and O in the normal state.

2. NORMAL STATE

The most remarkable expression of nonconventional physics of the high T_c cuprates is found in the "normal state," and particularly the transport properties. Most intensively studied is the dc resistivity, and through quantitative comparison and compilation of these data, some new insight can be gained and numerical values can be extracted for the various superconductors.

2.1 DC RESISTIVITY

The main observations are summarized here and will be discussed in the following:

- The resistivity is highly anisotropic: ρ_c, perpendicular to the Cu-O planes, is by ~ 2 to ~ 5 orders of magnitude larger than ρ_{ab}, parallel to the planes.

- ρ_{ab} and ρ_c have distinct and unusual temperature dependences: ρ_{ab} increases linearly (or very close to linearly) with temperature up to the highest measured temperature, while ρ_c has the opposite ("semiconducting") temperature dependence. This "semiconducting" characteristic is, however, not described by an activated law, but rather follows an inverse power-law T^{-n} with n being ~ 1 in best samples.

- The slope $d\rho_{ab}/dT$ of the linearly increasing in-plane resistance is surprisingly similar for several compounds.

- The transport scattering rate $\hbar/\tau$ is proportional to kT with a proportionality factor of order 1.

2.1.1 IN-PLANE RESISTIVITY ρ_{ab}

Already the first good-quality polycrystalline samples of $(La, Sr)CuO_4$ and $Ba_2YCu_3O_7$ exhibited a linear increase of resistivity between T_c and room temperature,[45] and in subsequent measurements the temperature range was substantially extended: to $\sim 1000K$ in $La_{1.85}Sr_{.15}CuO_4$, $\sim 600K$ in "123,", "Bi-2212" and $\sim 700K$ in "Bi 2201." In general, the temperature dependence is well described by $\rho(T) = \rho_0 + \propto T$. (Deviations from this form will be mentioned later). As the sample quality improves, the values for α and ρ_0 decrease, and ultimately ρ_0 vanishes in the best samples. $\rho(T)$ is then characterized only by the linearity-coefficient α. One goal of this summary is to extract the limiting values of α.

The fact that ρ is linear over such wide temperature intervals (e.g. from $\sim 10K$ to 700K in "Bi 2201") and that ρ_0 is zero rules out scattering off phonons as the dominant cause of resistance. Upper bounds of the electron-phonon coupling strength have been estimated for "214" and "123" and it was pointed out that attempts to describe $\rho(T)$ in the framework of el-phonon scattering lead to internal contradictions.[46] Using current parameters, these estimates would yield λ values in the range of 0.2 to 0.4 for both "214" and "123." The linearity of ρ up to the highest temperatures, and particularly the absence of a significant decrease of $\alpha = d\rho/dT$, reveals immediately that the mean free path is much longer than the intra-atomic space. This, together with the measured coherence length, (see 3.3) puts these superconductors in the clean limit.

In order to estimate the intrinsic values of the resistivity we need to employ several inter-comparisons involving polycrystalline samples, thin films and crystals. The data base is richest and most complete for "123." For each type of sample the lowest resistivity values are clustered in a narrow range ($\approx \pm 15\%$). This can be interpreted as a measure of the reproducibility from laboratory to laboratory and suggest that these values can be used for numerical analysis. For "123" with T_c at 90K, we find the following α values ($\mu\Omega cm/K$): sintered polycrystalline samples: 1.7-2.1; oriented thin films: 0.6-0.7; twinned crystals: 0.4-0.5. As we would expect, the crystals have the lowest resistivity, and the polycrystals the largest. A slope of $d\rho/dT$ close to 0.4-0.5 $\mu\Omega cm/K$ appears to be the intrinsic value for the resistivity averaged in the a-b plane of "123." Therefore, the mark of the highest quality crystal is a room temperature resistivity of 120-150 $\mu\Omega cm$, and no residual ρ_0.

In non-epitaxial thin films, the microscopic current distribution may not be homogeneous and the effective cross-section and path length can differ from what might be macroscopically expected. It is worth noting that quantitative studies of superconducting fluctuation effects required a re-scaling of the measured excess conductivities by a sample-dependent factor and resulted in "intrinsic resistivities" which gave $\alpha \approx 0.5$, indeed close to what we now consider the lower limit of $d\rho/dT$. The important results of those studies will be discussed in more detail later.

Polycrystals have an apparently four times higher resistivity than the best multi-domain crystals. A factor of 2 can be readily explained by effective medium theory for 2-dimensionally conducting grains and the other similarly large factor must be associated with the porosity of the samples. Well sintered samples have typically 70-80% of the theoretical density, a range where proximity to the percolation threshold requires significant corrections in the conversion of the measured resistance to conductivity.

For $(La, Sr)_2CuO_4$, the data were gathered almost exclusively from polycrystals. As in "123," the best data for α from many researchers compare well and fall in the range of 2.8-3.4, $\mu\Omega cm/K$. (The only exception, an α close to 2, might be due to preferred orientation of the crystallites.) The resistance slopes in $La_{1.85}Sr_{0.15}CuO_4$ polycrystals are thus 1.5-2 times larger than in "123" polycrystals. If the same correction factors as in "123" are applicable, then we can estimate an intrinsic slope α of $\sim 0.7 - 1\mu\Omega cm/K$. The lowest α's have been measured so far in epitaxial films, with $\alpha \simeq 2\mu\Omega cm/K$ for $La_{1.85}Sr_{0.15}CuO_4$. In general, $\rho(T)$ becomes less temperature dependent for higher Sr concentrations, and $\rho \sim T$ is found only near the optimal composition for superconductivity, $Sr_{0.15} - Sr_{0.20}$.

Single crystals of "Bi 2212" with T_c's of $\sim 85K$ have been studied in some detail.[47] Since these samples are single domain crystals, the anisotropy in the Cu-O plane could be studied: although the actual ratio α_a/α_b depends somewhat on the crystals, the average α_{ab} is always close to 0.45 ± 0.02, which is very close to the α for "123." The in-plane anisotropy is ~ 2, which is remarkable since the Bi-cuprates have no Cu-O chains.

High quality films of "Tl 2223" are reported with an α of 0.5, albeit with a residual resistance ρ_0 of $\sim 20\mu\Omega cm$.[48] The single Cu-O layer compound "Bi 2201," is remarkable with an α value of $\sim 0.7\mu\Omega cm$, which is constant within 10% from $\sim 20K$ to 700K.

For the oxygen-deficient "123" crystals with T_c's near 60K, the resistivity is close to but not perfectly linear in temperature, as it has been observed also for polycrystals. The approximate α below room temperature is $\sim 1\mu\Omega cm/K$, significantly larger than in the crystals before oxygen removal. Single crystal and polycrystal data show qualitatively different $\rho(T)$ characteristics around room temperature. While the former indicate an up-turn in $\rho(T)$, the latter show a slower-than-linear $\rho(T)$, reminiscent of saturation. (Clearly, more studies are necessary here).

The data are displayed in Fig. 3 as ρ vs. T (after a summary provided by S. Martin). The similarity of the $\rho(T)$ curves is striking, considering the various structures, and clearly point to the common Cu-O layers as the electronically active building block of all the compounds.

The slopes α vs. T_c are shown in Fig. 4. In those cases where $\rho(T)$ deviates from linearity, α is taken around room temperature. Added as open circles are the approximate α's for "123" crystals with two oxygen compositions that resulted in strongly suppressed T_c's around 30K and $\sim 50K$. These α values are from optically measured conductivities, extrapolated to dc, and more details will be discussed

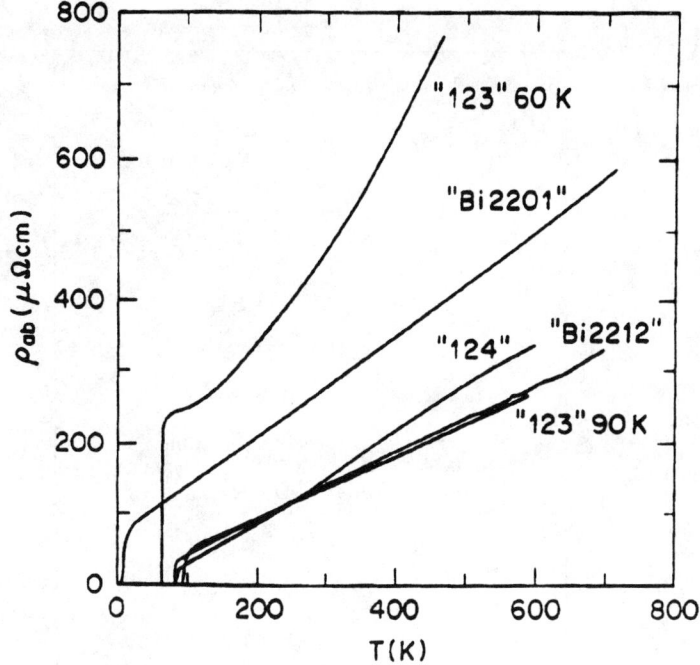

FIGURE 3 In-plane resistivity ρ_{ab} for various superconductors. ρ_{ab} is the average over ρ_a and ρ_b. (S. Martin)

later. The agreement between σ_{dc} and $\sigma_{opt}(\omega \to 0)$ is good for the 90K T_c and the 50-60K T_c sample; no dc data exist for the $\sim 30K$ T_c sample. The point worth mentioning is that a reduction of the carrier concentration (and T_c) within the same structure type increases the resistance, indicating that the dominant scattering rate remains roughly the same.

2.1.2 TRANSPORT SCATTERING RATES FROM DC MEASUREMENTS

Starting with best estimates for the intrinsic resistivities, we may proceed to calculate the characteristic scattering time τ and compare $\hbar/\tau$ with kT. In this analysis, the scattering time τ is interpreted as the parameter entering the Drude expression for the frequency-dependent conductivity $\sigma(\omega)$. At dc, $\sigma_{dc} = \omega_p \tau/4\pi$. For the plasmon frequency ω_p we have to rely on information from other experiments. In "123," we can relate ω_p either to the superconducting penetration length $\lambda(0)$, or use the (optically determined) plasma frequency. The former is associated with the number of carriers in the superconducting condensate and we call it ω_{psf}. Then through the definition $\lambda(0) = c/\omega_{psf}$, (and not incorrectly $\lambda(0) = c/\nu_{psf}$)

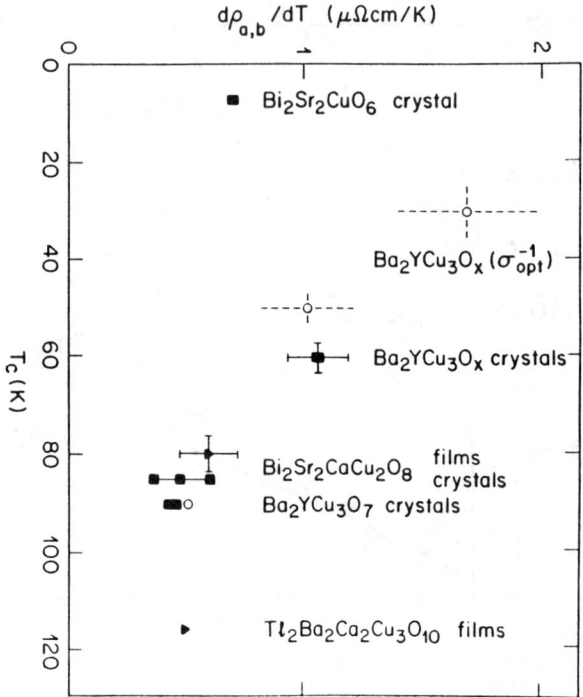

FIGURE 4 Temperature coefficient $\alpha(\mu\Omega cm/K)$ of the in-plane resistivity ρ_{ab} for various superconductors. Considering the different crystallographic structures, the similarity of the α values points to a common scattering mechanism characteristic for transport in the electronically active Cu-O layers. The values for the $Ba_2YCu_3O_x$ crystals with reduced T_c are approximations to the not strictly linear $\rho(T)$ curves. Squares: crystals; triangles: thin films, open circles: optically measured $\sigma(\omega)$ extrapolated to dc.

we obtain[49] for $\lambda(0) = 1400\text{Å}$ the plasmon frequency $\hbar\omega_{psf} = 1.4eV$. This is the same number as found in optical spectroscopy ($\omega_{psf} = 1.4eV$) for that part of the oscillator strength of $\sigma(\omega)$, which in the superconducting state collapses into the data function at $\omega = 0$. The agreement between these two numbers, incidentally, is one example of consistent results from different experiments. Using $\hbar\omega_{psf} = 1.4eV$ and $\sigma(100K) = 2 \times 10^{16}s^{-1}$, we calculate for the relaxation time $\hbar/\tau = 11.6meV$. This is an interesting result as it indicates a transport scattering rate $\hbar/\tau$ close to kT, (actually: $\hbar/\tau \cong 1.35kT$).

In an alternative way to estimate the transport relaxation rate, we replace in the above calculation ω_{psf} by ω_p associated with the total free carrier spectral weight, and not only the superfluid part of it. Applying sum rules up to energies considered to cover most of the relevant optical excitations, one can extract a plasmon energy $\hbar\omega$ close to 3eV.[50] The corresponding scattering rate $\hbar/\tau$ is then $\sim 53meV$ or $\hbar/\tau \simeq 6kT$. This second estimate of τ, however, is not appropriate, as discussed in the following.

Both of the above calculations are based on a particular simplified assumption about $\sigma(\omega)$ of the free carriers, i.e. a Drude-behavior with a single, energy-independent scattering time. The optical spectroscopy clearly reveals deviations from Drude behavior, and therefore the parameterization adopted here has its limitations. (See 2.4.1). The first type of parameterization using $\omega_{ps f}$, however, reflects more closely the actually measured spectral shape of $\sigma(\omega)$ at low frequencies, in that the coherent part of the free carrier conductivity resembles a Drude-type $\sigma(\omega)$. In any case, the scattering rate $\hbar/\tau$ is within a small factor of 1, given by the thermal energy kT. It will be interesting to compare this with the phase-breaking time derived from paraconductivity studies, (see 3.5).

The same numerical analysis can also be applied to the $Ba_2YCu_3O_x$ with an oxygen concentration to give a T_c between 50 and 60K, the only samples for which reasonably reliable dc and optical data are available. For this superconductor with an approximate $d\rho/dT \cong 1\mu\Omega cm/K$ and $\hbar\omega_{p,f} = 0.9eV$ (or $\hbar\omega_p \simeq 2.5eV$), we obtain the corresponding rates of $\hbar/\tau = 1.2kT$ (or 7-8 kT), which are very close to the rates for the 90K compounds. The difference in dc conductivity, therefore, is mainly due to a different carrier concentration, as mentioned before.

Actually, the assumption of a single scattering time is too simplistic as we know from optical spectroscopy studies. The basic conclusion, however, remains unchanged: $\hbar/\tau \approx kT$. High frequency resistance measurements on "123" crystals revealed no difference between dc and 1GHz. This is consistent with the above findings of τ and the fact that the optically measured conductivity (in the far-infrared range) extrapolates smoothly to the dc values. No additional scattering processes of significant strength in the GHz-THz range are present.[51]

2.1.3 HOW LINEAR IS $\rho_{ab}(T)$?

The linearity of $\rho_{ab}(T)$ in crystals of "Bi 2201," "Bi 2212," "123" and polycrystals is remarkable. An interpretation within the framework of conventional Bloch-Grüneisen scattering would require quite unusual parameters, particularly a very small characteristic phonon frequency and weak electron-phonon coupling. While one might argue for the presence of the required soft phonons in any one compound, it is not clear that such soft phonons are present in all of the compounds. The fact that $d\rho/dT$ is so similar, e.g. for "123" and for "Bi 2212" crystals with distinctly different crystal structures, would require essentially identical "soft" phonon frequencies.

Deviations from linearity, however, can be discerned in "248" and in "123" 60K T_c. In the former case, the resistivity is significantly smaller than in other compounds and a Bloch-Grüneisen law can be fit to slightly S-shaped data within a few percent, requiring a typical Debey-temperature of $\sim 500K$.[52] The origin of the smaller resistivity in "124" might be associated with the "Cu-O ribbons" which replace the "Cu-O chains" in the "123" compound and these might constitute parallel channels for conduction which "shunt" the in-plane transport.

The experimental situation with the "electron doped" superconductors i.e. $(Nd, Ce)_2CuO_4$, is not as well developed as with other materials. The resistivities are still very high and are even of a non-metallic nature. In crystals and thin films, $\rho(T)$ tends to saturate at low temperature, giving the whole $\rho(T)$ curve an upward-curving appearance. In this stage where materials complications are rampant, it would be too early to ascertain whether the intrinsic resistivity is linear in T or not. The situation is reminiscent of $La_{2-x}Sr_xCuO_4$ where good quality polycrystalline samples near x = 0.15-0.20 show linear $\rho - T$ behavior, but thin films and crystals show curvature in $\rho - T$.

2.1.4 CONDUCTIVITY PERPENDICULAR TO THE Cu-O PLANES

The linearity with temperature of the in-plane resistivity has become a hallmark of the superconducting cuprates, yet the electrical transport in the perpendicular direction is similarly intriguing and unusual.

Quantitative studies of ρ_c are quite difficult since most of the crystals are much thinner perpendicular to the planes than parallel to them. Further complications arise when ρ_c is larger than ρ_{ab} by many orders of magnitude and when crystalline defects provide a non-intrinsic current path across the Cu-O planes. Thus, the results for ρ_c differ somewhat on various samples.

Most data are published for "123" and examples are given in Fig. 5. Notable are the two "metallic-like" curves on crystals which the authors describe as either "fully oxygenated" or having the best "channeling characteristic."[53,54] It appears that this metallic-like characteristic is rather the exception than the rule and, at least for one example, we suggest that positive $d\rho/dT$ might be non-intrinsic. In this case, the contacting method involves a spark-attaching of gold leads to the crystals which "fuses" the wire onto the crystal. Not only does this method require reannealing to replenish the lost oxygen, but it will be detrimental to the crystalline quality and induce cross-links between the Cu-O layers. (The in-plane resistivity attests to the good, but not best sample quality.) Thus, the preponderance of semiconducting-like $\rho(T)$ curves suggests that electrical transport across the planes is indeed non-metallic in "123." Comparison with the other compounds supports this conclusion, but the definitive answer will come from further experimental studies.

Both the magnitude of ρ_c and the detailed temperature dependence in "123" depend on the samples and also on the numerical procedure to convert the measured 4 or 6-probe resistances into resistivities. The non-metallic character may be seen either as a shallow minimum at intermediate temperatures ($\sim 200K$) or as a pronounced 1/T-upturn. None of these curves can be well reproduced by a thermally activated process, but many of them can be described as $\rho_c(T) = A_cT + B_c/T$. The

Selected Experiments on High T_c Cuprates

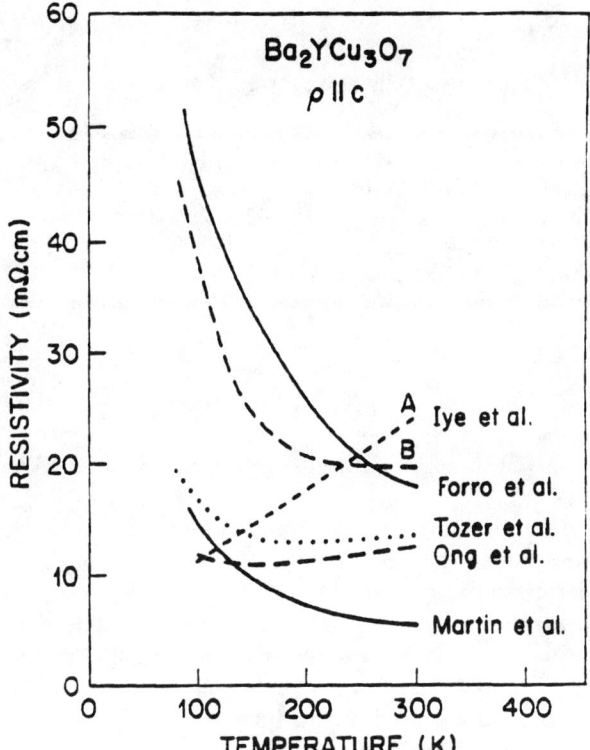

FIGURE 5 Examples of resistivities ρ_c perpendicular to the Cu-O planes in $Ba_2YCu_3O_7$ crystals. The $\rho(T)$ curves are not semiconductor-like (not thermally activated), and the ρ_c values are much larger than maximum metallic resistivities.

relative magnitude of A_c and B_c can be interpreted as a measure of electrical separation of the Cu-O double layers. In ideal crystals - without dislocations, stacking faults, twinning, etc. - one might expect A_c to vanish, and indeed, in some of the best samples the $1/T$ term dominates (e.g., lowest curve in Fig. 5).

The c-direction resistivities range from $5 - 50 m\Omega cm$. Such values are well beyond any minimum metallic conductivity estimate for these compounds (based, however, on 3d arguments) and highlight again the peculiar non-traditional transport mechanism. The ρ_c/ρ_{ab} anisotropies for "123" are 30 - 100 at room temperature and increase to $\sim$100-300 at 100K.

2.1.5 PERPENDICULAR RESISTIVITY ρ_c FOR OTHER COMPOUNDS

Several other superconducting compounds have been studied in "single crystal" form and one finds even much larger ρ_c/ρ_{ab} anisotropies than in "123." A selection

TABLE 2 Ranges of resistivity anisotropies in various cuprate superconductors

Compound	ρ_c/ρ_{ab} at 300K	ρ_c/ρ_{ab} at 100K
"123" 90K T_c	$0.3\text{-}1 \times 10^2$	$1\text{-}3 \times 10^2$
"123" 60K T_c	$1\text{-}7 \times 10^2$	$0.9\text{-}3 \times 10^3$
"Bi 2212"	$6\text{-}9 \times 10^4$	$1\text{-}4 \times 10^5$
"Bi 2201"	$0.4\text{-}3 \times 10^4$	$0.4\text{-}6 \times 10^5$

of ρ_c/ρ_{ab} data are given in Fig. 6, and numerically they are summarized in Table 2.

In "123" crystals with reduced oxygen content and a T_c of $\sim 60K$, the anisotropy is up to ~ 10 times larger than in the $90K$ T_c crystals. For this particular case, the microscopic details of oxygen removal and the influences on the local electronic structure are known: the oxygen vacancies are accommodated in the Cu-O chains with a strong tendency to form an ordered sequence of "full" and "empty" chains. The result is a reduced electronic coupling between the Cu-O double layers and the increased ρ_c/ρ_{ab} can be interpreted in this light.

The largest anisotropies are measured in the Bi double layer compound "Bi 2212," suggesting that the Bi-O secondary building blocks are a more effective barrier for charge transport than the "Cu-O chains" in "123." The values scatter significantly for the single layer Bi compound "Bi 2201". Interestingly, there is no correlation between the resistivity anisotropy (at least as it is measured macroscopically) and the occurrence of superconductivity. (A final answer requires more work. Double layer intergrowth is a concern here.)

Clearly, the observed perpendicular conductivities are by many orders of magnitude smaller than any estimate of minimum metallic conductivity. The functional dependence on temperature is not clear, except that $\rho_c(T)$ is not of a thermally activated form. In the case where $\rho_{ab} = \alpha_{ab}T$ and $\rho_c \sim 1/T$, i.e. no "cross-contamination" in the conductances, ρ_c/ρ_{ab} is proportional to T^{-2}. Such a behavior is observed for one example each of "123" (90K) and "Bi 2212."[47] The $\rho_c T$ vs. T^2 plot, which is useful to check the $\rho(T) = AT + B/T$ relationship, gives quite straight lines for many "123" samples and therefore, supports this particular form of $\rho(T)$. For the 60K "123," however, this temperature dependence is not observed.

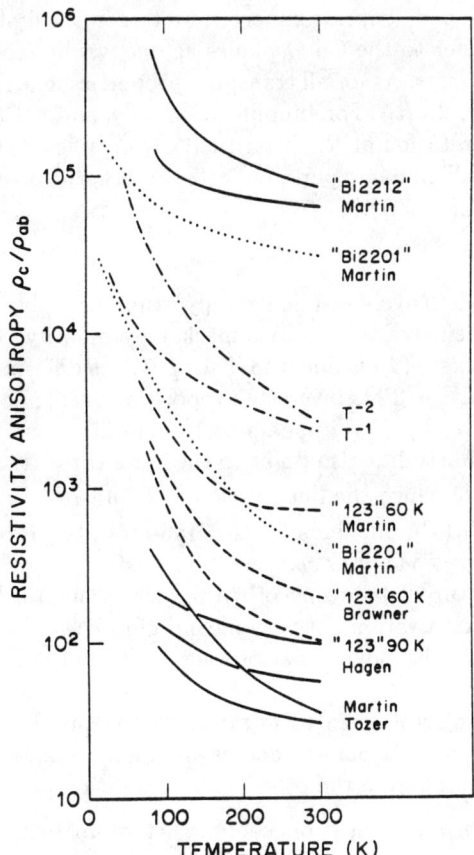

FIGURE 6 Resistivity anisotropy ρ_c/ρ_{ab} for various compounds.

2.2 HALL EFFECT

The comparisons in the previous section clearly point to the necessity of high quality single crystal work for quantitative analysis of transport properties. Many fewer Hall effect studies have been performed and thus, we are left with some ambiguities. Being aware that the resistivity of even the best polycrystalline "123" samples is about five times larger than the best single crystals, we are reluctant to put too much emphasis on transport data, particularly Hall effect, measured on polycrystalline samples. It turned out, however, that the sign of the Hall coefficient is usually the same for polycrystals and for single crystals with the magnetic field perpendicular to the Cu-O planes. It is this property that led to the terms "hole-doped" and "electron-doped" cuprate superconductors in the first place, confirming the sign of charge transfer to the Cu-O planes as one would expect it from simple electron counting arguments. As for all transport properties, one finds a wide scatter of reported Hall coefficients R_H. For simplification only, and not to imply that such a simple-minded interpretation of R_H is particularly significant, we have converted the R_H values in a Hall number per unit cell (or otherwise per Cu), using the relationship $n_H = V_{UC}/(e \cdot R_H)$.

$Ba_2YCu_3O_7$ "123"

This is the most intensively studied family, and the only one for which the anisotropy of R_H has been investigated. A remarkable similarity exists between $\rho(T)$ and $R_H(T)$ measurements: $\rho(T)$ is linear in T for ρ values differing by more than an order of magnitude, and $R_H(T)$ is inversely proportional to T, i.e. $1/R_H(T) \propto T$;[55] thus, the linearity of $n_H(T) \sim T$ appears to be another earmark of good "123" single crystals. Thin film studies also point to the close correlation between sample quality and the degree of which the linearities $\rho \propto T$ and $R_H^{-1} \propto T$ are observed.[56] Combining $\rho(T)$ and $R_H(T)$, one finds for the Hall motility $\mu_H \propto T^{-2}$, and in one example μ_H amounts to $\sim 34 cm^2/Vsec$ at 100K.

The strong temperature dependence of R_H renders a quantitative interpretation in terms of a carrier concentration not very meaningful. Nevertheless, R_H for $H \parallel c$ is positive, and is within the same order of magnitude as calculated from valence count.[53,55,56,57,58]

The Hall effect changes its sign to negative ("electron-like") when the external field is parallel to the a-b planes, and is distinctly smaller.[55] The temperature dependence is different from the other field orientation and dR_H/dT has even the opposite sign, i.e. n_H increases on cooling. At room temperature, $n_H^{H \parallel a,b}$ is larger than $n_H^{H \parallel c}$ by a factor of ~ 2 to ~ 5. The peculiar temperature dependence of R_H in $Ba_2YCu_3O_x$ is significantly altered in substituted compounds such as $Nd_{(1-x)-6}Ca_6Ba_{2-x}Cu_3O_y$, or $Nd_{1+x}Ba_{2-x}Cu_{3-z}Zn_zO_y$.[57,59] R_H becomes much less dependent on T and approaches a constant at lowest temperatures. The carrier concentration calculated from this last R_H is in good agreement with the hole concentration determined by wet chemical analysis (iodometric titration), and amounts to 0.1-0.15 mobile holes per in-plane Cu atom. Therefore, the unusual $R_H \sim T^{-1}$ relation appears to be most pronounced in $Ba_2YCu_3O_7$.

"Bi 2212"

In contrast to "123", the Hall coefficient in "Bi 2212" is only weakly T-dependent. The Hall number in crystals with $T_c > 80K$ is close to 0.3-0.4 per Cu ion in both cases.[35,59,60]

"Tl-compounds"

In "Tl 2223" polycrystals R_H is also positive ("hole-like") and $R_H(T)$ increases slightly at a lower temperature, corresponding to $n_H \cong 0.15$ holes per Cu. In "Tl 2212" the carrier concentration is lower, $n_H \cong 0.10$ holes/Cu.[57]

2.3 T_c AND CARRIER CONCENTRATION

For some time, it was thought that increasing the carrier concentration would necessarily lead to higher T_c, but this view seems not to be justified by more recent developments. Analysis of the "214" and the "123" systems using Hall effect and iodometric titration, and also of the Bi and Tl cuprates find no universal linear correlation between carrier concentration and T_c. The most striking counter-examples are "Tl 2223" ($T_c \simeq 124K$) and ($La_{1.85}Sr_{0.15}CuO_4$ ($T_c \simeq 38K$), which both have about 0.15-.02 carriers per Cu.[57,59,61] A summary of T_c vs carrier concentration is shown in Fig. 7, which is composed of data compiled by Penney et al.[59] and by Ong et al.[57] The symbols in Fig. 7 indicate the various substitutions and the names are the first authors of the corresponding publications listed by Penney et al.[59] Note that T_c is plotted vs. the number of mobile plane holes per plane Cu on the right panel and vs. the Hall number per plane Cu on the left panel. The former scale is based on (1) the determination of the total hole concentration by iodometric titration and (2) partition of the total number into mobile plane holes and localized holes, following Tokura et al.[61] The agreement between the two scales is generally good, but becomes less satisfactory when the Hall effect is strongly temperature dependent. Although the alleged $T_c - n_H$ correlation is absent, there appears to be a clustering of the highest T_c's for each compound family according to the number of closely-connected Cu-O layers, i.e. T_c is $\sim 120K$ for three layers, $\sim 80-90K$ for two layers, and $\simeq 40K$ for one layer.

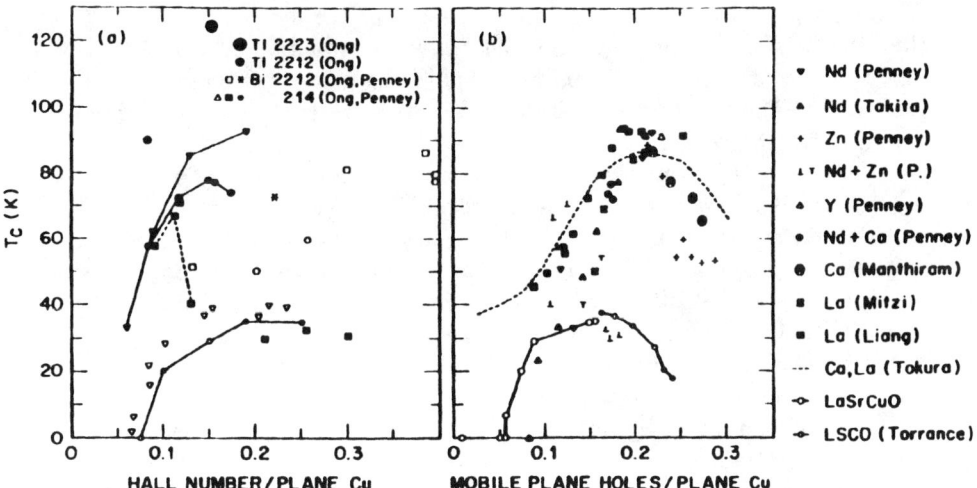

FIGURE 7 Relationship between carrier concentration and T_c for several compounds. The carrier concentration is deduced from Hall effect or from iodometric titration. (Penney et al.)

2.4 ELECTRONIC STATES NEAR E_F AND LOW FREQUENCY EXCITATIONS IN THE NORMAL STATE

Experiments that shed light on the electronic states near E_F and on low energy excitations include optical spectroscopy, inelastic light scattering, electron tunneling and photo-electron spectroscopy.

Electron tunneling spectroscopy is a powerful tool to investigate conventional superconductors, mainly to measure the superconducting gap and the electron-phonon coupling. Many attempts have been made to apply this technique to high T_c superconductors with more or less success. (See the recent paper by Kirtley[62]) Results pertinent to the superconducting state will be discussed later, only to mention here one of the most widely observed characteristics: the background dynamic conductance G = dI/dV is V-shaped with the minimum at zero bias. Single-particle tunneling is proportional to the density of states, and thus suggests that this quantity contains a term that increases linearly away from E.

2.4.1 OPTICAL SPECTROSCOPY

Optical spectroscopy provides us with the important information about the frequency dependent conductivity $\sigma(\omega)$ over a wide photon energy range from the far-infrared into the UV, i.e. from few meV to several eV. Of particular interest here are the variations of $\sigma(\omega)$ as the compounds change from insulators to superconductors. Many optical experiments have been performed both on single and polycrystalline samples. (For an almost up-to-date summary see T. Timusk and D. B. Tanner.[63]) As in other contexts, we chose here to discuss in some detail only the "123" system because a recent systematic investigation on crystals with varying oxygen concentration allows one to follow the spectral weight distribution in a clear way.[50] The experimental results from several research groups on thin films or crystals of "123" with $T_c \sim 90K$ are in close agreement.[64–67]

The essential information is given in Fig. 8 as the frequency dependence of σ, derived via Kramers-Kronig transformation from reflectivity data.[50] The spectra at 100K are shown for five crystals, one being non-superconducting. Increasing T_c is achieved by gradually filling up the vacant oxygen sites in the Cu-O chains.

For the interpretation of $\sigma(\omega)$, it is worthwhile to remember that charge introduced initially by oxygen is not transferred to the Cu-O planes, and is accommodated by electronic states associated with the Cu-O chains. The material becomes metallic and superconducting at a total oxygen content of $\sim 6.4 - 6.5$. The position of the bridging oxygen (between the chains and the planes) is a sensitive indicator of this charge transfer.[68] As a result, the marginally superconducting sample with $T_c \sim 30K$ is at this interesting composition where all the electronic states associated with the Cu-O chains are filled up and the new additional holes have to be accommodated in the Cu-O planes.

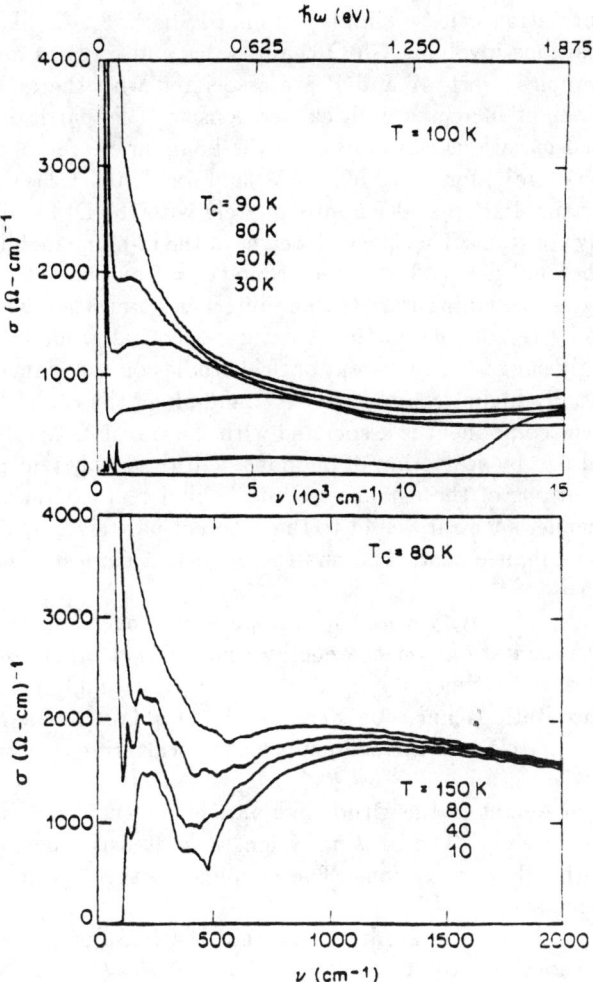

FIGURE 8 Optical conductivity spectra of $Ba_2YCu_3O_x$ crystals at different temperatures. (Orenstein et al.)

Inspection of Fig. 8 and analysis of all $\sigma(\omega)$ curves at various temperatures suggests that the additional spectral weight due to oxygen incorporation has three parts to it; Part A: this is the lowest frequency contribution to $\sigma(\omega)$, has a Drude-like shape and extrapolates very well to the dc conductivity, ($\pm 20\%$) which varies linearly with T; Part B: a contribution to $\sigma(\omega)$ which rises rapidly at low energies to peak in the 0.1-0.2 eV range; Part C: a broad and featureless conductivity which is best seen as approximately the difference between the two lowest curves (non-metallic vs. 30K T_c), extending up to the charge-transfer gap at $\sim 1.5 eV$.

The interpretation can be simply summarized: Part C is likely to be due to interband transitions involving Cu-O chain states and remains nearly constant in the metallic samples. Parts A and B are associated with the mobile carriers and their spectral weight increases with carrier density. This partitioning of $\sigma(\omega)$ can be substantiated in various ways although the boundaries are not strictly defined - for A: up to $\sim 70 meV$, for B: 70-500 meV, and for C: 0.5-1.25eV. Independent of the analysis details, Part B scales approximately with the Drude-like Part A, while Part C is nearly constant. The spectral weight of the in-plane mobile carriers is thus found in a "coherent" part (A) centered around $\omega = 0$ and an "incoherent" part (B) at higher energies. Assigning Part C to an interband transition within the O-Cu-O chains is not only reasonable on the structural-chemical grounds discussed above, it is also strongly suggested by recent optical studies on untwinned "123" crystals, where substantially higher conductivity is found along the chain direction.[54,69]

The coherent component is associated with the translational motion of the carriers influenced e.g. by scattering of phonons, and the incoherent part corresponds to optical transitions of the carriers leaving behind real excitations. The ratio of the total free carrier spectral weight to the coherent part gives by definition the low frequency mass enhancement. The analysis reveals a modest enhancement factor $1 + \lambda \cong 2 - 2.5$.

The optical conductivity $\sigma(\omega, T)$ is temperature dependent at 100K indicating interaction with excitations on the energy scale kT. In order to account for the spectral shape and the linearity in T at $\omega = 0$, the coupling of the carriers to the low-lying excitations must be weak. A fit to a Lorentzian gives a width of $\leq 2kT$ which is very close to the estimates for the scattering rate from dc-transport analysis. (2.1.2)

The spectral weight in the Drude-like part is not sufficient to account for the total free carrier weight. The missing weight is shifted to higher energies due to interactions with other excitations. The coupling to these excitations is stronger than to the processes at $\sim kT$.

In summary, the optical studies show that the free carriers interact with two classes of excitations: weakly to the ones with energies $\sim kT$ and moderately strong to others at higher energies in the range of kT to several tenths of an eV. Identifying the nature of the latter would be a big step toward understanding high T_c superconductivity in cuprates.

2.4.2 INELASTIC LIGHT SCATTERING

Unconventional low frequency excitation spectra are observed not only in optical, but also in inelastic light scattering spectroscopy. Most significant is the existence of a broad electronic excitation continuum which appears to be closely related to the free carrier conductivity discussed above. (For a recent review and references see Cooper and Klein.[70])

The broad excitation continuum extends at least up to the typical experimental limits of $\sim 100 meV$ and appears to be genuine to all high T_c cuprates. It was found in various types of $Ba_2YCu_3O_x$ samples, and in "Bi 2212" and "Tl 2223."

Such an intense scattering over a wide frequency range is unusual for metals and is most likely associated neither with intraband scattering (expected to occur up to $q.v_F \approx 50 cm^{-1} \sim 6 meV$) nor with interband transitions since it occurs in compounds with different crystallographic and electronic structures. Raman studies on untwinned $Ba_2YCu_3O_x$ crystals revealed slightly stronger electronic scattering when the exciting light is polarized along the Cu-O chains, indicating that excitations associated with the chains contribute to scattering both in the normal and superconducting states.[71] There is, however, no doubt about the electronic origin of the broad scattering because it interferes with selected Raman active phonons. These Fano-type interferences are most pronounced with the Ba modes ($116 cm^{-1}$) and the out-of-phase planar O modes ($340 cm^{-1}$), Their presence demonstrates a non-vanishing electron-phonon interaction, at least in $Ba_2YCu_3O_7$. One can study the redistribution of the electronic spectrum due to gap formation in the superconducting state by measuring the lineshape and position of the interfering phonon excitation as function of temperature.

The broad featureless Raman scattering background, as well as $\sigma(\omega, T)$ and other anomalous normal state properties, may well be linked to one particular dynamic process in this highly correlated electronic system.[72] The relative importance of spin and charge components remains to be elucidated.

2.4.3 PHOTO-EMISSION SPECTROSCOPY

High resolution and angle resolved photoemission electron spectroscopy has become an essential tool in studying the electronic properties of the cuprates. The experimental techniques have been improved enormously and energy resolutions of order 20-30 meV have been achieved. Progress in this field is so rapid that an up-to-date report is practically impossible.

It is hoped that photo-emission spectroscopy would reveal information on the central question in this field: Can the normal state be described as a Fermi liquid? No definite answer can be given yet, mainly because better experimental resolution is needed.

A set of angle resolved photoemission curves is shown in Fig. 9 for "Bi 2212." The variation of the angle allows one to probe along the $\Gamma - M$ direction in the Brillouin zone.[73-75] The authors interpret the data as follows: (1) There is no doubt about a band crossing E_F, and the crossings occur close to the points in k-space expected from band-structure calculations. (2) An analysis of cross-section as functions of photon energy indicates a mixture of 3d and 2p orbitals at E_F (approx ratio of 3:7). (3) The dispersive feature is sharpest at E_F and broadens as it moves away from E_F. (4) Only a portion of the broad background can be ascribed to secondary emission, suggesting perhaps the existence of a diffuse background of

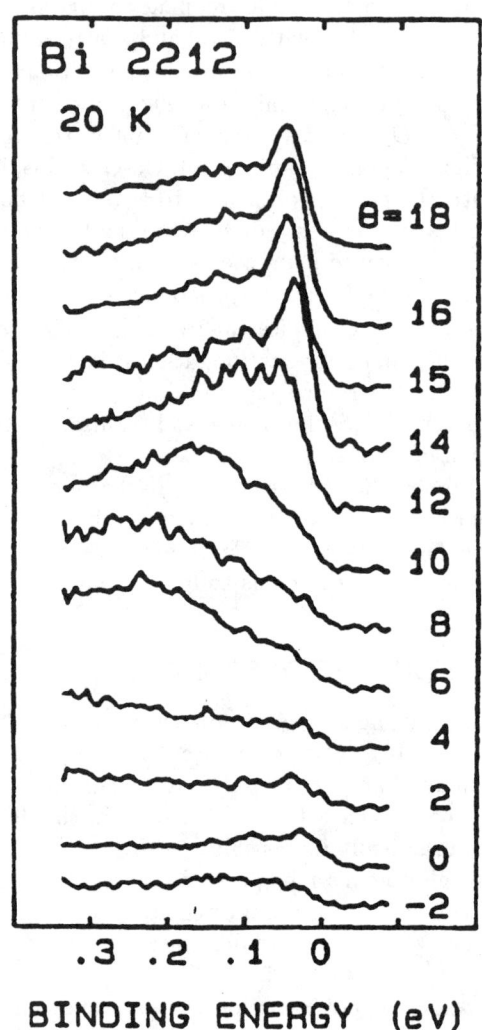

FIGURE 9 Set of angle resolved photo-emission curves from "Bi 2212" along the Γ-M direction. (Olson et al.)

electronic states. (5) The sharpness of the edge is identical to the edge measured on Pt metal within the 30 meV resolution. Any differences between them would have to be on an energy scale smaller than kT_c.

Further experimental progress will be necessary to search for the subtle differences predicted between Fermi-liquid and non-Fermi liquid theories.

3 SUPERCONDUCTING STATE

The phenomenology of high T_c superconductors may be summarized, with a gross simplification, as follows: the normal state properties are highly anomalous and hardly any anomalies are found in the superconducting state. Indeed, many probes find conventional, although anisotropic superconductivity, but a few point to significant non-conventional aspects. Much effort went into measurements of the anisotropy even of those properties that exhibit usual temperature dependence.

3.1 PAIRING AND FLUX QUANTUM ϕ_0

Within the first few months after the discovery of high T_c cuprates, measurements confirmed that the new superconductors behaved like the traditional ones in macroscopic measurements, indicating an effective charge e* = 2e. (See e.g., Gough et al.[76])

The flux line arrangement in $Ba_2YCu_3O_7$ single crystals was visualized through magnetic decoration in several external fields. From the area per line and the magnetic field strength, one finds for the flux quantum ϕ the "old" value of $\sim 2 \times 10^{-7} Oe cm^2$, i.e. hc/2e.[77]

Most recently, the Little-Parks experiments have been successfully performed on an array of μm^2-sized plaquettes ion-beam inscribed in $Ba_2YCu_3O_7$ thin films,[78] see Fig. 10. From the variation of T_c as a function of magnetic flux per square plaquette, the flux quantum can be calculated. Again, ϕ_o is within the accuracy of ±6% the same as hc/2e. From all these macroscopic measurements one concludes that the superconducting state involves pairing with total charge e* = 2e.

A more microscopic probe is Andreev reflection of electrons injected through a narrow metallic orifice into the superconductor. For energies smaller than the superconducting pair potential, the electron will be reflected or it can form a Cooper pair with another normal metal electron. In the latter case, conservation of momentum and energy leads to the reflection of a hole with opposite momentum. The resulting increase of total conductance at energies $< \Delta$ has been observed in several experiments, providing strong evidence for a paired superconducting state.[57,79,80]

3.2 MAGNETIC PENETRATION LENGTH λ

The magnetic penetration length λ and the coherence length ξ are the two important length scales to describe these extreme type II superconductors. Both of them are best known for "123" and the various measurements agree within $\sim 10-20\%$.

Already at first glance one can anticipate a large penetration length λ: the metallic cuprates are black compounds, and they lack the bright metallic luster. This means that the free carrier concentration is small, the plasma frequency ω_p is low and thus, the penetration length is long, since λ is given by c/ω_p.

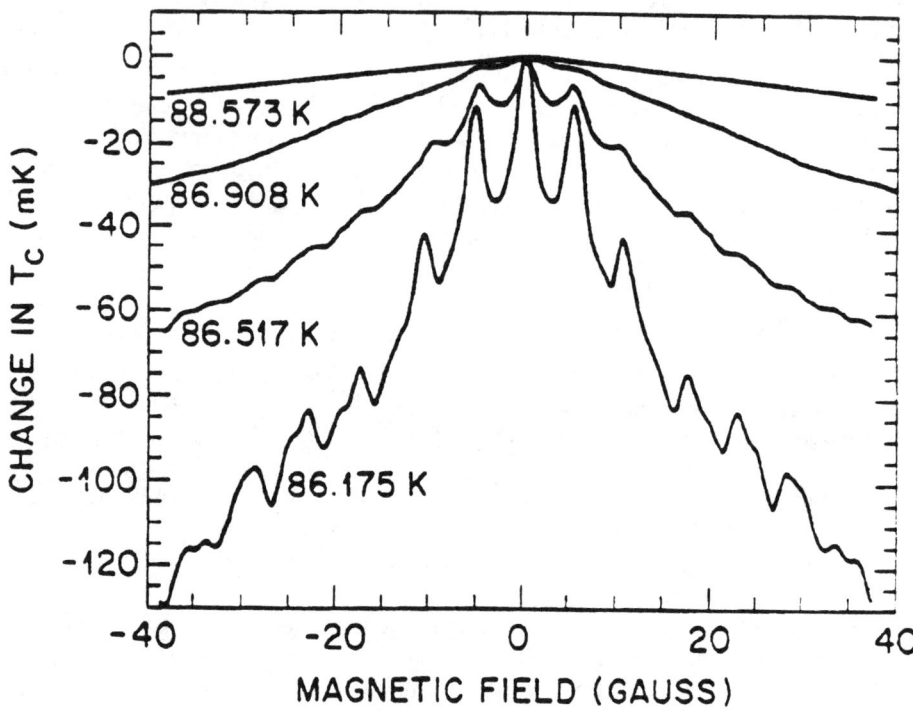

FIGURE 10 Little-Parks oscillations of T_c, from which the flux quantum is deduced to be $\phi_0 \pm 6\%$, in confirmation of earlier measurements of the ac-Josephson effect, flux jump, and flux lattice decoration experiments. (Gammel et al.)

While direct measurements of the coherence length are very difficult, λ can be measured rather directly in several ways: magnetization of thin crystals, muon spin rotation (μSR), kinetic inductance, and others. Anisotropies of λ can be estimated also from flux decoration and magnetic torque experiments. There is broad agreement among various studies at least for $Ba_2YCu_3O_7$ when the magnetic field is applied perpendicular to the Cu-O planes.

In $Ba_2YCu_3O_7$ crystals, $\lambda_{ab}(T \rightarrow 0)$ is $\sim 1400 \text{Å}$, from μSR measurements in excellent agreement with optical measurements of the superfluid density, as discussed earlier. Most significant is the temperature dependence at low temperatures, as it follows closely the conventional form. Although the data quality is not sufficient to distinguish unambiguously between the slightly different functional forms given by weak or strong coupling BCS, or by the two-fluid approximation, it is clear that $\lambda_{ab}(T)$ at the lowest temperatures is consistent with a superconducting gap opening all over the Fermi surface (if there is one). Or more specifically, $\lambda_{ab}(T)$ does not increase with temperature in a manner expected for a superconducting order parameter that vanishes at points or along lines in k-space, i.e. $\delta\lambda(T) \sim T^n$. (See also Fiory[81] for thin film measurements). In UBe_{13}, for comparison, a well-known non-S wave heavy fermion superconductor, the penetration length varies rapidly even at temperatures well below T_c. In Fig. 11, we have replotted the "123" penetration length as $\lambda^{-2}(T)$ from three μSR representative studies.[31,49,82] Such a plot over-emphasizes, of course, data scatter at low T, but it also facilitates comparison between weak-coupling and strong-coupling temperature dependences.

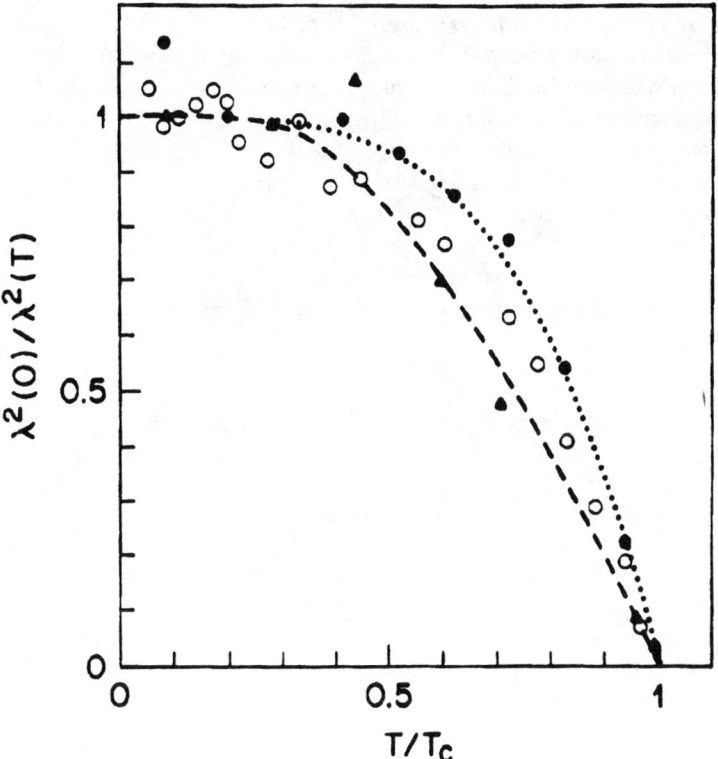

FIGURE 11 Penetration length in $Ba_2YCu_3O_7$ measured by μSR. Plotted is $\lambda^2(0)/\lambda^2(T)$ to indicate consistence with conventional T-dependence at low temperature and to facilitate comparison with weak-coupling (dashed line) and strong coupling. (The dotted line is given by the two-fluid model and serves as an approximation to strong-coupling T-dependence). Data by Harshman et al. (single crystals, triangles), Kiefl et al. (polycrystal, open symbols) and by Cooke et al. (polycrystal, closed circles).

Here the two-fluid approximation serves as a typical strong-coupling curve. The single crystal data follow more closely the "weak coupling" curve and the polycrystal data are better described by the two fluid model. This is typical for the uncertainty in many measurements, and the detailed T-dependence of λ has not been established so far. This point is critically discussed by Hebard et al.[83] From most recent H_{c1} measurements, however, one concludes that the T-dependence of λ is not weak coupling BCS-like.[84] These values of λ_{ab} of $\sim 1400 \text{\AA}$ are an average over λ_a and λ_b which are difficult to measure independently, even if large untwinned crystals are available. Their ratio λ_a/λ_b can be estimated from flux lattice decoration experiments. Instead of a regular triangular, a slightly distorted triangular lattice is observed. From the distortion, the λ-ratio can be calculated. Expressed in terms of an effective mass anisotropy, $m_a/m_b \approx 1.2 - 1.4$.[85,86] Interpratations of magnetic torque measurements rely on the validity of the effective mass approximation, which was found to describe the experimental data very well.[86] From such torque measurements, the anisotropy between in-plane and perpendicular-to-plane effective mass was found to be 5-6, in agreement with other experiments.

The various measurement techniques have their own strengths and weaknesses: μSR is a direct measurement of the field variation in a sample above H_{c1}. Calculations of λ depend on assumptions about the vortex distribution and details of the field profile within a vortex. Kinetic inductance and direct magnetization measurements, on the other hand, rely on exact knowledge of baselines in the experiments. (For details see Hebard et al.[83])

3.3 COHERENCE LENGTH ξ_{ab}, ξ_c

The superconducting coherence length is difficult to measure directly, but rather reliable estimates can be made from extrapolations to $T \to 0$ of the measured upper critical field (H_{c2}) slope at T_c. Alternatively, ξ can be extracted from fluctuation contribution to specific heat, susceptibility or conductivity. The best set of data is available for $Ba_2YCu_3O_7$ and some of the most recent results are given in Table 3. They agree in that the in-plane coherence length ξ_{ab} is $12 - 15 \text{\AA}$ and ξ_c is extremely short (only a few Å). The coherence volume is small and contains only a few Cooper pairs, which leads to pronounced fluctuation effects.

The Fermi velocity can be derived from the estimates of $\Delta(0)$ and ξ ($\hbar v_F = \pi \Delta(0) \xi_0$). For $\xi_{a,b} = 15 \text{\AA}$ and $2\Delta(0)/kT_c = 6$ (8) an in-plane averaged $v_F^{a,b}$ of 1.7 (2.2) $\times$ 10^7 cm/s follows. The same calculation would give a v_F^c of order $10^6 cm/s$, but its meaning is questionable because the transport perpendicular to the planes is not of a metallic nature ($\rho_c \gg \rho_{max.met.}, d\rho_c/dT < 0$).

TABLE 3 Estimates of the coherence length in $Ba_2YCu_3O_7$ from different studies: $\Delta\sigma$: magnetoconductance $\Delta\sigma(H)$ and $\Delta\sigma(T)$ [Reference 88], $\Delta\chi$: diamagnetic contributions to the normal state susceptibility [Reference 37]; H'_{c2}: extrapolations of the upper critical field to $T \to 0$ [Reference 89,90,91]

Experiment	$\xi_{a,b}(\text{Å})$	$\xi_c(\text{Å})$
$\Delta\sigma$	11.5	1.5
$\Delta\chi$	13.6±0.8	1.23±0.1
H'_{c2}	16.4	3.0

3.4 ENERGY GAP WITH ANISTROPY $\Delta(\vec{k},T)$

The quantity of prime interest is the superconducting order parameter Δ, its $\vec{k}$ - and temperature dependence $\Delta(\vec{k},T)$. Tunneling spectroscopy is the classical technique to study the superconducting gap and it is also widely applied in its various modifications to the cuprates. Since many physical properties are modified in the superconducting state, there are many other measurements which provide information either on the magnitude or the T-dependence of $\Delta(\vec{k},T)$. These are, e.g.: optical spectroscopy, photo-emission spectroscopy, inelastic light scattering, electrodynamic measurements, NMR-NQR, and others.

A consistent, but not yet complete, pattern emerges from the bulk of the experimental results, at least as $Ba_2YCu_3O_7$ is concerned. A redistribution of the density of states below T_c in a manner similar to traditional superconductors is found in various spectroscopic studies. The "gap" appears to be large within the Cu-O plane up to $6 - 7kT_c$, and smaller in the perpendicular direct $\sim 3 - 4kT_c)$. Most results suggest a gap function $\Delta(\vec{k})$ which has no zeroes at points or along lines in $\vec{k}$-space, i.e. $\Delta(\vec{k})$ is non-vanishing, in contrast e.g. to heavy fermion superconductors.

In the following, the various experiments will be discussed briefly with emphasis on agreed-upon features of the results.

3.4.1 TUNNELING SPECTROSCOPY

This technique was enormously successful in confirming the electron-phonon coupling theory in conventional superconductors. This proven track record makes it a prime candidate to shed some light also on the nature of high T_c cuprates. Under specific conditions, the dynamic tunneling conductance measures the single particle density of states. In reality, several complications conspire to give less than ideal results in the cuprates. Those may include: (1) The probing depth is given by the coherence length, which is very short. (2) The tunneling barrier height may be not much larger than the gap energy. (3) The composition of the surface can be different from the superconducting bulk

A sobering example are the Bi-cuprates which cleave nicely parallel to the Cu-O planes. The exposed surface is of good crystalline quality and has Bi-O as the top layers. Attempts to tunnel into this surface were only partly successful as the electronic density of states at E_F was found to be strongly suppressed, indicative of a non-metallic top layer in this electronically and crystallographically highly anisotropic material. These types of difficulties may ultimately limit our ability to study the cuprates.[92,93] Despite all the difficulties, a quite clear picture evolves when all the data are taken together. A comprehensive summary of tunneling studies is given in three recent publications.[62,94,95]

The various experimental methods differ in the nature of the tunnel barrier (vacuum, native surface barrier, controlled oxidation, artificial barrier) and the spatial extent of the tunneling contact area (point contact, large area planar junctions, break junctions, etc.). Much skill and experience are necessary to discern between tunneling spectra that reflect at least partly intrinsic behavior from artifact-dominated "spectra." Reproducibility is a necessary step in the right direction. (See e.g. for "123": Geerk et al.,[96] Fournel et al.,[97] and Gurvitch et al.[98])

Common to the tunneling spectra is a reduction of conductance below some characteristic energy Δ. Although reminiscent of the features due to the opening of a gap in traditional superconductors, several aspects are less well-developed in cuprate spectra: (1) broadened "gaps," (2) multiple "gap structures," (3) reduced density-of-states peaks at the gap edge, and (4) non-zero conductance well below the gap energy.

Here we focus first on the energy scales only, and particularly on the ratio of $2\Delta/kT_c$. In "123" there is essential agreement on a value of the gap-like structure around 18-20 meV and $2\Delta/kT_c \approx 4.5 - 5.5$. Examples of conductance vs. bias voltage spectra are shown in Fig. 12. An additional conductance peak is often observed around 30 meV. For $(La,Sr)_2CuO_4$ the gap values range from ~ 4.5 to 9.5 meV, corresponding to $2\Delta/kT_c \cong 3 - 6$.

Those gap values are found when no particular experimental precautions are taken to direct the tunneling current in a specific crystallographic direction. If this was possible, the gap anisotropy could be probed. An attempt in this direction involved the low-temperature fabrication of press-junction on freshly broken edges of oriented "123" films.[95] A clear indication for anisotropy was found, indeed, with gap values near $3.5kT_c$ for tunneling perpendicular to the Cu-O planes and $\sim 6kT_c$ in the Cu-O planes. Keeping in mind the sizeable opening of the cone in k space over which tunneling electrons probe $\Delta(\vec{k})$, the above anisotropy most likely underestimates the maximum $\Delta(\vec{k})$ anisotropy. In any case, these spectroscopic data clearly suggest considerable anisotropy of the gap function, with reduced gaps ranging

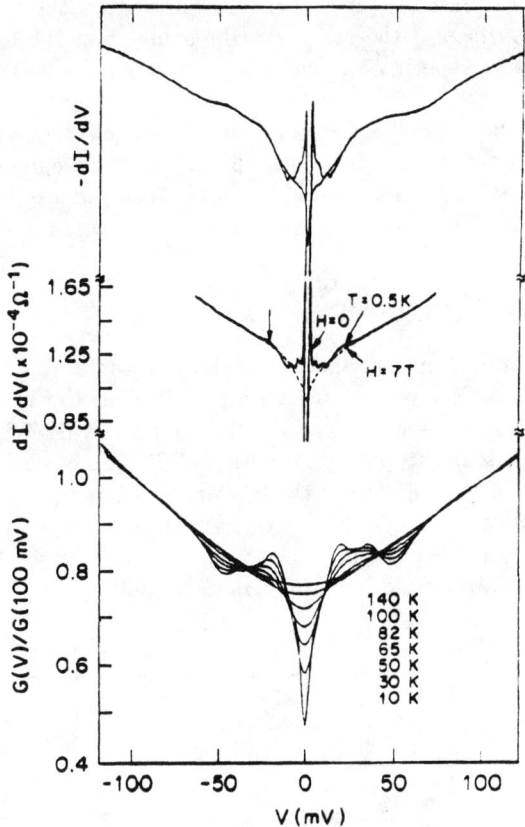

FIGURE 12 Examples of tunneling spectra on $Ba_2YCu_3O_x$. Remarkable is the V-shaped background at high bias voltage, suggesting a depression of density of extended electronic states. Below T_c the tunneling conductance peaks around $\pm 12 - 20 mV$ and is reduced in a range of $\sim \pm 10 - 15 mV$, but remains non-zero at zero bias. The sharp spikes in the two upper curves are due to superconducting counter electrodes with small energy gaps. From top to bottom: Fournel et al., Geerk et al., Gurvitch et al. 1989.

from the weak-coupling limit of ~ 3.5 to values of 6 and higher. Further support for this point is provided by a numerical simulation of tunneling spectra assuming an anisotropic gap $\Delta(\vec{k})$.[62] When averaged over the angle between junction and crystallographic orientation, the calculated curves closely resemble the point contact spectra in several non-trivial details.

In the above interpretation, it is tacitly assumed that the gap probed by tunneling at the very surface of the sample is the same as in the bulk. This, is however, questionable and experiments pertinent to it are discussed by Deutscher and Müller,[99] by Deutscher.[100]

A summary of the tunneling results is given in Fig. 13 as a plot of the "gap" value for compounds with various T_c's. (This figure is an extension of figures by Tsai et al.[95] and Kirtley[62]). The general trend is a systematic increase of Δ with T_c, and the gap-to-T_c ratio falls in a band between ~ 3.5 and ~ 6.

3.4.2 OPTICAL SPECTROSCROPY

The interpretation of optical spectra in the infrared and far-infrared is somewhat controversial. While there is a significant drop in the reflectivity in "123" around $500 cm^{-1}$, which corresponds to $\sim 8kT_c$ for the 90K T_c samples, its position in energy remains unaffected by reduction of T_c through oxygen removal.[49,63] This is evidence against being due to the opening of a superconducting gap, and is rather ascribed to an interband like excitation. At energies of $\sim 3.5kT_c$, however, the reflectivity of the single crystals reaches unity within experimental uncertainty which then might be associated with the superconducting gap.

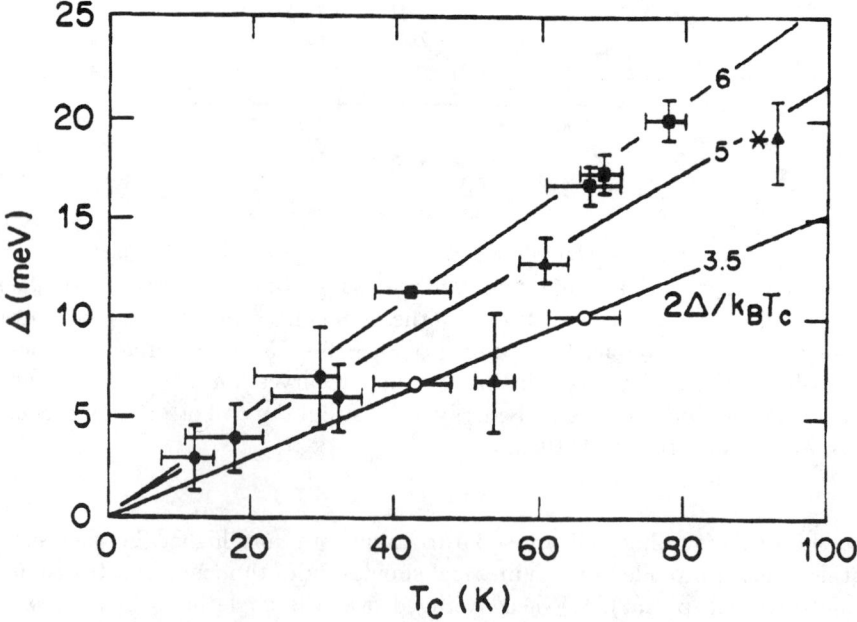

FIGURE 13 Summary of characteristic energies Δ from tunneling spectroscopy on samples with different T_c's. The ratio $2\Delta/kT_c$ ranges from ~ 3.5 to ~ 6. The former value is presumably associated with tunneling perpendicular to the Cu-O planes, and the latter with tunneling in the planes. (after Tsai et al., and Kirtley)

In a different study with the electric field vector either parallel or perpendicular to the Cu-O planes, temperature dependent reflectivity changes were interpreted as indications for an anisotropic gap with values of $2\Delta \sim 8kT_c$ in the planes and $\sim 3kT_c$ perpendicular to them.[66] In those disputes it is useful to remember that only very little modifications of the optical spectra around Δ are expected for a superconductor in the clean limit, i.e. $\tau^{-1} \ll \Delta, \xi_0 \ll l$.[63,67] While reflectivity measurements are not fully conclusive because the information is contained in the samll difference between the sample reflectivity and 100% reflectivity, absorption studies on thin films could be particularly useful to settle the question about the superconducting gap.

3.4.3 ANDREEV SCATTERING

When an electron traverses an ideal interface between a normal metal and a superconductor, and if its kinetic energy is less than the superconducting gap, then it can, together with another electron, propagate in the superconductor only as a Cooper-pair. This requires the re-emission of a hole into the normal metal in order to obey the conservation laws. Thus, Andreev scattering is seen as an increase of conductance below the gap. (Ideally, the conductance doubles).

In the experiments performed so far, the characteristics of Andreev scattering have been observed, but the values for Δ scatter somewhat. Fig. 14 shows the data with the most prominent Andreev reflection, together with the calculations for an ideal contact with a superconducting gap of 25 meV ($2\Delta/kT_c \approx 6.4$).[79] Other experiments yielded a smaller conductance enhancement ($\sim 3\%$) and also smaller gap values.[57,79] Smaller gaps might be due to reduced local T_c's, which have not been measured simultaneously.

3.4.4 PHOTO-EMISSION

A major progress in experimental capabilities, combined with the large gap values, makes it feasible to measure the superconducting gap with photo-emission techniques.

Angle integrated spectra on "Bi 2212" reflect a redistribution of density of states and gap values of $\simeq 8kT_c$.[101] More recent studies using angle resolved spectroscopy also found similar gap values in the same compound.[74] An example of an angle resolved spectrum[73] is reproduced in Fig. 15, together with the angle-integrated spectrum by Imer et al.[101] The instrumental resolution is as low as ~ 30 meV. In the angle resolved experiment, the gap is found to be the same for several points in the Brillouin zone independent of direction within the Cu-O plane, and amounts to $\sim (7 \pm 1)kT_c$.

Photo-emission, Andreev scattering and tunneling data are thus in essential agreement in observing gap parameters well in excess of the weak-coupling value of $3.5kT_c$, and the differences between $2\Delta \sim 5 - 6kT_c$ and $6-8\ kT_c$ remain to be

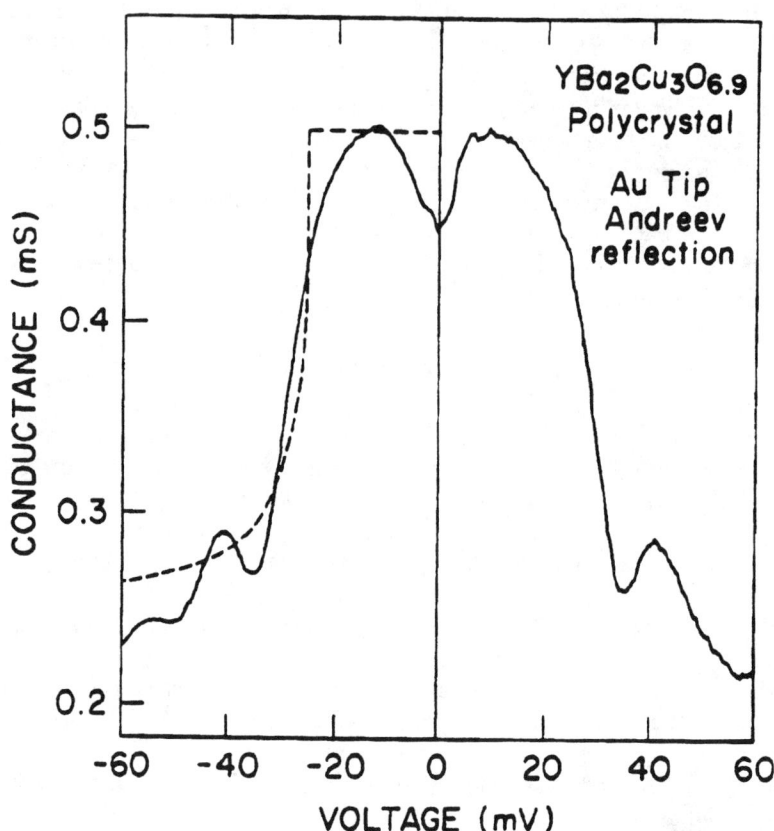

FIGURE 14 Andreev reflection of electrons at the interface between $Ba_2YCu_3O_{6.9}$ and a gold tip. The dashed line is the calculation for an ideal contact and a superconducting gap of $25meV (2\Delta/kT_c = 6.4)$. (K. Gray)

clarified. Since the probing depth of photo-emission spectroscopy is larger than that of tunneling spectroscopy - at least perpendicular to the Cu-O planes - one might speculate that the slightly smaller gap values found by tunneling are caused by the gap reduction at the surface of the superconductor.

3.4.5 INELASTIC LIGHT SCATTERING

Inelastic light scattering due to electronic processes is observed in all cuprate superconductors studied so far, giving rise to a broad scattering continuum. If these were traditional, isotropic BCS superconductors with a gap opening below

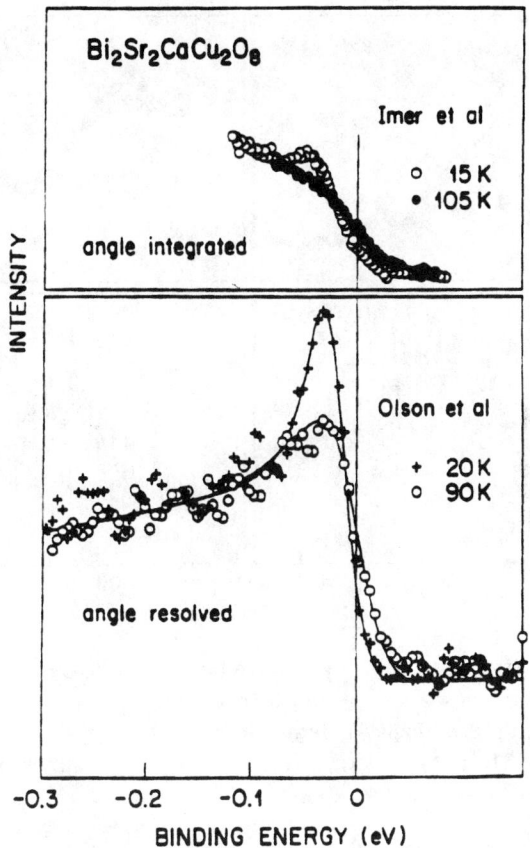

FIGURE 15 Photo-emission spectra from $Bi_2Sr_2CaCu_2O_8$ above and below T_c. The upper panel shows an angle-integrated spectrum with energy shifts of $2\Delta \simeq 8kT_c$, and the lower panel shows the results from angle resolved measurements, yielding $2\Delta = 7 \pm 1 kT_c$. (Imer et al., Olson et al.)

T_c, the scattering intensity below 2Δ would be suppressed to zero, and would be redistributed above 2Δ if the superconductor was in the dirty limit, i.e. if the scattering rate $\Gamma > 2\Delta$. Surprisingly, some aspects of Raman data are consistent with such a simple picture. Others, however, appear to contradict what we think we know about these compounds.

Examples of redistribution of scattering intensity are given in Fig. 16, showing A_{1g} and B_{1g} spectra for Ba_2YCu_3O, and equivalent spectra for "Bi 2212." Below T_c, the electronic scattering is reduced at low energies and increased at higher energies. (The dashed lines in the upper panels are 140K data.) Interestingly, the position

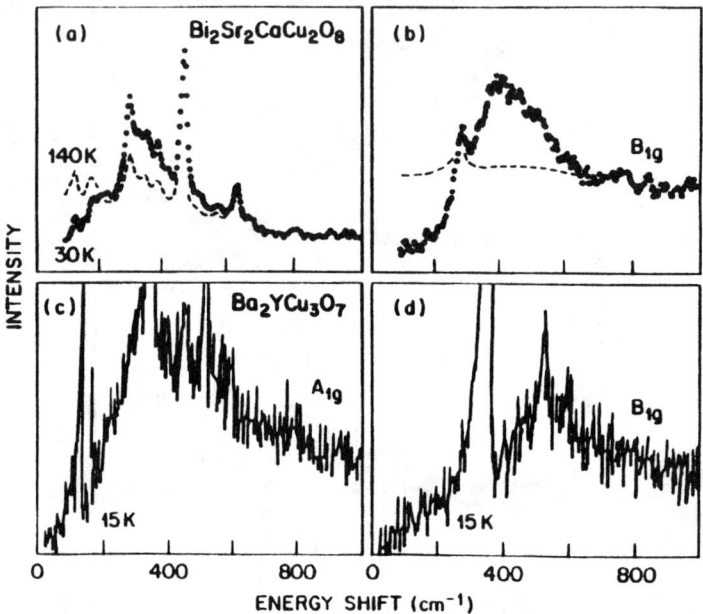

FIGURE 16 Inelastic light scattering spectra for $Ba_2YCu_3O_7$ and $Bi_2Sr_2CaCu_2O_8$. The sharp phonon lines are superimposed on a broad background. Below T_c the electronic background intensity is redistributed and peaks at different energies depending on the scattering geometry. Unlike in traditional superconductors, no sharp cut-off is observed below the peak energy. (Cooper et al., Yamanaka et al.)

of the scattering maximum depends on polarization, even when the electric field vector is in the Cu-O planes. For A_{1g} polarization ($E_I \parallel [110] \parallel E_S$), the maximum is near 42 meV (340 cm^{-1}), and for B_{1g} ($E_I \parallel [110] \perp E_S$) it is around 63 meV (510cm^{-1}). Both energies are close to gap values inferred from other measurements, corresponding to $\sim 5.5kT_c$ and $\sim 8kT_c$, respectively. A similar difference of peak positions has also been observed in $Bi_2Sr_2CaCu_2O_8$; i.e. the peak in the A_{1g} spectrum is lower than in the B_{1g} spectrum. The question whether the peaks can indeed, be interpreted as pair-breaking thresholds 2Δ is not settled.[70,102,103]

In conventional superconductors such as V_3Si and Nb_3Sn, the electronic Raman scattering well below T_c is completely suppressed at energies below 2Δ, creating a distinct onset at 2Δ. The high T_c cuprates behave differently, in that the electronic scattering decreases gradually below $\hbar\omega = 2\Delta$.

From the measured intensities corrected by the thermal Bose factor, one can estimate that $\sim 80\%$ of the normal state spectral weight below $\sim 30meV$ is removed in the superconducting state in "123." The line width and asymmetry of the 115cm^{-1} Ba phonon mode in "123" is affected by resonance interaction with the electronic continuum and thus, can serve as a probe of the spectral weight reduction below T_c. Even at T = 3K the shape of this 10 meV phonon line reflects residual density of electronic excitation; i.e. the gap is not complete.

The "continuum" of states below 2Δ is a general phenomenon and has been observed so far in $Ba_2YCu_3O_7$, in "Bi 2212" and in "Tl 2223." At least in the case of "123" it can be argued that the residual electronic scattering is not be due to surface effects, and that it is intrinsic to the compound. Furthermore, the fact that the Bi- and Tl- cuprates behave very similarly, strongly indicates a common origin. To what degree these and other experiments indicate nodes of the gap function, and thus non-S-wave pairing, will be discussed in the following section.

3.4.6 NODES IN THE GAP FUNCTION?

An important step towards understanding the coupling mechanism is to know the symmetry of the superconducting order parameter $\Delta(\vec{k}, T)$. Despite all efforts, the experimental information is still far from complete and even apparently controversial in some aspects. Leaving aside the clear indications of large gap anisotropies, probably due to differences of Δ in the planes and perpendicular to them, we look more closely at the temperature dependence of various quantities well below T_c. The expectations for various categories of order parameters are the following. If $\Delta(\vec{k})$ is finite for all $\vec{k}$'s, then the density of states is identically zero for $E < \Delta_0$ and the physical properties are at $T \to 0$ constant. When $\Delta(\vec{k})$ vanishes along lines (or at points) in $\vec{k}$ space , the density of states varies linearly or quadratically with energy and consequently the measured quantities vary in a power-law manner at $T \to 0$. (Those power-law temperature dependences were the first indications of non-conventional superconductivity in heavy fermion compounds).

The temperature dependence of the magnetic penetration length λ, the lower critical field $H_{c1}(T)$ and the spin susceptibility $K_s(T)$ are relevant to this discussion. The magnitude of $\lambda(T)$ has been discussed earlier. Thus, we mention here only that *all* data on high quality samples of "123" agree with respect to the low T behavior: up to $t \equiv T/T_c \sim 0.2 - 0.3$ the penetration length is essentially temperature independent, and does not follow a power-law in t.

Closely related to λ is H_{c1}, ($H_{c1} \sim \lambda^{-2}$ neglecting weakly T-dependent κ's) which is rather difficult to measure reliably over the entire temperature range. Using macroscopic magnetization measurements, H_{c1} identification relies on magnetic flux to enter or leave the sample, which can be drastically modified by the temperature dependence of thermally activated flux motion and other effects which are not related to H_{c1}. The resulting "$H_{c1}(T)$" curves can be grossly distorted, even showing upturns at low temperatures. A different technique (RF resonance cavity measurements) provided a complete $H_{c1}(T)$ curve (see Fig. 17 for "123"). Accordingly, $H_{c1}(T)$ is constant up to $t \sim 0.3$ and thus, confirms the $\lambda(T)$ results.[84]

The temperature dependence of the spin susceptibility below T_c, measured as NMR Knight shift K^s, provides further information on the superconductivity

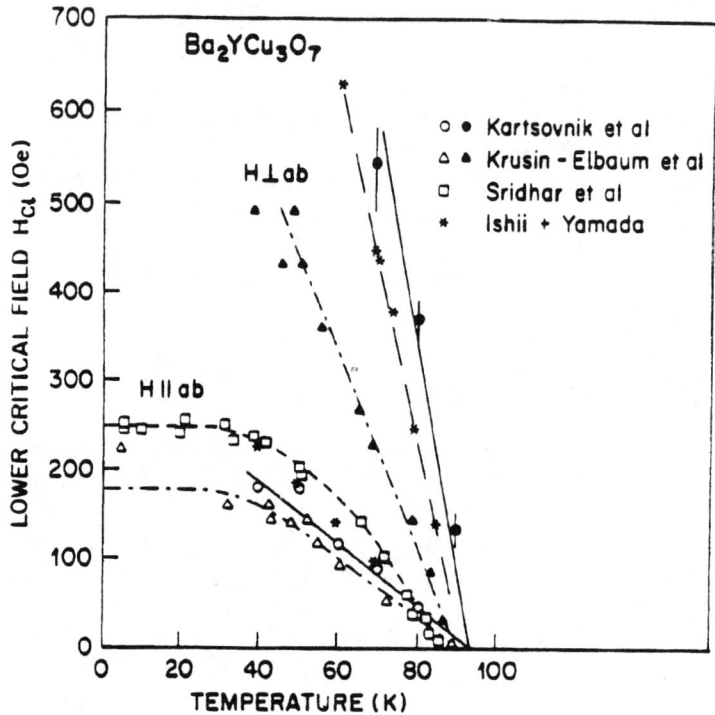

FIGURE 17 Lower critical field H_{c1} of $Ba_2YCu_3O_7$ measured with fields applied parallel or perpendicular to the Cu-O planes. Sridhar et al. used a RF resonance cavity technique, and the other studies employed dc magnetization measurements. H_{c1} remains constant up to $\sim 30K$ (open squares), in agreement with $\lambda(T)$ measurements.

state. Such studies were performed on the two Cu sites in "123".[40,41] The shift $K^s(T)$ for the chain-Cu site is well-described by a conventional BCS weak coupling Yosida function. The $K^s(T)$ data for the in-plane Cu site are influenced by both the Cu spins and the spins of the oxygen holes, which requires a decomposition of $K^s(T)$ into two parts. After this is done under some assumptions about symmetries, the resulting Cu spin contribution $\chi_s(T)$ is difficult to fit by a standard Yosida function. The measurements seem to follow closer theoretical curves for models for either strong coupling BCS s-wave or d-wave pairing. Considering the potential implications of these data, one would like to see these experiments repeated again. In any case, these data reveal different T-dependences of $K^s(T)$ at the two Cu sites and thus strongly suggest a non-uniform $\Delta(\vec{k})$.

3.5 TRANSPORT SCATTERING AND PHASE-BREAKING TIMES

A remarkable similarity is worth noting between the transport scattering time and the phase-breaking time at T_c in $Ba_2YCu_3O_7$. As discussed earlier in the context of dc resistivity and optical properties, the transport scattering time is essentially given by the thermal energy kT; i.e. $\hbar/\tau_p \cong 1.35kT$ when $\hbar\omega_p = 1.4 eV$ is used. Fluctuation effect studies give estimates for the phase-breaking time τ_ϕ. A detailed measurement of the magnetoresistance above T_c, combined with $\rho(T)$ in zero field, was analyzed in terms of superconducting fluctuations and τ_ϕ of $\sim 10^{-13} s$ was deduced at T=100K, or $\hbar/\tau_\phi = 1.0 - 1.3kT$.[88] Thus, the two characteristic times are the same within experimental uncertainties, and both are close to kT_c at T_c. The large value of $\hbar/\tau$ indicates pair-breaking must be intrinsic to this material, since T_c is highly reproducible from sample to sample. The short phase-breaking time entails (1) a large gap parameter $2\Delta_{00} \gtrsim 3.5kT_c$, (2) significant level broadening on a scale of $\hbar/\tau$ and smearing of the density of states distribution, and (3) a gapless region near T_c.

The close similarity between the charge transport scattering and pair-breaking rates strongly suggests that both are associated with the same microscopic mechanism. In analogy with strongly-coupled electron-phonon superconductors, one might speculate that the causes of pair-breaking and superconducting pairing are intimately related.

CONCLUDING REMARKS

The field of high T_c superconductivity is in an interesting phase. The early emphasis on reaching higher and higher T_c's led to the discovery of several compound families with quite complex unit cells and intricate crystal chemistry. Layers of Cu-O are the common, electronically active building blocks with adjacent charge reservoirs providing the proper number of carriers in the Cu-O layers. Guided by the idea that the essential physics is the same in all cuprate superconductors, the physicists have turned their attention to a few model systems, such as $Ba_2YCu_3O_7$, which can be prepared as crystals of sufficient size for most experiments. This has led to a significant development in the more recent past: experimental results from various research groups are in quantitative agreement and form a solid foundation for our emerging understanding of high T_c cuprates. Examples are transport, optical and spectroscopy and spin relaxation spectroscopy and spin relaxation measurements. Microscopic parameters like coherence length and penetration depth are now known with $\sim 10\%$ uncertainty, and progress in the experimental techniques have revealed information of unprecedented detail about electronic states and low energy excitations.

In near future we can expect answers to one of the central questions: To what degree is the conventional Fermi liquid picture applicable? It will be necessary to investigate more subtle aspects, such as anisotropies within the Cu-O planes or the significance of the Cu-O plane buckling, in order to identify the truly general properties, unaffected by compound specific details, e.g. the Cu-O chains in $Ba_2YCu_3O_7$ or the superstructure in the bismuth cuprates. Ultimately, we hope to be able to measure spin and charge excitations over a wide energy and momentum range. In the next few years these efforts are unlikely to produce headlines for the daily newspapers and the real excitement will be contributing to the solution of one of the most challenging problems in modern Condensed Matter Physics.

ACKNOWLEDGEMENT

I would like to thank many colleagues and friends with whom I have the privilege to share the excitement of the early development of a fascinating chapter in condensed matter physics. In particular, I am grateful to R. J. Cava for stimulating and fruitful collaborations, and to G. Aeppli, P. A. Fleury, P. Littlewood, S. Martin, A. Millis, P. Ong. J. Orenstein, T. Palstra, T. Penney, G. A. Thomas, C. M. Varma and R. Walstedt for helpful and stimulating discussions, and to L. W. Rupp, Jr. for excellent technical support.

REFERENCES

1. J. G. Bednorz and K. A. Müller, Z. Phys. **B64** 189 (1986).
2. P. W. Anderson, in "Strong Correlation and Superconductivity," p. 2, Ref. 7.
3. C. M. Varma, Int'l. J. of Modern Physics (1990).
4. "Physical Properties of High Temperature Superconductors I," Donald M. Ginsberg, (ed.) World Scientific Publishing Co., (1989).
5. "Mechanisms of High Temperature Superconductivity," H. Kamimura and A. Oshiyama (eds.), Springer Series in Materials Science, Vol. II, Springer-Verlag Berlin Heidelberg, 1989.
6. IBM J. Res. Devel. **33** No. 3, 197-404, (1989).
7. "Strong Correlation and Superconductivity," H. Fukuyama, S. Maekawa and A. P. Malozemoff (eds.) Springer Series in Solid State Sciences, Vol. 89, Springer-Verlag Berlin Heidelberg, 1989.
8. A. J. Millis, this volume.
9. S. Chakravarty, this volume.
10. R. J. Birgeneau and G. Shirane, in Ref. 4, p. 152-211.
11. P. B. Allen, Z. Fisk and A. Migliori, in Ref. 4, p. 213-264.
12. D. Vaknin, S. K. Sinha, D. E. Moncton, D. C. Johnston, J. M. Newsam, C. R. Safinya and H. E. King, Jr., Phys. Rev. Lett. **58** 2802 (1987).

13. S-W. Cheong, Z. Fisk, J. O. Willis, S. E. Brown, J. D. Thompson, J. P. Remeika, A. S. Cooper, R. M. Aikin, D. Schiferl and G. Grüner, Solid State Comm. 65 111 (1988).
14. T. Thio, T. R. Thurston, N. W. Preyer, P. J. Picone, M. A. Kastner, H. P. Jenssen, D. R. Gabbe, C. Y. Chen, R. J. Birgeneau and A. Aharony, Phys. Rev. **B38** 905 (1988).
16. S. Chakravarty, B. I. Halperin and D. L. Nelson, Phys. Rev. Lett. 60 1057 (1988).
17. G. Aeppli, S. M. Hayden, H. A. Mook, Z. Fisk, S-W. Cheong, D. Rytz, J. P. Remeika, G. P. Espinosa and A. S. Cooper, Phys. Rev. Lett. 62 2052 (1989).
18. T. Becher and G. Reiter, Phys. Rev. Lett. 63 1004 (1989).
19. K. B. Lyons, P. A. Fleury, L. F. Schneemeyer and J. V. Waszczak, Phys. Rev. Lett. 60 732 (1988).
20. R. R. P. Singh, P. A. Fleury, K. B. Lyons and P. E. Sulewski, Phys. Rev. Lett. 62 2736 (1989).
21. P. E. Sulewski, P. A. Fleury, K. B. Lyons, S-W. Cheong and Z. Fisk, Phys. Rev. B **41**, 225 (1990).
22. M. Sato, S. Shamoto, J. M. Tranquada, G. Shirane, B. Keiner, Phys. Rev. Lett. 61 1317 (1988).
23. M. J. Jurgens, P. Burlet, C. Vettier, L. P. Regnould, J. Y. Henry, J. Rossat-Mignod, H. Noel, M. Potel, P. Gougeou and J. C. Levet,Physica **156** and **157** 846 (1989).
24. J. M. Tranquada, G. Shirane, B. Keimer, S. Shamota and M. Sato, Phys. Rev. B **40**, 4503 (1989).
25. J. Rossat-Mignod *et al.*, Proc. Int'l. Conf. on the Physics of Highly Correlated Electron Systems, Santa Fe, NM, Sept. 1989.
26. R. A. Cowley, Phys. Rev. **B21** 4038 (1980).
27. K. Yamada, E. Kudo, Y. Endoh, Y. Hidaka, M. Oda, M.Suzuki and T. Murakami, Solid State Comm. **64** 753 (1987).
28. J. L. Budnick, B. Chamberland, D. P. Yang, Ch. Niedermayer, A. Golnik, E. Recknagel, M. Rossmanith and A. Weidinger, Europhys. Lett. 5 651 (1988).
29. D. R. Harshman, G. Aeppli, G. P. Espinosa, A. S. Cooper, J. P. Remeika, E. J. Ansaldo, T. M. Riseman, D. Ll. Williams, D. R. Noakes, B. Ellman and T. F. Rosenbaum, Phys. Rev. **B38** 852 (1988).
30. Y. J. Uemura *et al.*, Phys. Rev. Lett. 59 1045 (1987); Physica (Amsterdam) **153-155C** 769 (1988).
31. R. T. Kiefl *et al.*, Phys. Rev. Lett.63 2136 (1989).
32. N. Wada, H. Muro-Oka, Y. Nakamura and K. Kumagai, Physica C **157** 453 (1989).
33. H. Yasuoka, T. Imai and T. Shimizu, in Ref B, p. 254.
34. D. C. Johnston, S. K. Sinha, A. J. Jacobson, and J. M. Newson, Physica C **153-155** 572 (1988).
35. H. Takagi, Y. Tokura and S. Uchida, in Ref. 5, p. 238.
36. D. C. Johnston, Phys. Rev. Lett. 60 762 (1988).
37. W. C. Lee, R. A. Klemm and D. C. Johnston, Phys. Rev. Lett. 63 1012 (1989).

38. Y. Yamaguchi, M. Tokumoto, S. Waki, Y. Nakagawa and Y. Kimura, "Proc. of the Tsukuba Seminar on High T_c Superconductivity," p. 31, K. Masuda, T. Arai, I. Iguchi and R. Yoshizaki (eds)., University of Tsukuba (1989).
39. R. E. Walstedt and W. W. Warren, Jr., in Ref. 5, p. 137.
40. M. Takigawa, P. C. Hammel, R. H. Heffner and Z. Fisk, Phys. Rev. **B39** 7371 (1989)).
41. H. Alloul, T. Ohno and P. Mendels, Phys. Rev. Lett. **63** 1700 (1989).
42. D. J. Durand, S. E. Barret, C. H. Pennington, C. P. Slichter, E. D. Bukowski, T. A. Friedman, J. P. Rice and D. M. Ginsberg, in Ref. 7, p. 244.
43. F. Mila and T. M. Rice, Physica C **157** 561 (1989).
44. A. Junod, "Physical Properties of High Temperature Superconductors II," D. M. Ginsberg (ed.), World Scientific Publishing Co. (1990).
45. R. J. Cava, B. Batlogg. R. B. van Dover, D. W. Murphy, S. Sunshine, T. Siegrist, J. P. Remeika, E. A. Rietman, S. Zahurak and G. P. Espinosa, Phys. Rev. Lett. **58** 1676 (1987).
46. M. Gurvitch and A. P. Fiory, Phys. Rev. Lett. **59** 1337 (1988).
47. S. Martin, A. T. Fiory, R. M. Fleming, L. F. Schneemeyer and J. V. Waszczak, Phys. Rev. Lett.**60** 2194 (1988).
48. M. Hong, J. Kwo, C. H. Chen, A. R. Kortan, D. D. Bacon and S. H. Liou, Proc. 16th Int'l. Conf. Metallurgical Coatings, San Diego, CA, in "Thin Solid Films," (1989).
49. D. R. Harshman, L. F. Schneemeyer, J. V. Waszczak, G. Aeppli, R. J. Cava, B. Batlogg, L. W. Rupp, Jr., E. J. Ansaldo, D. Ll. Williams, Phys. Rev. **B39** 851 (1989).
50. J. Orenstein, G. A. Thomas, A. J. Millis, S. L. Cooper, D. H. Rapkine, T. Timusk, L. F. Schneemeyer and J. V. Waszczak, (Phys. Rev. B, 1990).
51. A. Zettl, A. Behrooz, G. Briceno, W. N. Creager, M. F. Crommie, S. Hoen, and P. Pinsukanjana, Ref. D, 249.
52. S. Martin, M. Gurvitch, C. E. Rice, A. F. Hebard, P. L. Gammel, R. M. Fleming and A. T. Fiory, Phys. Rev. **B39** 9611 (1989).
53. Y. Iye, in Ref. 5, p. 263.
54. Yu. A. Ossipyan, V. B. Timofeev and I. F. Schegolev, Physica C **153-155** 1133 (1988).
55. S. W. Tozer, A. W. Kleinasser, T. Penney, D. Kaiser and F. Holtzberg, Phys. Rev. Lett. **59** 1768 (1987).
56. H. L. Störmer, A. F. J. Levi, K. W. Baldwin *et al.*, Phys. Rev. **B38**, 2472 (1988).
57. N. P. Ong, T. W. Jing, Z. Z. Wang, J. Clayhold, S. J. Hagen and T. R. Chien; Ref. 7, p. 204.
58. L. Forro, M. Raki, C. Ayache, P. C. E. Stamp, J. Y. Henry and J.Rossat-Mignod, Physica C **153-155** 1357 (1988).
59. T. Penney, M. W. Shafer and B. L. Olson, Physica C, **162-164** 65 (1989). I am grateful to these authors for providing an updated version of the $T_c - n$ plot.
60. J. Clayhold, N. P. Ong, P. H. Hor and C. W. Chin, Phys. Rev. **B38** 7016 (1988).

61. Y. Tokura, J. B. Torrance, T. C. Huang and A. I. Nazzal, Phys. Rev. **B38** 7156 (1988).
62. J. R. Kirtley, Intl. J. of Modern Physics B, Feb. 1 (1990).
63. T. Timusk and D. B. Tanner, in Ref. 4, p. 339.
64. S. L. Cooper, G. A. Thomas, J. Orenstein, D. H. Rapkine, M. Capizzi, T. Timusk, A. J. Millis, L. F. Schneemeyer and J. V. Waszczak, Phys. Rev. **B40** 11358 (1989).
65. J. Schützmann, W. Ose, J. Keller, K. F. Renk, B. Roas, L. Schultz and G. Saemann-Ischenko, Europhys. Lett. **8** 679 (1989).
66. R. T. Collins, Z. Schlesinger, F. Holtzberg and C. Feild, Phys. Rev. Lett. **63** 422 (1989).
67. K. Kamaras, S. L. Herr, C. D. Porter, N. Tache, D. B. Tanner, S. Etemad, T. Venkatesan, E. Chase, A. Inam, X. D. Wu, M. S. Hedge and B. Dutta, Phys. Rev. Lett. **64** 84 (1990).
68. R. J. Cava, B. Batlogg, K. M. Rabe, E. A. Rietman, P. K. Gallagher and L. W. Rupp, Jr., Physica C **156** 523 (1988).
69. I. Bosovic, K. Char, S. J. B. Yoo, A. Kapitulnik, M. R. Beasley, T. H. Geballe, Z. Z. Wang, S. Hagen, N. P. Ong, D. E. Aspenes and M. K. Kelly, Phys. Rev. **B38** 5077 (1988).
70. S. L. Cooper and M. V. Klein, Comments on Solid State Physics (1990) and references therein.
71. F. Slakey, S. L. Cooper, M. V. Klein, J. P. Rice, D. M. Ginsberg, Phys. Rev. B.
72. C. M. Varma, P. B. Littlewood, S. Schmitt-Rink, E. Abrahams and A. E. Ruckenstein, Phys. Rev. Lett. **26** 1996 (1989).
73. C. G. Olson, R. Liu, D. W. Lynch, B. W. Veal, Y. C. Chang, P. Z. Jiang, J. Z. Liu, A. P. Paulikas, A. J. Arko and R. S. List, Physica C **162-164** 1697 (1989).
74. C. G. Olson, R. Liu, A. B. Yang, D. W. Lynch, A. J. Arko, R. S. List, B. W. Veal, Y. C. Chang, P. Z. Jiang, A. P. Paulikas, Science, **245** 731 (1989).
75. A. J. Arko, R. S. List, R. J. Bartlett, S-W. Cheong, Z. Fisk, J. D. Thompson, C. G. Olson, A-B. Yong, R. Liu, C. Gu, B. W. Veal, J. Z. Liu, A. P. Paulikas, K. Vandervoort, H. Claus and J. C. Campuzano, Proc. Int'l. Conf. on the Physics of Highly Correlated Electron Systems, Sept. 11-15, 1989; Santa Fe, NM.
76. C. E. Gough, M. S. Colclough, E. M. Forgan, R. G. Jordan, M. N. Keene, C. M. Muirhead, A. I. M. Rae, N. Thomas, J. S. Abell and S. Sutton, Nature **326** 85 (1987).
77. P. L. Gammel, D. J. Bishop, G. J. Dolan, J. R. Kwo, C. A. Murray, L. F. Schneemeyer and J. V. Waszczak, Phys. Rev. Lett. **59** 2592 (1987).
78. P. L. Gammel, P. A. Polakos, C. E. Rice, L. R. Harriot, D. J. Bishop, Phys. Rev. B (1990).
79. P. J. M. van Bentum, H. F. C. Hoevers, H. van Kempen, L. E. C. van de Leemput, M. J. M. F. de Nivelle, L. W. M. Schreurs, R. J. M. Smokers and P. A. A. Teunissen, Physica C **153-155** 1718 (1988); also Phys. Rev. **B36** 843 (1987).
80. K. E. Gray, Mod. Phys. Lett. **B2** 1125 (1988).

81. A. T. Fiory, Phys. Rev. Lett. **61** 1419 (1988);
82. D. W. Cooke et al. (preprint).
83. A. F. Hebard, A. T. Fiory and D. R. Harshman, Phys. Rev. Lett. **62** 2885 (1989).
84. S. Sridhar, Dong-Ho Wu and W. Kennedy, Phys. Rev. Lett. **63** 1873 (1989).
85. G. J. Dolan, F. Holtzberg, C. Feild and T. R. Dinger, Phys. Rev. Lett. **62** 2184 (1989).
86. L. Ya. Vinnikov, I. V. Grigorieva, L. A. Gurevich, Yu. A. Ossipyan, in "High Temperature Superconductivity from Russia," p. 171, A. I. Larkin and N. V. Zavaritsky, eds., World Scientific, 1989.
87. D. E. Farrell, C. M. Williams, S. A. Wolf, N. B. Bansal and V. G. Kogan, Phys. Rev. Lett. **61** 2805 (1988)).
88. A. G. Aronov, S. Hikami and A. I. Larkin, Phys. Rev. Lett. **62** 965 (1989).
89. U. Welp, W. K. Kwok, G. W. Crabtree, K. G. Vandervoort and J. Z. Liv, Phys. Rev. Lett. **62** 1908 (1989).
90. T. Ruf, C. Thomsen, R. Liu and M. Cardona, Phys. Rev. **B38** 11985 (1988).
91. M. M. Fang, V. G. Kogan, D. K. Finnemore, J. R. Clem, L. S. Chumbley and D. E. Ferrel, Phys. Rev. **B37** 2334 (1988).
92. C. K. Shih, R. M. Feenstra, J. R. Kirtley and G. V. Chandrashekhar, Phys. Rev. **B40** 2682 (1989).
93. M. Tanaka, T. Takahashi, H. Katayama-Yoshida, S. Yamazaki, M. Fujinami, Y. Okabe, W. Mizukami, M. Ono and K. Kajimura, Nature, **339** 691 (1989)).
94. M. Lee, A. Kapitulnik and M. R. Beasley, p. 220, Ref. 5.
95. J. S. Tsai, I. Takeuchi, J. Fujitu, T. Yoshitake, S. Miura, S. Tanaka, T. Terashima, Y. Bando, K. Iijima and K. Yamamoto, Physica C **153-155** .R 1385 (1988), and in Ref. 5, p. 229.
96. J. Geerk, X. X. Xi, and G. Linker, Zeitschrift für Physik **B73** 329 (1988).
97. A. Fournel, B. Oujia, J. B. Sorbier, J. Noel, J. C. Levet, M. Potel and P. Gougeon, Europhys. Lett. **6** 653 (1988).
98. M. Gurvitch, J. M. Valles, Jr., A. M. Cuculo, R. C. Dynes, J. P. Garno, L. F. Schneemeyer and J. V. Waszczak, Phys. Rev. Lett. **63** 1008 (1989).
99. G. Deutscher and K. A. Müller, Phys. Rev. Lett. **59** 1745 (1987).
100. G. Deutscher, in Ref. 6, p. 293.
101. J. M. Imer, F. Patthey, B. Darbel, W-D. Schneider, Y. Baer, Y. Petroff and A. Zettl, Phys. Rev. Lett. **62** 336 (1989).
102. R. Hackl, W. Gläser, P. Müller, D. Einzel and K. Andres, Phys. Rev. **B38** 7133 (1988).
103. A. Yamanaka, F. Minami, K. Inoue and S. Takegawa, ibid., p. 115.

DISCUSSION

D. J. Scalapino: Bertram, would you clarify how big the gap is? You said something to the effect that it might be 3.5 to 5 times $k_B T_c$.

B. Batlogg: That's what tunneling data tell us, up to about 6.0.

D. J. Scalapino: From your lecture I got a picture that in some of the data you like very much this ratio could be even larger, of order seven or eight.

B. Batlogg: Yes.

D. J. Scalapino: So, that's why I'm confused when you say 3.5 to 5 on that viewgraph.

B. Batlogg: What I mean is that these tunneling measurements, which supposedly tunneled perpendicular to the planes gave a ratio which is around 3.5. This was the only lower limit that I know of. The upper limit of eight, in the plane, comes from one photoemission experiment and to a lesser degree, of six to seven from tunneling. Remember that tunneling probes a "core" in q-space. In addition, there are not sharp edges in tunneling spectra. Kirtley and collaborators simulated tunneling spectra assuming a gap anistropy and smearing. One can get good agreement with experiment which picks up this anisotropy.

H. Ott: One other possibility could be that the short coherence length depresses the gap at the surface, and that tunneling gives the smaller gap.

B. Batlogg: Yes, as I mentioned.

N-P Ong: I've made a survey of Hall measurements on high T_c systems in the last three years (to appear in Don Ginsberg's Volume 2 on Physical Properties). There is remarkable convergence of the Hall results from various groups. One important point is that the temperature dependence is really there, even though it's very weak in certain systems, like the 2-plane BSCCO compound. It is very strong in the 3-plane bismuth compound, and somewhat weaker in the 3-plane Thallium compound, for reasons that we don't understand. But the new Nd superconductor is interesting because the Hall effect in the normal state is observable down to low temperatures (~ 1 K). Below 80K, the temperature dependence becomes very large, strikingly similar to YBCO.

B. Batlogg: Thank you for the comment.

P. W. Anderson: I wanted to make a comment on the comment. The comment on the comment is that the temperature dependence of the Hall effect that you eyeball is the Hall effect temperature dependence relative to the general magnitude of the Hall effect. If you measure that in terms of dn_H/dT, or if you measure even more in terms of the d/dT of R_H itself, because R_H's magnitude varies by enormous quantities, it tends to be very comparable from one system to another and to be very highly correlated with T_c. But in particular, dR_H/dT doesn't vary between the samples that look rather flat or the samples that look very steep ... doesn't vary by more than a factor of two. So it's a universal temperature dependence even though its size appears to be different.

C. M. Varma: What is the temperature dependence?

P. W. Anderson: It tends to be log squared.

D. Pines: I have a comment and a question. The comment is that as you showed, the NMR experiments seem to unambiguously say that the gap is quite different on the chains and on the planes in the $YBa_2Cu_3O_7$ as measured both on copper and oxygen. For the chains, it's not far from a BCS gap, while for the planes it seems to be a good deal larger and 2Δ is somewhere between 4 and 5 kT_c. My question is, given that, how can we attempt to unambiguously deduce gaps in experiments that do not distinguish between chain and plane of the sites?

B. Batlogg: That's experimentally very difficult, of course. If you knew what your top surface is in a photoemission experiment, and you know what escape depth is, approximately, from the photon energy, you can make a guess whether you're looking at the plains or by the chains. My guess is that ....the photoemission would pick up the gap in the plane, just simply because it's so large, but there's a assumption which I cannot justify. The data shown are for Bi-cuprate, however.

H. Ott: There is certainly this internal anisotropy, I think, from various NMR results, you would say that different sites see different gaps.

A. Malozemoff: I'd like to know your judgment about the gathering evidence for persistence of the gap, and maybe some kind of short range order, above T_c. I'm thinking for instance of a lanthanum-strontium neutron study of the suppression of the intensity of the magnetic fluctuations well below above the T_c in that material. You touched on it a bit in terms of the temperature dependence of the gap in tunneling. I know in their optic conductivity measurements Schlesinger *et al.* are talking about a temperature dependence of the gap which is not coming down once they take into account thermal smearing. Do you think this is a serious possibilty?

B. Batlogg: I would like to quote a good friend, Zachary Fisk, who said: "I'm here to present the data, not to think about them." (Laughter) But by leaving this data out I made a judgment already. I don't want to go into great depth about this apparent decrease of neutron scattering intensity in lanthanum strontium copper oxide. I'm not sure if that is relevant at all. These are measurements of a susceptibility in a certain q and a certain energy window.

Secondly, the question about the persistance of an optical gap to higher temperature is a bone of contention. We know that this 500 wavenumber-feature in reflectivity in some samples appears to persist to higher temperature. We are not sure if this is not an artifact, not of experimental nature, but rather of some coincidence in interband absorptions there. So there is no clear evidence for or against the persistance of any gap above T_c.

H. Ott: I just got the graph from where he points out that this decrease seems to be an artifact of not taking into account the Bose factor; if you take that out, the gap looks different as a function of temperature.

B. Batlogg: Gentlemen, I would respectfully prefer not to talk about data of authors who are not sitting here, and therefore I reserve judgment on that. But the statement is certainly right.

S. Chakravarty: I think what you said about scattering a few minutes ago is not right. They measure $S(q,\omega)$, and not just q equals (π,π). The q-integrated function $S(\omega)$ decreases as T is lowered below 100 degrees by a factor of two. Now whether that indicates a spin gap or not is a totally different story. But it is not correct to say that one is looking at a single q-vector. It's true that the analysis contains Bose occupation factors, but if you do a naive analysis using Bose occupation factors it will not agree at high temperatures. In other words, $S(\omega)$ is not going to saturate at high temperatures, it's going to rise linearly. So it's not fair to make that statement about Bose occupation factor either.

D. J. Scalapino: It depends very much on what you take for χ. For example, an RPA approximation for χ has a temperature dependence which leads to a saturation of $S(q,\omega)$. What I'm saying is that I think that it is uncertain whether data that we are talking about imply a spin-gap. The point is, there are approximations, such as RPA, in which you get a similar looking intensity without a spin-gap.

J. M. Tranquada: I have a question on the problem of whether there is a correlation between T_c and the density of holes in the planes. You showed a plot of your own unpublished data, for $YBa_2Cu_3O_{6+x}$ which showed a correlation between T_c and the formal valence of copper in the planes. Most people would consider the difference between 2+ and the formal valence on copper as a measure of the density of holes in the planes. So I would look at that as showing a correlation between T_c and the density of holes.

Then you showed the Hall effect data, and pointed out that the temperature dependence is not understood. Now if the temperature dependence isn't understood, it seems to me it's not clear how the Hall number is related to the density of holes. If you say it's directly related to the density of holes, then the density of holes is temperature dependent, so you have to be careful what temperature you choose in making the comparison. One should be cautious about drawing conclusions from Hall effect data concerning global trends of T_c versus hole density.

B. Batlogg: There's certainly no doubt that if you take the holes out, you don't have superconductivity. The drop between 90 degree and 60 degree is a rather precipitous drop. It was not my intention to say there is no relationship between the number of holes and T_c. What I am questioning is a general, universal relationship between T_c and the hole concentration, when comparing such compounds as $(LaSr)_2CuO_4$, $Ba_2YCu_3O_x$ with 90K and 60K T_c, the two and three-layer Bi and Tl compounds. To me it is more striking to see no clear correlation, although the estimates for the hole concentration are somewhat ambiguous. Results from the

Hall effect and iodometric titration may be different. Since we have fewer data for the thallium and bismuth systems, I reserve judgement on that.

J. M. Tranquada: Yeah, one question is of course how do you determine the hole number from Hall effect?

N-P Ong: Many complicated questions are intertwined in this question that John Tranquada just raised. First of all, I've examined all the Hall measurements, even for the systems that aren't very temperature dependent, like bismuth 2-plane, and you find that there's always a factor of two difference between the titration technique – which always gives a lower number for the hole density, compared with the idiot's way of analyzing the Hall data.

There's always this persistent factor of two which occurs in all the systems, except thallium. Now thallium is a problem because we're not sure how good the ceramic samples are, compared with BSCCO or YBCO. I think, until good single crystals become available, we should suspend judgment on thallium. The second point is that YBCO and the neodymium compound are the only ones in which, the temperature dependence is large enough to make this (determining the Hall density) an issue. But in the bismuth system, (the 2-plane and 1-plane), you still get this residual factor of two between titration and Hall. And the last point is, a lot of us are worried about this relationship, whether in fact T_c scales continuously with hole density, or undergoes an abrupt jump. You seem to make the point, that based on the oxygen doping experiments in YBCO, it's rather abrupt. I should point out that there's a movement of the apical oxygen, which we also see in cobalt doping experiments (on YBCO). The oxygen movement alters the chemical potential between the chains and planes, and that tends to make the hole count in the planes discontinuous rather than continuous.

B. Batlogg: Thank you for the comment. I'd like to clarify the last point.

I think that there is one system where we know what is going on, it is the 123-compound with varying oxygen content. In principle we know that there's a strong tendency for oxygen vacancies to order in these chains. At 6.5 they're full, empty, full, empty, etc. It's a new compound by itself. It has a T_c of around 50-60 degrees and therefore, it's not just filling in states, at all. And if we have new different orders, and we have evidence for that actually, if you have two full, one empty, you get a different T_c. So every homogeneous long-range ordered vacancy state has its own T_c, and therefore using any parameter continuously is probably highly misleading.

E. Abrahams: I just wanted to make a remark about trying to extract the gap from the tunneling measurements. If there's anybody who really knows how to do this, I don't think they've made themselves known. Most people like to do it just from the peaks in the tunneling spectrum, which would be fine if it were a BCS gap. On the other hand, in the presence both of temperature-dependent pair breaking and the fact that when you get close to T_c you can have important fluctuation effects make it completely unclear what the gap is just from where the lobes are. You might be off by factors of two.

B. Batlogg: Can you really be off by a factor of two? You can't make a peak if there's no density of states, but you can have the peak at higher energy.

E. Abrahams: You can have a peak in the tunneling density of states from fluctuations above T_c.

S. Trugman: I do not think that the gap near a surface is depressed more by a short coherence length than by a long one. It seems that the opposite should be true.

B. Batlogg: I would refer to at least one paper that I am aware of. Analysis of $J_c(T)$ across a grain boundary clearly indicates a reduced order parameter at the interface.

S. Trugman: The smaller ξ is, presumably, the larger the order parameter would be right at the surface.

B. Batlogg: I see many shaking heads, in the first row.

S. Trugman: Consider a statistical mechanics model like the (3d) XY model, in which ξ approaches zero as the temperature goes to zero. If the lattice has an edge, the order parameter at the edge approaches the full bulk value as the temperature (and correlation length) go to zero. The order parameter would, however, be significantly depressed at the edge if ξ were large.

H. Ott: I have also a remark now concerning the gap and the remaining states in the gap. In the paper of Imer *et al.* on photoemission they claim that the intensity is not conserved. What is going on?

R. S. List: It is true that the total integrated photoemission intensity does not appear to be conserved above and below T_c. Below T_c, there appears to be a greater gain in intensity 30 millivolts below the Fermi level than there is intensity lost in the gap formation. This however does not necessarily imply a non-conservation of the density of states since the angular resolved photoemission intensity measures the electronic density of states weighted by a photoemission cross-section at only a particular point in the Brillouin zone. Since both the photoemission cross-section and angular dispersion can be altered upon passing through T_c, the meaning of the non-conservation of intensity is not clear.

R. L. Martin: I was wondering, in the transition from the 90 Kelvin to the 60 Kelvin superconductor, do you see an abrupt change in the distance from the plane to the apical oxygen?

B. Batlogg: Yes, from 90K to 60K it tracks T_c one-to-one. But I think there's nothing mystic about that, other than the position of the oxygen, as you see in any diffraction experiment, reflects the total charge in the bond on both sides of this oxygen. All it says is that charge has been redistributed, even out of the planes, and the oxygen reflects that.

with the idea that a bismuth 1-layer has practically no T_c, while bismuth 2-layer has a very high T_c, in any theory, in other words, which gets the superconductivity from a third-dimensional interaction, your effective Lawrence-Doniach-Ginzburg-Landau theory then has only zero, or repulsive, in-layer terms, and the superconductivity is coming from the Lawrence-Doniach terms between layers, and in such a theory the superconductivity would tend to decrease, near the surface.

Another comment. If you look at the 90 degree infrared, everybody agrees there's a big change in σ, from 500 wavenumbers down, and everyone believes that the penetration depth pole at low frequencies comes from the integral of that σ over the entire range, up to 500. So in some sense, in the sense that there is missing conductivity up to 500 wavenumbers, that's a very clear statement that there is some gap up to 500 wavenumbers in the 90 degree material.

A. Millis: But it is equally clear that there are some states below 500 wavenumbers.

P. W. Anderson: Sure, it's not a perfect gap, it's just that it's mostly wiped out, and that it makes a very big contribution to the low-frequency behavior.

A. Millis: Yes.

C. M. Varma: May I ask where is the clear indication that there are low frequency states in the T tending to zero limit, much below the gap frequency?

A. Millis: Oh, the Regensberg reflectivity data [Schutzmann et al., Europhys. Lett. 8 679 (1989)] is clearly different from one in the range from 200 wavenumbers to 500. That's the clear indication that there is absorption.

C. M. Varma: Concerning the reflectivity, you have to know the absolute value of the reflectivity to .01, and you make different conclusions whether it is .01 or .005, I don't consider this a conclusive evidence.

A. Millis: The difference from unity is 0.02, which is outside the stated error.

J. R. Schrieffer: I would like to raise the question of the experimental determination of the quantum numbers of the elementary excitations in the cuprates. In the quasi one dimensional conductor, polyacetylene, NMR, transport and IR conductivity prove that charge and spin separation takes place via topological solitons. Can similar experiments be carried out in these quasi two-dimensional system in order to resolve this very fundamental question?

B. Batlogg: The main experimental problems have to do with the fact that these are metals that have very high reflectivity, and the optical probe, as it has been done in the past, is much more difficult to do here.

J. R. Schrieffer: But NMR, conductivity, and other measurements at least at the moment, don't seem to yield the desired resolution to this crucial question. Perhaps some creative thinking is required.

P. W. Anderson: Can I put my answer in the form of a question?

Question: Bertram, doesn't it strike you, looking through all these separate measurements of the gap, that there must be more than one different gap, and more than one different kind of excitation being gapped. That, therefore...in that we would find such a confusing picture in terms of the various kinds of measurements of gaps.

B. Batlogg: Absolutely. I couldn't agree more with you. There is a broad range, but it is a homogeneous range. Every experiment that they don't have to throw out for obvious nonsense has a lower limit of three-and-a-half or has a large limit up to six or eight.

J. R. Schrieffer: Responding to Phil's question, it is not a question of how many gaps there are, because even though there's only one fermion – let's take lead, or tin, or any low-temperature superconductor – there's a continuum of gaps. And as we know, if you make a dirty superconductor, as you originally pointed out, the spectrum of gap sizes decreases, and finally goes back to a single value. So the issue is not are there many gaps, but are there many independent quasiparticle excitations. Phil has advanced a very interesting suggestion today that there are two such Fermi surfaces. My own deep belief, and I believe Nature will agree with me ultimately, is that there is one, and only one, and you can divide it in two and call it two, but it's still just one. (Laughter)

C. M. Varma: I would like to come back again and ask Bob's question more specifically. We have this anyon version of RVB, which made a specific prediction. If the prediction is not seen, we know it does not work. We would similarly like to have a falsifiability criteria for Phil's ideas. Can he propose a specific experiment? If it is not seen, then we know that this thing does not work. If it is seen, then of course it only proves consistency. But we'd like to have at least one specific thing in which he says, "Do such and such experiment, if it is not observed, I abandon this idea."

B. Batlogg: It is amazing, I have so many viewgraphs, perhaps 200 with me. I'm sure I could find a viewgraph. He could tell me which one we have to look for. (Laughter)

P. W. Anderson: It's rather unfair that the real falsifiability experiment has already been done, and it came out OK. I was predicting photoemission a long time ago.

C. M. Varma: We predicted that, I think.

P. W. Anderson ...but there is one very specific remark, namely that there are additional places where there are zero energy spectra single particle in momentum space. With a Fermi liquid there's only one line where it has a zero energy. I am saying there are many places where it has zero energy, and there are even many accumulation points that are not on your Fermi surface. And the photoemission data show you that over quite a wide range there is a threshold that goes linearly to zero, which implies that there is a zero energy excitation there. There are, in addition,

$3k_F$ points, and they may or may not be observable, but they will eventually be observable...

G. Sawatzky: I think that Phil did make a prediction, indeed, saying there was a 3 k_F Fermi surface, and one should be able to measure that with photoemission. Now he also pointed out that there was some intensity in other regions in the Brillouin zone at the Fermi energy, and my question then would be: Isn't it rather strange that that intensity is terribly low, compared to the total spectral weight that you see at the Fermi level at 1 kF, and isn't that an indication that the holons and spinons are strongly coupled? Or do you have a way of explaining why the intensity would be so low, between k_F, shall we say, and $3k_F$, compared to the intensity at k_F?

P. W. Anderson: I don't have to explain it, I can calculate it, and it comes out that way. There is a coupling, in fact, but the point is that in terms of the quasiparticle Green's function, they start out in the same channel, and you only can spread them out to different channels by a rather complicated process, which does tend to make them couple. In addition, there are effects that I don't understand, that I can describe under the name of "gauge fields," that are rather tricky and appear in the one-dimensional case as well. Yes, they're coupled...they're coupled with a rather long-range overall effect, rather than specific scattering effects...

G. Sawatzky: This is what I was trying to get you to say, because if you predict as to what the relative intensity is going to be, at $3k_F$ as compared to k_F, then this is a very definite measurement that can relatively easily be done, with the beautiful equipment that's around now.

B. Batlogg: Returning to the question about the gap...I was digging out the viewgraphs, with the Raman scattering results, which I didn't show. First of all, it's an enormous experimental achievement. Secondly, in the bismuth, there's really not much left of the density of states near zero energy. One point that puzzles me about Raman scattering, is the following: Hackl and collaborators did single crystal work on "123" as function of temperature and with proper polarization analysis. With the electric field always in the plane, the redistribution of density of states depends on the orientation of the field. This is very similar to results of Cooper, Klein and collaborators on the same compound and to results by Yamanaka and collaborators on the Bi 2212 compound. At least it tells us that the maximum of density of states is not isotropic. That's something that I find remarkable.

F. Mueller: This goes back to a question which was raised with respect to photoemission, and it's really a kind of gedanken experiment with respect to this new kind of fermion, and also the questions that Bob Schrieffer was raising. Namely, what would happen in these kinds of materials, as we were hearing from Phil earlier, if one would take the system, put on a magnetic field, and then look for, perhaps, an oscillatory component of the susceptibility plotted, for example, as a function of 1/B? In other words, if one would do this gedanken experiment, would we see one Fermi surface, two Fermi Surfaces, many Fermi Surfi, or none.

P. W. Anderson: My prediction is you see one Fermi surface, but that it will damp out rather quickly.

F. Mueller: Would it be at 1 k_F, 2k_F, or 3k_F?

P. W. Anderson: 1 k_F. For that experiment, I suspect 1 k_F.

C. M. Varma: I would like to make a specific prediction about Fred's experiment. If you do that experiment, the effective mass will decrease logarithimically with the field.

D. Pines: I have a question for Bertram about the transparency that showed the Raman scattering at different symmetries. In all cases, you're really measuring the properties of the excitations in the plane, right? And you are below T_c.

B. Batlogg: Way below T_c. These are 30 degree measurements.

D. Pines: What is an alternative explanation other than the notion that one is looking at an anisotropic gap?

B. Batlogg: You're hard pressed coming up with an explanation.

D. Pines: It looks reasonably unambiguous, then, that the gap is anisotropic.

B. Batlogg: I'd like to quote a commentator by saying, if it looks like a duck, it quacks like a duck. And so on.

D. J. Scalapino: Could you comment on the low-temperature slopes on the same point. I'm curious about the slopes. It looks like the solid line is coming in quite differently than the other.

B. Batlogg: I'm not sure if we should push the data that far. In terms of details, I think that experimentally, that's the best data available. It is not easy to do this experiment, you know. But again, they have come a long way. Everyone is pushing the experimental art to the ultimate limit.

R. S. List: I have a comment and viewgraph relevant to the interpretations made this morning concerning the photoemission data. Specifically, I believe it is important to point out other, more conventional, explanations of the photoemission lineshapes other than does which were described by Prof. Anderson. Prof. Anderson relied heavily upon the asymmetry and sharp peaking of the photoemission spectra as they dispersed towards the Fermi level for support of his model. As Prof. Anderson pointed out, if we had infinite angular resolution and perfect Fermi liquid-like samples we'd expect to see a simple Gaussian peak. But in reality we have a finite acceptance angle of two degrees. If we account for this finite acceptance angle (dk) and experimentally measured dispersion of this photoemission peak near E_F (dE/dk), we obtain an effective energy resolution of not 25 millivolts, but 85 millivolts. That will give you a significant smearing of the peak. An important thing to point out is that in a Fermi liquid-like interpretation, the width of the peak varies as a function of the energy below E_F. So in fact instead of looking at a single Gaussian point, what we are in fact looking at is a summation of several Gaussians, to point out is that in a Fermi liquid-like interpretation, the width of the peak varies as a function of the energy below E_F. So in fact instead of looking at a single Gaussian point, what we are in fact looking at is a summation of several Gaussians, one very sharp here at E_F, another a little further down that's broader, and a successive series. If we do a summation of those, and scale them by the appropriate broadening, we get an asymmetry, an asymmetry, and also, the peaking effect near E_F is as would be expected for a Fermi liquid-like interpretation – the lifetime's much longer near E_F, and you get a much sharper peaking. So I think, I mean I'm very happy that Professor Anderson was interested in our data, and interpreted it the way he did, but I think we should keep in mind that there are other possible interpretations that are perhaps simpler.

P. W. Anderson: I'd like to comment on that. There are a couple of problems with the fit, although of course there is a theorem that you can fit any function with a set of Gaussians. As I kind of integrate that thing by eye, it's a little hard to understand how the intensity can peak up so strongly, and also how the one very close to the Fermi level can have so much intensity over here, when your angular acceptance is supposed to be finite. But I frankly believe that you are underestimating your angular acceptance, as I look at the sharpness of those peaks. Maybe you've got an acceptance angle that's not quite as wide as your detector. Of course I'm not an experimentalist, and it's hard for me to suggest this.

R. S. List: Well I'd sure like to believe that our data supports such interesting interpretations, but I see no compelling evidence.

P. W. Anderson: I'm a great believer in your data, I think it's better than you think...

B. Batlogg: We've moved far away from what Bob Schieffer told me to do: present the facts, only the facts, and perhaps all the facts.

D. J. Scalapino: Could you tell us whether the barium potassium bismuth oxide falls within the same framework as what you've been telling us about?

B. Batlogg: You probably noticed that I didn't mention that in my talk. Also, as you also know, it's one of my pet compounds. We simply don't have enough experimental data of similar quality that we can make this strong point as we can in the other ones, in the cuprates. That's related to the materials, the unavailability of very good crystals of large enough size to do decent transport, or any other measurement on them. If you ask for my gut feeling – and that's not relevant here – but I would try to point out that there's enormous crystal-chemical similarity between these two classes of compounds. Both of them are highly covalently bonded metal oxygen elements with degeneracies of the bare energies of the s, in the case of bismuth and oxygen, or the d in the copper and the oxygen. That's a similarity, that both of them have a very broad band, and in both cases the Fermi level sits, if you start from a band structure, up in the anti-bonding part of that. Very often the fact about isotope effects – I didn't bring up the question about the isotope effects at all, that alone would be a talk by itself – should not be mistaken as a proof of phonon-induced superconductivity. Actually Chandra Varma was pointing out the paper to me showing that if you have pairing due to high-energy excitations, and you have coupling to the phonons, there will be a strong phonon effect in T_c itself, and there will be an isotope effect. And therefore, one should not use the sizable isotope effect in the bismuth oxides in discussions about the pairing excitations.

P. W. Anderson: Could I add one answer to my answer to Chandra. Another thing we predict is that the fluctuations which are slow are the charge fluctuations, not the spin fluctuations, and that's quite different from any of the possible alternatives. Although I haven't calculated it yet and I don't know exactly what the functional form would be, I will predict that phonons are damped, they're damped in a way which is not compatible with conventional fermion damping, and that this is something that one may predict.

3. NORMAL STATE PROPERTIES

Chair — S. Doniach

Patrick A. Lee
Department of Physics
Massachusetts Institute of Technology

Normal State Properties of the Oxide Superconductors: A Review

I shall review experimental results which are relevant to the question of whether Fermi liquid theory applies to the normal state properties of the oxide superconductors. I shall then review theoretical attempts at describing the normal state, first the problem of a single hole in an antiferromagnetic background, and then the problem of a finite concentration of holes.

I. INTRODUCTION

In the three years since the discovery[1] of superconductivity in the copper oxide materials, we have accumulated a significant body of experimental observations which are universal, in the sense that they are observed in all oxide superconductors and the data are reproducible among many groups. In particular, the data for the normal state properties are anomalous and provide a challenge to the theories which have been put forward to model these materials. Much of the discussions have centered on the question of whether Fermi liquid theory applies.[2] In this paper I shall first briefly summarize the experimental results which are most relevant to this question. Since the magnetic properties will be reviewed by Chakravarty and Millis at this conference, I shall concentrate on the electronic properties, and in particular, I shall focus on the Hall effect and photoemission measurements as the two experiments which appear to provide conflicting information on the nature of the normal state. I shall then review the theoretical attempts at describing the normal state, first the problem of a single hole in an antiferromagnetic environment, and then the problem of a finite concentration of holes.

2. EXPERIMENTS ON THE NORMAL STATE PROPERTIES

We should first agree on the essential features of the Fermi liquid theory.
(1) There exist low lying single particle excitations which simultaneously add a charge and spin 1/2 to the ground state.
(2) These excitations are well defined in the sense that its decay rate to multiple particle-hole excitations is less than its energy.
(3) There exists a Fermi surface which is the locus in $\vec{k}$ space of the zero energy single particle excitations and the volume of the Fermi surface obeys Luttinger's Theorem, i.e. it accommodates all the electrons in the system.

The third condition is clearly much more specific and most experiments provide information only on the first two points. The most famous among these is the linear resistivity with temperature above T_c. This linear behavior seems to be universally observed and in one case[3] (single crystal $Bi_2 Sr_2 CuO_{6+\delta}$ with $T_c \sim 10K$) is linear down to 10K. This last observation provides strong evidence against a phonon mechanism, which will provide linear behavior down to only about one quarter of the characteristic phonon frequency. We should mention that recently Tsuei et al.[4] have claimed to find an exception to the linear law in the electron doped material $Nd_{1.85} Ce_{0.15}CuO_4$. They claimed that the published single crystal measurement[5] as well as their thin film data obey a T^2 law. Clearly, this potentially important result should be confirmed, particularly in view of the fact that data on the traditional hole doped materials often also show a deviation from linearity when the resistivity is high, indicating problems with samples or current homogeneity.

The optical measurements[6] reveal a Drude peak with a temperature dependent width, giving direct evidence that the linear term in resistivity is due to a scattering rate τ^{-1} which is linear and T, and not density of states effects. The measured rate is of order $\tau^{-1} \approx 2k_B T$. This suggests that the linear resistivity is due to inelastic scattering from some anomalous fluctuation in the correlated electron system. Indeed Raman scattering[7] has revealed the existence of a continuum of excitations which is described by the following response function.

$$R''(\omega) = S_o \omega / T \text{ for } \omega < T$$
$$R''(\omega) = S_o \qquad \text{ for } \omega > T \qquad (1)$$

This continuum extends up to several thousand cm^{-1} whereas in ordinary metals, scattering from particle-hole excitations extends only over $qv_F \approx 20$ cm^{-1} if q is identified with the inverse of the penetration depth. Klein et al.[7] and Varma et al.[8] have pointed out that the scattering of a single particle by this continuum will produce a scattering rate proportioned to ω or T whichever is greater.

Varma et al.[8] have taken this one step further and proposed that eq. (1) which is observed only for q$\approx$o in Raman scattering, be extended to all q as a phenomenological ansatz. They claim that a whole host of experiments can be explained by this single assumption. While I take issue with the success of their phenomenology, as I shall explain below, I find the points they made concerning the consequence of the linear ω or T behavior of the self energy interesting and significant. They observed that by Kramers-Kronig, an imaginary part of the electron self energy $\Sigma''(\omega)$, which is linear in ω implies a real part Σ' which is $\omega \ln \omega$. This produces a spectral weight $(1-\partial \Sigma'/\partial \omega)^{-1}$ which vanishes logarithmically. Thus the Fermi liquid description is at best marginal. They also pointed out that a large $\Sigma''(\omega)$ can lead to a tunneling density of states which is a constant plus a linear term as a function of voltage. Unfortunately, the ansatz Eq. (1) leads to a linear term with negative slope, in constrast with the positive slope observed experimentally.[9] This is because we can understand the tunneling density of states as being made up of a superposition of Lorentzians with widths increasing linearly with energy. Clearly the high energy Lorentzians contribute more to the low energy ones than vice versa, so that the total superposition is a decreasing function of energy. Varma et al. remarked that this problem can be fixed up by introducing additional momentum dependent broadening due to elastic scattering by impurities. However, this is incorrect because one should then consider tunneling into exact eigenstates of the impurity problem (as opposed to momentum eigenstates) and only the inelastic width will remain. Thus the linearly decreasing density of states is an inherent feature of this model. However, it may still be possible to explain the tunneling conductance with positive linear slope by postulating a tunneling matrix element which increases with energy and saturates beyond a certain energy scale. (C.M. Varma, private communication) It should be noted that the linear dependence of τ^{-1} on ω is crucial for the mechanism to work, and therefore is specific to the oxides. According to this view, the linear tunneling conductance is an extrinsic property depending on the barrrier and reflects only the ω linear inelastic scattering rate.

I should remark that Varma et al. also pointed to photoemission data which shows a rising background as a function of energy as evidence for a linearly increasing total density of states. We have seen that this is not possible within their phenomenological model, because the tunneling matrix element is obviously not involved. Apparently, the photoemission background depends on how the angular averaging is performed (C. Olson, private communications) and is probably not an intrinsic measurement of the total density of states.

Yet another consequence of the large inelastic rate is recently explored by Kuroda and Varma (preprint, and contribution by C.M. Varma at this conference) who carried out a detailed calculation of the superconducting state due to exchange of fluctuations given by Eq. (1). They produced a large $2\Delta/k_B T_c$ ratio[10] as well as an explanation of the absence of a peak below T_c in the nuclear spin relaxation rate. It is worth emphasizing that both these effects are consequences of the pair breaking effect of $\tau^{-1} \approx 2k_B T$ and should be independent of any specific mechanism for pairing.

Varma et al. successfully fitted the optical conductivity by a sum of a Drude term and a mid infra-red contribution coming from Eq. (1) with an upper cut-off. They also claim to fit the anomalously large copper T_1^{-1} in terms of excitation of the same spectrum given by Eq. (1) together with a Korringa type contribution. However, it is difficult to see how this picture can explain why T_1^{-1} on the oxygen and Y sites are Korringa-like whereas that on the copper site is not. I shall not go into the detailed interpretation of T_1^{-1}, which will be given by A. Millis at this conference. I would simply remark that he shows that a $R''(q,\omega)$ with a strong peak at $q=(\pi,\pi)$ to reflect the strong antiferromagnetic correlation[11] is required to explain the data. This picture explains why the oxygen and copper relaxation rates behave very differently above T_c because they are sensitive to different q average of $R''(q,\omega)$. Furthermore, it is remarkable that below T_c the copper and oxygen relaxation decreases in exactly the same way with temperature. If we interpret superconductivity as the opening of a gap in the charge degrees of freedom in the singlet spin channel, this implies that the magnetic fluctuations that relaxes the nuclear spin above T_c are strongly coupled to the charge degree of freedom, a requirement certainly satisfied by condition (1) for the Fermi liquid theory.

From the above discussion, we conclude that existing data are at least consistent with condition (1). Condition (2) appears to be violated, but only marginally, in the sense that the decay rate is comparable to the quasiparticle energy. Now we turn our attention to condition (3) and we are faced with data which seem to give us conflicting information. The most serious conflict is between Hall effect and angular resolved photoemission. Let us first examine the Hall effect and concentrate on the $La_{2-x}Sr_xCuO_4$ and $Nd_{2-x}Ce_x CuO_4$ system, where the interpretation is most clear-cut. The data from the University of Tokyo group[12] is reproduced in Fig. 1. For $La_{2-x}Sr_xCuO_4$ it has long been known that the Hall resistance is relatively temperature independent, and up to $x\approx 0.1$, it agrees in sign and magnitude with the simple formula $R_H = 1/nec$ where n proportional to x is the density of doped holes. Beyond $x = 0.15$ where the superconductivity is most stable, the deviation is only about a factor of two. Beyond $x = 0.25$, R_H becomes very small and changes sign beyond $x = 0.30$, together with the disappearance of superconductivity. The most natural interpretation of the data[13] is that the strong correlation produces a Mott-Hubbard gap (or a charge transfer gap in the sense of Zaanen, Sawatzky and Allen[14]) and that the doped holes go into the lower Hubbard band. If there is a Fermi surface (even if it exists only marginally, as our earlier discussion seems to indicate) this picture would seem to predict a Fermi surface volume of x holes per copper, as opposed to the Luttinger theorem predicting of $1 + x$ holes. There is also increasing evidence that the nature of the normal state changes dramatically beyond $x \approx 0.30$ and becomes a very ordinary Fermi liquid. In addition to the sign

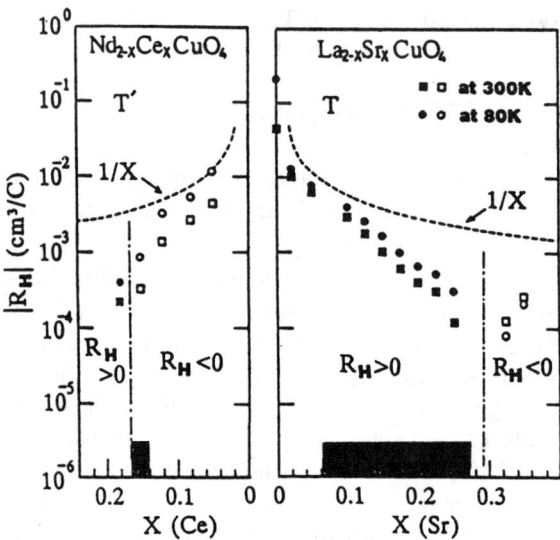

FIGURE 1 Hall constant vs. doping concentration for hole doped and electron doped materials from S. Uchida et al.[12]

change of R_H, we now have copper site $1/T_1$ data[15] which is linear in T, with a magnitude which is that expected for a Fermi liquid metal, ie. it is smaller than T_1^{-1} of $La_{1.85}Sr_{0.15}CuO_4$ by an order of magnitude. Thus the strong antiferromagnetic fluctuation believed to be responsible for the enhanced T_1^{-1} for x = 0.15 appears to have disappeared. Furthermore, the famous linear T resistivity at x = 0.15 acquires a significant curvature by the time we reach x = 0.3.[12]. All these data suggest a change from a Luttinger Fermi liquid for $x \geq 0.3$ to something else for smaller x.

This interpretation is further bolstered by the data on the electron doped material $Nd_{2-x}Ce_xCuO_4$. As seen from Fig. 1, there is symmetry between the electron and hole compounds. The Hall resistance is now negative for x < 0.16 and follows 1/nec very closely up to x = 0.1. The disappearance of superconductivity beyond x = 0.16 is again associated with a sign change and a small magnitude of R_H. While it can be argued that complicated band structure effect can cause a sign change to account for the Hall constant at an isolated value of x, the trend as a function of x for both electron and hole doping is so striking that it is difficult to imagine how the data up to x = 0.15 can be explained based on a band theory.

Qualitatively the Hall data is also universal in that all the hole doped copper oxide superconductors have positive Hall effect, with reasonable value (i.e. within a factor of two from what one might reasonably expect based on stoichiometry). This includes Bi 2212, Tl 2212, Tl 2223 which all have relatively temperature independent R_H.[16] Thus the linear T R_H observed in the 123 compound which received much attention earlier appears to be the exception rather than the rule, and is perhaps associated with the presence of chains. Furthermore, by doping Bi 2212 with [17]Y or [16]Tm, it is possible to drive the material insulating and at the same time R_H^{-1} linearly to zero. This is in agreement with the expectation that doping with trivalent Y or Tm add electrons to the Cu-O planes and drive the hole concentration back towards zero, another strong indication of the universality of the picture shown in Fig. 1.

An important set of experiments which can, in principle, determine the existence of quasiparticles and the shape of the Fermi surface is angular resolved photoemission . This technique measures the spectral density of a single hole state at momentum $\vec{k}$, i.e. Im G(k,ω). If the quasiparticle exists, it will appear as a pole in Im G(k,ω) and give rise to a peak in the electron spectrum. This peak will sit on top of a background which consists of a hole plus multiple particle-hole excitations or spin excitations. As $\vec{k}$ approaches the Fermi momentum, the quasiparticle peak is expected to sharpen as its energy approaches the chemical potential. As $\vec{k}$ continues to increase beyond the Fermi surface, the peak is expected to disappear. In this region, the technique of inverse photoemission (BIS) can provide information on the unoccupied single particle states. Several groups have reported angular-resolved photoemission experiments in the oxide compounds.[18,19,20] In Fig. 2 we reproduced high resolution data from Olson et al.[21] on Bi2212 which qualitatively shows the behavior expected for a quasiparticle approaching the Fermi surface. Further strong evidence that the sharp peak corresponds to a quasiparticle excitation comes from the observation that the peak sharpens considerably as the system undergoes the superconducting transition at 82K and in fact the sharpening can be fitted to an energy gap of 24 meV. These data are in support of the conditions (1) and (2) for the Fermi liquid discussed earlier. It seems that more data are needed to pin down conditions (3), but better and more data are coming in and we hope to obtain more information in the near future. In particular, the published data reproduced in Fig. 2 corresponds to a scan in $\vec{k}$ space roughly parallel to a direction from Γ to M which is tangential to the Fermi surface expected from band structure calculations. Recently, high resolution data along the line Γ to X are available[22] which shows qualitatively similar behavior in that a quasiparticle peak emerges as $\vec{k}$ approach a point somewhere less than half way between Γ and X. These and other data have been interpreted as indicating the existence of a Fermi surface which is in good agreement with band structure calculation which of course obeys Luttinger theorem.[23]

Theoretically, the dichotomy we face can be summarized by the schematic picture in Figure 3. For small doping with holes, the Luttinger Fermi surface can be expected to look like that shown in Fig. 3a. Indeed the band calculation Fermi surface can be interpreted as a hybridization of this Fermi surface with a small pocket centered at the M point.[24] On the other hand, in the Hubbard band picture, one might expect the locus of the low lying single particle excitations to look like that shown in Fig. 3-b. As discussed in the next sections, we have reasons to expect

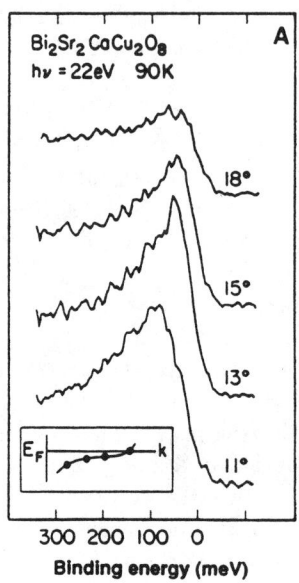

FIGURE 2 Angle resolved photoemission data from Olson et al.[20]

the hole "band" to have minima at $(\pm\pi/2, \pm\pi/2)$ and to be relatively flat along the direction towards the M point, leading to an elongated Fermi surface. The attractiveness of this band structure lies in the fact that the Hall effect can readily be explained. It is worth noting that the same picture is expected on the basis of weak coupling theory, where the appearance of an antiferromagnetic insulator at half filling is attributed to Fermi surface nesting. The square Fermi surface at half filling is gapped by the formation of spin density waves, and the introduction of holes can lead to the kind of Fermi surface shown in Fig. 3b.[25] Incidentally, since the Brillouin zone is halved due to magnetic ordering, the Fermi surface in Fig. 3b which contains x holes per copper actually satisfies Luttinger's theorem. According to this picture, as one scans along Γ - X, one would first encounter a Fermi edge which is located at a point which is approximately the same as that given in Fig 3a. However, beyond $(\pi/2, \pi/2)$, one expects a re-entrant behavior, where one encounters one Fermi surface again. This is in contrast with the Luttinger Fermi surface in Fig 3a, where one expect to see only unoccupied states via inverse photoemission (BIS) beyond the Fermi surface.

Available data seem to support Fig. 3a rather than Fig. 3b.[23] However, the high resolution data is available only up to the half way point between Γ and X[22] and so far do not distinguish between these two pictures. Of course, Fig. 3b is still a rather naive picture because it is not clear how it will be modified by complications

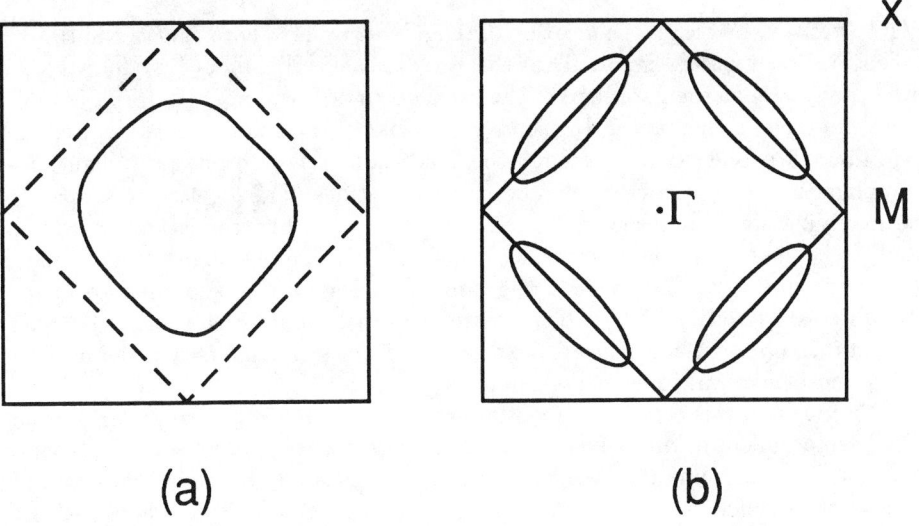

FIGURE 3 Schematic drawing of the locus of zero energy single-particle excitation in the a) Luttinger Fermi liquid and b) a possible alternative in the strongly correlated Hubbard band.

such as the absence of long range antiferromagnetic order or the appearance of incommensurate wave-vector.[26] However, it seems to me most important to establish firmly experimentally whether something resembling Fig. 3b is at all viable.

It is interesting to ask what is expected for a photoemission experiment in a one dimensional Hubbard model in the limit of strong correlation $U/t \gg 1$. The one dimensional Hubbard model exhibits a power law decay in the spin correlation function with a wave vector which is incommensurate and equals $2k_F$. The weak coupling limit $U/t < 1$ can be treated by renormalization group analysis.[27] A power law singularity is expected at $k = k_F$ for the single particle spectral density so that one may expect a quasiparticle like peak in photoemission when k approaches k_F where k_F is given Luttinger Theorem. The momentum distribution n_k does not show a discontinuity at k_F, but instead is predicted to have a power law singularity. The latter feature was found in a recent numerical study of the exact solution in the limit $U/t \to \infty$.[28] This confirms the expectation that the large U/t limit should behave qualitatively similar to the small U/t limit. Furthermore, a weaker singularity in n_k is found at $k = 3k_F$ and presumably a divergent power law also exists for $\mathrm{Im}G(k,\omega)$ near $k = 3k_F$ as well. The $3k_F$ singularly can readily be understood as a convolution of a single particle excitation at $k \approx k_F$ and a spin density excitation at a momentum near $2k_F$. The important point is that there are no singularity at $k = 2k_F$ which would correspond to the Fermi surface of spinless Fermions, or at $k =$

$\pi/2 + (\pi/2 - k_F)$ which is the location of the re-entrant hole Fermi surface in the one dimensional version of Fig. 3b. Thus the one dimensional Hubbard model, which is often thought of as the paradigm of the breakdown of Fermi liquid theory, actually obeys Luttinger's theorem in the sense of condition 3. Condition 2 is violated and conditional 1 is obeyed, even though the excitation is more properly thought of as a combination of separate charge and spin excitations. The location in k space of the low lying single particle excitations measured in a photoemission experiment would look close to the one dimensional version of Fig. 3a. As we mentioned before, this is not unexpected because a one dimensional system is supposed to behave smoothly as a function of U/t. The key unanswered question is whether the same smooth interpolation continues to hold in two dimensions and this is the question which photoemission experiments can help to answer.

We conclude this section by mentioning a very recent angle integrated photoemission experiment in which $La_{2-x}Sr_xCuO_4$ was compared with the electron doped material.[29] It was found that the Fermi energy lies at nearly the same energy within the charge transfer gap. This data was used to support a picture where impurity band-like states are created in the band gap upon doping which are then filled with carriers.[18] We would like to point out that within this picture, the discontinuous sign change of the Hall effect as shown in Fig. 1 would be a complete mystery. Furthermore, this behavior is in strong disagreement with the theory which does produce a Luttinger Fermi liquid, namely, that based on the Anderson lattice model used to describe heavy fermion materials. Within this theory, the photoemission is predicted to shift by an amount of the order of the gap upon changing from electron to hole doping.[30]

3. THEORY OF THE NORMAL STATE

I shall restrict most of my review to work on the one band Hubbard model, because as a model it is of intrinsic interest, and also there are reasons to believe that the low energy behavior of the full three band model may map onto the one band model.[2,31] For weak and intermediate U/t, the problem has been treated by Hartree Fock theory. If a spin density wave is assumed to condense at (π, π) the band minimum can occur either at the M point or at $(\pi/2, \pi/2)$, and in the latter case, the Fermi surface for a small but finite hole concentration will look like Fig. 3b.[25] This approach has recently been extended to the case where only short range order exists,[32] in which case there are only pseudo-gaps and presumably broadened quasiparticle peaks will remain in the vicinity of the "Fermi surface" shown in Fig. 3b. On the other hand, the Hartree-Fock ground state is found to be incommensurate with holes condensing on domain walls.[33,34] It is not clear how this picture can be extended to describe the short range order of the incommensurate spin structure or what its implication for photoemission will be.

In the strong correlation limit, $U/t \gg 1$, most work are based on the mapping to the t-J model, with nearest neighbor hopping t and exchange J, subject to the constraint of no double occupation. A convenient way to deal with the constraint is to formally write the electron operator $c_{i,\sigma}$ as a product of a Fermion and Bose operator. Two alternative methods have been used, depending on which species carry the spin label. In the slave boson method, one writes $c_{i\sigma} = f_{i\sigma} b_i^\dagger$ where the Fermion $f_{i\sigma}$ carries the spin label, and the constraint becomes $\sum_\sigma f_{i\sigma}^\dagger f_{i\sigma} + b_i^\dagger b_i = 1$ on every site i. Alternatively, one can use the slave Fermion (also know as the Schwinger boson method), where one writes $c_{i\sigma} = f_i^\dagger b_{i\sigma}$ and the constraint is $\sum_\sigma b_{i\sigma}^\dagger b_{i\sigma} + f_i^\dagger f_i = 1$. These constraints are then treated in the mean field (or large N) approximation. We shall discuss these methods in a little more detail later, but generally, since most theoretical efforts have focused on looking for a superconducting ground state, the normal state appears only as a finite temperature phenomenon and has not received too much attention, a deficiency which should be rectified. We shall begin by first reviewing the work on the simpler problem of a single hole moving in an antiferromagnetic background according to the t-J model.

3A. THEORY OF SINGLE HOLE IN AN ANTIFERROMAGNETIC BACKGROUND

It has long been appreciated that a single hole in a Néel background has difficulty hopping because each hop produces an overturned spin which is aligned parallel to its neighbors. In the Ising limit the hopping hole leaves behind a string of wrong spins, which gives rise to a discrete spectrum often referred to as the string states.[35,36] In the Heisenberg case, the spin flip term in the Hamiltonian can repair overturned spins provided the hole hops on the same sublattice. Thus one expect the spin to propagate coherently on the same sublattice, with a bandwidth of order J if $t \gg J$, since J is the term that limits the hopping rate. This expectation is confirmed by a self-consistent treatment of the coupling of the hole to spin excitations. It is found for both the Néel background[37,38] or a short range ordered RVB background[38] that the spectral function for the hole consists of an incoherent background and a pole on the low energy side at energy E(k). The spectral weight of the pole is of order J/t, leading to a dispersion E(k) with an effective hopping rate on the same sublattice of order J. Further development of the theory was made by Gros and Johnson,[39] who devised a different (but still uncontrolled) perturbation scheme which improves the treatment of the constraint of no double occupation. They also found a spectral weight which scales as J/t for very small J/t but found deviation for moderate J/t. On the other hand, Su et al.[40] have performed a self-consistent Hartree-Fock treatment which include the possibility of an average local spin distortion around the hole, in analogy with the polaron problem.

There is considerable numerical work on the problem ranging from exact diagonalization[41−44] to approximate calculations using a restricted basis set.[46,47] There is strong evidence that the hole minimum occurs at the $\left(\frac{\pi}{2}, \frac{\pi}{2}\right)$ point and the dispersion is almost flat towards the M point. If this band were filled with a small concentration of holes, it will lead to a Fermi surface as shown in Fig. 3b. The numerical studies are also consistent with a band width of order J. The recent work by Trugman[48] also make the interesting point that the mapping of the Hubbard model to the t-J model produces a second and third neighbor hopping term of order J. The direct hopping on the same sublattice is not subject to the problem of spin misalignment, and should not be renormalized downwards. Trugman found that this leads to an enhancement of the numerical prefactor in front of J for the effective bandwidth. This factor may improve the agreement with the effective mass for the Drude conductivity measured by analyzing optical reflectivity.[5,6] The experimentalists found an enhancement from the band mass of order 2 or 3 compared with an estimate of $t/J \approx 3$.

A number of theoretical issues remain unresolved, however. It is still not clear how to treat analytically the spin structure in the immediate vicinity of the spin, whether there is a core region of disordered spin and what the size of the core region is. In particular, it is not clear whether the string state in the Ising limit can describe the short distance or high frequency behavior. Trugman[48] has shown that his spectral function can be understood as remnants of string states. However, his basis set assumes a perfect Néel state in the background spin far away from the hole, and may be biased towards the Ising behavior. For example, it would seem that low frequency spin wave excitations are not allowed within this restricted basis set. It will be interesting to compare with the treatment of Sachdev,[46] which allows for quantum fluctuations in the background spin. Another unexpected feature in Trugman's result is the spectral weight of the hole scales as $(J/t)^{1/2}$, in contrast with the J/t expected based on the self-consistent self energy analysis.

We should remark that the situation appears to be qualitatively different in one dimension, in that the effective Fermi velocity v_F^* defined by the location of the cut in $G(k,\omega)$ for k near k_F, i.e. $G(k,\omega) \approx [\omega - v_F^* (k - k_F)]^{-\alpha}$, appears to be unrenormalized and given by approximately t, and not J.[49] In the renormalization scheme,[27] v_F^* does not have any logarithmic corrections. In the $U/t \to \infty$ limit, the wavefunction is such that the spin chain formed by the deletion of the hole site is described by the Bethe ansatz for the Heisenberg model.[28] We recall that Bethe ansatz can be understood as an RVB state with singlets between opposite sublattices only. Interpreted in this way, it is clear that the wavefunction for the hole state contains a large amplitude for a singlet bond between the two spins on each side of the hole. Note that these two spins are on the same sublattice in the physical lattice. It is now clear that that the hole can hop around without disturbing the spin wavefunction. The key for this possibility is the appearance of singlet bond between the same sublattice for spins located on opposite sides of the hole, so that the hole is functioning as a domain wall. This large scale re-arrangement of spins is of course not described by any self-consistent spin wave theory, so that these theories should not be applied to one dimension.[37] On the other hand, it is also clear that this wave function cannot be easily extended to two dimensions. Indeed, the short range RVB picture[50] retains the feature of having singlet bonds between opposite sublattices, and is therefore not particularly favorable as far as the kinetic energy of the hole is concerned. The numerical result that the hole spectral function and its dispersion is renormalized to J in two dimensions is a strong indication of the difference between one and higher dimensions.

3.B FINITE HOLE DENSITY

(i) Fermi Liquid Theory

This method proceeds along the line of the treatment of the Anderson lattice Hamiltonian as a model for the heavy Fermion materials.[51] The localized f electron operator is treated by the slave boson decomposition. In the mean field theory, the slave boson bose condenses and acquires a mean value, which serves to reduce the hybridization with the conduction electron. At the same time the constraint is treated on the average, and the Lagrange multiplier provides an energy shift which boosts the f level to the Fermi level, resulting in a narrow hybridized band which satisfies Luttinger Theorem, i.e. the local f moment is included in the Fermi surface volume. Kotliar, Lee and Read[52] applied the same procedure to a model for copper oxide where the copper hole orbital at energy ε_d is treated in an analogous way to the treatment of the localized f orbital and the oxygen hole orbital at energy ε_p plays the role of the conduction electron. This model exhibits a metal to insulator transition at half filling as the ratio of $\varepsilon_p - \varepsilon_d$ to t_{pd} is increased, where t_{pd} is the hybridization matrix element for finite doping. A heavy Fermi liquid theory result, with a Fermi surface which contains $1 + x$ hole according to Luttinger Theorem and a band width of order xt where $t \approx t_{pd}^2/(\varepsilon_p - \varepsilon_d)$ is the effective hopping matrix element.

This model has been further refined by the inclusion of the unhybridized oxygen orbital,[53] direct oxygen-oxygen hopping and realistic band parameters,[54,55] as well as the presence of an exchange term J.[56] One of the difficulties of the original model is that a bandwidth of order xt would predict a magnetic susceptibility χ proportional to x^{-1}, which is in disagreement with experiments which show χ increasing with doping. Similarly, this model predicts a Fermi velocity which is linear in x, a prediction which should be tested with angle-resolved photoemission experiments. Newns and Pattnaik[23] show that the inclusion of oxygen-oxygen hopping leads to a more complicated density of states which can increase as a function of χ. However, their model parameters are such that the system remains metallic at $x = 0$. Furthermore, it is not clear whether the same parameters would lead to a spectral weight in optical absorption which is linear in x, as observed experimentally.[6] Recently, Sa de Melo and Doniach[30] have emphasized that even with oxygen-oxygen hopping, the model can still exhibit a metal-insulator transition at half filling and it will be interesting to see how χ behaves as one dope the insulating phase.

Kotliar et al.[52] have argued that the heavy quasiparticles would interact with a residual interaction of order J. Thus the Fermi liquid picture is viable only if the bandwidth xt is greater than J. In the opposite limit of very low density of holes, it is more appropriate to bind the oxygen hole with a copper spin to form a singlet[31] and then consider the hopping of this singlet hole in the background of antiferromagnetically coupled spin 1/2 copper moments. In the extreme dilute limit this reduces to the problem of a single hole considered in the last section and the physics is clearly quite different from the Fermi liquid ground state. Thus one expects a phase transition as a function of x occurring at x≈J/t which is estimated to be 1/3. In light of the Hall effect, resistivity and the $1/T_1$ data discussed in section 2, it is tempting to identify this transition with the change in behavior at x = 0.3 for $La_{2-x}Sr_xCuO_4$. However, this is in apparent conflict with the photoemission data,[23] even though at present the understanding of the normal state for x < J/t is extremely limited, as we shall see below.

(ii) Slave boson RVB type theories

Early in the development, Anderson[2] has emphasized that in a strongly correlates system such as the copper oxides, the exchange interaction among the spin 1/2 local moments can lead to a novel quantum ground state which exhibits short range antiferromagnetic order, and that such a state may be induced by a small amount of doping.

In contrast with the heavy Fermion theory, this approach emphasizes the exchange coupling and the two methods clearly approach the interesting intermediate doping regime (x≈0.15) from opposite limits. Baskaran, Zou and Anderson[57] were the first to apply the slave boson method to the RVB theory to deal with the problem of strong correlation. They introduced the order parameter $\Delta_{ij} = \left\langle f_{i\uparrow}^\dagger f_{j\downarrow}^\dagger - f_{i\downarrow}^\dagger f_{j\uparrow}^\dagger \right\rangle$ to represent a resonating singlet bond on nearest neighbor sites. In their original treatment, Δ_{ij} was taken to be uniform, leading to a Fermi surface for "spinon" excitation which obeys Luttinger Theorem. Subsequently,[58,59] it was shown that at half filling Δ_{ij} is equivalent to $D_{ij} = \left\langle f_{i\downarrow}^\dagger f_{j\uparrow} + f_{i\uparrow}^\dagger f_{j\downarrow} \right\rangle$ by particle-hole symmetry and that one can associate a flux to each plaquette defined by

$$\phi = Im(ln \prod_{ij} D_{ij}) \qquad (2)$$

where the product is over the bonds around the plaquette. One of the large N mean field solution consists of complex D_{ij} such that the flux ϕ per plaquette is π and the spinon Fermi surface become Fermi points at $(\pm\frac{\pi}{2}, \pm\frac{\pi}{2})$.

There are many studies of the ground state for finite x using a variety of techniques such as mean field theory, renormalized mean field theory, Gutzwiller projected wave functions[60] etc. The interesting question is how the state at finite x is connected to the π flux state at half filling. Two interesting possibilities have emerged. The first is the uniform flux phase, where the flux per plaquette changes from π to $\pi(1+x)$ so that the flux per particle remains π. The motivation behind studying this state is that the kinetic energy of non-interacting particles in a tight binding model can be lowered if a commensurate flux is spontaneously generated, provided the energy for creating the flux itself is not included.[61] This state has an energy gap and is believed to be superconducting.[62] The state breaks time reversal symmetry so that its symmetry properties and the gap in the excitation spectrum is the same as those in the quantum Hall spin liquid state[63] and the resulting anyon superconductor.[64] These states are presumably intimately related. However, from renormalized mean field and Gutzwiller projection calculations,[62,65] it appears that the uniform flux state is stabilized only for $t \leq J$.

The other possibility to introduce holes to the π flux phase is to recognize that since π is the same as $-\pi$, the π flux phase can also be considered a staggered collection of $\pm\pi$ fluxes. Then it is natural to consider the staggered flux phase upon doping, where the fluxes become $\pm\pi$ $(1 + \phi)$ upon doping.[66-68] Poilblanc and Hasegawa[68] have concluded that this state is more favorable than the uniform flux phase for $t \leq J$. In the mean field theory the staggered flux phase does not have an energy gap, but instead has a Fermi surface which consists of small hole pockets centered around the $\left(\pm\frac{\pi}{2}, \pm\frac{\pi}{2}\right)$ points. Presumably this Fermi surface is unstable to superconducting pairing,[69] but the interesting point is that in the normal state the Fermi surface resembles that shown in Fig. 3b. It is likely that the "spinon" Fermi surface will lead to an incommensurate spin-spin correlation function so that this state is probably intimately related to the double spiral state we shall discuss in the next section. Another interesting feature of the staggered flux phase is that it possesses a staggered chirality, where the chirality[70,71] is defined as $\left\langle \vec{S}_1 \cdot (\vec{S}_2 \times \vec{S}_3) \right\rangle$ for three nearest neighbor spins forming a triangle with the same orientation. This is in contrast to the uniform flux phase which possesses a uniform chirality.

(iii) Slave Fermion Methods

The first application of the Schwinger boson (slave Fermion) method to the half filled case was given by Arovas and Auerbach,[72] who also gave an instructive comparison with the slave boson method. Here one introduces the order parameter defined on the ij bond as

$$D_{ij} = \left\langle b_{i\uparrow}^\dagger b_{j\downarrow}^\dagger - b_{i\downarrow}^\dagger b_{j\uparrow}^\dagger \right\rangle \tag{3}$$

to represent the formation of a singlet bond. It turns out that a solution of uniform nearest neighbor D_{ij} gives the best ground state energy, which is also better than the π flux phase energy in the large N limit. Furthermore, it has been argued that upon introduction of a single hole in the π flux phase, the hole which starts out as a slave boson acts like a Fermion.[73,74] Essentially a flux of π is bound to the hole, converting the statistics. This suggests that the slave Fermion method may be a more natural starting point for small doping. On the other hand, we have seen that the slave boson method leads naturally to the Fermi liquid solution in the large doping limit, and in the slave Fermion method it is not so clear how that can be accomplished. A more serious drawback of the slave Fermion method is that the nearest neighbor hoping term t is difficult to treat quantitatively. Physically we expect that for t > J, each hole can gain energy of order t by hopping around locally so that we expect a term in the ground state energy of order - xt. This term appears naturally in the slave boson method. In the slave Fermion method, it is necessary first to perform a self consistent treatment of the one hole problem as described in section 3a to produce a quasiparticle band with effective hopping on the same sublattice with matrix element J and located at energy of order -t. The energy -xt will then come from occupying these low lying quasiparticles. Since the major effect of the t term is to generate a hopping on the same sublattice, it makes sense to first study a simpler model, the t′-J model where t′ denotes hopping on the same sublattice only.[75][78] In this model there are two species of spinless fermions f_A and f_B which hop on the A or B sublattices. If the background spin is assumed to be in a quantum disordered state, i.e.it is characterized by $D_{ij} \neq 0$ but $\langle b_{i\sigma} \rangle = 0$, it has been shown that the phase of D_{ij} can be associated with a lattice gauge field A_{ij}.[79] The flux through a plaquette associated with the field A_{ij} has the physical meaning of being the solid angle subtended on the unit sphere by the *staggered* magnetic moment around the plaquette.[77,78] Furthermore, this gauge field is coupled to the hopping of f_A and f_B, but because $\vec{A}$ is related to a staggered flux, it naturally couples to f_A and f_B with opposite signs. Interaction via exchange of the gauge field leads to a pairing of f_A and f_B and a superconducting ground state.[75-77] The normal state property, being a finite temperature property, is much harder to describe. Naively the f_A and f_B Fermions form a degenerate Fermi surface identical to that shown in Fig. 3b. However, the f particles are a mathematical construct and cannot be identified with physical particles as they are not gauge invariant. The only gauge invariant objects are superconducting pairing $f_A f_B$, or a pairing of the f and the Schwinger bosons, which reproduce the physical electrons. The nature of this latter pairing at finite hole concentration has not been carefully discussed at this point. However, it seems clear that in order for this model to recover the phenomenology discussed in section 2 related to properties (1) and (2) of a Fermi liquid, the pairing between the Fermion and Boson must occur. We should mention that a related model, the quantum dimer model has been studied at half filling,[80,81] and it was also concluded that spinon excitations are confined. Extension of this work to finite hole concentration has also been made.[81]

If the t term is now re-introduced and treated in mean field theory,[82-85] it is found that the ground state favors a spiral distortion of the spins, so that the f_A and f_B fermions are now admixed by a term of order xt. This admixture gains an energy of order $x^2 t$. We repeat our caution that the mean field theory misses the more important energy term of order -xt and therefore the mean field theory can only be considered applicable to a quasiparticle with renormalized hopping, probably of order J. Thus the mean field phase diagram should not be trusted t > J, rather, the t>J limit should be renormalized down to t≈J. Thus the large regions of ferromagnetic phase found in the mean field phase diagram are probably spurious.

The slave Fermion method has the advantage of providing a mechanism for incommensurate spin distortion, even though in its simplest form, the spins retain long range ordering.[82] We note that the mechanism for a spiral distortion is different from that of Fermi surface nesting, in that the ground state remains metallic in the spiral state. Indeed the spiral leads to an admixture of the t_A and t_B band so that the degeneracy is lifted, resulting in a non-degenerate spinless Fermion type Fermi surface. Again, the implication of this for experiments such as photoemission is unclear, because one must consider the convolution with the Schwinger boson operators to recover the physical electron operator. Nevertheless it is doubtful that this method can produce a Luttinger type Fermi surface as shown in Fig. 3a.

Finally, we mention that an alternative to the single spiral state has been suggested, called the double spiral state which is isotropic and has short range order.[84] It has the additional feature that it possesses a staggered chirality.[84,86] This state seems to have many features in common with the staggered flux state described in the last section. For example, we already mentioned that the spin correlation of the staggered flux state is likely to be incommensurate, because the existence of a Fermi surface in the spinon excitation implies low lying spin 1 excitations at wave vectors shifted away from (π,π).

4. CONCLUSIONS

It seems to me that the experimental data are providing a rather clear set of requirements for any theory of the normal state properties. The angular resolved photoemissions data are still incomplete at this point, but will soon provide a rather complete determination of the Fermi surface and settle the choice between Fig. 3a and b. I shall proceed with the assumption that Fig. 3a is favored, i.e. Luttinger Theorem is obeyed. Then we have to deal with the Hall effect and ask how the Fermi surface emerges from the half-filled case. Let us consider the following scenario. For very small doping x, we have Néel order, and in the absence of disorder, the system will form a Fermi liquid with pockets as shown in Fig 4a. (For the purpose of this argument, we shall ignore complications due to the formation of an incommensurate state). From our discussion of the single hole problem we expect a mass given by J and a Hall number equal to x. We also introduce (N/m)* as the ratio of the effective carrier number per copper and the mass obtained from the spectral weight of the Drude part of the optical absorption. This Fermi liquid is a conventional one, and obeys Luttinger theorem in the reduced Brillouin zone. Upon further doping, the hole pockets will merge. If we assume that the system remains Néel ordered, the Fermi surface will look like Fig. 4b in the extended zone scheme. Let us introduce more holes and the system loses long range order. Then we may expect the Fermi surface to look like the solid line in Fig. 4c. If the spin coherence length is long,

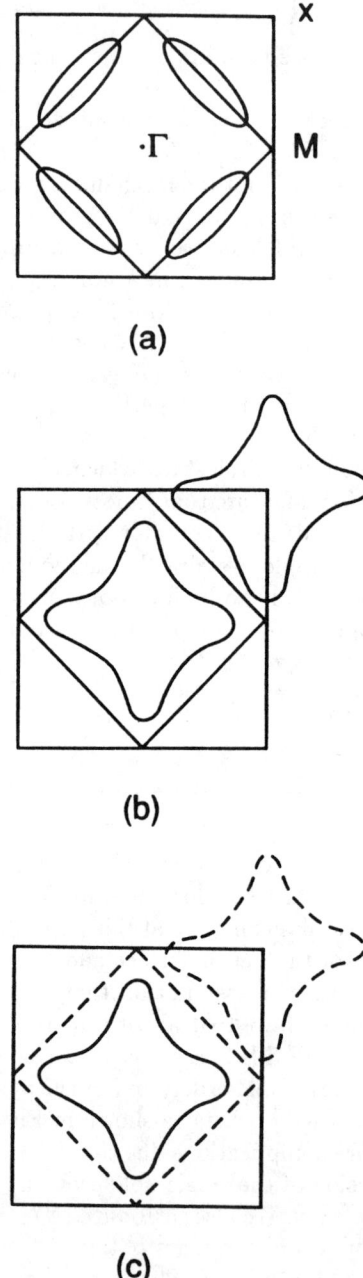

FIGURE 4 Schematic drawing of a possible evolution of the Fermi surface as doping is increased. a) small doping, long range ordered state, b) increased doping, LRO remains, c) Further doping, in the short range ordered state the dotted part of the Fermi surface disappears.

we may expect a "ghost Fermi Surface" indicated by the dotted line, indicating the locus of low lying and broadened single-particle excitation formed by the emission of a soft zone boundary magnetic excitation. Its Hall effect will be complicated due to the complex curvatures, and may have either particle or hole signs. But this Fermi liquid will be unconventional at least in the following sense. Experimentally, the optical sum rule is found to be proportional to x and given by $(N/m)^* = x/m_1$ where m_1 is typically twice the free electron mass and corresponds to a hopping parameter slightly larger than J. The same ratio is determined from London penetration depth measurement and is found to be consistent. The optics can only determine the ratio $(N/m)^*$ but now the effective mass m^* can be measured from the dispersion of the band in photoemission experiment. From the recent data[22] scanning from Γ to X or Y, we see that the mass enhancement from the band mass is only about a factor of two, leading to an $m^* \approx m_1$. We emphasize that if confirmed, this is a very significant and rather surprising result. The point is that very naively, one might have expected, based on looking at Fig. 3a, that the number of carriers is 1-x per copper and that $(N/m)^* = (1-x)/m^*$. With the measured $m^* \approx m_1$ this will greatly exceed the measured value of x/m_1. The heavy Fermion solution solves this problem by developing a heavy mass of order m_1/x. The fact that this mass enhancement is not seen shows that the heavy Fermion solution described in Section 3.b(i) is not viable. Of course, there is nothing inconsistent or even surprising with the observed phenomena, because we know that it is really the doped holes which are carrying the current. However, it is a challenge to come up with a theory which simultaneously satisfy Luttinger Theorem, the Hall effect, the optical sum rule, and a minimal quasiparticle mass enhancement. The challenge to the various theories are as follows:

(i) Theories based on doped holes in the slave Fermion representation will have the right Hall effect optical mass and quasiparticle mass, but will have difficulties satisfying Luttinger theorem.

(ii) Theories based on doped holes in the slave boson representations, such as the flux phases described in section 3.b. will also have difficulties with Luttinger theorem. The exception is the original BZA solution which contain a spinon Fermi surface (I thank G. Kotliar for reminding me of this fact). These theories may have difficulty with the Hall effect, even though no quantitative treatment has been attempted.

(iii) The weak coupling theory may have the best chance of realizing the scenario shown in Fig. 4. The challenge is to show that if the spin correlation is reduced to only 3 or 4 lattice constant such that the gap is largely filled in, that the optical sum rule remains of order x/m, and not revert back towards $(1-x)/m$.

(iv) Finally, we remark that that one dimensional Hubbard model manages to solve all the above problem of obeying Luttinger Theorem, no mass enhancement and optical sum rule x/m. However, it manages to do so by exploiting the possibility of domain formation and charge spin separation. Whether the two-dimensional Hubbard model will employ the same route to solve its problem, as Phil Anderson suggested, is an open question at this point.

5. ACKNOWLEDGEMENT

I am thankful to the NSF-MRL program under DMR 87-19217 for support.

REFERENCES

1. J. G. Bednorz and K. A. Müller, Z. Phys. B **64**, 88 (1986).
2. P. W. Anderson, Science **235**, 1196 (1987).
3. S. Martin, et al., AT&T Bell Labs preprint. For ceramic data, see G. Xiao et al., Phys. Rev. B **38**, 11824 (1988).
4. C. C. Tsuei, A. Gupta and A. Koren, IBM Yorktown preprint.
5. Y. Hidaka and M. Suzuki, Nature **338**, 635 (1989).
6. G. A. Thomas, J. Orenstein, D. H. Rapkine, M. Capizzi, A. J. Millis, R. N. Bhatt, L. F. Schneemeyer and J. V. Waszczak, Phys. Rev. Lett. **61**, 1313 (1988); R. T. Collins, Z. Schlesinger, F. Holtzberg, P. Chaudari and C. Field, Phys. Rev. B **39**, 6571 (1989).
7. For a review, see M. Klein et al. in *Strong Correlation and Superconductivity*, Ed. H. Fukuyama, S. Maekawa and A. Malozemoff, Springer-Verlag, 1989, p. 226, and S. Sugai, in Stanford Conference, Physica C, to be published, 1989.
8. C. M. Varma, P. B. Littlewood, S. Schmitt-Rink, E. Abrahams and A. E. Ruckenstein, Phys. Rev. Lett **63**, 1996 (1989).
9. M. Gurvitch, J. M. Valles, A. M. Cucolo, R. C. Dynes, J. P. Garno, L. Schneemeyer and J. V. Waszczak, Phys. Rev. Lett. **63**, 1008 (1989).
10. P. A. Lee and N. Read, Phys. Rev. Lett. **58**, 2891 (1987).
11. B. S. Shastry, Phys. Rev. Lett. **63**, 1288 (1989); F. Mila and T. M. Rice, ETH preprint.
12. S. Uchida, H. Takagi, Y. Tokura, N. Koshihara, and T. Arima, in *Strong Correlation and Superconductivity*, Ed. H. Fukuyama, S. Maekawa and A. Malozemoff, Springer-Verlag, 1989, p. 194.
13. N. P. Ong et al., Phys. Rev. B **35**, 8807 (1987); M. W. Shafer, T. Penney and B. Olson, Phys. Rev. B **36**, 4047 (1987).
14. J. Zaanen, G. Sawatzky and J.W. Allen, Phys. Rev. Lett. **55**, 418 (1985).
15. For a review, see H. Yasuoka, T. Imai and T. Shimizu, in *Strong Correlation and Superconductivity*, Ed. H. Fukuyama, S. Maekawa and A. Malozemoff, Springer-Verlag, 1989, p. 254.
16. N. P. Ong, T. W. Jing, Z. Z. Wang, J. Clayhold, S. J. Hagen and T. R. Chien, ibid, p. 204; Clayhold et al., Phys. Rev. B **39**, 7300 (1989).
17. T. Tamegai, K. Koga, K. Suzuki, M. Ichihara, F. Sakai and Y. Iye, Jpn J. Applied Phys. **28**, L 112 (1989).
18. T. Takahashi et al., Phys. Rev. B **39**, 6636 (1989).
19. J. Fink et al., preprint.

20. C. G. Olson, R. Liu, A. Yang. D.W. Lynch, A. J. Arko, R. S. List, B. Veal, Y. Chang, P. Jiang and A. Paulikas, Science **245**, 731 (1989).
21. C. G. Olson et al., Stanford Conference, 1989, Physica to be published.
22. C.G. Olson, et al., Ames Lab. preprint, 1989.
23. For a review, see D. M. Newns and P. C. Pattnaik, in *Strong Correlation and Superconductivity*, Ed. H. Fukuyama, S. Maekawa and A. Malozemoff, Springer-Verlag, 1989, p. 146.
24. See for example, M. Hybertson and L. Mattheis, Phys. Rev. Lett. **60**, 1661 (1988); H. Krakauer and W. Pickett, ibid. **60**, 1665 (1988); S. Massidda, J. Yu and A. J. Freeman, Physica C **152**, 251 (1988).
25. J. R. Schrieffer, X. G. Wen and S. Zhang, Phys. Rev. B **39**, 11663 (1989).
26. G. Shirane et al., Phys. Rev. Lett. **63**, 330 (1989) and references therein.
27. J. Solyom, Advances in Physics **28**, 201 (1979).
28. M. Ogata and H. Shiba, preprint 1989.
29. J. W. Allen et al., U. Michigan preprint, 1989.
30. C. A. R. Sa de Melo and S. Doniach, Stanford preprint 1989.
31. F. C. Zhang and T. M. Rice, Phys. Rev. B **37**, 3759 (1988).
32. A. Kampf and J. R. Schrieffer, Los Alamos preprint, 1989.
33. J. Zaanen and O. Gunnarsson, Phys. Rev. B **40**, 7391 (1989).
34. D. Poilblanc and T. M. Rice, preprint.
35. L. N. Bulaevskii, E. L. Nagaev and D.I. Khomskii, JETP **27**, 836 (1968).
36. B. Shraiman and E. Siggia, Phys. Rev. Lett **60**, 740 (1988).
37. S. Schmitt-Rink, C. M. Varma and A. F. Ruckenstein, Phys. Rev. Lett **60**, 2793 (1988).
38. C. Kane, P. A. Lee and N. Read, Phys. Rev. B. **39**, 6880 (1989).
39. C. Gros and M. D. Johnson, U. of Indiana preprint.
40. B. Su, Y. M. Li, W. Y. Lai, and L. Yu, Phys. Rev. Lett. **63**, 1318 (1989).
41. J. Bonca, P. Prelovsek, and I. Sega, Phys. Rev. B **39**, 7074 (1989).
42. E. Dagotto, J. R. Schrieffer, A. Moreo and T. Barnes, Santa Barbara preprint.
43. D. Poilblanc, Phys. Rev. B **39**, 140 (1989).
44. V. Elser, D. Huse, B.I. Shraiman and E.D. Siggia, Phys. Rev. B, to be published.
45. W. Stephan, K. von Szczepanski, M. Ziegler and P. Horsch, Stuttgart preprint 1989.
46. S. Sachdev, Phys. Rev. B **39**, 12232 (1989).
47. S. Trugman, Phys. Rev. B **37**, 1597 (1988).
48. S. Trugman, Los Alamos preprint, 1989.
49. P. W. Anderson, Kathmandu lectures (1989) and this conference.
50. S. Kivelson, D. Rokhsar and J. Sethna, Phys. Rev. B **35**, 8865 (1987).
51. N. Read and D. Newns, Solid State Comm. **52**, 993 (1984); A. Millis and P. Lee, Phys. Rev. B **35**, 3394 (1987); A. Auerbach and K. Levin, Phys. Rev. Lett. **57**, 877 (1986).
52. G. Kotliar, P. Lee and N. Read, Physica C **152-155**, 538 (1988).
53. C. Castellani, C. DiCastro and M. Grilli, Physica C **153-155**, 1659 (1988).

54. D. Newns, P. C. Pattnaik, M. Rasolt and D. A. Papaconstantopoulos, Phys. Rev. B **38**, 7033 (1988).
55. J. H. Kim, K. Levin and A. Auerbach, Phys. Rev. B **39**, 11033 (1989).
56. C. Castellani and G. Kotliar, Phys. Rev. B **39**, 2876 (1989).
57. G. Baskaran, Z. Zou and P. W. Anderson, Sol. St. Comm. **63**, 973 (1987).
58. I. Affleck and J. B. Marston, Phys. Rev. B **37**, 3774 (1988).
59. G. Kotliar, Phys. Rev. B **37**, 3664 (1988).
60. For a review see F. C. Zhang, C. Gros, T. M. Rice and H. Shiba, Supercond. Sci. Tech. **1**, 36 (1988).
61. Y. Hasegawa, P. Lederer, T. M. Rice and P. B. Wiegmann, Phys. Rev. Lett. **63**, 907 (1989); C. Montambaux, Phys. Rev. Lett. **63**, 1657 (1989).
62. P. Lederer, D. Poilblanc and T. M. Rice, Phys. Rev. Lett. **63** (1989).
63. V. Kalmeyer and R. B. Laughlin, Phys. Rev. Lett. **59**, 2095 (1987).
64. R. B. Laughlin, Phys. Rev. Lett. **60**, 2677 (1988), **61**, 379 (E).
65. S. Liang and N. Trivedi, Illinois preprint.
66. T. Dombre and G. Kotliar, preprint 1989.
67. A. B. Harris, T.C. Lubensky and E. J. Mele, Phys. Rev. B. **40**, 2631 (1989).
68. Y. Hasegawa and D. Poilblanc, ETH preprint.
69. F. C. Zhang, Cincinnati Preprint.
70. X. G. Wen, F. Wilczek and A. Zee, Phys. Rev. B **39**, 1143 (1989).
71. D. V. Kheshchenko and P.B. Wiegmann, preprint 1989.
72. D. Arovas and A. Auerbach, Phys. Rev. B **38**, 316 (1988).
73. N. Read and B. Chakraborty, Phys. Rev. B **40**, 7133 (1989).
74. F. D. M. Haldane and H. Levine, Phys. Rev. B **40**, 7340 (1989).
75. P. B. Wiegmann, Phys. Rev. Lett **60**, 821 (1988), Physica C **153-155**, 103 (1988).
76. X. G. Wen, Phys. Rev. B **39**, 7223 (1989).
77. P. A. Lee, Phys. Rev. Lett **63**, 680 (1989).
78. R. Shankar, Phys. Rev. Lett **63**, 203, 1029 (E), (1989).
79. N. Read and S. Sachdev, Phys. Rev. Lett. **62**, 1694 (1989).
80. L. B. Ioffe and A.I. Larkin, Phys. Rev. B **40**, 6941 (1989).
81. E. Fradkin and S. Kivelson, preprint 1989.
82. B. Shraiman and E. Siggia, Phys. Rev. Lett. **62**, 1564 (1989).
83. D. Yoshioka, J. Phys. Soc. Jpn. **58**, 1516 (1989).
84. C. L. Kane, P. A. Lee, T. K. Ng, B. Chakraborty and N. Read, Phys. Rev. B, to be published.
85. C. Jayaprakash, H. R. Krishnamurthy and S. Sarker, Phys. Rev. B **40**, 2610 (1989).
86. B. Shraiman and E. Siggia, Trieste lecture, preprint 1989.

DISCUSSION

P. W. Anderson: Listening to Patrick, and I think that Patrick was doing the best he could, although I don't think he actually believes in Fermi liquid theory either, I think it is true that there about two theories that have made falsifiable predictions. One of these is Fermi liquid theory which has made a great many, and I can hardly think of any one that has not been falsified. And yet we find that people go on trying to fit everything with Fermi liquid theory. Here's this theory that makes a lot of predictions, and you falsify them and then they try to squeeze out of it again and again. I think, we should think, falsifiability, and is there a theory which has made predictions. I don't think that anyone who has not made a theory that makes predictions should really have a right to talk about whether it's falsifiable or not. Fermi liquid theory has made many falsifiable predictions and they're almost all wrong.

Now, the second point. On those lanthanum strontium things – this is a very detailed point and really clarification – you jump suddenly from the highest curve to the lowest curve as you go from 15% to 30%. That is the point at which, if our band theory friends had been alert, they would have told you long since that your 3-band model is no longer valid and you should have been including the d_{z^2} band; and in fact the atoms move in such a way as to bring the dz-squared band more or less degenerate with the $d_{x^2-y^2}$ band, so this is exactly the point which you go into a two-band Hubbard model, and of course everything goes to hell. You've changed the structure, so this is a very strong argument in favor of the single-band Hubbard model.

Remarks about the Hall effect: One of the most fascinating things that could happen, and may even happen, is the de Haas-Van Alphen data can find you an area of the electron-like Fermi surface piece that is appropriate for Fermi liquid theory, but it will probably have the hole sign of the carriers around it. That's not an unequivocal prediction of the things I was talking about this morning, but it's a very likely prediction, and certainly the fact that the Hall effect is getting bigger as you go to lower temperatures makes it very likely that this thing will happen.

One other thing. Just saying about the one-dimensional thing that you change from singlets all coupling spins between the two sub-lattices to singlets coupling spins within a sublattice. One of the lovely thing about singlets, the singlet representation, is that it's INCREDIBLY overcomplete. It's overcomplete by some factor of N factorial, or something like that. So in fact you can describe it equally well, using singlets between both sublattices, and when you do that you actually shorten the range of the singlets. The wrongest thing about SU(N) is it destroys this overcompleteness, and makes it impossible to talk in terms of these various representations. That's probably the key to the two-dimensional thing too, right? It's much easier to break up this sub-lattice structure than you think.

D. J. Scalapino: I want to comment on the question of the doping dependence of the optical spectral weight. Baeriswyl and co-workers showed that the integral of the optical conductivity for the Hubbard model was proportional to the matrix element t_{eff} of the near-neighbor hopping (the kinetic energy). Now suppose one looks at t_{eff} for the half-filled Hubbard model. As U/t is increased, t_{eff} decreases, and for large values of U/t, t_{eff}/t varies as t/U. Now, at a fixed value of U, what happens when we begin to add more holes? For $U/t = 4$, we find from the Monte Carlo that t_{eff} initially increases linearly with x. This is also found at larger values of U, where the half-filled spectral weight is more suppressed. Thus it appears that the Hubbard model can lead to an optical spectral weight which varies as the doping x.

C. M. Varma: I want to go back to the basics, and I have question – it's almost a chemistry question – about this reduction of the three-band model to a one-band model. Now one of the possibilities of course is that if one could ever solve the three-band model exactly or very suitably that everything except the one-band will be irrelevant in some sense. My guess is that that's how Phil tends to think about it. And I have nothing to say about that. But my question really is about the way people tend to do perturbation theory to deduce the one-band model from the three-band model. Now in what I'm going to talk about I'm going be in the electron language, so my e_p and e_d will be in the electron language, and we know that e_p and e_d renormalized – mean field renormalizd, or density-functional renormalized, or Hartree-Fock renormalized – have to be pretty much the same, to within a volt. I think there is no argument about that. So people say $(e_p - e_d)_{bare}$ (as corrected by U's and what not) is a quantity of the order of 4 or 6 volts, and t is of the order one volt, and then start doing perturbation theory in $t/(e_p - e_d)_{bare}$. And then we do second order perturbation theory, and lo and behold, we have the one-band model. In fact, the parameter should be $t/(e_p - e_d)_{bare}$ but I won't quibble with that factor. But my real problem arises that if I start doing perturbation theory, if I go on working hard, I'm going to end up with energy denominators which are the renormalized energy denominators, which are zero. So I think in the sense of deriving one-band model from perturbation theory, I have a difficult time understanding how it can ever be derived. And I am talking about the metallic regime, and not in the insulating case. The basic point I'm trying to make is that what you might derive about the basic parameters of the problem for the insulating state may have absolutely nothing to do with the important parameters in the metallic state, where you have no new low-energy excitations and new interactions among those low-energy excitations.

E. Stechel: I would like to make it perfectly clear that in reducing the three-band Hubbard model to a single band model, no one recently has used perturbation theory. The work by Hybertsen, myself, Schluter and Jennison that Patrick Lee referred to consists of calculations on finite clusters with the three-band model solved EXACTLY. Reduction is accomplished by mapping onto the EXACT spectrum and exact wave functions. No perturbation theory has been applied.

C. M. Varma: That's a little more subtle point why you might still reduce the model to a one-band model in the insulating state, and yet not be able to do it in the metallic state. Thus, I am questioning perturbation theory in the metallic state.

T. M. Rice: I have a question for Patrick. You made a big point about the difference between the photoemission and the transport properties, but isn't that in some sense what Phil said happens in the Hubbard model. If you take n_k, which is the expectation value of the momentum that's the integral of the spectral weight of the photoemission, and it has its weight predominantly below k_F, and has a singularity at the Luttinger k_F, although even in that model, one knows that the transport properties and all that are proportional to the density of holes in one or two dimensions.

P. Lee: Yes, in one dimension, we have a case where things work out, and I'm just asking how are we going to produce a theory in two dimensions where all these features are preserved. That's presumably what Phil talked about.

V. Emery: I have another comment on the three-band, one-band discussion. There are numerical calculations by Gooding, Elsa, Shraiman, and Siggia with a much larger periodic lattice than the one you described, and they certainly don't verify the one-band model. They get a rather small overlap with the states in the one-band model. For my money, I think there are some rather special properties of the cluster you studied that are really not reflective of the full system. They give you a larger overlap than you'd really find if you had a larger system.

H. Ott: I had a question about the experimental facts that are predicted by models and falsified by models, or vice-versa. Now some of the normal state properties which are seen in these oxides are really very similar to a heavy electron system at high temperatures. Are those also not Fermi liquids at these temperatures?

P. W. Anderson: Well, no, actually. They are quite different from Fermi liquids. In the high temperature state you've got spins; the spin is not a Fermi liquid object, and so you're starting from a starting point which is very far from Fermi liquids. It tends to renormalize towards Fermi liquid. Some of us believe that in the meantime, in fact, it is going through a stage where it's doing something even more complicated, before it finally finds the Fermi liquid. I would answer that no, of course, that was the original problem...How does the Kondo model make a spin 1/2 particle, no charge particle out of a system that has no such particle? Or how does the Anderson model make the Kondo model? The Kondo model is a non-Fermi liquid kind of state.

G. Sawatzky: Maybe I could mention another observation that we should perhaps start considering, because a lot of photoemission data, on high T_c but also on model systems, show that when you actually dope the systems that you don't really move the Fermi level into the upper Hubbard band, or the lower Hubbard band, whether you hole-dope or electron-dope, but in fact the Fermi level seems to be rigidly fixed. So what you're doing is you're introducing new states, either above, or below, or at the Fermi level, spread out over a very large energy range.

P. W. Anderson: The answer to that is Yes, the ordered upper and lower Hubbard bands do have all these quite heavy effective masses, all these problems, yet in some magic way that I don't yet completely understand, the moment you introduce

introduce carriers, the holes are moving, with their full kinetic energy. That is the spectral densities move up, although I bet the Fermi level moves down more than a little.

C. M. Varma: Although the problems that you mentioned are taken care of by the small-U kind of theory, I think I would find it very remarkable if the other problems that were there in the experiments can really be taken care of by this.

P. Lee: I think you misunderstood me. I said that the small-U problem, as far as I see, may have difficulty taking care of the spectral weight.

C. M. Varma: But I want to say that the kinds of problems that you mentioned with the spectral weight, and optics, and London penetration depth, etc., at least on a phenomenological level, I do not have any of those problems with that thing which is not really a Fermi liquid – and indeed, has many of the crazy properties that you find in one dimension without having any charge and spin degrees of freedom separation.

P. Lee: I agree that on a phenomenological level we understand everything, because the experimentalists have told us the answer, and we look at it, and we know what is needed. And I think, listening to Doug's talk, Monte Carlo is getting close, also, to telling us what the answer is. I think this allows us to focus on what question to ask, now that the experiment is saying that the proper theory will have this conflicting property; looks like the Monte Carlo is going to tell us that the Hubbard model is going to force us into this dilemma as well. But the question I am raising is, microscopically, how do you construct such a theory.

C. M. Varma: I'm also saying that I don't think it is there in the repulsive Hubbard model at all.

P. Lee: Okay, well that's something that Doug would be able to tell us in a year or two when he gets his half-a-million dollars.

C. M. Varma: Yes, the half-million-dollar question.

G. Kotliar: I have a comment. If I look at the t-J model for large doping, in the large-N limit – that is, expanding around Anderson's uniform solution – then I don't have any problems with the Fermi surface. Now...It will be given by Luttinger's theorem. Now as far as the Hall effect, if you include the oxygen-oxygen overlap, then Levin, Auerbach, and Kim have calculated what the sign of the Hall effect should be, and it comes out hole-like. Now as far as the spectral weight if you compute the optical conductivity, you're going to find that the mass that changes the conductivity scales with δ, and you have the proper large-U behavior. On the other hand, if you focus for example on things like the specific heat, or quasiparticle properties, you're going to find that it's J that enters there. Now, since t/J is small, this is just a factor of three, if you're willing to live with a mass which is renormalized by a factor of three, this is not inconsistent with the photoemission. So, I don't see any real problem with that scenario.

P. Lee: Okay, let me make sure I understand you. Is this a heavy fermion-type theory?

G. Kotliar: No, it's not heavy fermion, in the sense that you have to include J. J cuts off, for example, the mass renormalization as entering the specific heat. In the Fermi liquid problem, one has different masses. One has a mass which enters in the transport, enters in the conductivity; one has a different mass which enters in the susceptibility; a different mass which enters in specific heat. Now, if you treat the t-J model in the large-N limit, you discover that all those masses are different, and that the mass, for example, that enters in the specific heat or in the susceptibility is not strongly renormalized. It's renormalized from t to J, which is a factor of three. On the other hand, the mass that enters in the conductivity is strongly renormalized; that scales with δ. On the other hand, the Fermi surface obeys Luttinger's theorem. As far as the Hall effect, there is a problem in the sense that in that model, if you take, strictly speaking the t-J model, you get the wrong sign of the Hall effect. So as far as that goes, there is a problem. But it's not clear to me that just by modifying slightly the shape of the Fermi surface, one can also fix that.

N-P Ong: Let me comment on the Kim-Levin-Auerbach model. Now, what they did was to approach the insulating state (as in D. Newns *et al.*) by letting m* go to ∞. They perform the calculation, essentially by, narrowing the band, and find that in two dimensions, again, they don't agree with experiment. You get a Hall resistivity that's flat with respect to x. (For the 3D case) they comment (paraphrased), "Well, if we could have hybridization in the c direction, the calculations are much harder. But let's assume that the conductivity remains metallic in the c direction, but becomes localized in the ab plane. Then under those circumstances, we could argue that it (the Hall coefficient) would diverge (as x goes to 0)." They hand-draw the curve to approximate experiments. However, this is a very strong assumption that goes against experiment. When we approach the insulating state in LaSrCu, in YBCO, and in BSCCO, the resistivity anisotropy becomes much larger. Localization sets in first in the c direction.

G. Kotliar: I guess you did not understand the comment I wanted to make. I mean, I think there's no hope in explaining the $1/\delta$ dependence of the Hall effect in this type of theory. That's definitely out of the question. But if you don't want to take seriously the low T_c materials, as far as things like transport, I think that what they showed, in that paper, is that the Hall effect is something which depends a lot on the details of the Fermi surface. So if you just want to think about the 1,2,3, compound and if you want to incorporate the oxygen-oxygen overlap, it's not inconceivable that the band structure, as far as that particular material is concerned, may come out with the right sign. And in fact they have. So that's the only thing I wanted to say.

A. Malozemoff: It might be useful, at least from my perspective, to comment on the lack of completeness in the experimental data that enforce Patrick Lee's picture. It seems to me that the spectral weight is hinging on those two optical studies from Japan, on the lanthanum strontium copper oxide (Suzuki) and neodymium cerium copper oxide (Uchida). But the photoemission data is mostly on the bismuth system.

P. Lee: But I can also get the ratio n/m^* from the London penetration depth. It's the same quantity that appears. And in all cases where it has been studied, like in the strontium-doped, and so on, they are in good agreement with what you get from the penetration depth.

Large Doping Approach to High T_c
J. R. Schrieffer

Recently, Arno Kampf and I[1] have been investigating the 2d one band Hubbard model for doping x far from half filling so that antiferromagnetic spin fluctuations are moderate in amplitude. This approach compliments the work of Wen, Zhang and myself[2] starting from the antiferromagnetic insulator which occurs for $x \ll 1$. In this small x regime, one finds hole like excitations that carry charge e and spin $\frac{1}{2}$ and are fermions. One finds that a hole locally depresses the antiferromagnetic order parameter over the coherence length $\xi_{SDW} \sim \hbar v_F/\pi\Delta_{SDW}$, where $2\Delta_{SDW}$ is the spin density wave gap in the electronic spectrum. These bag type excitations interact with the antiferromagnetic spin waves of the x=0 system. By direct calculation one finds a pairing attraction $V_{kk'}$ for momentum transfers $|\vec{k} - \vec{k}'|$ less than ξ_{SDW}^{-1}, while $V_{kk'}$ is repulsive[4] for $\vec{k} - \vec{k}' \sim \vec{Q}$, the nesting wave vector. This leads to s or d wave type superconductivity depending on the shape of the fermi surface and the relative strength of $V_{kk'}$ for small and large $\vec{k}-\vec{k}'$.

Since cuprate superconductivity occurs in the paramagnetic metal, where finite doping has destroyed antiferromagnetism, it is important to investigate whether similar bag like excitations occur in this regime, with the gap $2\Delta_{SDW}$ replaced by a pseudogap[5] induced by antiferromagnetic spin fluctuations. Consider the zero order hamiltonian to be band electrons, of energy ϵ_k coupled by an effective Hubbard U. This effective U includes screening and short range correlations making it considerably weaker than the bare U of the Hubbard model.

The essential point we wish to make is that much of the physics of this intermediate to large x regime is obtained in low order diagrams. For example, if one treats the one particle self energy Σ at the one loop level using the spin susceptibility χ, one has

$$\Sigma(k,\omega) = \frac{-3iU^2}{2N} \int G(k-q, \omega-\nu)\chi(q,\nu)d^2q d\nu/(2\pi)^3 . \tag{1}$$

for N cells per unit volume. Experimentally, neutron scattering[3] shows that $\chi(q,\nu)$ is large in the vicinity of $\vec{Q} = (\pm\pi, \pm\pi)$ for doping in the range corresponding to superconductivity. This result is also obtained in an RPA calculation, even though the RPA breaks down as long range spin order sets in. Since our results for Σ are found to be relatively insensitive to the detailed form of $\chi(q,\nu)$, we consider a simple parameterized form of χ,

$$\chi(q,\nu) = \sum_{Q=(\pm\pi,\pm\pi)} \frac{\Gamma}{(Q-q)^2 + \Gamma^2} \cdot \frac{2\omega_o}{\omega_o^2 - \nu^2 - i\delta} \tag{2}$$

As $\Gamma \to 0$, χ approaches the Bragg peak of the ordered antiferromagnet, while for $\Gamma \sim Q$, χ exhibits the broad q variation appropriate to the weakly coupled metal. We have carried out numerical RPA calculations of χ and confirmed as a function x and U that the essential features of Σ are quite insensitive to χ as long as χ is peaked in the vicinity of Q.

We recall that for the x=0 mean field SDW, Σ is given by

$$\Sigma_{SDW}(k,\omega) = \frac{\Delta_{SDW}^2}{\omega + \epsilon_k}, \qquad (3)$$

Thus, Σ_{SDW} has a pole at $-\epsilon_k$ and G(k,ω) has two poles, at energies $\pm E_k = \pm\sqrt{\epsilon_k^2 + \Delta_{SDW}^2}$, corresponding to the upper and lower bands arising from the doubling of the size of the unit cell. This result is in contrast with the one quasiparticle peak in G occurring in the weakly coupled large x metal.

Arno Kampf and I have studied[1] how one goes between these limits. While the calculations are carried out in diagram perturbation theory, there are reasons to believe that smooth behavior occurs over the physically relevant range of U/t, where t is the hopping. This smooth behavior was seen in the x=0 calculations of Wen and Zhang[2] for the spinwave velocity, the sublattice magnetization and the gap.

In the weakly correlated metal, $\Sigma(k,\omega)$ has a negative value of $\frac{\partial \Sigma}{\partial \omega}$ for $k \sim k_F$ and ω near the quasiparticle energy E_k. Therefore, the weight of the quasiparticle pole

$$Z_k = \frac{1}{1 - \frac{\partial \Sigma}{\partial \omega}}\bigg|_{k,E_k} \qquad (4)$$

is less than unity, consistent with the spectral function $A(k,\omega) = \frac{1}{\pi}|\text{Im}G(k,\omega)|$ satisfying the sum role

$$\int_{-\infty}^{\infty} A(k,\omega)d\omega = 1. \qquad (5)$$

However, as x decreases and $\chi(q,\nu)$ becomes peaked around Q, $\partial \Sigma/\partial \omega$ changes sign near $\omega=0$ (center of the gap) and Z_k as predicted by (4) becomes larger than unity violating (5). The resolution of the problem is that as $\frac{\partial \Sigma}{\partial \omega}$ becomes positive, there no longer exists a solution of the pole condition of G near $\omega=0$, even though $k \sim k_F$, i.e.

$$G^{-1}(k,\omega) \neq 0 \qquad (6)$$

This result is consistent with the x=0 antiferromagnetic insulator where for $k=k_F$, two poles, $\pm \Delta_{SDW}$, occur rather than one pole at $\omega=0$ as in the conventional fermi liquid.

However, if the chemical potential μ is less than $-\Delta$ as it is for $x \neq 0$, one must investigate states near $\omega \sim -\Delta$, the energy region where physically relevent excitation reside.

In Fig. 1, the density of states for the model χ is shown for U=4t and several values of Γ. One sees that as the spin-spin correlation length $L_{ss} \sim \frac{2\pi}{\Gamma}$ lengthens

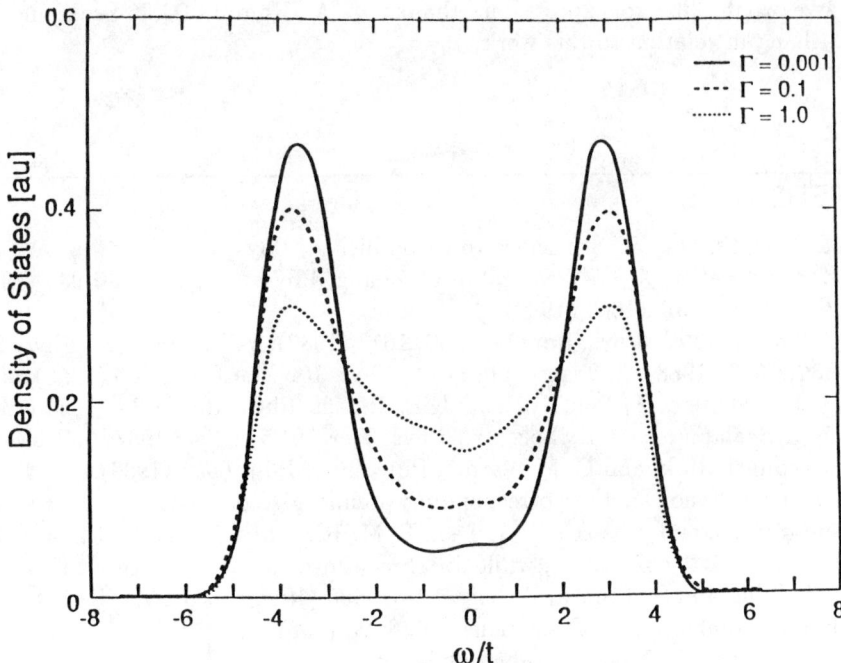

FIGURE 1 Density of states as a function of energy in units of the hopping t for several values of the width Γ of the peak of the spin susceptibility χ near $Q=(\pm\pi,\pm\pi)$.

the pseudo gap becomes better defined. Similar results are obtained with χ calculated within the RPA as well as that of Pines and Monien.[6] We note that the two peaks in $A(k,\omega)$ as a function of ω arise not from a splitting of the quasiparticle peak but from the sharpening up of the incoherent electron and hole parts of $A(k,\omega)$, with the disappearance of the conventional quasiparticle peak.

Having obtained the pseudo gap, one can get bag like excitations by working at the two loop level, i.e. one loop makes the pseudo gap and the other locally suppresses it through the Pauli principle. The relevant diagram in Σ is the two crossed line graph, an effect which reduces the pseudogap somewhat but more importantly giving internal structure to the excitation.

The most important property of the bags for high T_c is that they attract for $|\vec{k}-\vec{k}'| \lesssim 1/\xi$, where ξ is the coherence length of the pseudo gap. Explicit calculation of the pairing potential shows results similar to the $x\simeq 0$ interaction.[2]

Currently we are studying the symmetry of stable solutions Δ_k of the pairing gap equation as well as investigating the normal phase properties arising from the pseudo gap.

We would like to express our thanks to A. Kampf, D. J. Scalapino and S.-C. Zhang in relation to this work.

REFERENCES

1. A. Kampf and J. R. Schrieffer, to be published, Phys. Rev. B.
2. J. R. Schrieffer, X.-G. Wen and S.-C. Zhang, Phys. Rev. Lett. **60** 944 (1988); Phys. Rev. **B39** 11663 (1989).
3. G. Shirane et al., Phys. Rev. Lett. **59**, 1613 (1987); Y. Endoh et al., Phys. Rev. B**37**, 7663 (1988); J. Tranquada et al., Phys. Rev. Lett. **60**, 156 (1988).
4. D. J. Scalapino, E. Loh, Jr., and J. E. Hirsch, Phys. Rev. B**34**, 8190 (1986); D. J. Scalapino and E. Loh, Jr., Phys. Rev. B**35**, 6694 (1987); K. Miyake, S. Schmitt-Rink, and C. M. Varma, Phys. Rev. B**34**, 6554 (1986).
5. For a discussion of the corresponding pseudo gap in CDW systems, for the one dimensional case see P. A. Lee, T. M. Rice, and P. W. Anderson, Phys. Rev. Lett. **31**, 462 (1973); while for three dimensional cases see K. Kitazawa, S. Uchida, and S. Tanaka, Physica **135B**, 505 (1985); S. Uchida, K. Kitazawa, and S. Tanaka, Phase Transitions Vol. 8, 95 (1987).
6. D. Pines and H. Monien, to be published.

DISCUSSION

A. Millis: I'm confused by the fact that the quasiparticle renormalization Z is $(1-\partial \sum /\partial \omega)^{-1}$ to the minus one, and yet you said that in this pseudo-gap region your $\partial \Sigma/\partial \omega$ was positive. So Z appears to be larger than one.

J. R. Schrieffer: Right. One has to be very careful about what the spectral function looks like. The standard formula you mention only works, as you know, when the real part of $\sum$ satisfies the pole condition, $\omega - E_k = \sum(k,\omega)$. That usual formula breaks down when you have $\partial \sum /\partial \omega > 0$ since the pole does not occur, i.e. the quasiparticle peak has zero weight, not Z>1.

A. Millis: But I also thought I always got these analytic properties provided perturbation theory was in some sense OK. And yet you're telling me you've done what amounts to a perturbative calculation.

J. R. Schrieffer: In fact, our analysis is highly non-perturbative. But this is basically that of a very small fraction of nearly degenerate sets coupled to each other by a very small interaction. We are using Brillouin-Wigner perturbation theory, the standard method for treating such problems, i.e. Feyman the usual way, diagrams. If you had a very large fraction of the states involved, with a large range of q's coupled, rather than only a small range of γ's around the nesting, then there could be more serious problems. A nice example is the SDW gap where second order

Brillouin-Wigner gives the exact result. Quantum fluctuations of the exchange field is given by the one loop level diagrams.

P. W. Anderson: This comment really doesn't have to do with modified spin bags, it has to do with these old, unmodified spin bags. This is a really fundamental reason for having some problems with the original spin bags. The fundamental reason has to do with the fact that the philosophy of the spin bag is that one wants to somehow leave the Neel state commensurate, and to make a local region in which the order parameter is decreased. Unfortunately the Neel state in this case has no commensurability energy, it has no desire to be of any particular wavelength. There is no singular term, no cusp term, coupling the register of the Néel state to the underlying lattice. The terms which appear to be commensurability terms, are CHARGE commensurability terms. The Umklapp terms are telling you that there is a Mott-Hubbard gap, and that you want to have charge commensurability of the state with the lattice, and then the Neel state has one particle per lump, and so you gain charge commensurability by having that state. But the minute you add a carrier there's no longer any reason why the order parameter should have any register with the lattice. And so one must very generally have a twist, and in fact Shraiman and Siggia have shown that it will twist in strong coupling, We worked it through in weak coupling, although I've only done it in one dimension, and weak coupling in one dimension also twists. I see no reason why the general mean field techniques that you use for that shouldn't work equally well in two dimensions. I think that the point is the minute you have overcome the Mott-Hubbard gap you no longer have any strong predilection to have a particular wavelength.

J. R. Schrieffer: I do not agree with that statement. If one studies the 2d Hubbard model when an SDW is present, one can readily see that the exchange energy is reduced in magnitude when the phase of the SDW increases from zero. This is because the peak of the SDW moves from on site to between sites and the self exchange is much stronger than the interatomic exchange. Concerning the splitting of the neutron scattering peak, it is possible that the SDW remains commensurate but a correlated array of discommensurations modulates the spin density, producing the double peak in certain cases.

P. W. Anderson: I disagree. And in fact, you can explain the observed structure on the basis that you get localization of the carriers and the spins twist only where the carriers are localized, and of course where there are not carriers they do not.

J. R. Schrieffer: That is a gedanken observation...

P. W. Anderson: That is a gedanken observation...

J. R. Schrieffer: ...and my gedanken tells me your gedanken ain't gedankt right.

D. Pines: There is an additional piece of experimental evidence which tends to support Bob's argument about commensurability. That is, the NMR experiments on at least the simplest interpretation suggest very strong antiferromagnetic correlations which are of reasonably short range and which must be very close to the commensurability wave vector in order to explain the Korringa behavior of the oxygen, and the non-Korringa behavior of the copper. It certainly suggests that that basic picture is the right one.

Recent Angle Resolved Photoemission Results from $Bi_2Sr_2CaCu_2O_8$ Single Crystals

R. S. List

Los Alamos National Laboratory, Los Alamos, NM 87545

This paper presents a summary of the recent photoemission results which are primarily due to the efforts of C. G. Olson.[1,2,3] Reviews of the Fermi surface topology, gap isotropy and detailed peak lineshape will be presented. All the measurements discussed have exceptionally good energy and momentum resolution, i.e. $|\Delta E| = 32$ meV and $|\Delta K_\parallel| = 0.075$ Å^{-1} ≈ 5% of the square Brillouin zone width.

In angular resolved photoemission, the energy versus momentum dispersion of an electron state can be mapped by measuring at fixed incident photon energy the energy distribution of photoemitted electrons as a function of their emission angle from the crystal. This is due to the fact that the component of photoelectron momentum parallel to the crystal surface is conserved as it is excited from the crystal into the vacuum. In the simplifying case of a two dimensional crystal, the measured kinetic energy and emission angle of an electron in vacuum completely specify the E and **k** of the electron in the crystal. A point on the Fermi surface is defined as a value of **k** at which a peak disperses across the Fermi level. Since photoemission only measures occupied electron states, there is also a complete loss of intensity in the peak as it crosses the Fermi level. So far we have measured dispersion of peaks along $\Gamma(0,0) \to M(1/2,1/2)$, $\Gamma(0,0) \to X(1,0)$, $(1/8,0) \to (1/8,1)$ and $(0,3/8) \to (1,3/8)$. The measured Fermi surface crossing points are $(3/8,3/8)$, $(3/8,0)$ and $(1/8,3/8)$. These points are consistent with either a Fermi surface calculated from band structure[4] or a more simple rounded square at 1/4 filling.[5] The differences between these two Fermi surfaces is only evident near M(1/2,1/2) where our finite angular resolution and relatively complex band structure make a distinction between the two difficult. Either Fermi surface appears to be consistent with Luttinger's theorem.[6] It should also be noted that the measured E versus k dispersion of the bands near the Fermi level is substantially flatter than predicted by band structure calculations.[4] Along the $(1/8,0) \to (1/8,1)$ direction, an effective mass enhancement of 2 can be estimated. Along $\Gamma \to M$, the measured band is exceedingly flat, but unfortunately the complexity of the calculated band structure preclude a mass enhancement estimation.

Measurements of an anisotropy in the value of the superconducting gap is of course essential in establishing the presence of d-wave pairing. High energy resolution, angle resolved photoelectron spectroscopy is ideally suited for such measurements since it can measure the gap directly at any k point along the Fermi surface. To carefully determine a value for the gap, consecutive high resolution spectra of the density of states near the Fermi level must be taken well below and just above the critical temperature. These spectra may then be fit to a Lorentzian convoluted with a Fermi-Dirac function and also convoluted with a BCS gap function in the case of the superconducting state. From such fits using 32 meV resolution spectra, differences in the gap of less than 2 meV can be determined. So far, such measurements and fits have been obtained for four points along the Fermi surface: (1/8,3/8), (3/8,0) and two distinct points near (3/8,3/8).[2] For each of these points we obtain the same value of Δ to within experimental resolution, i.e. $\Delta = 18\pm2$ meV. This yields a value of $2\Delta = 5kTc$ which is clearly larger than the simple BCS value. From the apparent isotropy, we may conclude that a simple d-wave pairing gap anisotropy such as $|\cos\theta|$ can be ruled out. Smaller amplitude or more localized anisotropies however have not been removed from consideration.

The final and more controversial topic to be discussed is the significance of the detailed lineshape and background of the photoemission peaks near the Fermi edge in the normal state. Figure 1 shows the dispersion of a photoemission peak as a function of angle away from the surface normal. These spectra span approximately linearly across the Brillouin zone from (1/8,1/8) to (1/8,7/16), i.e. roughly along the $\Gamma \to Y$ symmetry direction. Two inequivalent interpretations of these spectra have been proposed. The first, more conventional description sees a broad peak which sharpens as it disperses towards the Fermi level. At 12° the peak has just crossed the Fermi level and by 14° it has already passed through. The broadening

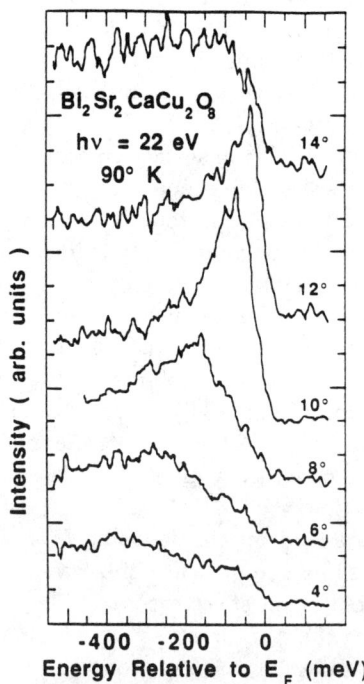

FIGURE 1 Angle resolved photoemission spectra at several angles approximately along the $\Gamma \to Y$ direction in the Brillouin zone.

of the peak as it disperses below the Fermi level is observed in most materials and is due to the shortened photohole lifetime for states far from the Fermi level.[7] The fact that this broadening is so rapid is merely a reflection of the enhanced electron mass due to correlation effects.[8] The slight asymmetry of the lineshape can be explained by the finite angular resolution of the analyzer.[1] If one accounts for the 2° acceptance angle of the analyzer ($|\Delta_{||}| = 0.075$ Å^{-1}) and the experimentally measured energy dispersion of the band near the Fermi level (dE/dk $= 1.1$ eV/Å^{-1}) one obtains an effective energy smearing over 85 meV. Since this smearing sums with equal intensity over broad, low peaks away from the Fermi level and narrow, high peaks near the Fermi level, the resultant lineshape is naturally asymmetric with the sharper edge being near the Fermi level. Detailed fittings to these lineshapes with backgrounds subtracted give very good fits to the data if a broadening proportional to E-E_F is assumed. The featureless background apparent in the spectra can be attributed to scattering from impurities, phonons or other low energy ($\approx$ 100 meV) excitations. This interpretation is strongly supported by the shape of the spectrum for k $>$ k_F. In the 14° spectrum, the shape of the background is very rounded and featureless, just as one would expect for a scattered secondary background. Furthermore the absolute lack of intensity in this spectrum at the Fermi level suggests that the background is in fact made up of scattered secondaries. It is important to note that if we assume a background at any given energy to be proportional to the sum of primaries at higher kinetic energies, then the background calculated for the 12° spectrum is exactly the 14° spectrum.

The second inequivalent interpretation of the data is that they demonstrate the separation of spin and charge in the normal state of high temperature superconductors into spinons and holons.[9] The evidence for such an identification is the presence of the large featureless background and the asymmetry of the peaks in the spectra. The background and asymmetry are intrinsically present in a model using a two dimensional analogy of the one dimensional Hubbard model.[9] While such an interpretation is interesting and elegant, it has trouble explaining the spectra along $\Gamma \rightarrow$ M for k $>$ k_F.[4] Experimentally the featureless background drops off very rapidly as k is increased past k_F, even though the band dispersion is very flat. This is consistent with a scattering interpretation, but is inconsistent with an intrinsic background. It is also important to point out that the size of the featureless background appears to be sample dependent. For poor cleaves which exhibit no evidence for any dispersing peaks, only the featureless background, similar to that in an angle integrated spectrum, is visible. This strongly suggests that the background is extrinsic and related to scattering or sample inhomogeneities. Finally a smaller intrinsic background is found in nearly all metals[10] and in no way verifies the separation of spin and charge. Clearly proof for the existence of a novel state of matter requires more than the observation of a structureless photoemission background in an imperfect crystal.

REFERENCES

1. C. G. Olson, R. Liu, D. W. Lynch, R. S. List, A. J. Arko, B. W. Veal, Y. C. Chang, P. Z. Jiang and A. P. Paulikas, Phys. Rev. B (submitted 12/15/89).
2. C. G. Olson, R. Liu, D. W. Lynch, R. S. List, A. J. Arko, B. W. Veal, Y. C. Chang, P. Z. Jiang and A. P. Paulikas, Phys. Rev. B (submitted 11/15/89).
3. C. G. Olson, R. Liu, D. Lynch, B. W. Veal, Y. C. Chang, P. Z. Jiang, A. P. Paulikas, A. J. Arko and R. S. List, Proceedings of the Mechanisms and Materials of High Temperature Superconductivity Conference, Stanford, CA, July 25 to 28, 1989.
4. S. Massidda, J. Yu and A. J. Freeman, Physica C **52**, 251 (1988).
5. Patrick A. Lee, these proceedings.
6. J. M. Luttinger, Phys. Rev. 119, 1153 (1960).
7. "Photoemission in Solids I," eds. M. Cardona and L. Ley, Springer-Verlag, New York (1979) p. 77.
8. Kevin S. Bedell, NATO ASI Series, Vol. B, Interactions of Electrons in Reduced Dimensions, October 1988, Torino, Italy.
9. P. W. Anderson, Nature (submitted 10/18/89).
10. S. Doniach and M. Sunjic, J. Phys. C. Solid State Phys. 3, 285 (1970).

DISCUSSION

P. Lee: Can you put back the global Fermi surface figure. The scan from Γ to M is actually *not* consistent with the band structure. The band structure predicts an electron pocket around the M point whereas the data do not show a re-entrant behavior after the first crossing of the Fermi surface from Γ to M. It seems to me that the data is more consistent with a single square Fermi surface around the Γ point as expected for a two-dimensional square lattice near half filling.

R. S. List: All right. It's hard to say exactly what happens near the Γ-M crossing. As I said, there are two bands that are very close to each other in the band structure calculations, one of copper-oxygen character and another of bismuth- oxygen character which are both close to crossing the Fermi surface. I cannot definitively say whether the copper-oxygen band which we watch disperse close to E_F from Γ towards M actually crosses the Fermi surface, but the crossing points on the Fermi surface are the same as predicted by the band structure calculations.

P. Lee: Do you have any additional evidence that this so-called band associated with the bismuth oxide exists?

R. S. List: Well, there is some Fermi Surface crossing there. Whether or not it is in fact the bismuth-oxygen derived band, I can't say.

P. W. Anderson:: There are many band calculators who do not believe that that band exists. The Freeman calculations were made with the bismuth on the nominal positions, but they are in fact a very large distance away from them, and also such band calculations would be incompatible with the physical properties of the compounds, particularly the fact that it cleaves between the bismuth layers, and there can not be the bismuth to bismuth covalent bond which that would tend to imply. So then that hole can't be there, and I think that probably Patrick's implied construction is reasonably correct.

R. S. List: I've seen at least two other band structure calculations that have similar structures with a pocket, the bismuth-oxygen type pocket. This is a more complicated structure, it's not a cubic structure, so it's somewhat of an unusual Fermi surface.

R. M. Martin: I'd just like to comment on that, that it really looks like that pocket should be sensitive to what's going on, so you shouldn't believe the band structures very much. So either you find that you just don't see that band or it's not there. You couldn't say that the crossing point is the same, because you see something that has the opposite direction from the calculation.

R. S. List: Yes, and then the electron-hole sense seems to be opposite to that predicted. I agree with that, yes.

B. Batlogg: I'd like to bring up the point discussed this morning about gap anisotropy within the plane. I think you do have this selective point only in the Brillouin zone, and I'd like to bring up this viewgraph. These Raman scattering experiments were done in the plane, rotating polarization in the plane. Both in the bismuth and the copper compounds one has a distinct difference between different polarizations. I wonder whether we are fortunate in picking up with Raman just that part in the Brillouin zone, which apparently has a larger gap, could you have missed that, just by not going in the 45 degree direction.

R. S. List: Right, well we do have a finite number of points, but at least the points we looked at, it seems to be constant.

D. Pines: Did you use strong coupling BCS theory, that is BCS theory modified with strong coupling corrections to get out a gap, because the gap is larger than the BCS value?

R. S. List: Yes, although I did not do the fitting myself, so I cannot definitively answer that. I believe – I could look at the paper, I have it with me, it probably has a reference to it – I believe it was not strong coupling, but I should really check on it.

D. Pines: Then I would treat with extreme caution any gaps that you deduce.

4. SPIN FLUCTUATIONS IN THE INSULATING AND METALLIC PHASES

Chair — D. Pines

Sudip Chakravarty
Department of Physics
University of California at Los Angeles
Los Angeles, CA 90024-1547

Magnetic Properties of La_2CuO_4

A theoretical review of the magnetic properties of stoichiometric La_2CuO_4 is presented.

I. INTRODUCTION

In this review I shall discuss the magnetic properties of stoichiometric La_2CuO_4. More specifically, I would like to explore the extent to which a S=1/2 nearest-neighbor Heisenberg model on a square lattice with an antiferromagnetic exchange (QHAF) adequately describes these properties.

In order to compare with experiments it is necessary to develop a theory which is valid at non-zero temperatures. Thus the main emphasis is the behavior of this model at finite temperatures. I shall assume that the ground state of QHAF is ordered and deduce, based on this assumption, the properties of this model at finite temperatures. The correctness of our theory, and our assumption, will then be tested by comparing with experiments. Of course, by now considerable numerical evidence exists that QHAF is ordered at T=0 (cf. below).

The main theoretical framework is based on a continuum model, the quantum mechanical non-linear σ-model (QNLσM), which correctly reproduces the long wavelength, low temperature behavior of the QHAF. Of course it is also important to understand the short wavelength properties of the QHAF, particularly because spin-pair excitations in light scattering experiments probe precisely these energetic spin fluctuations. These short wavelength excitations will be treated separately using the recently obtained series expansion results and the results obtained from exact diagonalization studies of finite lattices.

Although this is a theoretical review, I shall place considerable emphasis on experiments, and, in particular, I shall discuss neutron and light scattering experiments in some detail. I shall also present some new results on electron and nuclear magnetic relaxation rates. I believe that measurements of the nuclear spin-lattice relaxation rate, $1/T_1$, can provide valuable information regarding the low frequency spin dynamics and thus it would be very interesting to perform these experiments.

Two review articles which discuss the experimental results in some depth are a relatively long presentation by Birgeneau and Shirane,[1] and a short presentation by Birgeneau.[2]

II. WHY HEISENBERG MODEL?

A simple valence counting argument quickly leads to the fact that Cu must be in a 2+ state. Because the electronic configuration of Cu in the atomic state is [Ar] $3d^{10}4s^1$, the Cu^{2+} ion in the solid must have a 3d hole. The five-fold degeneracy of the $3d^9$ orbital is lifted in the crystal field, and the only remaining orbital of interest is the $3d_{x^2-y^2}$ orbital. Thus it is not difficult to imagine that the intervening oxygen ions could mediate an antiferromagnetic coupling via the superexchange mechanism,[3] leading to a S=1/2 Heisenberg model. A simple estimate[4] then leads to an exchange constant $J \sim 1000$ K. As we shall see later, this picture is entirely consistent with experiments.

To some, the arguments in the last paragraph may appear to be deceptively simple. Indeed, the single electron picture would predict La_2CuO_4 to be a nonmagnetic metal, in complete disagreement with experiment. This well known but less understood dilemma is the Mott phenomenon[5] in which strong electronic correlation effects are of crucial importance. Thus La_2CuO_4 is an insulator, not because single particle bands are filled (in fact, a description in terms of single particle bands does not exist) but the charge carriers collectively organize themselves in a localized state which cannot carry current. We often associate such an insulating state with a broken symmetry -- a spin density wave, Wigner crystal, etc. What is perhaps not obvious is that this need not be the case. The transition to the insulating state as the electronic correlations get stronger may signify the restoration (not a breaking) of a symmetry, a local gauge symmetry. Although a precise *characterization* of Mott insulators in terms of a local gauge principle exists,[6] it has thus far not proven possible to elevate this principle to a truly dynamical principle. Attempts to establish such a dynamical principle lie at the heart of a number of fascinating recent developments in this field.[7]

From the structure of La_2CuO_4 it is not difficult to guess that the ratio of the interplanar coupling J' to intraplanar coupling J should be extremely small. The current experimental estimate for this ratio, as we shall see later, is about $J' \approx 10^{-5} J$. Thus, to a good approximation, it is sufficient to consider a two-dimensional model. However, if with decreasing temperature the spins become well correlated in the plane, even a tiny interplanar coupling can drive a three-dimensional phase transition at a finite temperature. Qutite generally, one can argue that in the disordered phase one needs to go to a temperature exceedingly close to the this three-dimensional ordering temperature to discern the effect of the interplanar coupling. To see a significant deviation from two-dimensionality the two-dimensional fluctuating energy scale has to be of the order of the interplanar coupling J' and therefore the two-dimensional correlation length has to be very large.

The discussion in this section would not be complete if I did not say a few words about the role of anisotropy. In one respect La_2CuO_4 is very different from K_2NiF_4. Until recently this isostructural (ignoring the small orthorhombic distortion in La_2CuO_4) material was known to be the best experimental realization of a two-dimensional Heisenberg model. Indeed, many elegant neutron scattering measurements[8] were carried out on this material nearly twenty years ago. However, K_2NiF_4 differs from La_2CuO_4 in an important way. The spin on Ni is 1 as opposed to 1/2 on Cu. Thus, single-site Ising anisotropy can be important in K_2NiF_4 but not in La_2CuO_4. This is important because simple considerations involving symmetry and energetics dictate that the finite temperature phase transition is a two-dimensional Ising transition in K_2NiF_4 (very close to the transition there would be a crossover to a three dimensional transition) but not in La_2CuO_4. Consequently, the fluctuation dynamics in these two materials could be drastically different. What about other possible anisotropies? The most obvious candidate is of course the dipolar anisotropy. Another possiblity could be the pseudo-dipolar anisotropy arising from second order spin-orbit coupling.[9] What is a bit surprising is that neither of these anisotropies are dominant. The dominant anistropy turns out to be the antisymmetric Dzyaloshinski-Moriya interaction,[10] as is evident in many properties involving the three-dimensional ordering in this material.[11] Moreover, without this understanding it would be difficult to explain why the electron paramagnetic resonance signal is not seen in this material at any reasonable temperature. I shall discuss these topics later, but for the most part of the article I shall assume that the anisotropy energies are sufficiently small that they can be neglected.

III. GROUND STATE PROPERTIES

Let us first consider the ground state of the Heisenberg model. For a ferromagnet it is easy to prove that the spin rotational symmetry is spontaneously broken in the ground state.[12] But what about an antiferromagnet? Is there a staggered order parameter?

At this point I should define my model more carefully. I shall first restrict myself to bipartite lattices, i.e. hypercubic lattices such that all the nearest neighbors of a site in sublattice A are sites in sublattice B. Furthermore, the Hamiltonian will be assumed to be isotropic in spin space, and the interactions between the spins will be assumed to extend only to nearest neighbors. Both of these restrictions are physically important. Clearly, the symmetry group of the Hamiltonian must be one of the important factors determining the nature of the ground state. However, less obvious is the role of further-neighbor interactions. Such interactions can have important effects on the ground state. Although this is not the subject of the present article, I would like to remark that recently considerable progress has been made on this subject.[13] Thus the model that I shall consider is,

$$H = J \sum_{(i,j)} S_i \cdot S_j ,\qquad(1)$$

where the sum is over all distinct nearest neighbor pairs.

When many spins are coupled together there is the possiblity that, *on the average*, the spins may be antiparallely aligned in the ground state. In this state $<S_i> = \pm N_0 \hat{\Omega}$, where $\hat{\Omega}$ is an arbitrary unit vector. The staggered order parameter is $\pm N_0$, depending on the sublattice on which the site i is located. Such a state, whatever its wavefunction, I shall call a *Néel state* as opposed to a *classical Néel state* in which the spins are strictly up or down. It is easy to see that the classical Néel state, which is the ground state of an Ising antiferromagnet, cannot be an eigenstate of the Heisenberg Hamiltonian in any dimension. Moreover, it is not difficult to convince oneself that the Néel state cannot be the ground state in 1 dimension, irrespective of what the spin is. Indeed, the ground state wavefunction for S=1/2 discovered by Bethe,[14] many years ago, has the property that $N_0=0$.

Evidently, if we could turn off quantum mechanics, the ground state would be a classical Néel state, because the energy is minimized in this configuration. This we could achieve by a *gedanken* experiment in which the value of the spin S is increased to ∞ in such a way that JS^2 remains finite, while S/S tends to a classical unit vector $\hat{\Omega}$. Thus 1/S plays the role of $\hbar$. As we increase 1/S from 0 there are two possibilities: either the broken symmetry state persists up to a finite value of 1/S, or the symmetry is restored as soon as we turn up 1/S. We shall see that it is the latter situation that prevails in 1 dimension, while it is highly plausible that the former is the case for dimensions greater than 1. In fact, it is now rigorously known[15] that the nearest-neighbor, antiferromagnetic Heisenberg model on hypercubic lattices has long-range antiferromagnetic order for $S \geq 1/2$ in $d \geq 3$, and $S \geq 1$ in d=2. No such theorems exist for S=1/2 and d=2. It seems to me that these rigorous theorems are not quite adequate yet from the physical point of view. For example, although the theorem can be proved for a three-dimensional cubic lattice, it fails for a body centered cubic lattice! A careful examination of these theorems shows that probably a competely new set of techniques would be required to improve upon the existing set of results.

A convenient way to parametrize our *gedanken* experiment is to replace the spin operators by Boson operators invented by Holstein and Primakoff[16] which are:

$$S^+ = (2S)^{1/2}\sqrt{1 - a^\dagger a/2S}\, a, \tag{2a}$$

$$S^- = (2S)^{1/2} a^\dagger \sqrt{1 - a^\dagger a/2S}, \tag{2a}$$

$$S^z = S - a^\dagger a \tag{2c}$$

This is an exact transformation in spite of the fact that the Fock space for the Boson operators is infinite dimensional, while the dimensionality of the space of spin operators is finite. This is because the operators cannot connect the physical space of the spin operators with the unphysical part introduced by this representation. The advantage of the Holstein-Primakoff transformation is obvious. The Bosons satisfy simpler commutation rules which are preserved under cannonical transformations. The disadvantage is equally obvious. The transformation is non-linear and any practical calculation involves an expansion of the square root, which leads to many-Boson interactions in the Heisenberg Hamiltonian. Moreover, once the square root is expanded, the physical and the unphysical spaces must mix. The best one can hope for is that the mixing is small. Many other transformations are available in the literature. For some calculations Dyson-Maleev[17] or Schwinger-Boson[18] formalisms may be more effective. Each of these transformations have their own difficulties when it comes to practical calculations. In other words, there is no free lunch in this world. Fortunately, for what I am about to discuss, which is the absence of long range order in the ground state in one dimension, it would make no difference as to which transformation I choose. I shall therefore try to illustrate my point using the Holstein-Primakoff transformation.

As I said before, I need to expand the square root. When is this justified? Clearly, S must be large, and $a^\dagger a$, in some suitable sense, should not be large, i.e. the number of bosons excited at a given site must not be large. But whether or not $a^\dagger a$ is large or small is a dynamical question and cannot be settled *a priori*, i.e., without actually solving the Hamiltonian. If we turn this argument around we could equally well say that the expansion could be justified even when S is small if $a^\dagger a$ happens to be small. The expansion is therefore justified if $(a^\dagger a/2S)$ has a small matrix element acting on the "relevant" portion of the Hilbert space.

If we now Holstein-Primakoff transform the Heisenberg Hamiltonian and expand all square roots in sight, we generate an expansion in powers of $1/S$. For an antiferromagnet we choose the spin quantization axis to be opposite on opposite sublattices. This conforms to our intuitive notion of an antiferromagnet. If we keep only the leading term in this expansion, we find the classical Néel state, and the expectation value of S^z is $\pm S$ depending on the sublattice. The leading term contains no quantum fluctuations. We begin to see something interesting when we also include the next term in the expansion. The Hamiltonian at this level is quadratic in Boson operators, or equivalently harmonic. It is therefore easy to diagonalize by a Bogoliubov transformation. The result is a set of two normal modes for each wave vector k in the Brillouin zone. These modes are the quantum mechanical precession waves, or spin waves, which are gapless and are in accordance with Goldstone's theorem. The expectation value of S^z on a given sublattice is now given by,

$$S- <S^z> = \text{const.} \int_0^\Lambda dk \frac{k^{d-1}}{k}, \qquad (3)$$

where I have made some inessential simplifications to exhibit the potential infrared divergence of the integral. Λ is a short wavelength cutoff, of the order of the inverse lattice spacing. In dimension d=1 the integral is divergent and the spin deviation from the maximal value is infinite due to quantum fluctuations. Hence, the long wavelength zero-point spin waves must destroy long-range antiferromagnetic order and the correct ground state must have zero staggered magnetization and no broken spin rotaional symmetry. Surely, if the inclusion of the harmonic term already destroys long-range order, the situation would get worse if we include the neglected higher order anharmonic terms. Thus, the ground state cannot have long range antiferromagnetic order in one dimension.This is the analog of the argument due to Peierls[19] for the absence of long range order at finite temperature, in d=2, for problems involving continuous symmetry. I fully believe that the argument can be legitimized in a fashion similar to the Hohenberg-Mermin-Wagner theorem.[20] However, to the best of my knowledge, such a rigorous theorem is not known.[21]

What happens in higher dimensions? Basically, nothing. The spin deviation calculated in this manner is just a number. There are no crisp theorems to prove. How good is this number ? Well, that I cannot tell you. I can only tell you what the next term is. For d=2 the result is[22]

$$<S^z> = S - 0.197 + 0 \times (1/S) \qquad (4)$$

As of writing, the term of order $(1/S)^2$ is not known. Does this series converge? I would be very surprised if it did. I think that it is probably a decent asymptotic expansion. The third term on the right hand side of Eq. (4) is identically zero for any bipartite hypercubic lattices -- one of those miracles that you learn to live with.

Although it leaves a queasy feeling in your stomach, you learn to say that in dimensions greater than or equal to 2 quantum fluctuations do not destabilize the Néel state. It is at least a local minimum in the configuration space. This is how things stood for a long time until P. W. Anderson (who was the first[23] to notice everything that I said above) revived the horse that was considered dead. He raised the serious possibility that the Néel state may not be the ground state in d=2, particularly for triangular (non bipartite) lattices.[24] This point of view was also empasized by him in the context of high temperature superconductivity in cuprates.[4] We shall see that the finite temperature properties of the stoichiometric La_2CuO_4 can be well understood on the basis of a Néel ordered ground state, dispelling any doubts that one might have harbored. For the moment, let me enumerate what the spin wave theory predicts for the various zero temperature properties[22] in d=2 that we shall need later.

The spin wave velocity, c, can be written in the form,

$$c = \frac{2S\sqrt{2}Ja}{\hbar} Z_c(S) , \qquad (5)$$

where the renormalization factor Z_c is

$$Z_c = 1 + 0.158/2S + O(1/2S)^2 . \qquad (6)$$

Although it is not obvious, within spin wave theory, the renormalization of the spectrum is independent of the wave vector, **k**, in the Brillouin zone.

The uniform magnetic susceptibility $\chi_\perp(T=0)$, in the direction perpendicular to the staggered magnetization, is given by (in units where $g\mu_B/\hbar = 1$),

$$\chi_\perp(T=0) = \frac{\hbar^2}{8Ja^2} Z_\chi(S) , \qquad (7)$$

where the renormalization factor $Z_\chi(S)$ can be expressed as

$$Z_\chi(S) = 1 - 0.552/2S + O(1/2S)^2 . \qquad (8)$$

One of the most useful variables that I shall be using over and over again is the spin stiffness constant, $\rho_s(T=0)$ ($\equiv \rho_s(0)$), which is a bending energy required to twist the system away from the staggered direction. Using the relation[25] $\rho_s(0) = c^2 \chi_\perp(0)$, we can write,

$$\rho_s(0) = JS^2 Z_c^2(S) Z_\chi(S)$$

$$= JS^2 Z_{\rho_s} \qquad (9)$$

The multiplicative renormalization factors, the Z-factors, contain the effect of quantum fluctuations present in the Néel state.

The T=0 spin wave theory discussed above should not be construed as a proof that the ground state is ordered. In fact, it is quite likely that the spin wave expansion is asymptotic, and misses[26] out terms of the type $e^{-\alpha S}$, where α depends on the dimension of the lattice. These terms cannot arise in a 1/S expansion. However, if α happens to be large, such an expansion can still be useful even if S is small. But we must first convince ourselves that the ground state is ordered. I believe that recent quantum Monte Carlo simulations[27,28] have forcefully demonstrated that this is so for S=1/2. Of course, some care must be exercised in interpreting the numerical data. In particular, because one is forced to extrapolate to the infinite system based on results obtained on rather small lattices, it is important that there is a well justified rationale for this extrapolation. As emphasized in Ref. 28, different assumptions with respect to the size dependence can lead to drastically different extrapolations for the value of the staggered order parameter. However, the possible size dependencies that can be justified lead to a value of the staggered magnetization which is astonishingly close to the spin wave result quoted earlier. Of course this is sheer luck, but this gives us some legitimacy to use the spin wave expressions for all values of the spin, at least numerically.

In my opinion, the best estimates of the Z-factors, for S=1/2, are those given by a careful analysis by Singh.[29] The results are:

$$Z_c = 1.18 \pm 0.02, \qquad (10a)$$

$$Z_\chi = 0.52 \pm 0.03, \qquad (10b)$$

$$N_0 = 0.302 \pm 0.007 . \qquad (10c)$$

These numbers are very close to the results obtained from spin wave theory outlined above. The spin wave results for S=1/2 are:

$$Z_c = 1.158, \qquad (11a)$$

$$Z_\chi = 0.448, \qquad (11b)$$

$$N_0 = 0.303 . \qquad (11c)$$

IV. QUANTUM NON-LINEAR σ-MODEL (QNLσM)

In this section I shall argue that there exists an effective continuum theory which correctly reproduces the low-energy, long-wavelength behavior of QHAF. This continuum theory is a generalization of the classical O(3) non-linear σ-model[30] and will be called the quantum non-linear σ-model. It is well known that the classical model correctly captures the long-wavelength physics contained in the classical Heisenberg lattice models where the spins are represented by three-component unit vectors.

QNLσM was first derived from the Heisenberg model by Haldane[31] using a large-S expansion. It is therefore natural to associate this model with a large S expansion. However, the applicability of this model is not confined to the restrictions imposed by the large S expansion. It is very useful to think about it in a manner similar to Landau-Ginzburg model. As long as the ground state is ordered, it is the simplest continuum model with the correct symmetry and the correct spin wave spectrum at long wavelengths. Moreover, because the interactions between the goldstone modes at long wavelengths are entirely determined by symmetry, they are also correctly given by QNLσM.

The model is parametrized by two phenomenological parameters defined on a scale which sets the short wavelength cutoff in the theory. The local spin stiffness constant, ρ_s^0, determines the stiffness against local spatial fluctuations, and the local uniform susceptibility, $\chi_\perp^0$, determines the stiffness against local temporal fluctuations. These parameters are the input parameters of the model and are *not* the physical macroscopic parameters.

Within the theory, one could fix these constants in two different ways. Firstly, if one knew the macroscopic physical values, one could work backwards and determine these local prameters. Clearly, these local parameters would depend on their defining length scale. Alternately, a sufficiently accurate microscopic calculation could also be used to obtain these parameters. For example, $\rho_s^0 = JS^2 \Lambda^{d-2}$, and $\chi_\perp^0 = (\hbar^2/4dJ) \Lambda^d$, in the limit $S \to \infty$, where $\Lambda = 1/a$, a being the lattice constant of the square lattice. Because these input parameters are scale dependent, they depend sensitively on the precise manner in which the underlying microscopic theory is coarse grained to arrive at this contnuum theory. Similarly, because of the coarse graining operation, the spin dependencies of these parameters are expected to be complicated. It is only in the limit $S \to \infty$ do they have a simple S dependence.

The justification of the QNLσM along these lines has been discussed in detail by Chakravarty, Halperin, and Nelson[32] (CHN). In effect, as long as the continuous O(3) symmetry is spontaneously broken, the interaction between the goldstone modes at long wave lengths is precisely given by QNLσM, regardless of the actual value of the spin. This is reminiscent of a theorem due to Weinberg[33] in the context of pion physics as was pointed out by Rosenstein and Warr.[34]

There is one remaining point that deserves some further thought. QNLσM does not distinguish between integer and half integer spins. Haldane[35] has recently argued that due to quantum fluctuations an intrinsic dependence of the quantized value of S will appear through the creation of certain topological singularities. However, it is highly unlikely that these topological singularities have any perceptible effect on the low-energy, long- wavelength properties if the ground state is ordered. Because these topological defect states are separated by a gap, they would play an important role only when the gap, which is proportional to macroscopic spin stiffness ρ_s, is very small. This is likely to happen if the ground state is disordered but not if the ground state is ordered. As mentioned earlier, I shall assume that the ground state is ordered and derive the consequences of this assumption at finite temperatures. Therefore I shall not discuss this question further.

The effective Euclidean action of the QNLσM may be written in the form,

$$S_{\text{eff}}/\hbar = \frac{\rho_s^0}{2\hbar} \int_0^{\beta\hbar} d\tau \int d^d x \left(|\nabla \hat{\Omega}|^2 + \frac{1}{c_0^2} |\frac{\partial \hat{\Omega}}{\partial \tau}|^2 \right), \quad (12)$$

where $\hat{\Omega}$, a three-component *unit* vector field, is to be interpreted as the local staggered magnetization. Although a short-distance cutoff, Λ^{-1}, is assumed for the spatial integrals, no such intrinsic cutoff is assumed to exist for the time integral. Because of this non-Lorentz invariant cutoff procedure, c_0 may differ from the actual spin wave velocity c at long wavelengths by a finite factor. In a Lorentz invariant theory c_0 must equal c. However, to one-loop accuracy, this equality also holds in the present model. Here, $\rho_s^0 c_0^{-2} = \chi_\perp^0$. By changing variables one can also recast the integrand of the effective action in terms of the dimensionless variables u and y and write the action as follows:

$$S_{\text{eff}}/\hbar = \frac{1}{2g_0} \int_0^{\beta \hbar c_0 \Lambda} du \int d^d y \left(|\nabla_y \hat{\Omega}|^2 + |\frac{\partial \hat{\Omega}}{\partial u}|^2 \right) . \qquad (13)$$

The upper limit of the imaginary time integral is the ratio of the thermal de Broglie wavelength to the short wavelength cutoff of the model. The coupling constant g_0 is given by,

$$g_0 = \frac{\hbar c_0 \Lambda^{d-1}}{\rho_s^0} . \qquad (14)$$

The action in Eq. (13) is equivalent to a classical (d+1)-dimensional nonlinear σ-model in which one of the dimensions is finite for non zero temperatures but becomes infinite at T=0.

V. A RENORMALIZATION GROUP ANALYSIS

In this section we consider the renormalization group equations for the QNLσM. The emphasis here is to show how, *at finite temperatures*, these equations can be used to map the quantum model to an effective classical non-linear σ-model (CNLσM) in the regime in which the ground state of the quantum model has long-range order. It is *only* in this regime that we have a *precise* control over our calculations. This mapping is quite useful because we can make use of the known results for the classical model in d=2 to obtain results for the spin-spin correlation length and the staggered susceptibility of the quantum model, and therefore of the S=1/2 QHAF. We of course need to establish that the mapping is sufficiently accurate for our purposes.

At this point the reader might be wondering as to why we could not obtain the same results from a direct simulation of the S=1/2 *quantum* Heisenberg model. The answer is that in principle we could, but despite considerable recent progress,[36] in practice, it is extremely difficult to obtain reliable results for temperatures of interest, i.e., $T \leq J/2$. On the other hand, it is precisely the low temperature regime in which the renormalization group method is expected to be most effective.

(a) **The Classical non-linear σ-model (CNLσM)**

Before I discuss the quantum problem I would like to illustrate a part of the logic by considering the classical problem. The classical non linear σ-model is defined by the Hamiltonian

$$\beta H = \frac{1}{2t_0} \int d^d x |\nabla \hat{\Omega}|^2 \, , \qquad (15)$$

which is the naive continuum limit (as opposed to statistical continuum limit[37]) of the classical Heisenberg model defined by,

$$\beta H = \frac{J_{cl}}{T} \sum_{(i,j)} \hat{\Omega}_i \cdot \hat{\Omega}_j \, . \qquad (16)$$

The coupling constant t_0 in Eq. (15) is $T\Lambda^{d-2}/J_{cl}$, corresponding to a square lattice of lattice constant a, and $\Lambda=1/a$. It is advantageous to introduce a slightly more general notation and define t_0 to be $t_0 = T/\rho_s^0$.

This model does not have a finite temperature phase transition in two dimensions. Its critical temperature is at T=0 and its correlation length diverges exponentially with decreasing temperature.[38] Thus, in the limit $T \to 0$, a description in terms of a continuum field theory becomes meaningful. The renormalization group equation in this limit is given by (e^l is the length rescaling factor),

$$\frac{dt}{dl} = \beta(t) \, , \qquad (17)$$

where the renormalization group β-function is now known up to four loops.[39] For our purposes it would suffice to quote results only up to two-loops.[40] Thus,

$$\beta(t) = \beta_2 t^2 + \beta_3 t^3 \, , \qquad (18)$$

where $\beta_2=1/2\pi$, $\beta_3=(1/2\pi)^2$. This loop expansion is a low temperature expansion. Each loop adds an additonal power of temperature to the β-function. It is important to note that this two-loop β-function is universal,[41] i.e., it does not depend on the precise regularization scheme. However, higher order terms do not enjoy this property.

It is now natural to ask if this β-function allows us to compute physical quantities such as the correlation length. At first, one is tempted to say yes. All we have to do is to integrate out the recursion relations up to a scale l* at which $t(l^*) \approx 1$. At this scale the correlation length is of the order of lattice spacing, and working backwards, we can find out what the physical correlation length is at the physical temperature T. This is almost correct except that there are some genuine ambiguities. The first concerns the precise value of the running coupling constant at which to stop the integration, i.e., what should we choose t(l*) to be --- 1, 2π, or something else? The second is more subtle. This renormalization group equation is perturbative, and is only valid for low temperatures. It certainly has a fixed point at T=0 but at any finite temperature t(l) grows. Thus it it is plausible that the system is disordered, but we cannot deduce this from this equation alone. As we integrate the recursion relation we quickly move away from the region of validity of the equation because t(l) grows with iteration. Therefore, we must have an idependent piece of information that the system is indeed disordered. Only then can we use this equation to compute say the spin-spin correlation length. Suppose that we know this, then it is easy to see that the correlation length must be[40]

$$\xi = \Lambda^{-1} C_\xi^{cl} \left(\frac{t_0}{2\pi}\right) \exp(2\pi/t_0) , \qquad (19)$$

where Λ^{-1} is the short distance scale at which the theory is regularized.

C_ξ^{cl} is an integration constant which *cannot* be determined from this renormalization group equation; it does not matter how many terms in the β-function we know. In order to determine C_ξ^{cl} we try to match this expression to a suitable calculation which is valid in the strong coupling regime (i.e., in the high temperature regime). Such a calculation, for example, may be a Monte Carlo calculation. This matching, or eqivalently this interpolation, if successful, determines this constant. However, we must remember that this expression cannot be legitimately used for temperatures above the matching temperature.

To summarize, this technique is not simply a result of a perturbative renormalization group calculation. It combines the results obtained from a perturbative renormalization group β-function with those obtained from a non-perturbative method and can therefore give an accurate description of the crossover from the weak coupling to the strong coupling regime. To the best of my knowledge, this clever analysis involving a judicious matching of a method which works best at high temperatures with the low temperature renormalization group method for the classical Heisenberg model in d=2 was first carried out by Fisher and Nelson.[42] However, the details of their calculation was a bit flawed. Subsequently the problem was solved more fully by Shenker and Tobochnik.[43]

How many terms do we need in the loop expansion? As many as we need to produce a smooth matching. It happens to be the case that for the model we are considering the two-loop β-function is sufficient. Higher loop corrections, although not universal, could be used to refine the calculation (analytic corrections) but this possibility has not been explored in detail.

Recall that Λ^{-1} is the short distance cutoff in the problem, and that C_ξ^{cl} is an integration constant that is independent of the coupling constant but depends on the regularization scheme. In principle, one can choose an infinite number of regularization schemes; each will have their own Λ^{-1} and their own C_ξ^{cl}. However, because the strong coupling expansion to which we are supposed to match is often carried out on a lattice model, it is essential that we use the quantities appropriate to the lattice regularization scheme. It turns out that there is a simple trick by which we can convert from one scheme to the other. Amusingly, only a perturbative *one-loop* calculation is necessary to find this conversion factor. This trick is well known in the context of lattice gauge theory.[41] I have found the discussion by Kogut[44] to be particularly pedagogic and I shall closely follow him; see also Ref. 32.

From Eqs. (17) and (18), we can write,

$$1^* = \frac{1}{\beta_2 t_0} + \frac{\beta_3}{\beta_2^2} \ln t_0 , \qquad (20)$$

and inverting it, we get,

$$\frac{1}{t_0} = \beta_2 1^* + \frac{\beta_3}{\beta_2} \ln[1^* + O(t_0)] , \qquad (21)$$

where t(l=0) is t_0, and l* is an integration constant. We can also write this equation in terms of the correlation length, ξ, to get

$$\frac{1}{t_0} = \beta_2 \ln(\Lambda\xi/C_\xi) + \frac{\beta_3}{\beta_2} \ln[\ln(\Lambda\xi/C_\xi) + O(t_0)] , \qquad (22)$$

This equation has the property of asymptotic freedom, and shows that we must decrease t_0 if we increase Λ, if we were to keep the physical correlation length ξ fixed. An important feature of this equation is that the *two-loop correction* (the second term on the right hand side of Eq. (22)) *does not redefine the constant* C_ξ. Thus only a one-loop calculation is necessary to define C_ξ. Now imagine that the same calculation is carried out in a different regularization scheme. Then we should get

$$\frac{1}{t_0'} = \beta_2 \ln(\Lambda'\xi/C_\xi') + \frac{\beta_3}{\beta_2} \ln[\ln(\Lambda'\xi/C_\xi') + O(t_0)] . \qquad (23)$$

Because we are holding the physical correlation length fixed, t_0' must in general be different from t_0 we had earlier. Thus,

$$\frac{1}{t_0} - \frac{1}{t_0'} = \beta_2 \ln\left(\frac{\Lambda C_\xi'}{\Lambda' C_\xi}\right) . \qquad (24)$$

The next step in the calculation is to find an independent relation between t_0 and t_0'. For this we simply need a one-loop perturbative calculation. We calculate the renormalized coupling constant, t_R, for the bare couplings t_0 and t_0' and demand that t_R be identical. This is of course equivalent to holding the physical correlation length fixed. Thus, we must have

$$t_R = \frac{t_0}{Z_1} = \frac{t_0'}{Z_1'} \qquad (25)$$

The definition of the renormalization factor is the same as that of Brézin and Zinn-Justin.[40] A simple calculation shows that

$$t_0' = t_0 - [(A - A') + \beta_2 \ln(\Lambda'/\Lambda)]t_0^2 \qquad (26)$$

It is not difficult to guess that this must be the case. Quite generally, to one loop order, one must have

$$Z_1 = 1 + B(\Lambda)t_0 , \qquad (27)$$

where $B(\Lambda)$ must be given by,

$$B(\Lambda) = -\beta_2 \ln\Lambda + A . \tag{28}$$

A is the finite part of the counter term. That this must be so follows simply from the observation that the condition $\Lambda(\partial/\partial\Lambda)t_R = 0$ must yield the β-function. A similar expression holds for Z_1'. Combining Eqs. (24) and (26) we get,

$$\frac{C_\xi{}'}{C_\xi} = \exp((A'-A)/\beta_2) . \tag{29}$$

Thus, all we need to do is to calculate the finite part of the counter term in one-loop order to find the conversion factor. The conversion factor between the momentum cutoff scheme employed in Ref. 32 and the lattice regularization scheme necessary for the matching caluculation, if we want to match to a Monte Carlo calculation on a square lattice, is given by,

$$C_\xi^{\mathrm{mom}} = C_\xi^{\mathrm{lat}} \sqrt{32} e^{\pi/2} . \tag{30}$$

This is identical to the conversion factor between the Pauli-Villars regularization scheme and the lattice regularization scheme calculated by Parisi.[45]

We now define the static structure factor S(**k**) to be

$$S(\mathbf{k}) = \Lambda^{-d} \sum_i e^{-i\mathbf{k}\cdot\mathbf{r}_{ij}} N_0^2 <\hat{\Omega}_i \cdot \hat{\Omega}_j> , \tag{31}$$

where, for convenience, we have introduced a normalization factor N_0 for the order parameter, i.e., the order parameter is $N_0\hat{\Omega}_i$. Of course N_0 is 1 in the classical problem.

The renormalization group analysis[40] predicts that

$$S(k=0) = B_s \xi^2 t_0^2 N_0^2/(2\pi)^2 , \tag{32}$$

where B_s is another dimensionless constant, and like C_ξ^{cl} is not determined by the two-loop renormalization group. B_s may be determined, as explained above, by matching Eq. (32) to a strong coupling calculation. Interestingly, for the low temperature behavior, it is sufficient to calculate the product $B_s(t_0/2\pi)^2$ to leading order in the bare coupling -- higher order corrections due to different regularizations are unimportant as $t_0 \to 0$. The reason we were so fussy when we were calculating ξ is because these corrections appeared in the exponential and we could not drop them without loosing accuracy.

Without giving you the details let me state a number of results that follow from the renormalization group analysis of CHN.[32]

The dimensionless running coupling constant, t(k), in one loop approximation, is given by,

$$t(k) = \frac{1}{[1 + \frac{1}{2} \ln(1 + k^2\xi^2)]} \tag{33}$$

The static structure factor, S(k) can be expressed as

$$S(k) = S(k=0)f(k\xi) \tag{34}$$

with

$$f(x) = \frac{1 + \frac{1}{2}\ln(1+x^2)}{1+x^2} . \tag{35}$$

The scaling function f(x) in Eq. (35) is a one-loop result. It can be generalized to include corrections of two-loop order and beyond. However, the form of S(k) given in Eq. (34) should remain valid. Thus, with S(0) given by Eq. (32), we expect that the temperature dependence of S(k) to be correctly given by Eq. (32) in the limit $t_0 \to 0$ for any fixed $k\xi$. Although a more accurate form of the scaling function can change the coefficient multiplying the logarithmic term of f(x), when x is large f(x) should be given by,

$$f(x) \approx \ln x / x^2 , \tag{36}$$

so that, classically

$$S(k) \approx \frac{2t_0 N_0^2}{k^2} \tag{37}$$

as $T \to 0$.

If we define the angular momentum susceptibility tensor by,

$$T\chi_{\alpha\beta} = \Lambda^{-d} \sum_j <M_{i\alpha}M_{j\beta}> , \tag{38}$$

where M_i is given by,

$$\mathbf{M}_i = \chi_\perp^0 \Lambda^{-d} \hat{\Omega}_i \times \frac{d\hat{\Omega}_i}{dt} , \tag{39}$$

then

$$\chi_{\alpha\beta} = \frac{2}{3}\delta_{\alpha\beta}\chi_\perp^0 [1 + O(t_0^3)] \tag{40}$$

for a fixed finite $k\xi$. This result should be valid, to leading order in t_0, for $t_0 \to 0$, when higher loop corrections are included.

The spin stiffness constant at wave vector k can also be calculated, and we find that,

$$\rho_s(k, t_0) = \frac{t_0}{2\pi} g(k\xi) , \tag{41}$$

in the limit $t_0 \to 0$, for any fixed value of the product $k\xi$. A one-loop calculation gives

$$g(x) = 1 + \frac{1}{2}\ln(1+x^2) \,. \tag{42}$$

We expect the form of Eq. (41) to remain valid when the calculation is extended to include higher loop corrections. However, the scaling function g(x) given in Eq. (42) should be correct for large x and only approximately valid for x~1.

(b) The QNLσM

I now turn to the discussion of QNLσM using the renormalization group equations derived in Ref. 46. These equations are,

$$\frac{dg}{dl} = (1-d)g + \frac{K_d}{2}g^2 \coth(g/2t) \,, \tag{43}$$

and

$$\frac{dt}{dl} = (2-d)t + \frac{K_d}{2} gt \coth(g/2t) \,. \tag{44}$$

Here, $K_d^{-1} = 2^{d-1}\pi^{d/2}\Gamma(d/2)$, and the initial values of the dimensionless coupling constant g(l) and the temperature scale t(l) are $g_0 = \hbar c \Lambda^{d-1}/\rho_s^0$ and $t_0 = T\Lambda^{d-2}/\rho_s^0$. At T=0, there is a non-trivial fixed point at $g = g_c$, given by,

$$g_c = \frac{2}{K_d}(d-1) \,. \tag{45}$$

For g<g_c the ground state is ordered, whereas for g>g_c the state is disordered. At finite temperature there are no non-trivial fixed points for d≤2. For d>2 there is a non-trivial finite temperature fixed point at $t_c = (d-2)/K_d$.

These one-loop equations have been discussed in detail in Refs. 32 and 46. In this section I shall show that these equations can be used to map the quantum problem to a classical problem at any finite temperature provided that g<g_c. I shall concentrate on the case d = 2, and follow a method due to Kopietz and Chakravarty.[47] An alteranate method can be found in Ref. 32.

The recursion relations can be explicitly integrated in d=2, and we find:

$$\frac{1}{t} = \frac{1}{t_0} + K_2[\ln\left(\sinh\left(\frac{g_0 e^{-1}}{2t_0}\right)\right) - \ln\left(\sinh\left(\frac{g_0}{2t_0}\right)\right)] \,, \tag{46}$$

and,

$$\frac{g}{t} = \left(\frac{g_0}{t_0}\right)e^{-1} \,. \tag{47}$$

As we integrate the recursion relations, i.e., we coarse grain the problem on longer and longer length scales, these equations should tend to the solution of the classical problem which is

$$\frac{1}{t} = \frac{1}{t_0} - K_2 \mathbf{l} \ . \tag{48}$$

Well, not quite, because we started from the quantum problem. We get instead

$$\frac{1}{t} = \frac{1}{t_0}\left(1 - \frac{K_2 g_0}{2}\right) + K_2 \ln\left(\frac{g_0}{t_0}\right) - K_2 \mathbf{l} \ , \tag{49}$$

provided the length scale is larger than the thermal de Broglie wavelength which is proportional to $\hbar c/T$ for an antiferromagnet. Thus the recursion relations are those of an effective classical problem where,

$$\frac{1}{t_0^{eff}} = \frac{1}{t_0}\left(1 - \frac{K_2 g_0}{2}\right) + K_2 \ln\left(\frac{g_0}{t_0}\right) \ . \tag{50}$$

Substituting the explicit values of the parameters we get,

$$\frac{1}{t_0^{eff}} = \frac{\rho_s(0)}{T}\left(1 + \frac{T}{2\pi\rho_s(0)}\ln\left(\frac{\Lambda\hbar c}{T}\right) + O(T^2)\right) \ , \tag{51}$$

where $\rho_s(0)$ is the zero temperature spin stiffness renormalized by quantum fluctuations.[32] While it is necessary to keep the logarithmic term, since we expect it to appear in the argument of an exponential, one can drop the term of the order of T^2 in Eq. (51) along with higher loop corrections. Although we have shown that the first term is the renormalized T=0 spin-stiffness only to one loop order, it is obvious that this result should hold to all orders. Thus, it is not necessary to go beyond a one-loop calculation to map the QNLσM to an effective CNLσM.

We can now take over the results we obtained earlier for the classical problem. If we substitute the expression for t_0^{eff} (Eq. (51)) in Eq. (19) for the correlation length, and make use of the conversion factor in Eq. (30), we find that the spin-spin correlation length for the QHAF is given by,[32]

$$\xi = \sqrt{32} e^{\pi/2} (2\pi C_\xi^{\text{lat}}) \left(\frac{\hbar c}{2\pi \rho_s(0)}\right) \exp\left(\frac{2\pi \rho_s(0)}{T}\right) \ . \tag{52}$$

Note that the explicit dependence on Λ has disappeared from this equation and the final result contains only quantities which are long wavelength properties of the system. This lends further support to the fact that a consistent continuum theory exists.

It is as though we have made the replacements:

$$\Lambda^{-1} \to \sqrt{32} \, e^{\pi/2} \left(\frac{\hbar c}{T}\right) \ , \tag{53}$$

$$t_0 \to \frac{T}{\rho_s(0)} \ . \tag{54}$$

Similarly, the staggered susceptibilty, χ, for the QHAF can be obtained from Eq. (32), and we get,

$$T\chi = S(k=0) = \frac{B_s \xi^2 T^2 N_0^2}{(2\pi \rho_s(0))^2}, \tag{55}$$

where, unlike the classical problem, N_0 is not 1, but is the T=0 staggered magnetzation of the QHAF. This is natural because in order to obtain the effective classical problem we had to integrate out the quantum fluctuations up to a length scale given by Eq. (53), and therefore the magnitude of the order parameter must be reduced. In Eq. (55) we have ignored logarithmic and higher order corrections which vanish in the limit T→0 (see Eq. (51)).

The static structure factor $S(k)$ is still given by Eq. (34) but $S(0)$ is now given by Eq. (55) and the correlation length ξ by Eq. (52). This expression for $S(k)$ should be valid for $k \ll (T/\hbar c)$. A physically plausible extension of this formula to wavelengths of the order of the thermal wavelength, $\hbar c/T$, is possible. Within a distance of the order of the thermal wavelength we can use non-interacting spin wave theory in which the parameters are renormalized by the T=0 quantum fluctuations; thermal fluctuations are relatively unimportant in this regime. Thus, we simply need to multiply by a factor $(\hbar ck/2T)\coth(\hbar ck/2T)$. One can check that in this way one recovers the correct long wavelength limit at T=0, i.e.,

$$\lim_{T \to 0} S(k) = \frac{\hbar c N_0^2}{k \rho_s(0)}. \tag{56}$$

This result is a consequence of Goldstone's theorem if we remember that $S(k)$ is an equal time spin-spin correlation function.

VI. STATIC PROPERTIES OF QHAF

Explicit results for the correlation length and the static structure factor can be written down from the expressions given in the previous section. We need, however, the values for C_ξ^{cl}, B_s, N_0, $\rho_s(0)$, and $\hbar c$.

C_ξ^{cl} and B_s have been extensively studied in the literature;[48] see also Kopietz and Chakravarty.[47] However, within 30% one can safely assume that $2\pi C_\xi^{cl}$ is 0.009 (scaled to our definition; see below), as determined by Shenker and Tobochnik.[43] Thus the correlation length ξ is given by

$$\xi = 0.24 \left(\frac{\hbar c}{2\pi \rho_s(0)} \right) \exp\left(\frac{2\pi \rho_s(0)}{T} \right). \tag{57}$$

The spin wave expressions for $\hbar c$ and $\rho_s(0)$ are given in Sec. III. More accurate results for S=1/2 can be obtained from Ref. 29. Similarly, B_s (scaled to our definition; see below) form Ref. 43 is 148; the spin wave expression for N_0 was given in Sec. III. We therefore get,

$$S(k) = 148 \left(\frac{\xi T N_0}{2\pi\rho_s(0)}\right)^2 f(k\xi) . \tag{58}$$

The accuracy of the fits at finite values of T improves if the classical expressions, Eqs. (19) and (32) are modified to

$$\xi = \Lambda^{-1} C_\xi^{cl} \frac{\exp(2\pi/t_0)}{(2\pi/t_0) + 1} , \tag{59}$$

and

$$S(k=0) = \frac{B_s \xi^2 N_0^2}{[(2\pi/t_0) + 1]^2} . \tag{60}$$

These modifications are not a result of the renormalization group analysis but are only plausible correction terms introduced by Shenker and Tobochnik. In these expresssions $2\pi C_\xi^{cl}$ should be 0.01 instead of 0.09 used earlier in Eq. (57), and B_s should be 180 instead of 148 used earlier in Eq. (58). A more recent calculation by Tyč, Halperin, and Chakravarty (THC)[49] which also uses the forms given in Eqs. (59) and (60) yields $2\pi C_\xi^{cl}$ to be 0.01 but B_s to be 125.

If we use Eqs. (59) and (60), we get, for the quantum problem,

$$\xi = 0.27 \left(\frac{\hbar c}{2\pi\rho_s(0)}\right) \frac{\exp\left(\frac{2\pi\rho_s(0)}{T}\right)}{1 + (T/2\pi\rho_s(0))} , \tag{61}$$

and

$$S(k=0) = \frac{B_s \xi^2 N_0^2}{[(2\pi\rho_s(0)/T) + 1]^2} . \tag{62}$$

Recently Gomez-Santos, Joannopoulos, and Negele[50] have calculated the correlation length and the staggered susceptibility using a direct quantum Monte Carlo simulation of S=1/2 QHAF on a square lattice of lattice constant a; see also Manousakis and Salvador.[51] From a fit to their numerical results Gomez-Santos et al. find that the correlation length, ξ, is given by,

$$\xi = 0.32 \, a \, \exp(J/T) . \tag{63}$$

However, one should keep in mind that it is difficult to obtain reliable numerical results at low temperatures.

Making use of the spin wave expressions for the spin wave velocity, c, and the T=0 spin stiffness constant, $\rho_s(0)$, in Eq. (57) for the correlation length, we get for S=1/2,

$$\xi = 0.42 \, a \, \exp(0.94 J/T) . \tag{64}$$

As explained earlier, this is an asymptotic result which is valid at sufficiently low temperatures.

We are thus confident that Eqs. (61) and (62), with the parameters N_0, c, and $\rho_s(0)$ determined from the T=0 spin wave theory, are accurate for all S. Later, I shall compare these theoretical results with experiments performed on La_2CuO_4 (S=1/2) and K_2NiF_4 (S=1).

There is a further interesting test that one can perform on this procedure of mapping a quantum Heisenberg model to an effective classical non-linear σ-model. Because it is considerably easier to simulate a S=1/2 quantum ferromagnet, it is interesting to ask if the procedure can be tested on quantum ferromagnets. The answer is yes; the corresponding calculation was carried out by Kopietz and Chakravarty.[47] The renormalization group results are in excellent agreement with simulations.

In the past two years numerous papers have been written on the exponential temperature dependence of the correlation length in QHAF. Most notable amongst these are the papers by Arovas and Auerbach[52,53] in which they developed a new method which is now widely used in this field.

With respect to the regime in which the ground state is ordered at T=0, it predicts that the spin-spin correlation length should be given by,

$$\xi = \text{const.} \frac{\hbar c}{T} \exp\left(2\pi\rho_s(0)/T\right), \tag{65}$$

which is precisely of the form given by the one-loop renormalization group approach presented in Refs. 32 and 46. Of course, the prefactor in Eq. (65) is not correct. Similar calculations have also been reported by Takahashi.[54] That the self-consistent gap equation used to determine the correlation length in these approaches is precisely equivalent to the one-loop renormalization group equation can be easily checked. This is shown explicitly in the case of a ferromagnet in Ref. 47 (see, in particular, Appendix B); the proof for the antiferromagnetic case is identical.

VII. DYNAMIC PROPERTIES OF QHAF

The dynamic properties are more difficult to calculate reliably. It is therefore worth while to begin with a brief qualitative discussion of what one expects. These properties are most conveniently discussed in terms of the dynamic structure factor $S(q,\omega)$, which is the Fourier transform of the time dependent spin-spin correlation function. The dynamic structure factor contains information about the elementary excitations of the system. Inelastic neutron scattering experiments provide sensitive measurements of the correlation of the staggered order parmeter. *Thus, in this section we shall be mainly concerned about the region of the wave vector space close to the incipient antiferromagnetic wave vector $(\pi/a,\pi/a)$.*

In two dimensions the Heisenberg model does not have long range antiferromagnetic order. Nonetheless, as discussed earlier, the spin-spin correlation length diverges exponentially as $\exp(2\pi\rho_s(0)/T)$ in the limit T→0. It is therefore reasonable to assert that at low temperatures and for wave vectors k such that $k\xi \gg 1$ there is a large degree of local order. These local regions should therefore support propagating spin waves. In fact, for such large wave vectors renormalization due to thermal fluctuations should be negligibly small and the spin waves will have, predominantly, the character of T=0 spin waves which are renormalized by quantum fluctuations.

In order to make these statements more precise one of course needs, at the very least, an estimate of the damping of these elementary excitations. If damping happens to be small compared to frequency of these short wavelength spin waves, there should be single spin wave peaks in the dynamic structure factor. However, for longer wavelengths, when $k\xi \sim 1$, this picture should break down. As we have seen, in two dimensions, the dimensionless running coupling constant which measures the interaction between spin waves grows with increasing length scale. This means that not only the damping of the spin waves increases but there should be a sizable increase of multi-magnon excitations as we approach the length scale of the order of ξ^{-1}.

For $k\xi \ll 1$, the system appears to be truly disordered and there cannot be any propagating modes. At these wave vectors the behavior should be diffusive and we expect a overdamped mode centered at $\omega=0$. Since the staggered order parameter is not a conserved quantity, the relaxation rate at k=0 should be finite and cannot be very different from the relaxation rate at the scale ξ^{-1}. This relaxation rate should set the scale of the width of this overdamped mode. This overdamped peak is known as the quasielastic peak in the literature.

At intermediate wave vectors, where $k\xi \sim 1$, the situation is more complicated. One may particularly worry about the dynamics of Belavin-Polyakov solitons[55] which could make important contribution to the dynamic structure factor since the spin stiffness constant must vanish at this length scale. Not much is known about the dynamics of these objects. I shall pretend as though these objects did not exist. Nor would I be terribly concerned about multi-magnon excitations which for phase space reasons should give rise to a broad background.

If the above picture is correct, one should be able to establish: (a) that for $k\xi \gg 1$ there are propagating spin waves which are weakly damped, and (b) that there is a characteristic frequency scale $\bar{\omega}_0$ for the diffusive motion when $k\xi \ll 1$. In addition, in order to compare with experiments, it is important to provide a functional form of the dynamic structure factor.

The following discussion relies heavily on the arguments given by CHN, the numerical analysis by THC, and recent microscopic calculations by Tyč and Halperin (TH)[56] and Kopietz.[57]

CHN has argued that the dynamic properties of the QHAF at sufficiently low temperatures and low frequencies, and wave vectors very close to the antiferromagnetic Bragg peak, may be related directly to the low-frequency long-wavelength behavior of a *classical* lattice rotor model (CLRM), which can be studied by molecular dynamics simulations.

The CLRM is defined by the Lagrangian:

$$L = \frac{\chi_\perp^0 \Lambda^2}{2} \sum_i |\partial_t \hat{\Omega}_i|^2 + \rho_s^0 \Lambda^2 \sum_{(i,j)} \hat{\Omega}_i \cdot \hat{\Omega}_j \;, \tag{66}$$

where $\{\hat{\Omega}_i\}$ are a set of three-dimensional unit vectors located on the sites of a two-dimensional square lattice with the lattice constant $1/\Lambda$. The lattice constant of this model should be chosen small compared to macroscopic length scales of physical interset but it need not be simply related to the lattice constant of the quantum antiferromagnet whose properties we wish to reproduce. The parameters in the Lagrangian are defined at the scale Λ^{-1}. Each site represents a rotor which can be considered as a point mass constrained to lie on the surface of a unit sphere.

The quantization of the classical system described by the Lagrangian given in Eq. (66) is simple. We simply functionally integrate over all possible complexions of the set of unit vector fields $\{\hat{\Omega}_i\}$, and weight these by the action. The resulting quantum problem is none other than the lattice-regularized quantum mechanical non-linear σ-model that we have been discussing.

Although there are not yet dynamic renormalization group calculations appropriate to CLRM in d=2, CHN supposed that in the limit of low temperatures, *where the system is close to criticality*, the dynamic structure factor $S(\mathbf{k},\omega)$ defined by,

$$S(\mathbf{k},\omega) = \Lambda^{-d} \int_{-\infty}^{\infty} dt \sum_i e^{i\omega t - i\mathbf{k}\cdot\mathbf{r}_{ij}} N_0^2 <\hat{\Omega}_i(t) \cdot \hat{\Omega}_j(0)> \tag{67}$$

satisfied a dynamic scaling hypothesis[58], which may be expressed as

$$S(k,\omega) = \bar{\omega}_0^{-1} S(k) \Phi(k\xi, \omega/\bar{\omega}_0) \;, \tag{68}$$

where $\bar{\omega}_0$ is the scaling frequency, and Φ is assumed to satisfy the normalization condition:

$$\int_{-\infty}^{\infty} \frac{dy}{2\pi} \Phi(x,y) = 1 \;. \tag{69}$$

Recall that the wave vector k is defined with respect to the antiferromagnetic Bragg point.

It is this assumption of a *scaling* form that allows us to relate the dynamics of CLRM to the dynamics of QHAF. As discussed before, the knowledge of the static structure factor, S(k), for the classical model allows us to determine the structure factor for the corresponding quantum problem. This correspondence involves no adjustable parameters provided that the T=0 macroscopic spin stiffness, $\rho_s(0)$, the uniform susceptibility, $\chi_\perp(0)$, and the staggered order parameter N_0 of the QHAF are known. The scaling function f(kξ) which involves the product kξ can be carried over to the quantum model. In the case of La$_2$CuO$_4$, where the microscopic coupling constant is not precisely known, it is useful to consider $\rho_s(0)$ as an adjustable parameter which sets the energy scale. The remaining parameter can then be determined from spin wave theory[22] or from other microscopic calculations.[29]

Similarly, the scaling function Φ which involves the product $k\xi$ and the ratio $\omega/\bar{\omega}_0$ can also be carried over after we have suitably generalized the definition of the scaling frequency $\bar{\omega}_0$ to include quantum fluctuations. We must of course use the quantum mechanical expression for the correlation length ξ (see Eq. (61)).

Therefore we must determine the the scaling frequency $\bar{\omega}_0$ and the scaling function Φ. Finally, we must justify the dynamic scaling hypothesis.

CHN argued that the scaling frequency must be of the form

$$\bar{\omega}_0 = c\xi^{-1}\left(\frac{T}{2\pi\rho_s(0)}\right)^{1/2}, \qquad (70)$$

where c is the long wavelength spin wave velocity at T=0 which is renormalized by quantum fluctuations.

I shall now turn to a discussion of the scaling frequency and the scaling function. Consider first the CLRM. As mentioned earlier, for length scales shorter than ξ the system has a reasonably well defined local order. Thus, from the hydrodynamic arguments given by CHN there should be propagating spin wave modes for $k\xi \gg 1$. The frequency, ω_k, of the spin waves will be given by,

$$\omega_k = \left(\frac{\rho_s^0(k)}{\chi_\perp^0(k)}\right)^{1/2} k, \qquad (71)$$

where I have used scale dependent values for the spin stiffness constant and the uniform susceptibility. For spin wave hydrodynamics to apply the fluctuations which renormalize the spin stiffness constant, $\rho_s^0(k)$, and the uniform susceptibility, $\chi_\perp^0(k)$, must occur on a time scale much shorter than ω_k^{-1}. But this is likely to be true because dominant renormalizations of these parameters come from fluctuations with wave vectors, k', short compared to k, and therefore their frequencies are likely to be larger.

Although the short wavelength spin wave frequencies can be determined from hydrodynamic arguments, the damping of these excitations cannot be determined from hydrodynamics. The damping in two dimensions is mainly due to interactions between excitations at long wavelengths. As mentioned earlier, the dimensionless coupling constant, which is a measure of interactions between spin waves increases at long wavelengths. By contrast, in a three dimensional system the spin wave hydrodynamics applies for all temperatures below the Néel temperature[59] and the damping $\gamma_k = Dk^2$. As emphasized by TH, this hydrodynamic damping is given by short wavelength fluctuations and neglects completely the damping due to interaction of long wavelength spin waves. This may be a good approximation in higher dimensions since the phase space is proportinal to $k^{d-1}dk$, but, in the absence of long-range order, it breaks down in d=2. From the renormalization group point of view the effective interaction between magnons, below the Néel temperature, *decreases* with increasing length scales in d=3.[60]

Recently TH have used the Dyson-Maleev formalism[61] to calculate the damping of spin waves in the d=2 QHAF at asymptotically low temperatures and long wavelengths, both in the the quantum and the classical regimes. They have concluded that, as long as $k\xi \gg 1$, damping is much smaller than the frequency of spin waves and thus the spin waves are well defined excitations. In particular, they have shown that, for any positive x, if $k\to 0$ and $T\to 0$ with $ka \propto (T/\rho_s)^x$ then the ratio of damping to frequency vanishes. They have also taken into account fluctuation renormalizations of the damping rate at low but finite temperatures and have found good agreement with the simulations of CLRM. However, there are as yet no such rigorous theories available for the regime in which $k\xi \sim 1$. Therefore, it seems worth while to go through the qualitative arguments which prompted CHN to suggest a form for the scaling function.

Because it is likely that the short-wavelength, high-frequency fluctuations would adiabatically follow the modes at the intermediate length scales k^{-1}, it is possible to argue that the dominant damping should come from modes with wave vectors close to the wave vector k. Thus, these short wavelength modes can shift the frequency of the modes at the wave vector k, but cannot lead to a large damping of the modes at wave vector k. It is therefore plausible to postulate that the ratio of the damping to the frequency at wave vector k is some positive power, w, of the dimensionless coupling constant t(k) which characterizes the interaction between the modes at the length scale k^{-1}. We can therefore write,

$$\frac{\gamma_k}{\omega_k} = t(k)^w . \qquad (72)$$

Considering Fermi's golden rule it is tempting to guess that w should be 2, modulo logarithmic corrections that may arise from phase space considerations in two dimensions.

Therefore, interestingly, static renormalization group analysis can be used to estimate damping. Because t(k) increases with increasing length scale, we expect that the ratio γ_k/ω_k will also increase. In particular, this ratio must reach ~ 1 when $k=\xi^{-1}$ (The correlation length ξ is defined to be the scale at which $t(k)\sim 1$.). At this length scale multiple spin wave excitations become important, but since the short wave length excitations can still be assumed to follow the motion adiabatically, the structure factor, $S(k,\omega)$, could, in principle, be determined from a self consistent calculatuion at the scale $k=\xi^{-1}$, corresponding to a frequency scale $\omega_{\xi^{-1}}$. This characteristic frequency must match the frequency ω_k for $k > \xi^{-1}$. Moreover, because the staggered order parameter is not conserved, the relaxation rate at $k=0$ cannot be be very different from the rate at ξ^{-1}. Thus we finally arrive at a characteristic frequency $\bar{\omega}_0$, given by,

$$\bar{\omega}_0 = \omega_{\xi^{-1}} = \left(\frac{\rho_s^0(\xi^{-1})}{\chi_\perp^0(\xi^{-1})}\right)^{1/2} \xi^{-1} \qquad (73)$$

From the renormalization group equations obtained earlier we get,

$$\bar{\omega}_0 = \left(\frac{\rho_s^0}{\chi_\perp^0}\right)^{1/2} \xi^{-1} \left(\frac{t_0}{2\pi}\right)^{1/2}, \qquad (74)$$

using the results $\rho_s(\xi^{-1}) \approx t_0 \rho_s^0/2\pi$ and $\chi_\perp(\xi^{-1}) \approx 2\chi_\perp^0/3$.

The above formula for the scaling frequency can be immediately generalized to the case of QHAF: t_0 should be replaced by t_0^{eff} (Eq. (51)), ρ_s^0 should be replaced by $\rho_s(T=0)$, and $\chi_\perp^0$ by the T=0 uniform susceptibility $\chi_\perp(T=0)$. For the correlation length ξ we must use the quantum expression given in Eq. (61). This leads to the expression for the scaling frequency given in Eq. (70).

According to the hydrodynamic analysis given above the spin wave frequency ω_k must satisfy

$$\omega_k \sim \sqrt{3/2}\,\bar{\omega}_0(k\xi) \ln(k\xi), \qquad (75)$$

in the limit $k\xi \gg 1$. In deriving this equation we have made use of the renormalization group expression for the wave vector dependent spin stiffness constant given earlier. In the same limit we expect that the damping γ_k/ω_k should be proportional to some positive power w of $(1/\ln(k\xi))$, where we have made use of the renormalization group expression for the the dimensionless coupling constant $t(k)$. More precisely, from the work of TH we now know that the dependence is given by $\gamma_k/\omega_k \sim \ln(\ln(k\xi))/(\ln(k\xi))^2$.

These considerations led THC to suggest a scaling function:

$$\Phi(q,\nu) = \frac{\gamma_q}{(\nu - \nu_q)^2 + \gamma_q^2} + \frac{\gamma_q}{(\nu + \nu_q)^2 + \gamma_q^2}, \qquad (76)$$

where we have defined $q=k\xi$, $\nu=\omega/\bar{\omega}_0$, $\nu_q=\omega_k/\bar{\omega}_0$, and $\gamma_q=\gamma_k/\bar{\omega}_0$. They suggested the following parametrized expressions for the dimensionless spin wave frequency ν_q and the width γ_q:

$$\nu_q = \sqrt{3/2}\,q[\delta + \frac{1}{2}\ln(1+q^2)]^{1/2}, \qquad (77)$$

$$\gamma_q = \frac{\gamma_0(1+\mu q^2)^{1/2}}{[1+\frac{1}{2}\theta \ln(1+q^2)]^{3/2}}. \qquad (78)$$

Note that this scaling function is analytic in q, for q→0. The parameters γ_0, μ, θ and δ were determined from a molecular dynamics simulation of the CLRM model to be 0.8, 2.0, 0.15, and 1.7, respectively. THC found that attempts to fit with a form which allowed $\nu_0 \neq 0$ led to a value which was not distiguishable from zero.

More recent calculations of TH also provides us with a low temperature scaling function derived from the analytic damping calculations and is complementary to the scaling form given in Eqs. (77) and (78). These scaling functions have been tested by them using molecular dynamics simulations of the CLRM. The agreement is quite good.

$\gamma_0=0.8$ compares reasonably well with the value 0.96 obtained by Grempel[62] from a mode-coupling theory. By contrast, the approximation of Auerbach and Arovas[53] gives a characteristic frequency whose temperature dependence differs from Eq. (70), and a line shape which is quite different from the one discussed above. Nonetheless, it is important to note that their analysis also gives well defined spin waves at short wavelengths.

These results are in disagreement with Becher and Reiter[63], and Reiter.[64] According to Becher and Reiter[63] the damping rate in a d=2 QHAF should be proportional to k^2 and independent of temperature, in the limit T→0 for fixed small k. By contrast, according to TH, damping should be zero in this limit, provided that the T=0 magnon spectrum has a negative curvature at small k. This is certainly true for large S and is also likely to be true for S=1/2, if the spin wave expansion has at least asymptotic validity for small S. A negative curvature of the magnon spectrum at long wavelengths insures the stability of long wavelength magnons against spontaneous decay.[61] Reiter[64] has also suggested that the damping is proportional to T^2, and independent of k when k vanishes faster than T, but $k\xi \gg 1$. This also disagrees with the calculations of TH, and is inconsistent with the above discussion.

Some years ago a diagrammatic calculation of the damping for S=1/2 QHAF using Dyson-Maleev formalism was given by Kosevich and Chubukov.[65] Their results are qualitatively consistent with the calulation of TH, and is also consistent with the picture given by CHN. However, since they did not consider in detail the effect of the scattering surface, it is difficult to compare their results with those of TH. Roughly speaking, their long wavelength result does not contain the logarithmic corrections found by TH in their regime A.

The extreme short wavelength results, the regime D according to TH, agrees with an independent calculation by Kopietz.[57]

VIII. NEUTRON SCATTERING EXPERIMENTS

Neutron scattering measurements[66,67,68] have contributed a great deal to our understanding of the magnetic properties of La_2CuO_4 and related cuprates. In this subsection I shall present a selective review which ties in with the theoretical discussion given above.

The crystal structure of La_2CuO_4 at high temperature has the tetragonal symmetry, space group I4/mmm. At a temperature of about 500 K, there is a structural transition to the orthorhombic symmetry, space group Cmca. At this transition the CuO_6 octahedra undergoes a staggered tilt. However, because of the equivalence of the two tilting directions in the CuO_2 plane, a twin structure is formed. In general, this structural transition is inessential for a basic understanding of the magnetic properties, and is ignored quite frequently (however, see below). Thus, the CuO_2 plane can be assumed to be square-planar with a lattice constant of 3.8Å. However, in the orthorhombic notation, the room temperature lattice constants are a=5.354 Å, b=13.153 Å, and c=5.401 Å.

The system undergoes a three-dimensional (3d) antiferromagnetic transition at the Néel tempertuure T_N, at about 200 K. The staggered order parameter has the characteristic temperature dependence of a 3d transition and is very different from the 2d Ising transition seen previously in K_2NiF_4.[8] This transition was first discovered by Vaknin et al.[69] in their powder diffraction study. From their analysis they deduced that the spin is along [0 0 1] (this turns out not to be quite right; see below) while the antiferromagnetic modulation is along [1 0 0], using the notation of Cmca orthorhombic unit cell. In the notation of the tetragonal unit cell there would be lines of spins perpendicular to the [1, 1] direction which are alternately parallel and antiparallel. In the notation of orthorhombic Cmca unit cell, the axes are rotated by 45°. The point (1,1) becomes (1,0), in suitable units.

It is also known that the 3d transition temperature is extremely sensitive to oxygen concentration.[70] Although the maximum possible T_N may be ~300 K, a small change in the oxygen concentration can reduce T_N dramatically.

Vaknin et al.[69] also found that in the limit T→0 the ordered moment is 0.5±0.15 μ_B considerably smaller than the saturation value for Cu^{++} which is ~1 μ_B. This is not difficult to explain. As mentioned earlier, quantum fluctuations reduce the value of the staggered order parameter at T=0. From spin wave theory (or from more recent numerical work[29]) the ordered moment would be ~$0.6\mu_B$. As shown in Ref. 32 (see, in particualr, Appendix F), the negligibly small interplanar coupling, J', present in the system, has very little effect on the T=0 sublattice magnetization if the system is well into the ordered regime at T=0. This would not be the case if the system at T=0 is extremely close to the boundary between the ordered and the disordered phase.

The three dimensional transition in this system is a crossover phenomenon. Because the intraplanar coupling, J, is so much larger than the interplanar coupling, J', the system behaves as though it were two-dimensional until it gets to a temperature very close to T_N when the in-plane correlation length ξ becomes sufficiently large such that $(N_0/S)^2(\xi/a)^2 J' \sim T_N$.[32,46] At this temperature the system crosses over to the three dimensional behavior. Because (J'/J) is very small, of the order of 10^{-5}, (ξ/a) has to be very large for this crossover to take place and therefore the *3d* critical region is expected to be extremely narrow and the behavior above T_N should be largely unaffected by J'. Thus, above T_N, it should be legitimate to consider the system to be two dimesnional, with no transition at any finite temperature, the critical temperature being T=0. It is this two dimensionality that has been the focus of considerable attention.

The two-dimensional spin correlations in La_2CuO_4 have been studied in some detail in a series of neutron scatttering measurements.[66,67,68] In these experiments one integrates over the energy without changing the in-plane momentum transfer by collecting all outgoing neutrons in a direction parrallel to the 2d magnetic rod, the so-called two-axis focusing scan. If the spin correlations are two-dimensional, the dynamic structure factor must be independent of the momentum transfer perpendicular to the CuO_2 plane.

The static structure factor $S(q_{2d})$ is given by,

$$S(q_{2d}) = \int_{-\infty}^{\infty} d\omega\, S(q_{2d},\omega) . \tag{79}$$

It is important to note that, experimentally, this energy integration is carried out not between $-\infty$ and $+\infty$, but between $-T$ and the incident neutron energy E_i. Clearly a neutron cannot loose an energy greater than its incident energy. However, it can gain any amount of energy from the available excitations but the integration will be effectively cutoff by the thermal occupation factor. Thus, if the dominant contribution is contained within this range, we can safely assume that this procedure gives the true $S(q_{2d})$. The correlation length x is then extracted by fitting $S(q_{2d})$ to a Lorentzian, i.e.,

$$S(q_{2d}) \propto \frac{1}{q_{2d}^2 + (1/\xi)^2} . \tag{80}$$

Of course, in principle, one could use a more sophisticated line shape formula given Eqs. (34), (35), and (62), but this refinement appears to be outside the scope of the current set of experiments. Finally, this formula has to be convoluted with the instrumental resolution function.

A plot of the spin-spin correlation length as a function of temperature[68] is shown in Fig. 1, where the data obtained earlier[67], labelled NTT#2, are superposed with the more recent data,[68] labelled NTT#7. The dashed line is the theoretical formula evaluated for S=1/2 (Eq. (61)) with the spin wave velocity fixed at $\hbar c = 0.85$eV-Å; the remaining parameters are determined from the T=0 spin wave theory. The agreement between theory and experiment appears to be quite good. Using T=0 spin wave theory to relate the spin wave velocity to the nearest neighbor exchange constant J, we get J$\sim$0.13eV.

In a three-axis scan one measures both the energy as well as the momentum transfer. The three-axis constant-E_f scans (the energy is varied by varying the incident neutron energy) across the magnetic rod (1,K,0), or its orthorhombic twin (0,K,1), are shown in Fig. 2.[68] Note that the scattering from a two-dimensionally ordered system yields magnetic Bragg rods instead of magnetic Bragg points observed in the scattering from a three dimensionally ordered system. The intensity of scattering is seen to be independent of K, and exhibits a sharp peak at h=1 (corresponding to the wave vector $(\pi/a,\pi/a)$ in the tetragonal notation) confirming our expectations that the spin fluctuations are two-dimensional in character, and are dominantly located at the incipient antiferromagnetic wave vector.

The temperature evolution of constant energy scans across the magnetic rod, above T_N, are shown in Fig.3.[68] The dashed lines are the results of calculations according to the dynamic structure factor given by CHN and THC. These calculations were performed with $\hbar c$ set to 0.85eV-Å, and the correlation length, ξ, set to experimentally measured values. The theoretical formula was also convoluted with the experimental resolution function. The only adjustable parameter, the overall

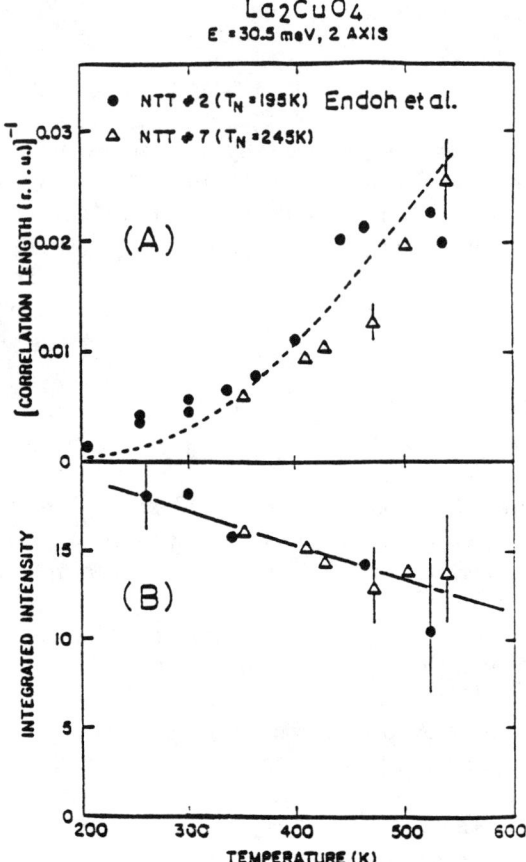

FIGURE 1 Temperature dependences of (a) inverse correlation length and (b) the integrated intensity measured by two-axis focusing scan from Ref. 68. The dashed curve in (a) is a fit to the formula by CHN.

intensity scale, was fixed by matching theory and experiment at the energy transfer of 6 meV, at 290 K. The agreement appears to be excellent.

The energy dependences of the integrated intensities of constant energy scans across the magnetic rod at different temperatures are shown in Fig. 4.[68] The inset shows the results obtained from the dynamic structure factor proposed by CHN with the parameters determined in the simulations of THC. The agreement between theory and experiment is quite good except in the quasielastic regime in which the cross section rises steeply. It is plausible that this steep rise arises primarily from

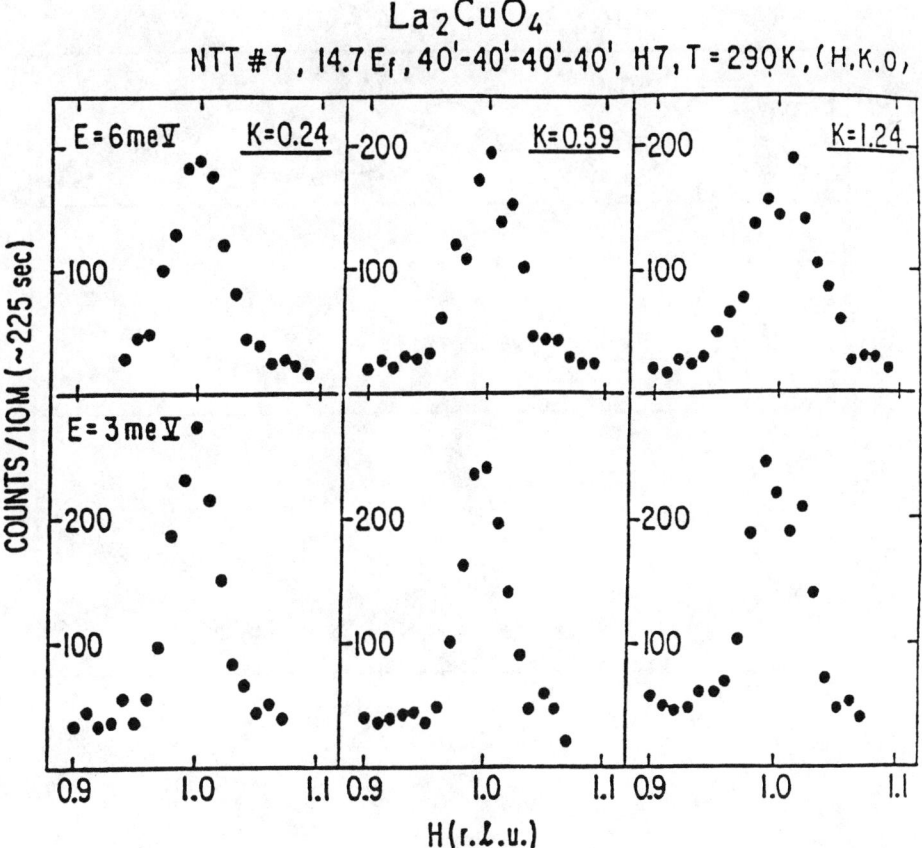

FIGURE 2 Three-axis constant-E_f scans across the magnetic rod $(1,K,0)$ [or $(0,K,1)$] at different K values above T_N from Ref. 68.

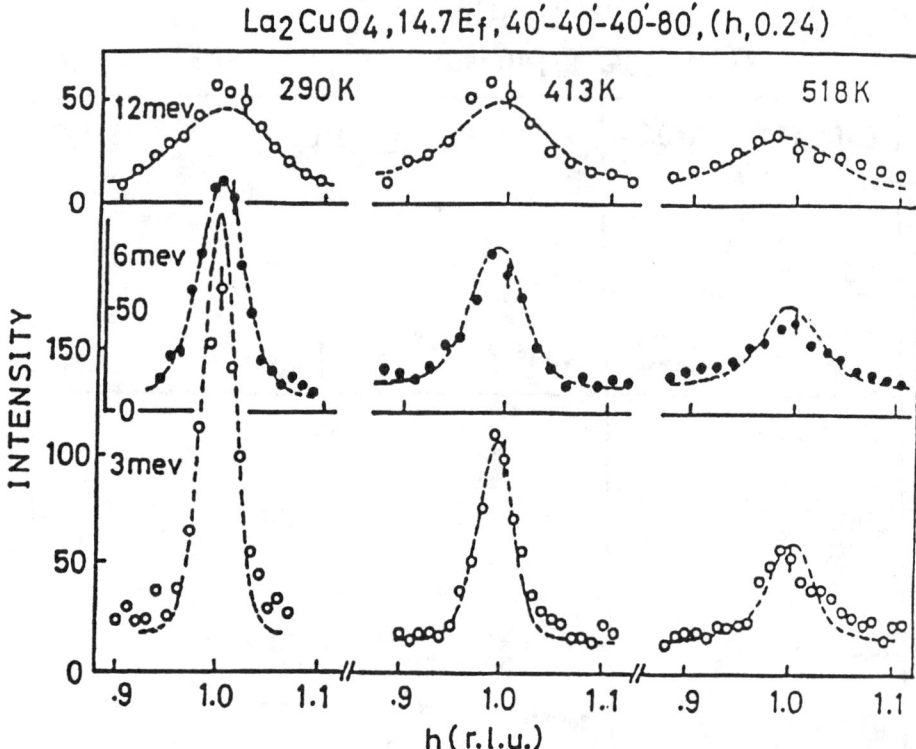

FIGURE 3 Temperature evolution of constant energy scans across the magnetic rod above T_N from Ref. 68. The dashed lines are the results of calculations utilizing the formula of CHN and THC for the dynamical structure factor convoluted with the experimental resolution function.

magnetic defects and is similar to the central peak phenomenon observed in structural phase transitions.[71] Further work is necessary to elucidate the true nature of this quasielastic feature. The temperature dependence of the integrated intensities is shown in Fig.5.[68] Once again, the agreement between theory and experiment is excellent except in the quasielastic regime. Here the word quasielasticity is used in a sense different from the earlier usage on page 21. The quasielasticity discussed here is due to magnetic defects and is non-intrinsic. This unfortunate choice of the

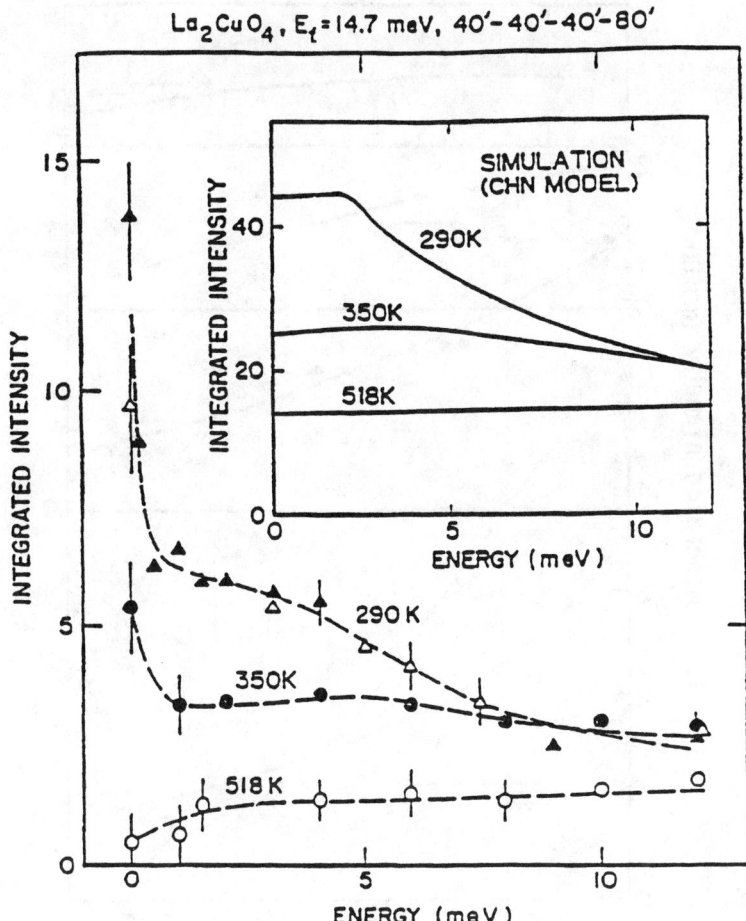

FIGURE 4 Energy dependences of integrated intensities of constant energy scans across the magnetic rod at different temperatures from Ref. 68. The data at 290 K are taken for two crystals NTT-7 (open triangle) and NTT-8 (solid triangle). The inset is the results of calculations utilizing the model due to CHN and THC.

same word to describe both intrinsic and non-intrinsic quasielasticity has led to much confusion in the current literature.

Recently[72] high-energy inelastic neutron scattering has been successfully used to resolve spin waves in La_2CuO_4. The measured spin wave velocity was found to be 0.85±0.03 eV-Å.

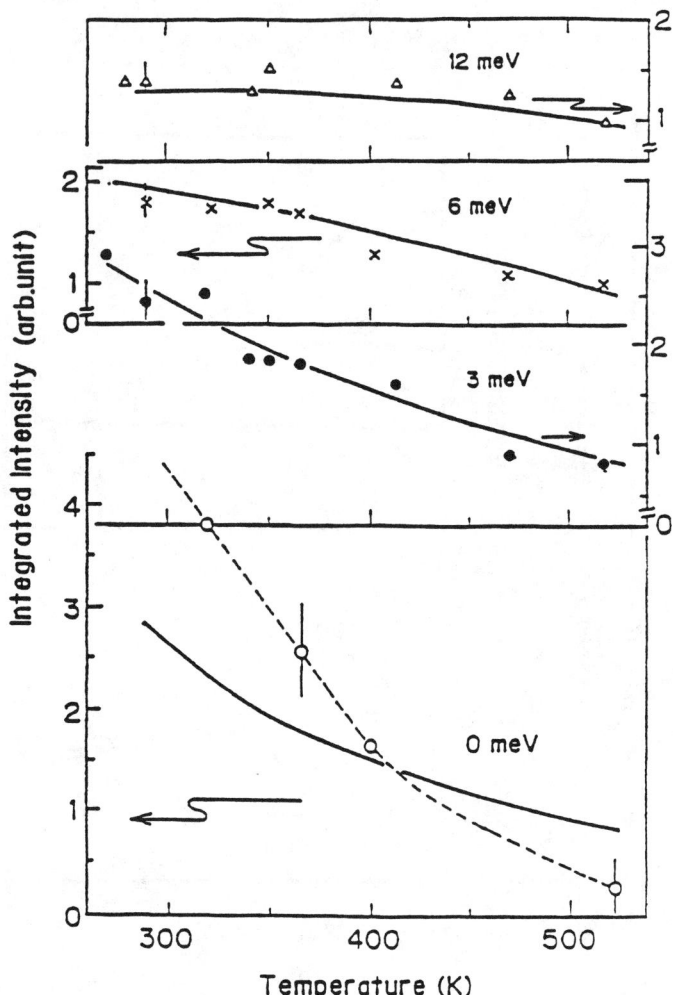

FIGURE 5 Temperature dependences of the integrated intensities of constant energy scans across the magnetic rod at different energies from Ref. 68. The solid lines are the results of calculations utilizing the formulas of CHN and THC.

Nearly twenty years ago Birgeneau, Skalyo, and Shirane[8] measured the 2d antiferromagnetic structure factor in K_2NiF_4 over a wide range of temperatures. K_2NiF_4 is a well chracterized layered antiferromagnet which can be described by a 2d S=1 quantum Heisenberg model. However, as mentioned earlier, the onsite Ising anisotropy in this material can be substantial, and the behavior in this model crosses over from a 2d Heisenberg behavior to a 2d Ising behavior close to T_N. The finite temperature transition is a two-dimensional Ising transition. The crossover to 3d behavior occurs immeasurably close to T_N.

Recently Birgeneau[73] has compared the experimental measurements of the correlation length and the staggered susceptibility in K_2NiF_4 with the theory of CHN.

The theory developed by CHN should apply for temperatures above the Heisenberg-Ising crossover. For S=1, we obtain from Eq. (61),

$$\frac{\xi}{a} = 0.16 \frac{e^{5.3J/T}}{1+T/5.3J} \;. \tag{81}$$

In the case of K_2NiF_4, the exchange constant J is known rather well and is 104 ± 1 K. The agreement between theory and experiment is excellent and the experimentally determined prefactor 0.123 ± 0.02 is in remarkable agreement with the theoretical value. As pointed out by Birgeneau,[73] Heisenberg-Ising crossover takes place when $\xi/a \sim 23$, corresponding to $T\sim 102$ K. This is consistent with the previous estimate[8] of the range of Ising criticality, which is $|T/T_N - 1| < 0.05$. The agreement between theory and experiment for the staggered susceptibilty was also found to be equally good.[78]

I would like to end this section with a few brief remarks about anisotropy. The staggered tilt of the CuO_6 octahedra at the orthorhombic transition induces an antisymmetric Dzyaloshinski-Moriya[10] interaction which turns out to be the dominant anisotropy in this system. This antisymmetric exchange arises from first order spin-orbit coupling to excited state orbitals. The correct nearest-neighbor Hamiltonian is then given by,

$$H = \sum_{(i,j)} \mathbf{S}_i \cdot \mathbf{J} \cdot \mathbf{S}_j \;, \tag{82}$$

where the matrix $\mathbf{J}$ is given by,

$$\mathbf{J} = \begin{pmatrix} J^{aa} & 0 & 0 \\ 0 & J^{bb} & J^{bc} \\ 0 & -J^{bc} & J^{cc} \end{pmatrix}, \tag{83}$$

where $J^{aa} \approx J^{cc}$.

As pointed out by Moriya,[74] J^{bc} is of the order of $J(\Delta g/g)$, multiplied, roughly, by the overlap between crystal-field-split levels on neighboring sites with $J=(1/3)(J^{aa}+J^{bb}+J^{cc})$. Δg is the deviation of the g-value from the free electron value. Thio et al.[75] have argued that $J^{bc} \sim \phi_0 J(\Delta g/g)$, where ϕ_0 is the angle, in radians, of the rotation of the CuO_6 octahedra. Note that $J^{bc} \to 0$ as $\phi_0 \to 0$, and therefore J^{bc} must be proportional to some power of ϕ_0 for small ϕ_0. But symmetry allows a first power of ϕ_0 and therefore it is plausible that J^{bc} is proportional to ϕ_0. Using the experimentally measured values of ϕ_0 (~ 0.049) and $(\Delta g/g)$ (~ 0.1) we get $J^{bc} \sim 0.6$ meV.[77] Actually, because electron paramagnetic resonance has not been observed in either La_2CuO_4 or $YBa_2Cu_3O_6$, Δg is not known precisely. A rough estimate can be obtained from Ref. 76. This value of J^{bc} is in agreement with the most recent experimental measurements[77] which yield $J^{bc}=0.7\pm0.1$ meV. Moriya[74] has also shown that the exchange anisotropy, which is a symmetric correction to the superexchange, is second order in spin-orbit coupling, and is given by the order of magnitude estimate $(\Delta g/g)^2 J$, multiplied, very roughly, by a complicated overlap integral. The naïve estimate $(\Delta g/g)^2 J \sim 1.3$ meV is more than two orders of magnitude larger than the experimentally measured value: J^{cc}-J^{bb}=0.003±0.002 meV.[75,77] However, there is no reason to believe that this estimate is correct, as it sets a complicated multiplicative factor to unity. Because exchange anisotropy does not necessarily vanish in the limit $\phi_0 \to 0$, this overlap factor bears no relation to the one that determines J^{bc}. In fact, it is quite surprising that the simple estimate for J^{bc} given above comes out so close to the experimental value. Approximately the same value of (J^{cc}-J^{bb}) was also obtained in a completely independent measurement involving zone-center spin wave excitations.[78] Thus the value 0.003 meV, for the exchange anisotropy, should be reiable.

The antisymmetric exchange leads to a number of interesting consequences. Firstly, it creates a spin canting of about 0.2°, in a direction perpendicular to the basal plane, below T_N. The canting angle derived from an analysis of the uniform susceptibility[75] agrees with that derived from the magnetic field dependence of the structure factor,[79] which, as a function of the applied field, exhibits a transition from the antiferromagnetic ordering of the canted component of the spins to the ferromagnetic ordering. Secondly, the antisymmetric exchange couples the uniform order parameter to the staggered order parameter and thus drives a peak in the uniform susceptibility at T_N, a rather dramatic effect. The calculated bulk susceptibility using the in-plane staggered order parameter correlation length, ξ, as given by CHN leads to a very good agreement with experiments.[75]

It is also interesting to note that the interplanar coupling, $J' \sim 0.003$ meV, obtained from the argument given above (i.e., $(N_0/S)^2 (\xi/a)^2 J' \sim T_N$) is also consistent with the experimentally derived value obtained from the finite field studies of the canting due to Dzyaloshinski-Moriya interaction discussed above.[75,77]

IX. SPIN-PAIR EXCITATIONS IN LIGHT SCATTERING EXPERIMENTS

So far I have concentrated on low frequency, long wavelength magnetic behavior of La_2CuO_4. However, the short wavelength, high frequency spin fluctuations are equally interesting, but far more demanding from a theorists point of view. Microscopic details can no longer be ignored. However, if one holds the point of view that the spin fluctuations play a major role in high temperature superconductivity, it is likely that a thorough understanding of the short wavelength behavior could be quite rewarding. In this section I shall therefore turn my attention to inelastic light scattering experiments which directly probe these energetic spin fluctuations.

Inelastic light scattering experiments from spin-pair excitations have a long and distiguished history. However, I shall entirely devote myself to recent developments. Shortly after the new superconductors were discovered, Lyons et al.[80] looked for the signature of the "two-magnon" Raman scattering from La_2CuO_4, and subsequently from the doped material $La_{2-x}Sr_xCuO_4$.[81] Just as in previously studied[82] K_2NiF_4, which is well described by a two-dimensional S=1 Heisenberg model, the Raman spectrum in La_2CuO_4 exhibited a broad peak centered around 3000 cm^{-1} and satisfied the selection rules appropriate to magnetic spin-pair excitations. Subsequently, scattering was found to extend to wave numbers as high as 8000 cm^{-1}.[83] Although qualitatively the experiments could be interpreted in terms of "two-magnon" Raman scattering, previously well understood in the context of K_2NiF_4, at first sight there appeared to be a number of intriguing problems.

The same line-shape analysis, developed by Parkinson[84], and applied previously to the S=1 case, yielded a theoretical spectrum in considerabe disagreement with experiment, although the peak position gave a reasonable value of the exchange constant, J.[85] In fact, in spite of more refined experiments, to be discussed later, this initial estimate turns out to be surprisingly accurate once the simple spin wave renormalization of the spectrum due to Z_c is taken in to account. Recall that, within spin wave theory, the renormalization of the spectrum is independent of the wave vector in the Brillouin zone.

However, subsequent measurements[83] also revealed significant scattering in the A_{1g} and B_{2g} symmetry, and not only in the B_{1g} symmetry. The symmetry notation corresponds to the tetragonal D_{4h} structure and ignores the negligibly small orthorhombic distortion in this material. Moreover, as remarked above, the scattering extends to energies well beyond the noninteracting magnon cutoff which is 4J, in fact a sizeable scattering in the B_{1g} symmetry exists upto 8J. These facts are particularly perplexing if one recalls that the agreement between the existing theory[84] (spin-wave plus random phase approximation) and the experiment[82] in K_2NiF_4 was found to be spectacular. As we shall see, the problems have been considerably clarified by Singh et al.[86] and Singh.[87] Nonetheless, it seems worth while to discuss this problem in some detail.

There are several possible sources of photon-magnon interactions. In addition to direct magnetic-dipole coupling and indirect electric-dipole coupling via spin-orbit interaction, it was shown some time ago, by Fleury and Loudon[88], that there could be a relatively strong "two-magnon" process (I shall soon switch over to a better terminology and call this process a spin-pair process) due to a mechanism caused by excited-state exchange interaction which is not simply an iteration of a first order process, say due to spin-orbit scattering, taken to 2nd order. Because the excited-state exchange scattering bears no relation to possible first order scattering mechanisms, *a priori*, the strength of the first order process has nothing to do with the the strength of the second order process as was noted by Fleury and Loudon.[88] In particular, the spin-pair spectrum in an antiferromagnet is often stronger than the first-order spectrum.

Because the spin-pair process flips two spins, there is no excited-state exchange-scattering mechanism for one magnon scattering. Of course spin-pair process is inoperative for ferromagnets because there is no two-magnon state of 0 spin and positive parity.[88]

The scattering Hamiltonian describing the interaction of the spin-pairs with light is given by,[88,89]

$$H_R = \sum_{(i,j)} (\mathbf{E}_{inc}.\sigma_{ij})(\mathbf{E}_{sc}.\sigma_{ij}) M_{ij} \mathbf{S}_i.\mathbf{S}_j , \qquad (84)$$

where $\mathbf{E}_{inc}$ and $\mathbf{E}_{sc}$ are the incident and the scattered electric field vectors of the photons, and σ_{ij} is a unit vector connecting the sites i and j. If we further restrict ourselves to nearest-neighbor pairs, i.e., M_{ij} is 0 unless (i,j) is a nearest-neighbor pair, it can be shown that the scattering is exclusively in the B_{1g} symmetry.

The scattering Hamiltonian arises in third order perturbation theory.[88] It consists of two dipole matrix elements involving incident and scattered electric field vectors, one matrix element involving exchange, and two energy denominators. Although the final form of the Hamiltonian is written down in terms of only spin operators, the intermediate states involve charge excitations which are high energy states. The nature of the intermediate states will determine the form the Hamiltonian, a point that I glossed over in writing down the above Hamiltonian. However, because an exchange matrix element is expected to fall off rapidly as a function of distance, it is plausible to set M_{ij} to 0 unless (i,j) is a nearest neighbor pair. In the presence of resonance enhancement it is no longer possible to restrict ourselves to a sum over nearest neighbor pairs.[86] In fact, we shall see that at the very least a diagonal next nearest neighbor interaction is necessary to explain the scattering seen in A_{1g} and B_{2g} symmetry (I shall have more to say about this point).[86] However, the magnitude of this diagonal next-nearest neighbor interaction bears no relation whatsoever to the magnitude of the possible further ranged interactions in the spin Hamiltonian.[86] Presently I shall assume that only nearest neighbor terms dominate.

The scattering intensity at T=0 is given by,

$$I(\omega) = \sum_i \delta(\omega - (E_i - E_0)) |<0|H_R|i>|^2 , \qquad (85)$$

where $|0>$ is the ground state and $|i>$ are the excited states of the Heisenberg Hamiltonian. The generalization of this formula to $T \neq 0$ is simple. One simply has to multiply within the sum by the probability to find the system in a state $|m>$ (not just the ground state), modify the delta function accordingly, and also sum over all possible initial states $|m>$.

The scattering geometries are defined as follows.[83] We define the axes x and y along the Cu-O bond directions and x',y' rotated by 45° with respect to x,y. Then the scattering with the polarization of the electric field vectors along $(x'x')$ corresponds to $A_{1g}+B_{2g}$, (xy) to B_{2g}, $(x'y')$ to B_{1g}, and $(x'x'-xy)$ to A_{1g}. The scattering Hamiltonian for B_{1g} symmetry simplifies to

$$H_R = A \sum_i \mathbf{S}_i \cdot [\mathbf{S}_{i+\hat{x}} - \mathbf{S}_{i+\hat{y}}] \ . \tag{86}$$

where A is an undetermined constant.

Before I discuss any actual calculation I would like to make a few qualitative remarks. The Raman Hamiltonian flips a pair of nearest neighbor spins in the ground state. If the ground state were a classical Néel state, 6 nearest neighbor bonds would be broken and the energy cost would be 3J, as opposed to 4J, if the spins flipped were not nearest neighbor pairs. Thus, one might say that we have a "bound pair."

Of course the classical Néel state is not the ground state of the Heisenberg Hamiltonian, but the ground state of the isotropic Heisenberg Hamiltonian can be obtained by perturbing around the Ising ground state, i.e., we start with the anisotropic model,

$$H = J_z \sum_{(i,j)} S_i^z S_j^z + J_{xy} \sum_{(i,j)} (S_i^x S_j^x + S_i^y S_j^y) \ , \tag{87}$$

and determine the ground state in an expansion in powers of the parameter x= (J_z/J_{xy}). Thus, the true ground state would contain a linear combination of states with many spins flipped with respect to the classical Néel ground state. The effect of the Raman Hamiltonian cannot be as simple as before. In particular, the spectrum cannot have a cutoff at 4J any more and should, in principle, extend up to ∞. Of course if such high energy states played an important role, we would no longer have any reasons to trust the Heisenberg model to correctly describe such processes. Recall that the Heisenberg model describes only the low energy sector of the Hubbard model.

There is another way to think about this problem. Consider first the lowest order spin wave approximation. In this picture the ground state is a condensate of non interacting zero-point spin waves. The Raman Hamiltonian acting on this state flips two-nearest neighbor spins. The flipped spins can be described by two wave packets of magnons which are initially created at a distance of a lattice spacing with 0 total momentum (ignoring the momentum of the light in comparison). If the magnons were non interacting, the spectrum would be nothing but the joint density of states weighted by the symmetry factor appropriate to the B_{1g} symmetry. From simple phase space considerations it is not difficult to see that the spectrum will be dominated by zone boundary magnons. In fact, it has a logarithmic singularity at the zone boundary.[84] But, nonetheless, the spectrum cuts off at 4J because that is the maximum energy a pair of non interacting magnons can have.

However, because the pair was created only a lattice spacing apart, it is clear that one cannot ignore interactions. The pair is likely to undergo repeated scattering and the problem, as formulated, is approximately equivalent to a Bethe-Salpeter equation for a bound state in which the sum of the wave vectors of two magnons $k_1+k_2=0$.[90] In the approximation in which the propagation of the magnons is treated within non interacting spin wave approximation and the repeated scattering effect is assumed to give rise to a local field correction (usually replaced by the classical Néel value), the solution of the Bethe-Salpeter equation is nothing but a random phase approximation. The spectrum still cuts off at 4J because this is the maximum possible energy a pair of non-interacting magnons can have, but the peak of the intensity is shifted downwards to 2.7J due to interaction effects instead of 3J as obtained in the classical Néel limit. It is possible to improve this treatment a bit further and incorporate the self energy correction because the magnons are not moving in vacuum but in the presence of other magnons. But note that since the spectrum is dominated by short wavelength magnons, we need self energy corrections for short wavelength magnons. *A priori*, there is no reason to believe why this self energy correction should be purely real. However, within the spin wave expansion, for large S, the T=0 magnon spectrum has a negative curvature for *all* k, and therefore the magnons are stable against spontaneous decay.[61] Therefore, it appears to me, that *within the context of spin wave theory*, it is unjustified to use a phenomenological damping parameter in the Green's function of the single magnon to account for the Raman width.[91] However, the absence of the decay of a magnon within the large-S spin wave theory cannot be a general result because single magnon states are not exact eigenstates of the Hamiltonian, and therefore it is likely that for small S, and for *short* wave lengths, the self energy may have a sizeable imaginary part. In contrast, if the ground state has long range antiferromagnetic order, it is possible to argue (see below) that the damping of the *long* wavelength magnons is negligibly small.

The magnon treatment described above seems to work spectacularly well for K_2NiF_4 (S=1), but, as mentioned earlier, does not work well for La_2CuO_4. The question is why? Also, note that the traditional magnon treatment is far from being classical. The very fact that we are talking about magnons at T=0 means that we are considering a ground state which has quantum fluctuations.

I shall now turn to some additional questions of principle. The long wavelength magnons are well defined elementary excitations at T=0 if the ground state of the two-dimensional S=1/2 Heisenberg antiferromagnet is ordered. And therefore, in treating these long wavelegth magnons one need not worry, at least qualitatively, about the actual value of the spin. In Dyson's[17] language these long wavelength magnons can essentially float over each other and the kinematic restrictions due to the finiteness of the spin are not important. This picture also follows from the renormalization group treatment of the (2+1)-dimensional non-linear σ-model. As can be seen from Eq. (43), at T=0 the effective coupling representing the interaction between magnons scales to zero at long wavelengths, as long as we are in the regime such that the ground state has long range order. This flow towards weak coupling is equivalent to a flow towards the large effective spin limit, i.e., as though the effective spin gets bigger at longer wavelengths. Note that the coupling constant of the non-linear σ-model is dimensionful in (2+1) dimensions. Thus, it is easy to see that the long wave length magnons should be well defined elementary excitations with negligible damping, similar to the behavior of the 3d system[59] for temperatures below the Néel temperature, and that their properties can be quantitatively calculated within spin wave theory once we take into account quantum renormalizations, the actual value of the spin does not impose any severe kinematic restrictions.

The difficulty of course is the fact that the Raman intensity is primarily dominated by short wave length magnons where the above arguments do not apply. If we turn this observation around, it is quite possible to deduce that a detailed theory of the Raman line shape can yield the valuable information that we have been looking for in this field, i.e., what is so special about S=1/2? In particular, it is interesting to ask, are there well defined short wave length excitations which are *essentially different* from spin wave excitations? For S=1/2 there could be interesting possibilities. Because for S=1/2 one cannot put more than one spin deviation at a site, the kinematic restrictions are quite severe. It is possible that the calculations of the Raman spectrum by Coppersmith[92] and Hsu[93], using Anderson's non-interacting spinons[4], lead to a much better agreement with the experiment than the traditional spin wave treatment is significant in this respect.

What seems to emerge from this discussion is the fact that the nature of the elementary excitations can change with the length scale, perhaps from fermionic spinons at short wavelengths to bosonic spin waves at long wave lengths, and that perhaps Raman scattering in this material can tell us if there are surprises to discover. It would therefore be valuable to develop a detailed theory of the *line shape*, and needless to say, it is also essential to develop a proper understanding of possible resonant enhancements observed in in La_2CuO_4.

Suppose, for a moment, that we do not want to invoke the language of spinons, then how should we think about Raman scattering in the conventional language of spin waves and what calculations should we be doing? First of all note that the Raman Hamiltonian produces two spin flips adjacent to each other. Each spin flip can be represented by a wave packet made out of wave vectors close to the zone boundary. It is therefore plausible that this wave packet should contain components from one-magnon, three-magnon,...sectors and that quantum fluctuations present at T=0 allow for such superpositions. If we believe that this is the case, we expect to see 4-magnon, 6-magnon,...peaks in the Raman spectrum, provided that the intensities are sufficiently large and provided that the kinematics does not wash them out. Of course, as mentioned earlier, at very high energies the Heisenberg model description is no longer valid. At the very least we expect a broad asymmetric peak with significant intensity well above 4J. This conclusion also follows from the analysis based on the Ising perturbation theory in x, as discussed above, and was noted by Singh et al..[86]

The spectral moments of the Raman intensity have been calculated by Singh et al.[86] in an expansion in powers of x. If we define,

$$I_T = \int d\omega I(\omega) = <0|H_R^2|0> , \tag{88}$$

$$\rho_1 = \frac{1}{I_T}\int d\omega\, \omega I(\omega) = \frac{1}{I_T} <0|H_R[H,H_R]|0> , \qquad (89)$$

$$\rho_2 = \frac{1}{I_T}\int d\omega\, \omega^2 I(\omega) = -\frac{1}{I_T} <0|[H,H_R]^2|0> , \qquad (90)$$

$$\rho_3 = \frac{1}{I_T}\int d\omega\, \omega^3 I(\omega) = \frac{1}{I_T} <0|[H,[H,H_R][H,H_R]|0> , \qquad (91)$$

etc., then for S=1/2 they obtain

$$\rho_1/J = 3 + 0\ x + \frac{1}{2}\ x^2 + 0\ x^3 + 0.0286\ x^4 + 0.0372\ x^5 + 0.0159 x^6 + \ldots \qquad (92)$$

$$\rho_2/J = 9 + 0\ x + \frac{10}{3}\ x^2 + 0\ x^3 + 0.719\ x^4 + 0.290\ x^5 + 0.0828\ x^6 + \ldots \qquad (93)$$

$$\rho_3/J = 27 + 0\ x + 17.16\ x^2 + 0\ x^3 + 6.474\ x^4 + 1.663\ x^5 + 0.711\ x^6 + \ldots \qquad (94)$$

This allows us to define the cumulants, for n>1, by,

$$(M_n)^n = \frac{1}{I_T}\int d\omega\, (\omega - \rho_1)^n\ I(\omega) . \qquad (95)$$

The first cumulant is defined to be $M_1 = \rho_1$.

These expansions are, in fact, convergent despite long wavelength spin wave modes (Goldstone modes); the presence of Goldstone modes slows down the convergence but does not destroy it. In any case, because light scattering intensity is dominated by short wavelength excitations, the effect of long wavelength singularities should be negligible. Thus, Padé approximants can be safely used to extrapolate to x=1. In this manner Singh et al.[86] obtain $\rho_1/J = 3.58 \pm 0.06$, $M_2/J = 0.81 \pm 0.05$, and $M_3/J = 1.00 \pm 0.14$, where the uncertainities reflect the spread in the Padé approximants. These cumulants allow for a parameter free test of the theory. The predicted ratio M_2/M_1 relates the linewidth to the central frequency, independent of J. This ratio, 0.23 ± 0.02, is in excellent agreement with the experimental value 0.27 ± 0.03. Moreover, this theoretical value is almost a factor of 2 larger than that obtained in the traditional spin wave treatment for S=1/2. This certainly sorts out the initial puzzle regarding the discrepancy between the spin wave result and the experimental linewidth seen in La_2CuO_4. Similarly, the ratio M_3/M_1 is a prediction for the skewness and helps to put a bound on the high frequency tail of the spectrum. The experimental and the theoretical values for M_3/M_1 are 0.27 and 0.27. The agreement in the third cumulant can be interpreted to mean that there is no appreciable scattering beyond the experimental range ~ 8000 cm^{-1}. Equating the experimental first moment to the theoretical value one obtains $J = 1030 \pm 50$ cm^{-1}.

Because Raman intensity is primarily dominated by short wavelength excitations, one can argue that finite lattice calculations could give us useful information. From exact diagonalization studies of a 4×4 lattice Scharf and Chakravarty[94] have found that for S=1/2, $M_1 = 3.24$ J, $M_2/M_1 = 0.245$, $M_3/M_1 = 0.352$ ($M_2 = 0.797$ J, and $M_3 = 1.14$J). These exact diagonalization studies also reveal that there is appreciable intensity, about 10% of the two magnon peak at 2.976J, at an energy range where one expects a 4-magnon "bound state." If we flip 4 spins in a row we break 10 bonds, and we expect a peak at 5J. Similarly, if we flip all 4 spins around a square plaquette, we break 8 bonds, and we expect a peak at 4J. Exact diagonalization studies give a peak at 5.48 J and 4.47 J of appreciable intensity beyond the strongest peak at 2.976 J.

Singh et al.[86] have put forward an interesting explanation of the observed scattering in A_{1g} and B_{2g} symmetry by invoking an appreciable diagonal next nearest neighbor coupling in the Raman Hamiltonian as mentioned earlier. That this is true has also been verified in the finite size studies by Scharf and Chakravarty.[94] Clearly, more theoretical work is necessary to make any definitive statement. Also, more experimental work is necessary to explicate fully the role of resonant enhancement in La_2CuO_4.

This still leaves us with the responsibility to explain why the traditional spin wave theory agrees so well with experiments in K_2NiF_4? Singh[87] has recently examined this S=1 case in detail. His moment calculation similar to the ones outlined above shows that $M_1 = 7.22\pm0.02$ J. The spin wave calculation yields instead $M_1 = 6.63$ J. However, if one uses the (1/S) corrected Z_c, M_1 would be 1.08×6.63J, i.e., 7.16 J. How about M_2/M_1? The spin wave theory gives 0.09 whereas the calculation by Singh gives 0.12±0.03. Keep also in mind that the spin wave theory now has a cutoff at 8J but the exact calculation by Singh does not have such a cutoff, and neither does the experimental spectrum[82], as far as I can see. However, the experiments in K_2NiF_4 have a line shape which is in excellent agreement with the spin wave theory. The experimental second cumulant, M_2, obtained by *truncating* the spectrum at 8J produces the ratio M_2/M_1 to be ~0.1, may be a little smaller.[95] This is understandable if the scattering beyond 8J is weak enough. Thus, although the first moment may not change appreciably, the second cumulant obtained by truncating at 8J can be consistently lower than Singh's exact result. As pointed out by Singh,[87] since the width of the spectrum is about a tenth of its peak position, even 1% integrated intensity around twice the peak position can easily account for a 30% difference. Such a weak scattering, spread over a large energy range, may be difficult to determine experimentally, although more refined experiments should shed further light on this question. However, it is amusing to note that the value of J determined previously in light scattering experiments[82] is still unchanged.

What about scattering in A_{1g} and B_{2g} symmetries? Probably because there are no optically active states near the incident laser energy in K_2NiF_4, one need not invoke a substantial diagonal-next-nearest-neighbor Raman term. In the absence of resonant enhancement any possible diagonal-next-nearest-neighbor term would be extremely small compared to the nearest neighbor term. This would then imply that there could, in principle, be appreciable scattering in A_{1g} and B_{2g} symmetries in La_2CuO_4 but not in K_2NiF_4. In fact, if one looks carefully at the experimental data presented in Ref. 86, one finds a systematic variation with respect to the incident laser energy which may be significant.

The story, as it stands, does hold together quite well. What is missing is a real understanding of the *line shape*. First 3 moments of a spectrum rarely define a line shape completely. In particular, it appears to me that the B_{1g} spectrum in La_2CuO_4 could be fitted well by a two peaked distribution, signifying a 2-magnon and a 4-magnon peak. In fact, we must ask: Is it possible to use Raman scattering to obtain information about the exciting possibilities regarding the short wavelength fluctuations in La_2CuO_4 beyond what we expect from spin wave theory? Could the elementary excitations be fermionic at short wavelengths but bosonic at long wavelengths? And, is there something special about S=1/2? It seems to me that a proper understanding of the short wavelength magnetic properties of undoped La_2CuO_4 may be extremely valuable in understanding the superconducting properties of the doped material if magnetism has anything to do with high temperature superconductivity.

X. ELECTRON AND NUCLEAR MAGNETIC RELAXATION RATES

In this section I shall develop a theory of electron and nuclear magnetic relaxation phenomena in order to explore more fully the low frequency spin fluctuations in La_2CuO_4. The discussion follows closely the recent work by Chakravarty and Orbach.[96] I shall show that the EPR signal in La_2CuO_4 is severely broadened by the antisymmetric Dzyaloshinski-Moriya interaction, making it unlikely that EPR can be observed at room temperature, but may in fact be observable at higher temperatures. The EPR linewidth given here disagees seriously with that of Mehran and Anderson[97] by a factor of $(\xi/a)^2$. It will be shown that in spite of critical fluctuations the NMR spin-lattice relaxation time T_1 should be long enough to measure at room temperature and above, over a substantial range of temperature. Thus, given the validity of this theory, NMR should provide an accurate determination of the correlation length ξ.

The nuclear spin-lattice relaxation rate for the Cu-nuclei, $1/T_1$, when the external magnetic field is in the z-direction is given, quite generally, by,

$$\frac{1}{T_1} = \frac{A^2}{2\hbar^2} \int_{-\infty}^{\infty} dt \, \cos\omega_N t \, <\{S_i(t), S_i(0)\}> , \qquad (96)$$

where A is the hyperfine coupling constant and ω_N the resonance frequency. Here the expectation value is taken with respect to the thermal ensemble, and the electronic spin system is assumed to be in the disordered phase, i.e., $<S^\alpha>=0$. Also note that $\{a,b\} = (ab+ba)/2$.

In the disordered phase one can also rewrite the above formula in the form,

$$\frac{1}{T_1} = \frac{A^2}{3\hbar^2 N} \sum_\mathbf{q} S(\mathbf{q}, \omega=0) , \qquad (97)$$

where, as usual, $S(q,\omega)$ is the dynamic structure factor and N the total number of spins in the two dimensional plane. We have assumed that the resonance frequency, ω_N, is negligibly small compared to all important dynamical frequency scales in the problem and can therefore be neglected.

If the fluctuations of the electronic spin system are in the critical region, the dominant contribution to the sum will arise from the region of q close to the incipient antiferromagnetic wave vector, $(\pi/a, \pi/a)$.[98] Thus the $S(q,\omega)$ proposed by CHN and THC can be applied over the entire Brillouin zone with impunity and we obtain,

$$\frac{1}{T_1} = \frac{1}{T_{1\infty}} \left(\frac{\xi}{a}\right) \left(\frac{0.8}{Z_c}\right) \left(\frac{T}{2\pi\rho_s}\right)^{3/2} \left(\frac{1}{1+T/2\pi\rho_s}\right)^2, \qquad (98)$$

where $T_{1\infty}$ is defined to be,

$$\frac{1}{T_{1\infty}} = \frac{A^2}{\hbar^2} \left(\frac{\sqrt{\pi}}{4}\frac{\hbar}{J}\right). \qquad (99)$$

A word of caution is necessary here. In 2d there is also an important contribution to the sum (Eq. (97)) from wave vectors q close to 0. This is the diffusive contribution given by,

$$\left(\frac{1}{T_1}\right)_{\text{diffusive}} = \frac{A^2}{3\hbar^2 N} \sum_{q\approx 0} \frac{T\chi_{q=0}}{Dq^2}. \qquad (100)$$

The sum as it stands is divergent. However, the divergence is only logarithmic and would be effectively cut off by any small anisotropies or interplanar coupling present in the system. This, by itself, does not make this contribution negligible. However, as shown by CHN, the diffusion constant D diverges as $(T/\chi_\perp)^{1/2}\xi$ as T→0. Thus the relative importance of the diffusive contribution compared to the $(\pi/a,\pi/a)$ contribution is down by a factor of $(1/\xi)^2$ and is negligibly small in the limit $\xi \to \infty$.

It is important to state the regime of validity of the formula for T_1 (Eq. (98)) in the scaling limit. An estimate can be obtained by examining the range of validity of the formula for the correlation length obtained by CHN. As pointed out by THC, for S=1/2, the expression for the correlation length should be valid for $2\pi\rho_s/T>2$. Using the T=0 spin wave expression for the spin stiffness constant, this translates to 0.94J/T>2, i.e., T<J/2. For S=1, a similar analysis leads to the estimate $2\pi\rho_s/T>3$, or 5.3J/T>3; for S=5/2, we have instead $2\pi\rho_s/T>4$, or 37.2J/T>4.

T_1 as predicted by Eq. (98) exceeds $T_{1\infty}$ at about 400K. This, in principle, could mean a break down of the formula because $S(q,\omega=0)$ may not be dominated by the region close to the antiferromagnetic wave vector. However, at this temperature $(\xi/a)\approx 17$ and we do not expect this to happen. It is therefore reasonable to assert that the behavior of T_1 is non-monotonic. This assertion follows from the consideration of the high temperature series expansion result obtained by Moriya[99] many years ago. When applied to S=1/2, and to square lattice, the high temperature expansion yields,

$$T_1 = T_{1\infty}\sqrt{1+J/4T}\,\exp((J/2T)^2(1+J/4T))\,. \tag{101}$$

Note that this formula implies that T_1 must rise above $T_{1\infty}$ as the temperature decreases. Because in the scaling limit T_1 decreases as the temperature decreases, there must be a maximum. A guessed interpolation between this high temperature formula and the low temperature scaling formula has been given by Chakravarty and Orbach.[96]

However, it is important to note the limitations of this high temperature formula for T_1. In deriving this formula it was assumed that the local field spectra obeys a Gaussian distribution at sufficiently high temperatures. In view of what was said about the diffusive pole, or equivalently the long time tail, this cannot be quite correct. However, an extensive high temperature series expansion far beyond Moriya's work is in progress.[100] In this work high temperature series expansion for the short time behavior is combined with an analysis of the long-time tail. Initial analysis suggests that the qualitative behavior implied above will remain unchanged for temperatures T>J/2, while the results in the scaling limit are not affected at all. Thus, T_1 is indeed non-monotonic and that the crossover takes place around J/2.

An estimate of T_1 for a Cu^{2+} nucleus in La_2CuO_4 can be obtained setting A to be 220 kG/μ_B,[101] and taking Cu moment to be $0.6\mu_B$. Thus one finds that $T_{1\infty}$ is 440μs, using J to be approximately 1300K (We have used[29] ρ_s=0.18J, and[68] $2\pi\rho_s \approx$1500K). We should also note that Eq. (98) holds as long as exchange narrowing is valid, implying the condition $A/\hbar \geq 1/T_1$. Using the same parameters one finds a temperature well below the three dimensional ordering temperature, T_N, in this material.

Following Kawasaki[102] and Huber[103], a similar calculation for the singular contribution to the EPR linewidth was made by Chakravarty and Orbach.[96] This singular contribution at low temperatures arises due to critical fluctuations present in the 2d Heisenberg spin system. In contrast to the NMR relaxation rate which involves a spatially local correlation function, the EPR response involves excitations of arbitrary long wavelengths. Thus, interactions such as dipolar, pseudo-dipolar, or antisymmetric Dzyaloshinski-Moriya interaction which do not commute with the total magnetization can contribute significantly even if they are small compared to the isotropic exchange J.[104]

Unfortunately, the expression for the EPR linewidth involves a four-spin correlation function. Thus, one cannot directly take over the results of CHN and THC. While dynamic scaling can be used to determine the form of this four-spin correlation function, the actual function is not known. However, if we use the random phase approximation to express the wave vector dependent four-spin relaxation function in terms of products of two-spin relaxation functions, then we can once again use the dynamic structure factor proposed by CHN and THC in the scaling regime. As noted by Halperin and Hohenberg[105] and Kawasaki,[102] the random phase approximation probably overestimates the effects of critical fluctuations. Nevertheless, this factorization allows us to use the results of THC to calculate the four-spin relaxation function without further approximation.

Denoting the singular part of the EPR linewidth by $\Gamma_\perp^2$, we find that

$$\Gamma_\perp^s = \left(\frac{0.94}{Z_c^3 Z_\chi^2}\right)\left(\frac{A_K a^4}{\hbar^2}\right)\left(\frac{\hbar}{J}\right)\left(\frac{\xi}{a}\right)^3 \frac{(T/2\pi\rho_s)^{5/2}}{(1+T/2\pi\rho_s)^4}, \qquad (102)$$

where K is the antiferromagnetic wave vector $(\pi/a, \pi/a)$. A_K is a complicated form factor which sets the scale.[96] Note that both the secular and nonsecular terms contribute to broadening as long as the electronic precession frequency is small compared to the dynamic scaling frequency, $\bar{\omega}_0$, discussed earlier. In this case, as pointed out by Huber,[103] the resulting linewidth is given by the angular average of the zero field rate in the plane perpendicular to the applied magnetic field.

The important point to note is that the largest anisotropy in La_2CuO_4 is the Dzyaloshinski-Moriya interaction, which has been determined[75] to be 0.55meV; a more recent and precise measurement gives 0.7 ± 0.1meV.[77] This anisotropy, as was pointed out earlier, is more than two orders of magnitude larger than the pseudo-dipolar (exchange anisotropy) or dipolar anisotropy. Using the value 0.55meV, Chakravarty and Orbach[96] estimated ω_p^2/ω_{ex} to be approximately 920G, where $\omega_p^2 = A_K a^4/\hbar^2$ and $\omega_{ex} = J/\hbar$. These values mean that the above equation for the linewidth is valid only above 380K where $\Gamma_\perp^s = 190$kG (by implication, $\Gamma_\perp^s$, is of this value for lower temperatures). This is because Eq. (102) is valid only if $\Gamma_\perp^s < (A_K a^4/\hbar^2)^{1/2}$. Use of Eq. (102) leads to values of $\Gamma_\perp^s$ of 100kG at 400K, 34kG at 450K, and 13kG at 500K. This enormous value for the room temperature EPR linewidth, and its decrease to more reasonable values only at very high temperatures, may explain its lack of ovservability.[97]

In the scaling limit the contribution from the diffusive pole is even less significant in the case of the EPR linewidth. It is not difficult to see that the diffusion contribution is down by factor of $(a/\xi)^4$ compared to the singular contribution in this case, and is entirely negligible in the limit $\xi \to \infty$. Of course at sufficiently high temperatures it can not be neglected.

An analysis of the nuclear relaxation rate similar to the one given above, but only qualitative, was also given by Shastry.[106] As was pointed by Shastry himself, there is no particularly compelling reason to believe that the dynamics of the spherical model used by him is appropriate to the problem under consideration, except for the qualitative purpose of exhibiting the effect of critical slowing down at low temperatures. Shastry does not, however, discuss the high temperature behavior.

Recently, Bulut et al.[107] have also calculated $1/T_1$ using the dynamic structure factor of Arovas and Auerbach.[53] They find that $1/T_1 \propto T^2\xi$. The factor T^2 is surely incorrect, as it arises from the incorrect temperature dependences of the prefactors for the correlation length and the susceptibility, and the incorrect temperature dependence of the scaling frequency obtained by Arovas and Auerbach (cf. the discussions in Secs. VI and VII). Moreover, they predict a contribution similar to the diffusive contribution discussed above, which is proportional to $\ln^3(J/\omega_0)/\xi$, where ω_0, in their notation, is a cutoff frequency. This is in disagreement with the behavior predicted according to the diffusive pole in the spectral function. Because spin diffusion follows from rather general considerations involving conservation laws, I am tempted to conclude that this result is incorrect. As mentioned earlier, the prefactor of the asymptotic behavior (the long distance behavior) of the spin-spin correlation function calculated by Arovas and Auerbach does not appear to be correct and this may well be the source of the problem.

XI. CONCLUDING REMARKS

Although magnetic properties of La_2CuO_4 are largely well understood, there is considerable room for further work. More work seems to be necessary to understand fully the spin dynamics at finite temperatures. In this respect measurements of the NMR relaxation rate, $1/T_1$, may be helpful. It would also be extremely valuable to develop a theory of the *line shape* for the spin-pair excitations seen in Raman scatttering to see if there are any surprises here.

As is well known, it is difficult to write a review article in a field which is developing so rapidly, especially under a tight deadline. In particular, I apologize to many authors whose work was not discussed here.

XII. ACKNOWLEDGEMENTS

First and foremost, I would like to thank B. I. Halperin, D. R. Nelson, and R. Orbach for many interesting discussions and collaborations. The origin of my interest in the subject is an inspiring talk given by R. J. Birgeneau at Harvard in the fall of 1987. I would also like to thank him for many valuable discussions on the subject. I am deeply indebted to W.W. Warren, Jr., and M. Takigawa for teaching me much about NMR relaxation rates. Special thanks are also due to P. A. Fleury for getting me interested in the problem of light scattering, and Rajiv Singh for teaching me much about his work in this area. I also owe much to my students P. Scharf and P. Kopietz for collaborations and for keeping me on my toes. Last but not least, I am grateful to M. Gelfand for a careful reading of the manuscript, and for many suggestions which have helped me to improve the manuscript.

This work was supported by grants from the National Science Foundation: DMR-86-01908 and DMR 89-07664.

Note added:
Recently, in what appears to be an extensive and impressive piece of work, Ding and Makivić[108] have calculated correlation functions of the $S=1/2$ two-dimensional Heisenberg antiferromagnet at low temperatures via a large scale quantum Monte Carlo simulation on lattices upto 128×128. The correlation length was found to be

well described by the form derived by CHN. In their simulations they have been able to reach tempeartures as low as J/4, considerably lower than the previous work mentioned in the text. From their fit to the neutron scattering data up to temperatures of the order of J/4, they conclude that J=1450±30 K. They find that the correlation length can be fitted well by the formula,

$$\xi = 0.276a \exp(1.25J/T).$$

They also find that if the prefactor is assumed to be of the form T^x, then x must be less than 0.03, i.e., to an excellent approximation the prefactor is independent of temperature. They have also calculated the asymptotic form of the staggered spin-spin correlation function, C(r), given by,

$$C(r) = A \left(\frac{e^{-r/\xi}}{r^\lambda} + \frac{e^{-(L-r)/\xi}}{(L-r)^\lambda} \right),$$

where L is the linear dimension of the system. They find that at higher temperatures λ is given quite closely by the Ornstein-Zernike value of 1/2. At lower temperatures λ crosses over to a value very close to 0.4. By contarst, Schwinger boson mean field theories[52,53] and the theory due to Takahashi[54] give λ=1. The result of CHN, as described above, is close to the Ornstein-Zernike result, except for logarithmic corrections (see Eq. (35)), and is consistent with the above result.

XIII. REFERENCES

1. R. J. Birgeneau and G. Shirane, in, "*Physical Properties of High Temperature Superconductors*," edited by D. M. Ginsberg (World Scientific, Singapore, 1989).
2. R. J. Birgeneau, Am. J. of Phys., to be published.
3. P. W. Anderson, Phys. Rev. **115**, 2 (1959).
4. P. W. Anderson, Science **235**, 1196 (1987).
5. N. F. Mott, "*Metal Insulator Transitions*" (Taylor and Francis, London, 1974).
6. W. Kohn, Phys. Rev. **133A**, 171 (1964); also in "*Problème à N Corps (Many Body Theory)*," edited by C. DeWitt and R. Balian (Gordon and Breach, New York, 1968).
7. See, for instance, G. Baskaran and P. W. Anderson, Phys. Rev. **B37**, 580 (1988) and L. B. Ioffe and A. I. Larkin, Phys. Rev. **B39**, 8988 (1989).
8. R. J. Birgeneau, J. Skalyo, Jr., and G. Shirane, Phys. Rev. **B3**, 1736 (1971), and references therein.
9. T. Nagamiya, K. Yosida, and R. Kubo, in "*Advances in Physics*," edited by N. F. Mott (Taylor and Francis, London, 1955) Vol. 4, p. 1.
10. I. Dzyaloshinski, J. Phys. Chem. Solids **4**, 241 (1958). T. Moriya, Phys. Rev. **20**, 91 (1960).

11. See Ref. 2, for example.
12. N. W. Ashcroft and N. D. Mermin, *"Solid State Physics"* (Saunders College, Philadelphia, 1976).
13. M. P. Gelfand, R. R. P. Singh, and D. A. Huse, Phys. Rev. B, in press; E. Dagotto and A. Moreo, Phys. Rev. Lett. 63, 2148 (1989) and references therein.
14. H. Bethe, Z. Physik **71**, 205 (1931).
15. F. J. Dyson, E. H. Lieb, and B. Simon, J. Stat. Phys. 18, 335 (1978); E. J. Neves and J. F. Peres, Phys. Lett. **114A**, 331 (1986); I. Affleck et al., Comm. Math. Phys. 115, 477 (1988) (see in particualr the appendix); T. Kennedy, E. H. Lieb, and B. S. Shastry, J. Stat. Phys. **53**, 1019 (1988).
16. T. Holstein and H. Primakoff, Phys. Rev. **58**, 1098 (1940).
17. F. J. Dyson, Phys. Rev. **102**, 1217 (1956); 102, 1230 (1956); S. V. Maleev, Zh. Eksp. Teor. Fiz. **30**, 1010 (1957) [Sov. Phys., JETP **64**, 654 (1958)].
18. J. Schwinger, U. S. Atomic Energy Commission Rpt., NYO-3071 (1952), reprinted in L. Biedenhern and H. Van Dam (eds.), "Quantum Theory of Angular Momentum" (Academic, New York, 1965). See also Refs. 52 and 53.
19. R. Peierls, H. Phys. Acta **7**, Suppl. 2, 81 (1934); Ann. Inst. H. Poincaré **5**, 177 (1935).
20. P. C. Hohenberg, Phys. Rev. **158**, 383 (1967); N. D. Mermin and H. Wagner, Phys. Rev. Lett. **17**, 1133 (1966).
21. A theorem by S. Coleman, Comm. Math. Phys. 31, 259 (1973) is often quoted in this context. However, one should bear in mind that the theorem only applies to Lorentz-invariant field theories. To the extent that the one dimensional antiferromagnetic Heisenberg model can be replaced by a Lorentz invariant (1+1)-dimensional field theory at T=0, i.e., by a (1+1)-dimensional non-linear σ-model, the theorem should apply. But to the best of my knowledge a rigorous theorem for the lattice model is not known.
22. T. Oguchi, Phys. Rev. **117**, 117 (1960).
23. P. W. Anderson, Phys. Rev. **86**, 694 (1952).
24. P. W. Anderson, Mater. Res. Bull. **8**, 153 (1973); P. Fazekas and P. W. Anderson, Philos. Mag. **30**, 432 (1974).
25. B. I. Halperin and P. C. Hohenberg, Phys. Rev. **177**, 952 (1969).
26. R. Kubo, Rev. Mod. Phys. **25**, 344 (1953); S. J. Miyake, Prog. Theo. Phys. **74**, 468 (1985).
27. J. D. Reger and A. P. Young, Phys. Rev. B37, 5978 (1988); M. Gross, E. Sanchez-Velasco, and E. D. Siggia, Phys. Rev. B39, 2484 (1989).
28. J. D. Reger, J. A. Riera, and A. P. Young, J. Phys. Cond. Matter 1, 1855 (1989).
29. R. R. P. Singh, Phys. Rev. B39, 9760 (1989); R. R. P Singh and D. A. Huse, Phys. Rev. B40, 7247 (1989).
30. See, for instance, D. R. Nelson and R. A. Pelcovits, Phys. Rev. B 16, 2191 (1977), and D. J. Amit, in *"Field Theory, the Renormalization Group, and Critical Phenomenon,"* 2nd ed. (World Scientific, Singapore, 1984).
31. F. D. M. Haldane, Phys. Rev. Lett. **50**, 1153 (1983); Phys. Lett. **93A**, 464 (1983).

32. S. Chakravarty, B. I. Halperin, and D. R. Nelson, Phys. Rev. B39, 7443 (1988).
33. S. Weinberg, Phys. Rev. Lett. 17, 616 (1966); C. G. Callan, S. Coleman, J. Wess, and B. Zumino, Phys. Rev. 177, 2247 (1969).
34. B. Rosenstein and B. J. Warr, preprint.
35. F. D. M. Haldane, Phys. Rev. Lett. 61, 1029 (1988).
36. See Refs. 50, 51.
37. K. G. Wilson, Rev. Mod. Phys. 47, 773 (1975).
38. A. M. Polyakov, Phys. Lett. B59, 79 (1975).
39. W. Bernreuther and F. J. Wegner, Phys. Rev. Lett. 57, 1383 (1986).
40. E. Brèzin and J. Zinn-Justin, Phys. Rev. B14, 3110 (1976).
41. M. Creutz, "*Quarks, Gluons, and Lattices*" (Cambridge Univ. Press, Cambridge, 1983).
42. D. S. Fisher and D. R. Nelson, Phys. Rev. B16, 2300 (1977).
43. S. H. Shenker and J. Tobochnik, Phys. Rev. B22, 4462 (1980).
44. J. Kogut, Rev. Mod. Phys. 55, 775 (1983).
45. G. Parisi, Phys. Lett. 92B, 133 (1980).
46. S. Chakravarty, B. I. Halperin, and D. R. Nelson, Phys. Rev. Lett. 60, 1057 (1988).
47. P. Kopietz and S. Chakravarty, Phys. Rev. B40, 4858 (1989).
48. M. Fukugita and Y. Oyanagi, Phys. Lett. 123B, 71 (1983); B. Berg *et al.* Phys. Lett. 126B, 467 (1983); B. Berg *et al.*, Nucl. Phys. B239, 149 (1984); M. Falconi *et al.*, Nucl. Phys. B225, 313 (1983); A. Hasenfratz and A. Margaritis, Phys. Lett. 133B, 211 (1983); S. Itoh *et al.*, Nucl. Phys. B250, 312 (1985).
49. S. Tyč, B. I. Halperin, and S. Chakravarty, Phys. Rev. Lett. 62, 835 (1989).
50. G. Gomez-Santos, J. D. Joannopoulos, and J. W. Negele, Phys. Rev. B39, 4435 (1989).
51. E. Manousakis and R. Salvador, Phys. Rev. Lett. 60, 840 (1988); Phys. Rev. B39, 575 (1989).
52. D. P. Arovas and A. Auerbach, Phys. Rev. B38, 316 (1988).
53. A. Auerbach and D. P. Arovas, Phys. Rev. Lett. 61, 617 (1988).
54. M. Takahashi, Phys. Rev. B40, 2494 (1989).
55. A. A. Belavin and A. M. Polyakov, JETP Lett. 22, 245 (1975).
56. S. Tyč and B. I. Halperin, preprint.
57. P. Kopietz, preprint.
58. B. I. Halperin and P. C. Hohenberg, Phys. Rev. 188, 898 (1969).
59. B. I. Halperin and P. C. Hohenberg, Phys. Rev. 177, 952 (1969).
60. This follows directly from the scaling flows of the (3+1)-dimensional non-linear σ-model at temperatures below the Néel temperature.
61. A. B. Harris, D. Kumar, B. I. Halperin, and P. C. Hohenberg, Phys. Rev. B3, 961 (1971).
62. D. R. Grempel, Phys. Rev. Lett. 61, 1041 (1988).
63. T. Becher and G. Reiter, Phys. Rev. Lett. 63, 1004 (1989). [Note added: The claims made in this paper have now been withdrawn. See T. Becher and G. Reiter, Phys. Rev. Lett. 64, 109(E) (1990)]

64. G. Reiter, in *"Proceedings of Nato Advanced Research Workshop on interacting Electrons in Reduced Dimensions,"* Torino, October 1988, edited by D. Campbell, to be published in Plenum Press.
65. Y. A. Kosevich and A. V. Chubukov, Zh. Eksp. Teor. Fiz. **91**, 1105 (1986)[Sov. Phys. JETP **64**, 654 (1986)].
66. G. Shirane *et al.*, Phys. Rev. Lett. **59**, 1613 (1987).
67. Y. Endoh *et al.*, Phys. Rev. **B37**, 7443 (1988).
68. K. Yamada *et al.*, Phys. Rev. **B40**, 4557 (1989).
69. D. Vaknin et al., Phys. Rev. Lett. **58**, 2802 (1987).
70. K. Yamada *et al.*, Solid State Commun. **64**, 753 (1987).
71. B. I. Halperin and C. M. Varma, Phys. Rev. **B14**, 4030 (1976).
72. G. Aeppli *et al.*, Phys. Rev. Lett. **62**, 2052 (1989).
73. R. J. Birgeneau, preprint.
74. T. Moriya, Phys. Rev. **120**, 91 (1960).
75. T. Thio et al., Phys. Rev. **B38**, 905 (1988).
76. G. J. Bowden *et al.*, J. Phys. **C20**, L545 (1987).
77. T. Thio *et al.*, Phys. Rev. B, in presss.
78. C. J. Peters et al., Phys. Rev. **B37**, 9761 (1988); R. T. Collins *et al.*, Phys. Rev. **B37**, 5817 (1988).
79. M. A. Kastner *et al.*, Phys. Rev. **B38**, 6636 (1988).
80. K. B. Lyons *et al.*, Phys. Rev. **B37**, 2353 (1988).
81. K. B. Lyons *et al.*, Phys. Rev. Lett. **60**, 732 (1988).
82. P. A. Fleury and H. J. Guggenheim, Phys. Rev. Lett. **24**, 1346 (1970).
83. K. B. Lyons *et al.*, Phys. Rev. **B39**, 9693 (1989).
84. J. B. Parkinson, J. Phys. **C2**, 2012 (1969).
85. See Refs. 80 and 81.
86. R. R. P. Singh *et al.*, Phys. Rev. Lett. **62**, 2736 (1989).
87. R. R. P. Singh, preprint.
88. P. A. Fleury and R. Loudon, Phys. Rev. **166**, 514 (1968).
89. R. J. Elliot and M. F. Thorpe, J. Phys. **C2**, 1630 (1969).
90. J. Sólyom, Z. Physik **243**, 382 (1971).
91. W. H. Weber, Phys. Rev. B and G. W. Ford **40**, 6890 (1989).
92. S. Coppersmith, private Communication.
93. T. Hsu, Phd. thesis, unpublished, Princeton University (1989).
94. P. Scharf and S. Chakravarty, unpublished.
95. R. R. P. Singh, private communication.
96. S. Chakravarty and R. Orbach, Phys. Rev. Lett. **64**, 224 (1990).
97. F. Mehran and P. W. Anderson, Solid State Commun. **71**, 29 (1989).
98. T. Moriya, Prog. Theo. Phys. **28**, 371 (1962).
99. T. Moriya, Prog. Theo. Phys. **16**, 23 (1956).
100. R. R. P. Singh, S. Chakravarty, M. Gelfand, A. Jaggannathan, and R. Orbach, unpublished.
101. F. Mila and T. M. Rice, Physica **157C**, 561 (1989).

102. K. Kawasaki, Prog. Theo. Phys. **39**, 285 (1968); also in "Phase Transitions and Critical Phenomenoa," Vol. 5A, edited by C. Domb and M. S. Green (Academic, New York, 1976).
103. D. Huber, Phys. Rev. **B6**, 3180 (1972).
104. Z. Soos *et al.*, Phys. Rev. **B16**, 3036 (1977).
105. B. I. Halperin and P. C. Hohenberg, Rev. Mod. Phys. **49**, 436 (1977).
106. B. S. Shastry, Phys. Rev. Lett. **63**, 1288 (1989).
107. N. Bulut, D. Hone, D. J. Scalapino, and N. E. Bickers, preprint.
108. H.-Q. Ding and M. S. Makivić, preprint.

DISCUSSION

P. W. Anderson: I would like to say a little bit on Ted Hsu's work. In the first place you made it sound as though it had been done by Hsu and myself, but that was very primitive, and Ted actually showed that we were quite wrong, and that you can do it much better.

Hsu shows in the thesis that if you start from spinons in the flux phase, there is an instability, which in fact our field theorists friends tell us is a theorem, that the spinon Hamiltonian for the Heisenberg model is unstable to formation of Néel ordering. And he follows out the Néel ordering process and shows that in fact the Néel ordering, is about the right order of magnitude – in fact a very accurate value for the mean sub-lattice magnetization – follows by starting from spinons and forming spin-waves as collective modes of the spinon Hamiltonian. And so he does exactly the structure that you talked about, taking an absolutely continuous interpolation between spinons and spin waves. And it's quite clear that what you said is right, that the spin waves are completely correct at long wavelengths, but that there's very little perturbation on the spinon Hamiltonian at short wavelengths, and he has some pretty good fits (they're still not as happy as they would have liked, but I think perhaps he's missing the polarization questions, i.e. they're not perfect fits) to the Raman spectra.

P. Lee: Phil, this is a spinon with no Fermi surface?

P. W. Anderson: Right, with no Fermi surface; point-like Flux phase. It seems to give a good description.

S. Chakravarty: Thank you Phil. I did talk to him on the phone but I haven't seen his thesis yet. I'm beginning to believe that the short wavelength excitations in the Heisenberg model spin-1/2 can be quite different. And there is a way of checking it. I think that Raman is a pretty good way of checking it, if you do a good job in the line shift analysis, in my opinion.

E. Abrahams: I have a question about Bob Schrieffer's first viewgraph, which anyone can answer. If you take this beautiful agreement that you have of spin wave theory with the experimental neutron scattering spectra, then you might be led to conclude, since the theory is basically the Heisenberg model, that reality is to the strong. That is, the Heisenberg model is essentially in the strong-coupling sector of the Hubbard model. So my question is, is there any hope that from a spin-density wave ground state that one could in any way get similar agreement?

S. Chakravarty: Again, I think I know what Bob is going to say. Bob actually has done a calculation, starting from the weak coupling limit, in collaboration with Zhang and Wen, and in particular they have looked at the sub-lattice magnetization which agrees quite well with the strong coupling results; that's the only thing I think, as far as I know...

J. R. Schrieffer: Also the spin wave spectrum...

S. Chakravarty: Spin wave stiffness...

J. R. Schrieffer: We obtain the known weak and strong coupling mean field results and a smooth interpolation between them.

S. Chakravarty: OK. Those things, you get rather well.

J. R. Schrieffer: We get the exact results in both limits, according to Phil's 1972 paper, that is, the spin wave approximation to the Heisenberg limit.

S. Chakravarty: But let's be careful about it. You get $1/s$ correction as well? Is that correct? Or you get the leading term...

J. R. Schrieffer: I'll have to check.

S. Chakravarty: Yeah, because Phil's expression is S minus $.197$ in two dimensions. That's his old result, which is still the result for spin wave theory as far as I know, because the next term is zero, and the following term has not been calculated. So it's still his '53 results...Now do you get that? You must, because you get the same sub-lattice reduction...

J. R. Schrieffer: I think that's right.

S. Chakravarty: OK. The answer is that small U calculation extrapolated to large U checks quite well for some properties after you do resummation of diagrams, random-phase type. You seem to need very little vertex corrections.

J. R. Schrieffer: Can I comment on that. I completely agree with what you say. The surprise to me was that we did get the large-U limit correctly, in that sense, of the spin wave limit to both the sub-lattice magnetization and the spin wave velocity. It's not all that surprising, however, in the sense that in the large-U limit, you have in one description, the mean-field gap, the other you have the Mott-Hubbard gap, and you're looking at quadratic fluctuations, Gaussian fluctuations, about, in essence, the same state. And in fact you can make a two-line proof that these have to be the same. The RPA is different from what one might think because of, as you say, the vertex corrections. The basis states are those in the staggered order parameter, and these change the matrix elements in the RPA. In fact, that's critical to the result.

S. Chakravarty: Maybe I'll make one more comment. Yeah I think Bob you're right, but you have to understand that when you calculate sub-lattice magnetization, you're summing over all q's. And quite a lot of this reduction is coming from the long q behavior. And as I mentioned before, the long q behavior is taken care of very well by the Random Phase Approximation, but I'm not entirely convinced, though, that, for example, if you go ahead and compute Raman intensity, which is dominated by short wavelengths, that you'll come out right. You might...

J. R. Schrieffer: What I was referring to was the spin wave spectrum.

J. R. Schrieffer: I believe we get the right Gaussian fluctuation corrections to the spin wave spectrum for all values of q.

R. Walstedt: Just a question about your T_1 calculations, and how they vary with the spin quantum number in the quadratic layer antiferromagnet. Because in fact there are several isomorphs, K_2NiF_4 as you mentioned as well as others.

S. Chakravarty: Were you the referee?

R. Walstedt: No..No... (laughter). Anyone who has dealt with this subject knows about these things. Back in the early '70's these quadratic layer antiferromagnets became very popular, they also occur with Mn^{2+} and with rubidium in place of potassium. They are the same except for $S = 5/2$. Do you expect a strong contrast between $S = 1/2$ and $S = 5/2$.

S. Chakravarty: Right, yes, indeed the referee had pointed out that we should take a look at K_2NiF_4 and K_2MnF_4 measurements done in the '70's. (If the referee's over here, please identify yourself, I'll wring your neck). We did look at them. The problem is a little bit different, because of the Heisenberg-Ising crossover. But if you look above the Heisenberg-Ising crossover – by the way, the quality of the data is not very good, OK – our results are consistent with the data. We're doing a lot of extensive work now. In the case of K_2NiF_4, the S=1 case, the measurements were not done up to temperatures, where we predict a peak in the T_1, so this showed only the critical slowing down; K_2MnF_4, if I remember it right, they were done up to temperatures a little bit higher than the temperatures at which we predict a peak, and sure enough you see a peak...Now, I don't want to imply that the agreement is quantitative. The height of the peak, if you draw by hand, can be off by factors of two. But the old data is plotted on a logarithmic scale, and if you plot it out carefully on a linear scale it's not unusual, in fact it's quite consistent.

You can see it more clearly that I think it would be interesting to repeat these experiments on K_2NiF_4 and K_2MnF_4.

C. M. Varma: What is the latest number, for the effective magnetic moment and, with what error bars, and also I want to know the best current theoretical value for a Heisenberg model, and whether the experimental number is really with the error bars, at the very upper edge.

S. Chakravarty: John Tranquada can certainly answer this question, but one has to be a little bit careful because the g-factor is really not known because ESR has not been seen in La_2CuO_4. I don't know what the current error bounds are, but I can tell you that the spin wave result agrees spectacularly well with Monte Carlo calculation, and the beautiful series expansion that Rajiv Singh has done. So we know that number from the Heisenberg model quite well.

C. M. Varma: I want to know the best theoretical estimate, where does it lie within the best experimental numbers, including the error bars...

S. Chakravarty: Okay, the best theoretical estimate I would say is .603 for the reduction, and I don't know what the experiment...is there a better experimental estimate now, John?

J. M. Tranquada: Well it should be qualified, in that to get the order parameter from experiment you have to know the form factor accurately, and that hasn't really been treated well enough, I think, in most of the experiments. Assuming a Cu^{2+} form factor as was done by Vaknin *et al.*, values of 0.6 μ_B for the order parameter have been seen in lanthanum copper oxide single crystals, and 0.65 μ_B in yttrium barium copper oxide single crystals, but you have to allow for significant uncertainties in extracting those values from the experiment.

C. M. Varma: I guess I wanted to know what your best value was, what's your estimate...

S. Chakravarty: Estimate for the uncertainty. He wants a number. It's .6 plus or minus...

C. M. Varma: ...value lie in the upper edge, the middle, the lower edge, where...

J. M. Tranquada: I think the experimental and theoretical values are consistent within experimental uncertainty, but I would put the uncertainty at 10 or 15 per cent, and possibly bigger. It depends. There are some people here who think there should be big hybridization effects in the form factor, which would change the order parameter that's pulled out

P. W. Anderson: Is that correcting for hybridization, or without the correction?

J. M. Tranquada: No, the values of the order parameter I quoted were obtained with little consideration of hybridization. You can't determine the order parameter without knowledge of the form factor.

C. M. Varma: There is an assumption about the form factor when you make a Heisenberg model. That's basically one of the things to check.

D. Pines: There is an indirect argument about the size of μ_{eff}, based on a comparison of the results for the antiferromagnetic resonance frequency as measured in an NQR experiment, and the size of the hyperfine quantities, the direct coupling A and the transfer hyperfine coupling constant B. If one believes, as seems to be the case, that those quantities do not change in going from the insulator to the metal, one winds up with a μ_{eff} which is very close to two-thirds. So it's not direct, but it is within the experimental accuracy, and within the current theoretical estimates, I believe.

S. Doniach: I have a quick question about the second neighbor coupling. In the two band theory, you have reason to believe there's some finite J_2, since that's frustrating, it could strongly affect the results like this.

S. Chakravarty: I think the best person to ask is Richard M. Martin, because he has done these calculations.

R. M. Martin: Only to give an idea what the J_2 is, but not what all the results are, the consequences are. Although J_2 is very sensitive to one number, that's the poorest number known – the E_p-E_d energy that, we would say J_2 is aroung ten percent of J_1.

S. Chakravarty: It's like ten percent or less of the largest J, but that's VERY uncertain.

E. Stechel: We calculate J' to be 3 meV (or about 2% of J). This is very small, and as far as affecting any of the properties, such as the Raman spectrum, it cannot be seen.

S. Chakravarty: J_2 in Raman that has been invoked, by Singh *et al.* is different. That comes from resonant enhancement. And that strength has nothing to do with the J_2 that Seb is talking about.

E. Stechel: No, calculating the B-2 component with the nearest, and next-nearest neighbor Heisenberg model makes no significant difference.

H. Ott: With respect to the saturation moment, our measurements on the T' phase Praeseodymium-Copper O_4 gives even a smaller saturation moment than this .5, it's even less than that.

S. Chakravarty: Well, you can get a very small saturation moment even in lanthanum copper oxide, you just change the oxygen concentration. The important point is to find out what's the largest one that you can get.

H. Ott: It was a fully oxygenated sample, with the highest T_n we could get...

S. Chakravarty: I believe anything you experimentalists say.

D. Pines: Has the antiferromagnetic resonance been done on the La_2CuO_4, or on the praeseodymium?

M. Takigawa: The experiment has been done in 2,1,4 compounds. The internal field at the Cu site in Nd_2CuO_4 is 104 k Oe, which is significantly larger than those in La_2CuO_4 (79 k Oe) and in $YBa_2Cu_3O_6$ (77 k Oe). But the hyperfine coupling could also change.

E. Stechel: Could you make some comments on the Raman spectra, where in the clusters, as you know, you get peaks around 3J, 4J, 5J, and 6J. Do you believe that this structure should survive in the infinite system?

S. Chakravarty: Ellen, I don't know. We are doing a calculation of exact diagonalization on an 18-site lattice now...

E. Stechel: I already calculated the spectrum for an 18-site cluster.

S. Chakravarty: Oh you have already done the 18?

E. Stechel: The structure persists, that is no new peaks arise.

S. Chakravarty: If I remember it right, the first peak was at 2.976, the next peak was at 4.47, and the third peak was out at 5.3 or 5.4, something like that, I have my numbers somewhere around – that is for a 16-site lattice. Now when you go to 18-site lattice how do the peaks shift?

E. Stechel: They move around a little bit. They shift to slightly higher energies.

S. Chakravarty: That's what we're also getting...

E. Stechel: ...but you still get the same number...

S. Chakravarty: ...number of peaks...

E. Stechel: But if you average clusters, the structure washes out except for a tiny little bump that's in the vicinity of 6J.

S. Chakravarty: Well you have more information right at this point than I do, on this problem.

E. Stechel: Well I thought you had some strong feelings on it, so I was asking for your comment...

Anisotropic Antiferromagnets from the Hubbard Model
Zlatko Tešanović
The Johns Hopkins University and Los Alamos National Laboratory

I discuss here several recent results on the two-dimensional Hubbard model obtained using a Hartree-Fock mean-field theory plus one-loop self-consistent quantum fluctuations. I will concentrate on the strong coupling regime and the half-filling situation but it is the advantage of this approach that it can be used to study the Hubbard model at any filling and for any value of U/t-thus, it could have a considerable bearing on the understanding of high temperature superconductivity.[1]

Since the mean-field theory and the ensuing loop expansion are formally valid only in the weak coupling it is not obvious that the procedure should work for $U/t \gg 1$. However, the mean-field solution, resulting in two bands split by $2\Delta_{SDW} = U$, removes the charge degrees of freedom in a way equivalent to the Schrieffer-Wolff canonical transformation. The new low energy scale J ($= 4t^2/U$), characterizing quantum spin fluctuations, is thus correctly reproduced despite its non-perturbative character. Consequently, at the one-loop level, the effective low energy Hamiltonian is given by the spin-wave description of the quantum Heisenberg antiferromagnet and the physics of the two is equivalent.[2] In particular, the spin-wave dispersion in the strong coupling limit is given by

$$\Omega(\vec{Q}) = 2J\sqrt{1 - \gamma_{\vec{Q}}^2} \qquad (1)$$

where $\gamma_{\vec{Q}} = (1/2)(\cos Q_x a + \cos Q_y a)$. This also implies that the quantum fluctuations reduce the ground state magnetization to 60% of its mean-field value of unity.[1,2] In this approach, however, one can also evaluate the spin-wave dispersion and the magnetization for any value of U/t.[2] Overall, the agreement with the exact and Monte Carlo numerical results on the two-dimensional Hubbard model at half-filling is very good for $U/t > 5$.[3]

We have further considered the effect of a weak coupling between CuO_2 planes on the dynamics of the antiferromagnet. In the La_2CuO_4 compound, the effective coupling between layers is due to the orthorhombic distortion. This distortion renders unequal the two pairs of couplings by which a Cu spin in a plane is coupled to its four out-of-plane nearest neighbors. The effective coupling between planes is thus governed by the larger of the two couplings. We shall represent the coupling between planes by an effective interplanar hopping term, and later discuss how it relates to the structural features of La_2CuO_4.

If r denotes the ratio of an effective interplanar to planar hopping strength, then the free-particle energy dispersion relation is

$$\varepsilon_{\vec{k}}^{\sigma} = -2t[\cos k_x a + \cos k_y a + r \cos k_z c] , \qquad (2)$$

Carrying out the evaluation of the transverse spin susceptibility we find that the energy of the spin-wave mode is given by[4]

$$\Omega(\vec{Q}) = 2J\sqrt{1-\gamma_{\vec{Q}}^2}\,, \tag{3}$$

where J is related to the planar exchange energy, $J_p = 4t^2/U$, by $J = J_p(1+r^2/2)$ and $\gamma_{\vec{Q}} = (\cos Q_x a + \cos Q_y a + r^2 \cos Q_z c)/(2+r^2)$.

We first consider the correction due to the zero-point, spin-wave excitations to the sublattice magnetization at zero temperature.[2] The HF sublattice magnetization is lowered due to these zero-point excitations. The resulting magnetization is monotically increasing as a function of the hopping ratio r and goes from ~ 0.6 in the strictly $2D$ case to ~ 0.85 in the isotropic $3D$ case.

Extending the zero-temperature result for the self-energy correction to finite temperatures, we obtain,

$$-\delta M(T) = \frac{1}{N}\sum_{\vec{Q}}\left(\frac{1}{\sqrt{1-\gamma_{\vec{Q}}^2}}\frac{2}{e^{\beta\Omega_{\vec{Q}}}-1}\right) \tag{4}$$

For $k_B T \ll J$ only spin waves with small [mod ($\pi/a, \pi/a$)] planar wavevector will be excited and thereby give a significant reduction in the sublattice magnetization. Therefore, retaining terms up to quadratic order in $\cos Q_x a$ and $\cos Q_y a$, and denoting $\{(Q_x a)^2 + (Q_y a)^2\}^{1/2}$ by θ_p and $Q_z c$ by θ_z, we obtain,

$$\delta M(T) = 2\int_{-\pi}^{\pi}\frac{d\theta_z}{2\pi}2\int_0^{\infty}\frac{\theta_p d\theta_p}{2\pi}\left(\frac{\theta_p^2}{2}+r^2(1-\cos\theta_z)\right)^{-1/2}\frac{1}{e^{\beta\Omega_{\vec{Q}}}-1} \tag{5}$$

where the spin-wave energy in the quadratic approximation for the cosines of in-plane momenta is $\Omega_{\vec{Q}} \approx 2J[\theta_p^2/2 + r^2(1-\cos\theta_z)]^{1/2}$. Integrating over θ_p yields,

$$-\delta M(T) = \frac{2}{\pi^2}\left(\frac{k_B T}{J}\right)\int_Q^{\pi} d\theta_z \ln(1-e^{-\frac{2Jr}{k_B T}(1-\cos\theta_z)^{1/2}})^{-1} \tag{6}$$

The above equation is applicable outside the critical region around T_N. There are two different temperature regimes. First, for $k_B T \gg 2Jr$, we obtain,

$$-\delta M(T) = \frac{2}{\pi}\left(\frac{k_B T}{J}\right)\ln\left(\frac{k_B T}{2Jr}\sqrt{2}\right)\,. \tag{7}$$

In this regime the spin-waves are quasi two-dimensional. For low temperatures, however, when $k_B T \ll 2Jr$, the significant contribution comes from the spin-wave modes with long-wavelength in the z-direction. In this case we obtain,

$$-\delta M(T) = \frac{\sqrt{2}}{3}\left(\frac{k_B T}{J}\right)\left(\frac{k_B T}{2Jr}\right)\,. \tag{8}$$

$M(T)$ obtained from Eq. (5) is in an excellent agreement with the available data on La$_2$CuO$_4$ if we choose $M(0) = 0.5$, $J = 1600$ K and r = 0.011.

We now discuss how is the effective interlayer hopping term related to the structural characteristics of La$_2$CuO$_4$. The important feature in the La$_2$CuO$_4$ structure is the orthorhombic distortion, because of which the two pairs of exchange terms by which a Cu moment is coupled to its out-of-plane nearest neighbors are not equal. If J_1 and J_2 denote two out-of-plane nn couplings the effective interplanar exchange energy is $2(J_1 - J_2)$. If we express the various exchange energies in terms of the respective hopping strengths ($J = 4t^2/U$), then the effective interplanar hopping, t'_{eff}, is related to the average out-of-plane nearest-neighbor hopping, t', and the difference, $\Delta t'$, in the out-of-plane nearest-neighbor hoppings (due to the orthorhombic distortion) by $(t'_{eff})^2 \approx 4t'\Delta t'$. Dividing by t^2, where t is the planar nearest-neighbor hopping, we obtain:

$$r^2 \equiv \frac{(t'_{eff})^2}{t^2} \approx 4 \left(\frac{t'^2}{t^2}\right)\left(\frac{\Delta t'}{t'}\right) \qquad (9)$$

The ratio of the conductivities $\sigma_\perp/\sigma_\|$ should be $\propto t'^2/t^2$. Therefore, r^2 is seen to be related not only to the anisotropy in the conductivities, but also to the fractional anisotropy in the out-of-plane nearest neighbor hoppings arising from the orthorhombic distortion.

The Sr$_2$CuO$_2$Cl$_2$ compound differs from La$_2$CuO$_4$ in that it stays in the tetragonal phase down to the lowest temperatures. This leads to a frustration between planes. Most likely, magnetic dipole interactions break this frustration and introduce a very weak coupling between layers. The magnetic behavior in Sr$_2$CuO$_2$Cl$_2$ should therefore be expected to be even more two dimensional in nature.

We have obtained J and r for the Sr$_2$CuO$_2$Cl$_2$ system by fitting our results for the measured sublattice magnetization.[5] The theoretical curve fits very well with the data in this case also and best fit yields $J = 800/M(0)$ K and $r/M(0) = 0.004$. Using $M(0) = 0.34$ in Sr$_2$CuO$_2$Cl$_2$[5] we obtain $r = 0.0014$. Thus r^2 is about two orders of magnitude smaller than that in La$_2$CuO$_4$. This lends support to the view that it is a much weaker (than exchange type) interaction (possibly a magnetic dipole interaction) which is responsible for the coupling between layers in the absence of any orthorhombic distortion.

The work reported here was done in collaboration with A. Singh (JHU).

REFERENCES

1. J. R. Schrieffer, X.-G. Wen, and S.-C. Zhang, Phys. Rev. Lett. **60**, 944 (1988); J. R. Schrieffer, X.-G. Wen, and S.-C. Zhang, Phys. Rev. B **39**, 11663 (1989).
2. A. Singh and Z. Tešanović, Phys. Rev. B **41**, 1 January (1990).
3. D. Scalapino, elsewhere in this volume.

4. A. Singh and Z. Tešanović, (to be published).
5. D. Vaknin, S. K. Sinha, C. Stassis, D. C. Johnston and L. L. Miller, (to be published).

DISCUSSION

P. W. Anderson: To some of us, the hopping integral between planes is a very important parameter, and it also appears, from things people were saying, that the sub-lattice magnetization – at least as measured in the perfectly ordered state – seems a bit big. And, I noticed when you had hopping between planes, you got a little bit more sub-lattice magnetization.

Z. Tesanovic: Yes.

P. W. Anderson: This is one of many ways in which perhaps the magnetic measurements could be getting at the amount of interlayer hopping which is a very important parameter.

Z. Tesanovic: That's a very good point. What we tried to do, Phil, is we tried to correlate this parameter r with conductivity, but...

P. W. Anderson: It won't.

Z. Tesanovic: It won't, we had a little bit of a problem with that. But we also have a problem with Hans' data, because his data shows magnetization which is .45, and that's quite a bit below...And I think the most natural thing to assume, in the Hubbard model, is that there's some itinerance.

P. W. Anderson: It think it's quite important in the YBCO series to look separately. This is maybe the one way in which you could actually find the hopping between the two planes, distinct from the long-range hopping...a very important distinction, and a very important number.

T. M. Rice: Did you calculate the Raman spectrum, and would the particle-hole excitations across the Hubbard gap show up in Raman (because that goes out to much higher energies)? Is that a test?

Z. Tesanovic: Well, that depends on what you believe is a gap. You see, when you go to such high energies of course, then you may ask a question whether you really have a Hubbard model, or you have some effective one-band model. So I would have to assign the meaning to this gap.

T. M. Rice: Within your Hubbard model, can you put some bounds on U/t from the ordered AF moment.

Z. Tesanovic: That's an excellent question. Let me take the following attitude, that the one-band model description is good, is valid, and that whatever moment is in oxygen is negligible. If I assume that most of the moment is in copper, if I do that, that may be a wrong assumption, but if I do that, I can then use the magnetization curve that I've just shown to you to try to extract what is U/t. By reading off what is .45 here, I can use this curve to extract what is U/t, and that would give me around ten or twelve. So that would be roughly the energy that I would associate with the gap. But it's perfectly possible, that you have other bands...as you open this big gap that there are other bands, which have a lower gap, consistent with optical measurements.

R. M. Martin: I just wanted to make a comment to follow up on what Phil Anderson was saying, and John Tranquada could really say this better. It's the coupling between the nearer planes in the 1,2,3 materials that does show a significant correlation between the moments on those two planes, experimentally, and that could affect both the magnitudes of the spins, and also the signature, to get at sort of what's happening.

J. R. Schrieffer: Just to follow-up on Maurice's comment. I very much like what I hear, of course. The possibility of looking at the $\sum(\omega)$, and at the threshold for what sometimes is called interband transition, charge transfer transition, but reinterpreted in terms of the intersite transition in the staggered lattice, gives you numbers for, say, the lanthanum cuprate in the range of 1.5 to 1.8 volts for the gap, compared to a bandwidth which is somewhat larger. So one might say that the effective U compared to the bandwidth is of order one in that interpretation. Whether that's right or not of course is another question. But I think this gives more information.

Z. Tesanovic: I just wanted to say one word. It is important to emphasize that this procedure is working in the limit of U/t very large, and also to try to understand whether it would work when we have a finite density of holes. We have some results for a single-hole case where we seem to be getting similar results as those Patrick was talking about yesterday. The overall hole bandwidth is given by $4t$ and its coherent mass arises through an interaction with spin waves with energies of order J. And it seems the reason is that once you have a very large gap you get rid of charge degrees of freedom and then the only question is whether your spin degrees of freedom, your spin flips, come out with the right scale, and they do. And once that happens, then you are in the game, because your effective low-energy Hamiltonian is the same. It turns out that this low-energy spin wave scale comes out to be identically what it is in the Heisenberg model.

J. R. Schrieffer: Yes, the reason for that is provable and well understood.

Z. Tesanovic: It happens, it happens.

Z. Tesanovic: The advantage of this approach is that you can do things at finite U, and also at finite doping. When you put holes in, something happens. Even if U is very large, that you cannot get out from the t-J model. There is actually the rearrangement of charges of the spins that in the t-J Hamiltonian you assume belong to a spin background.

A. J. Millis
At&T Bell Laboratories
Murray Hill, NJ 07974

Spin Dynamics of Superconducting Cuprates

Presently available experimental evidence for the existence of anomalous spin dynamics in the normal state of superconducting cuprates is summarized.

I. INTRODUCTION

High temperature cuprate superconductors are closely related to magnetic insulators. The simplest example is the $La_{2-x}Sr_xCuO_4$ system. At $x = 0$ the material is an antiferromagnet insulator, while at $x = 0.15$ the material is a high temperature superconductor with a $T_c \sim 40K$. The notion that high temperature superconductivity may be produced by modest doping of magnetic insulators has led to intense theoretical interest in models involving a small number of itinerant charge carriers moving in a background of ordered or fluctuating magnetic moments. A variety of models have been proposed; all involve two crucial assumptions: (i) that the spin dynamics in superconducting materials are in some way unusual and (ii) that the coupling of the unusual spin dynamics to the charge degrees of freedom is essential for high temperature superconductivity.

It is not at present known whether assumption (ii) is correct. It could be that at a doping of 0.15 carriers per Cu atom (as in the $La_{1.85}Sr_{.15}CuO_4$ system) or at the still larger dopings believed to exist in the $Bi_2Sr_2CaCu_3O_8$ and $YBa_2Cu_3O_7$ systems, the important physics (superconductivity) is due to some non-magnetic interaction. There is however some experimental evidence of non-trivial spin dynamics in cuprate superconductors. This review is a summary of this evidence with emphasis upon characterization of the way in which the magnetic properties in the normal (nonsuperconducting) state are unusual. Section II is an overview. It contains a qualitative discussion of the physics expected or conjectured for doped antiferromagnetic insulators. Section III is an analysis of bulk measurements such as dc susceptibility, and contains some consideration of issues of sample quality and homogeneity, especially in reference to the doping dependence of physical properties. Section IV treats neutron and Raman scattering experiments. Section V treats nuclear magnetic and quadrupole resonance experiments. Section VI is a summary and conclusion.

II. OVERVIEW

This section reviews the physics expected for doped antiferromagnetic insulators. Model Hamiltonians are discussed, with attention to the question of whether a one band or a two band model is appropriate for the hole doped cuprate superconductors. Theoretical scenarios for variation of physical properties with doping are considered.

a. Model Hamiltonians

The cuprate superconductors have various complicated crystal structures with large unit cells. The different structures share a common feature: CuO_2 planes. Superconductivity is associated with carriers moving in the planes; coupling between different planes is apparently weak. Most theories focus on two-dimensional motion of carriers within a plane and this review will focus on experiments which provide information concerning the two-dimensional spin dynamics of electrons in a CuO_2 plane.

In the stoichiometric compounds such as La_2CuO_4 and $YBa_2Cu_3O_{6.0}$ the CuO_2 planes are insulating antiferromagnets. When doped, they are metallic and superconducting. One way of thinking about the metal-insulator transition is based on band theory. Local density approximation (LDA) band structure calculations for, e.g., La_2CuO_4 predict that one band per CuO_2 plane crosses the chemical potential.[1] This band is nearly dispersionless in the direction perpendicular to the planes and is composed of strongly hybridized Cu $d_{x^2-y^2}$ and O $p\sigma$ orbitals. For the stoichiometric compound La_2CuO_4 this band is half filled and a finite strength but relatively weak repulsive interaction added to the LDA Hamiltonian will lead to a spin-density wave (SDW) ground state. The repulsive interaction must be added to the LDA Hamiltonian because "spin polarized" LDA calculations predict a non-magnetic ground state. Modest doping will destroy the SDW and yield a one-band fermi liquid with strong antiferromagnetic correlations. Further doping will weaken the antiferromagnetic correlations leading to a situation well described by band theory. There is a well established weak coupling theory of spin-density waves. Although La_2CuO_4 is not a weak coupling SDW (for example, the gap to charge excitations $E_g \gtrsim 1.5\ eV$ much larger than the maximum spin-wave energy $E_{SW} \sim 0.4\ eV$, while the weak coupling assumption implies $E_{SW} > E_g$) it is possible that one may gain insight into high temperature superconductivity by extrapolating the weak coupling theory to stronger couplings.

This line of thinking leads to the hypothesis that the high temperature superconductors are most appropriately modeled by a theory involving one band of fermions moving on a lattice and subject to a repulsive interaction which promotes antiferromagnetism and further suggests that the model may be studied by conventional perturbation techniques. The simplest such theory to write down is the Hubbard model, which is described by the Hamiltonian

$$H = \sum_{ij,\sigma} t_{ij} c^\dagger_{i\sigma} c_{j\sigma} + \sum_i U n_{i\uparrow} n_{i\downarrow} \qquad (2.1)$$

Here $c^\dagger_{i\sigma}$ creates an electron of spin σ in a Wannier function made of states in the Cu-O antibonding band and centered in unit cell i. t_{ij} is a hopping matrix element connecting sites, i and j (not necessarily nearest neighbors), and U is the on-site repulsion which causes the SDW instability at half filling. U is the dominant interaction for doping sufficiently near half filling. Other interactions may become important at larger doping.

Some authors have considered multiband models in which e.g. Cu-d and O-p orbitals are explicitly retained. Within weak coupling approximations,[2] the low energy physics turns out always to be describable by a one-band model. The interactions may in general be more complicated than the simple on-site repulsion displayed in Eq. 2.1. If, however, the model parameters are such that at half filling (one hole per CuO_2 unit) the ground state is antiferromagnetically ordered, then sufficiently near half filling the appropriate Hamiltonian is found in weak coupling theory to be Eq. 2.1. At larger dopings other interactions may become important; these promote "charge-transfer" rather than magnetic instabilities, and should have little effect on magnetic properties. Thus arguments based on perturbing about the band theory calculations lead to the conclusion that the one-band Hubbard model is the appropriate model with which to analyse the metal-insulator transition and the magnetic properties of the metallic state at those dopings where they are significantly different from those of a non-interacting Fermi gas.

A somewhat different point of view holds that band theory is not directly relevant and that the stoichiometric materials such as La_2CuO_4 are "Mott" or charge transfer insulators in which the insulating behavior is due to a strong repulsive interaction which prevents electrons from hopping from site to site. In this point of view one describes the low energy physics ($\hbar\omega < E_g$) in the insulator (e.g. La_2CuO_4) as involving one localized hole per CuO_2 unit cell. The hole resides primarily upon the Cu site (although because of Cu-O hybridization the density on the O site is non-zero). Each hole has a spin degree of freedom; these are coupled by an antiferromagnetic exchange interaction, J, leading to an antiferromagnetically ordered state at $T = 0$.

If the CuO_2 planes are doped by electrons (as is believed to happen in the new $Nd_{2-x}Ce_xCuO_{4-\delta}$ materials[3]) the added electrons reside on the Cu sites. Each added electron eliminates one spin degree of freedom, and the resulting picture is one of spinless entities (in this case, Cu sites with no holes) moving in a background of spins. The Hamiltonian describing this situation is believed[4] to be the "t-J" model

$$H_{t-J} = \sum_{ij} t_{ij}\tilde{c}_i^+\tilde{c}_j + J\sum_{\langle ij\rangle} \vec{S}_i \cdot \vec{S} , \qquad (2.2)$$

Here the object $\tilde{c}_i^+$ creates a spinless particle on site i, subject to the constraint that no more than one such particle may occupy any given site. t_{ij} is a hopping matrix element which need not only couple nearest neighbors. In the term proportional to the antiferromagnetic coupling J, $\vec{S}$ is a spin operator and the sum is understood to run over only those sites i and j which are (i) nearest neighbors and (ii) contain none of the spinless particles created by $\tilde{c}_i^+$. The t-J model has been derived from the Hubbard model (Eq. 2.1) in the limit of large on-site repulsion U and a nearly half filled band.[5]

The t-J model has been intensively studied recently because of its possible connection with high-temperature superconductivity. We briefly discuss some results of this study in part (b) of this section. Here we simply note that because it is a model with non-trivial strong interactions, the low-lying excitations need not have a simple representation in terms of the operators in Eq. 2.2. In particular, even though the charge carriers are described by spinless fermions in Eq. 2.2, the physical low-energy charge excitations are believed to have spin.[6]

In the more widely studied "hole-doped" superconductors such as $La_{1.85}Sr_{.15}CuO_4$ and $YBa_2Cu_3O_{6+x}$, the situation is more complicated. It is generally accepted that the charges added when the material is hole-doped reside primarily on the O sites (although the change in the charge density on the Cu sites is non-zero). This raises the important question whether in the hole doped materials the effective Hamiltonian describing the low energy degrees of freedom involves more than one spin degree of freedom per unit cell (e.g. the original $s = 1/2$ on the Cu site and a new $s = 1/2$ on a fraction of the oxygen sites) or less than one, as in the t-J model introduced above.

The "heavy electron" metals provide an apt analogy.[7] These materials are alloys composed of a rare earth element with a partially filled f-shell, such as Ce or U, and a metallic element. At temperatures of order room temperature the magnetic (and other) properties are clearly those of a two component system: a band of itinerant electrons and a set of local moments. At temperatures T and energies $\hbar\omega$ below an energy scale k_BT_K (which in heavy electron metals is of order 10-100K) the local moments and itinerant electrons become strongly coupled and behave in most respects as a one-band Fermi liquid (albeit with an enormously enhanced effective mass). The small value of the energy scale $k_B T_K$ is due to the weak hybridization between the localized and itinerant electron states. The temperature T_K is called the Kondo temperature. It is believed that any system of itinerant electrons hybridized with orbitals containing local moments will have a positive Kondo temperature unless some other instability (such as a magnetic ordering of the local moments) occurs at a temperature $T_c > T_K$.

The relevant question for hole-doped cuprate superconductors is thus the magnitude of the "Kondo temperature" below which the dopant spin is strongly coupled to the local moment. It is natural to expect that the Kondo temperature depends upon the strength of the hydridization between the orbital in which the local moment resides and the orbital to which the dopant hole is added. If the dopant wave function has a large admixture of $O_{p\sigma}$ states, the hybridization will be large. Indeed the overlap t_{pd} between Cu $d_{x^2-y^2}$ and $O_{p\sigma}$ orbitals is believed on the basis of band calculations to be $t_{pd} \sim 1.3\ eV$.[1,8] If however the dopant hole wave function is e.g. of primarily $O_{p\pi}$ or O_{pz} character, the hybridization will be very small (although it has not been estimated precisely). Band theory[1,8] and photoemission[9] experiments and (as will be explained below and in section V) NMR favor the $O_{p\sigma}$ orbitals. Early indications from diagonalization of small clusters[10] that dopant holes go into $O_{p\pi}$ states have not been confirmed by later calculations, however, it is apparently still unclear whether $O_{p\sigma}$ or O_{pz} orbitals are favored by cluster calculations.[11] Because the present experimental consensus favors $O_{p\sigma}$ orbitals we shall primarily consider theories involving these in what follows.

A number of theories exploiting the analogy with heavy fermion physics have appeared.[12-14] As expected from the qualitative discussion presented above, all apparently lead to effective "Kondo temperatures" $T_K \gtrsim 0.5\ eV$; for energies less than $k_B T_K$, a one-band model describes the spin dynamics.

The notion of a coupling between localized an itinerant electron moment leading to an effective one band model at low energies as introduced into the high-T_c literature by Zhang and Rice.[14] They used reasoning similar to that presented by Lacroix[15] in a study of the strong coupling limit of the heavy fermion problem to argue that the low-energy physics of one hole introduced on to an $O_{p\sigma}$ orbital in a CuO_2 plane would be described by the one-band "t-J" model discussed above. A more systematic derivation was given by Shastry.[16]

The conclusion that at low energies a one band model suffices has been disputed by Emery and Reiter,[17] who have presented a solution to the problem of an added oxygen hole moving in a background of ferromagnetically aligned Cu spins. They argue that because the charge-carrying excitation in this model has a spin degree of freedom, a one-band model cannot be correct. However, in both the heavy fermion materials[7] and (at least in some limits) the t-J model,[6] the charge-carrying excitations have been shown to carry spin, so the argument of Emery and Reiter is not conclusive.

In sum, it seems likely that the low energy spin dynamics of the cuprate superconductors can be described by a "one-band" model involving (at most) one $s = 1/2$ spin degree of freedom per unit CuO_2 cell. In the low-energy theory at least at low dopings the dominant interaction is a repulsion leading to formation of local moments and to a magnetically ordered state at half filling. Thus sufficiently near half filling it seems plausible that the model would seem to reduce essentially to the Hubbard or $t - J$ models discussed above.

Result of Studies of Model Hamiltonians

In this section we review what is known or conjectured about the dependence of various magnetic properties upon doping and interaction strength within the one-band Hubbard and "t-J" models discussed in the previous section. We will not discuss the results in detail, but will focus on the qualitative behavior of such observables as the static and dynamic spin correlations.

We begin with weak coupling analyses of Eq. 2.1. Most workers have studied the canonical two dimensional Hubbard model in which the hopping matrix element t_{ij}, connects only nearest neighbor sites. The two relevant dimensionless parameters are the normalized interaction strength U/t and the doping δ. If the number of holes per unit cell is n_h then $\delta = n_h - 1$. Within the weak coupling expansion formalism at $T = 0$ two types of phases are found. For sufficiently large U ($U > U_c(\delta)$) at fixed δ or sufficiently small $\delta(\delta < \delta_c(U))$ at fixed U the ground state is antiferromagnetically ordered, while for smaller U and larger δ a Fermi liquid with a weak d-wave superconducting instability results.[18] For $\delta \neq 0$ in the magnetic phase the magnetic order is found to be linearly polarized and incommensurate.[19] Specifically, within the Hartree-Fock approximation, and for $\delta_c \gg (\delta_c - \delta) > 0$ the spin density $\vec{S} = \langle c_\alpha^+ \vec{\sigma}_{\alpha\beta} c_\beta \rangle$ takes the form

$$\vec{S}(r) = \vec{S}_0 e^{i\vec{Q}\cdot\vec{r}} \tag{2.3}$$

where $\vec{Q}$ is a vector nearly equal to the commensurate value $\vec{Q}_0 = (\pi/a, \pi/a)$. The vector $\vec{Q}$ is found not to be parallel to $\vec{Q}_0$. In the limit $\delta \to 0$, the incommensurability is claimed to take the form of sharp domain walls separating commensurate regions.[19,20] Different authors disagree on the orientation of the domain walls. As discussed below, this form of the magnetic order is different than that found in the $U \to \infty$ limit using the "t-J" model, suggesting that a phase transition between different magnetic states occurs at finite U and fixed δ. Comparison with quantum Monte-Carlo calculations[18,21] suggests that the leading order (Hartree-Fock) calculation overestimates the tendency toward magnetism. Properly self-consistent calculations carried to the next order beyond Hartree-Fock apparently agree reasonably well with Monte-Carlo calculations for $U/t \lesssim 2$; the effect of these higher order terms on the magnetic order parameter has not been studied in detail.

For $U < U_c(\delta)$ or $\delta > \delta_c(U)$ the Hartree-Fock/RPA calculation predicts an antiferromagnetically correlated Fermi liquid with a weak d-wave superconducting instability. The charge fluctuation properties of this state, in particular the f-sum rule spectral weight, are essentially doping independent and gives by their band theory values. The antiferromagnetic correlations may be characterized in several ways. One is via the dynamic susceptibility $\chi''(q,\omega)$. Relatively little detailed information concerning the doping, frequency and momentum dependence of this quantity has been published so far. Indirect information concerning $\chi''(q,\omega \to 0)$ calculated for $U/t = 2$ was presented in the context of a theory of magnetic relaxation experiments.[27] These results suggest that the antiferromagnetic fluctuations are only appreciable for $(\delta - \delta_c) \gg \delta_c$.

The antiferromagnetic fluctuations lead to d-wave superconductivity. Most calculations indicate that the superconducting T_c is too low to be relevant for cuprate superconductors.[18,21,23] It has been suggested[24] that more complicated diagrams, not considered in the usual RPA calculation, made lead to a higher T_c, but quantitative results are not available.

In sum, the weak coupling approaches based on expansion about the Hartree-Fock solution lead to an incommensurate, linearly polarized spin density wave at low doping. As the doping is increased, a second order transition to a Fermi liquid state with antiferromagnetic correlations occur. The correlations rapidly dwindle as the doping is increased. In this picture the mechanism for superconductivity is unclear. Many authors have argued that another interaction must be added to the Hubbard model to obtain high-temperature superconductivity.

We now turn to the strong coupling limit of the Hubbard model. In this limit at half filling, the model is an antiferromagnetic insulator with a gap to charge excitations E_g which is large compared to the maximum spin wave energy, which is of order the exchange constant J. The strong constraints inhibiting charge motion imply that when the material is doped, the electrical conductivity is to be thought of as due to a small number of charge carriers doped into an insulating background. Specifically, the f-sum rule spectral weight $\omega_p^2 = 8 \int_0^{E_g} \sigma(\omega)d\omega$ (where $\sigma(\omega)$ is the frequency dependent conductivity) will be much smaller than the band-structure value and will be proportional to the dopant concentration δ and to the hopping matrix element t.[25] Thus, loosely speaking, the energy scale for charge motion is δt. However, there are two possible energy scales for spin fluctuations. One is δt; the other is the magnetic exchange constant J. In the limit $\delta t \gg J$ the ground state of the Hubbard or $t - J$ model is either a Fermi-liquid (with a renormalized Fermi temperature $\sim \delta t$) or, at sufficiently small δ, a metallic ferromagnet. The signature of this regime is that static susceptibility $\chi \sim 1/\delta$. The critical δ beyond which the ferromagnet is unstable is not known even for $J = 0$, although variational bounds are available. As discussed in section III, the limit $\delta t \gg J$ is probably not relevant to the separate superconductors. In the other limit, $\delta t \lesssim J$, the physics is much less well understood. The essential physics is supposed to be well represented by the "t-J" model, Eq. 2.2. Many interesting proposals have been made concerning the behavior of this model.[26] However, theory has not reached the point where it can be quantitatively compared with experiment. Indeed much theoretical work is based on uncontrolled approximations, so it is difficult to compare theory with theory. We therefore simply summarize various scenarios mentioning consequences for experiment. At zero doping and finite J, the ground state of the $t - J$ model is expected to have commensurate antiferromagnetic order. At sufficiently small doping and $J = 0$, the ground state is, as previously mentioned, expected to have ferromagnetic order. At finite J and sufficiently small doping, the commensurate antiferromagnetic state has been shown to be unstable to a metallic phase with long ranged spiral magnetic order.[6] The pitch of the spiral is small, proportional to the doping. The issue of the stability of this state to e.g. phase separation into regions of commensurate magnetic order and no charge carriers coexisting with regions containing charge carriers is not settled.[27] Note that the spiral order is apparently not found in the weak interaction limit. In this limit the dopant-induced incommensurability in the magnetic order is claimed to take the form of domain walls separating regions of commensurate order.

In the commensurate antiferromagnetic phase the spin excitations are spin waves. A gap, typically of order the exchange constant J, is present except near the zone center and the zone corners, where the spin-wave energy vanishes. In the spiral phase there are two sorts of spin excitations.[6] One class of excitations may be thought of as the spin waves of the spiral structure. These excitations have a gap, typically of order the exchange constant J, except near the tone center and the zone corners. The spin wave energy vanishes at a wavevector Q_0 near (π, π) determined by the pitch of the spiral. In the vicinity of Q_0 a new low-lying Goldstone mode related to changing the orientation of the spiral arises. The second class of spin excitations in the spiral phase are the incoherent, relaxational spin fluctuations due to the dopant holes. These are fermions and have a small magnetic moment, proportional to the doping, δ, and therefore presumably give a Fermi-liquid like contribution of small spectral weight to the dynamic susceptibility $\chi''(q,\omega)$. This contribution has not been calculated in detail.

As the doping is increased, a transition occurs to a state with no long range magnetic order. This may be Fermi-liquid phase, or a "spin liquid" phase with various interesting properties may intervene. Various models for the spin liquid phase have been proposed, including "double spiral" phase and uniform and staggered flux phases.[26,28] It is not known whether any of these are the ground state of the $t - J$ model for any values of doping, although many approximate calculations exist. It is presumed that at sufficiently large δ the ground state becomes Fermi liquid like, but the critical value of δ is not known. The order parameter for these phases is $I = \vec{S}_1 \cdot \vec{S}_2 \times \vec{S}_3$ where the $\vec{S}_i$ are spin operators on lattice sites i. The three spins must be on the same sublattice. In the staggered flux and double spiral phases the sign of I alternates between sublattices. In the uniform flux phase, it does not. The dynamical susceptibility for these phases has not been calculated in detail. Qualitative arguments suggest that as in the single spiral phase, the dynamical susceptibility in the spin liquid phases may be separated into two parts. One is the Fermi-liquid like contribution due to dopants; this apparently does not exist in the uniform flux phases; in the other phases it should be qualitatively similar to the dopant contribution in the spiral phase. The other contribution, from the spins, is also expected to similar to that found in the spiral phase, except that a gap opens up even near $q = (\pi, \pi)$ where the spectrum in the spiral phase had gapless modes. The size of the gap is expected to be small, at least in the "double spiral" phase.

To summarize this section, we have argued that a one component model of the spin dynamics should suffice for the cuprate superconductors, and that as far as the spin dynamics are concerned, this model may be taken to be the Hubbard or $t - J$ model. We have distinguished between the weak coupling limit, in which the spectral weight in the low ω optical conductivity, (a measure of the behavior of charge degrees of freedom) shows only a weak doping dependence, while the dynamic susceptibility (a measure of the behavior of the spin degrees of freedom) as displays a rapid buildup of low ω spin fluctuations as the doping is decreased to a critical value. In the strong coupling limit the spectral weight in the low ω optical conductivity scales with the doping, δ, while the dynamical susceptibility at most wavevectors exhibits a relatively doping-independent spin gap. The details of the magnetic order in the $\delta \to 0$ regime are different in weak and strong coupling.

III. STATIC SUSCEPTIBILITY

In this section we compare the doping dependence of the experimental magnetic susceptibility and the effective low frequency plasma frequency. We argue that this dependence shows that in reduced T_c materials the charge and spin excitations are governed by different energy scales, although these energy scales tend to converge as the doping (and T_c) are raised.

Studying the doping dependence of physical properties is dangerous because there is evidence that samples of $La_{2-x}Sr_xCuO_4$ with $x\ <\ 0.14$ and $YBa_2Cu_3O_{7-\delta}$ with $\delta\ >\ 0.1$ are inhomogeneous on a length scale $> 1000 \text{\AA}$.[29] (Also, of course, both $La_{1.85}Sr_{.15}CuO_4$ and $YBa_2Cu_3O_{7-\delta}$ for $\delta > 0$ have short-range disorder because, e.g. in the La-Sr material not every unit cell contains an Sr ion). The possible presence of large-scale inhomogeneities renders interpretation of data difficult, and much published data has been obtained from inadequately characterized samples. However, μSR[30] and magnetic[31] relaxation measurements suggest that it is possible to obtain samples of $YBa_2Cu_3O_{7-\delta}$ ($T_c = 90K$), $YBa_2Cu_3O_{6.6}$ ($T_c = 60K$) and $La_{1.85}Sr_{.15}CuO_{4-\delta}$ which are homogeneous on length scales greater than 1000 Å and are not too badly disordered on shorter length scales. Mutually consistent measurements on these materials have been obtained by many groups, and we shall discuss these data in what follows.

Interpreting susceptibility data on cuprate superconductors is difficult because the measured susceptibility includes diamagnetic contributions from ionic cores and paramagnetic (van Vleck) contributions from virtual transitions between Cu d-levels, as well as the contribution from the low energy spin degrees of freedom in which we are interested. Because the core diamagnetism and van Vleck terms must be estimated and subtracted from the data, uncertainty in the spin susceptibility remains.

Values for the core diamagnetism and the van Vleck paramagnetism are given in the Table caption. The core diamagnetism values are quoted by all workers as coming from "standard tables." The van Vleck term has been estimated in several ways;[32-34] the values are roughly consistent, although the doping dependence of this contribution in the $YBa_2Cu_3O_{6+x}$ system has not been clearly established.

In the $YBa_2Cu_3O_{6+x}$ system an additional difficulty exists: this system contains Cu-O chains as well as CuO_2 planes. Any bulk measurement will involve both plane and chain contributions, which must be deconvolved. In $YBa_2Cu_3O_{6.0}$, neutron scattering experiments[35] suggest that the chain Cu site has no magnetic moment. This is consistent with chemical arguments that two-fold coordinated Cu should be in the magnetically inert d^{10} state. As oxygen (x) is added to the $YBa_2Cu_3O_{6+x}$ system, an ordered magnetic moment appears on the chain Cu sites for $0.1 < x < 0.4$. For $x > 0.45$ there is no ordered moment on the chains. In the O_7 material, knight shift measurements[36] clearly indicate a spin susceptibility on the chain Cu site comparable in magnitude to that on the planar Cu sites. Thus as a rough estimate for the susceptibility associated with one CuO_2 cell in the plane for $YBa_2Cu_3O_7$ one may divide the measured susceptibility (corrected for core and orbital effects) by three. In the $YBa_2Cu_3O_{6.6}$ $T_c = 60K$ material there is an alternation of full and empty chains, and the interpretation is less clear. We simply divide the corrected χ by 3 for all metallic $YBa_2Cu_3O_{6+x}$ samples, recognizing that it leaves some uncertainty.

In the $La_{2-x}Sr_xCuO_4$ material the measured bulk susceptibility is sensitive to oxygen content,[37] so the intrinsic value is not clear.

Despite the difficulties it is of interest to attempt to determine the spin susceptibility and to compare it with the value predicted by band-structure calculations and the value in the insulating antiferromagnetic compounds. The enhancement over band-structure is readily calculable within the RPA or weak coupling theories referred to in the previous section and the change from the insulating value ought to be calculable within a strong-coupling "t-J" model framework, although as yet neither approach has been carefully compared with experiment.

The band-structure density of states is (within a rigid band approximation) strongly doping dependent.[1] For example, because of a van Hove singularity the density of states of $La_{1.85}Sr_{.15}CuO_4$ is calculated to be almost a factor of two higher than that of La_2CuO_4. In this paper I compare the experimentally determined susceptibilities with the band theory prediction for La_2CuO_4 which we take to be a representative band theory density of states for CuO_2 planes. I believe that the van-Hove singularity is unlikely to be relevant in a strongly correlated material or over a wide range of dopings.

The values of the static spin susceptibilities obtained for $La_{1.85}Sr_{.15}CuO_4$ and $YBa_2Cu_3O_{6+x}$ using the procedures described above are given in the Table at temperature $T = 300K$. Note that (i) all susceptibilities are greater than the band structure value by factors of order 2-3 and greater than that expected from the quantum Heisenberg model with $J = 0.15$ eV by factors of 1-2, (ii) the magnitude of the susceptibility at any temperature T, increases with T_c and doping and (iii) at fixed doping the magnitude of the susceptibility increases with T, except for the $T_c = 90K$ $YBa_2Cu_3O_7$ material. These trends have not been explained by any microscopic theory. D. C. Johnston[37] has interpreted the data phenomenologically in terms of a Heisenberg model with a varying J.

It is interesting to compare the doping dependence of the bulk susceptibility with a measure of the energy scale associated with charge fluctuations. One such measure is the square of the quasiparticle plasma frequency, which may be determined either from optical reflectivity[38] or from measurements of the superconducting penetration depth.[30] The plasma frequency squared may be compared with the band structure prediction. This comparison is discussed in more detail in Ref. 14. The results are again given in the Table. It is clear that the size of the enhancement and the trend with doping are different for the charge than for the spin. The scaling of the quasiparticle plasma frequency differs from what is expected on the basis of RPA calculations but is in qualitative accord with expectations for the large U Hubbard model discussed in the previous section. The behavior of the magnetic susceptibility is opposite to that of the plasma frequency. This suggests that different physics governs the charge and spin degrees of freedom as might be expected within a "t-J" model description. However, the difference in energy scales is least apparent in the $YBa_2Cu_3O_7$ material where T_c is highest.

TABLE 1 Column 1 identifies the material. Column 2 gives an estimate of the doping δ. For $La_{2-x}Sr_xCuO_4$ we assume $delta = x$. For the $YBa_2Cu_3O_{7-\delta}$ materials we use valence counting arguments such as those given in ref. 61. Column 3 gives the enhancement of the susceptibility over the band theory values. The susceptibility is obtained from Refs.[32-34] corrected for van Vleck and core diamagnetism. For $La_{1.85}Sr_{.15}CuO_4$ we assumed that $\chi_{obsd} = \chi_s - 75 \times 10^{-6} cm^3/mole$;[37] for $YBa_2Cu_3O_{6+x}$ we assumed $\chi_{obsd} = \chi_s - 46 \times 10^{-6} cm^3/mol$ planar Cu.[36] The fourth column gives the ratio of the band structure plasma frequency ω_p^2 to the quasiparticle plasma frequency ω_p^{*2} obtained from penetration depth[30] or optical[38] measurements. The fifth column gives a typical spin fluctuation frequency at a wavevector q not near zone center or corner obtained from oxygen or yttrium NMR[58] as explained in section V. The sixth column gives the zone corner spin fluctuation frequency Γ^* obtained from copper NMR and NQR[58] as explained in section V. The data represented here suggest that the charge fluctuation energy scale and the antiferromagnetic spin fluctuation energy scale are doping dependent, but that the spin fluctuation energy scales for momenta far from the zone corner are not.

Material	δ	χ^2/χ_s^{band}	ω_p^2/ω_p^{*2}	$\Gamma(eV)$	Γ^*
$La_{1.85}Sr_{.15}CuO_4$	0.15	3	10	—	—
$YBa_2Cu_3O_{6.6}$	0.1-0.2	2	10	0.5	0.01
$YBa_2Cu_3O_7$	0.3-0.5	2.5	5	0.4	0.02

IV. NEUTRON AND RAMAN SCATTERING

Neutron and Raman measurements on magnetic insulating cuprates have provided detailed information on the magnetic excitations in these materials. The experimental data have been quantitatively fit to accurate theoretical calculations based on the quantum Heisenberg model. The agreement between experiment and theory is very good. For a review, see the article by Professor Chakravarty in this volume. The situation in the superconducting cuprates is much less satisfactory. Because of the difficulty of obtaining good samples and because the magnetic scattering is weaker, the presently available data are much less precise than in the case of the insulating materials. Also, no detailed comparison of experiment and theory is available. However, the experiments suggest that interesting magnetic behavior occurs. In the balance of this section, we review the available data, first for neutron scattering and then for Raman scattering.

Neutron scattering studies of superconducting cuprates are difficult to perform because of the scarcity of sufficiently large single crystals and are difficult to interpret because of doubts concerning the homogeneity of available samples. Sample quality has proven difficult to characterize. Neutron measurements are performed on single crystals of linear dimension $\sim 1\ cm$. Measurements of magnetic screening or (because muons penetrate only a few microns) muon spin rotation probe only the the surface region. Other measurements such as specific heat and magnetic resonance require that a small piece of the sample be removed, preferably from the interior. In the $La_{2-x}Sr_xCuO_4$ sample the temperature and sharpness of the orthorhombic-tetragonal phase transition (which may be probed with neutrons) have been used to characterize samples.[41] A systematic comparison of this method of sample characterization with other methods has not been made.

A further difficulty is that the magnetic scattering in superconducting cuprates is much weaker than in La_2CuO_4, in part because of the absence of long range order. Because the scattering is weaker, it is more difficult to distinguish from background.

Neutron scattering experiments have been performed on several $La_{2-x}Sr_xCuO_4$ crystals with $T_c \lesssim 20K$.[40] These experiments suggest the presence of incommensurate antiferromagnetic correlations characterized by a weakly doping and temperature dependent coherence length of the order of 3-5 lattice constants. However, many samples of $La_{2-x}Sr_xCuO_4$ with $T_c < 30K$ or x too different from 0.15 are known to be inhomogeneous;[30] thus the extent to which the data is representative of superconducting material is not clear.

More recently, data on a $T_c = 33K$ nominally $La_{1.85}Sr_{.15}CuO_4$ crystal have become available.[41] This crystal exhibited a sharp structural phase transition, suggesting it is microscopically homogeneous. Plots of the structure factor $S(q,\omega)$ with $q = h(\pi/a, \pi/a)$ have been presented. As h is varied at fixed ω a clear peak above background is seen near $h = 1$ (i.e. $q = (\pi/a, \pi/a)$). This peak suggests that antiferromagnetic fluctuations are present. The fluctuations are incommensurate, with a maximum at $h \cong 0.85$. One qualitative measure of the intensity of the magnetic scattering is $I(\omega) = \int_{h_{min}}^{h_{max}} dh\, S(h,\omega)$. Here the integral over momentum is over that part of the zone diagonal for which the spin fluctuations are distinguishable from the background. At $\omega = 6\ meV$, this quantity has an interesting temperature dependence, shown in Fig. 1 which is reproduced from ref. 4. Now $S(q,\omega)$ is related to the dynamical susceptibility $\chi''(q,\omega)$ via

$$\chi''(q,\omega) = (1 - e^{-\beta\omega})\, S(q,\omega) \qquad (4.1)$$

We have suggested in section 2 that $\chi''(q,\omega)$ is the quantity of relevance to microscopic theories. The temperature dependence of $I_\chi(\omega) = \int_{h_{min}}^{h_{max}} dh \chi''(h,\omega)$ is also plotted in Fig. 1. The points are obtained from Fig. 1 and Eq. 4.1. It is clear that the general trend is for $I_\chi(\omega)$ to slowly increase with decreasing T, as expected from a temperature dependent build-up of antiferromagnetic correlations. The significance of the structure in $I_\chi(\omega = 6\ meV)$ near $T = 100K$ is not clear at present.

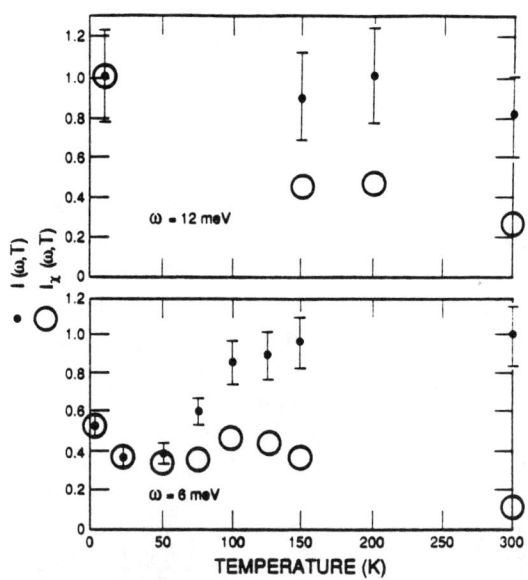

FIGURE 1 Temperature dependence of normalized integrated scattering intensity obtained as described in the text from the dynamical structure factor $I(\omega, T)$ (solid points and error bars) and from the dynamical susceptibility $I_\chi(\omega, T)$ (open circles, error bars omitted), for $La_{2-x}Sr_xCuO_4$ at two energy transfers: $\hbar\omega = 6mev$ (lower box) and $\hbar\omega = 12mev$ (upper box). The solid points and error bars are from ref. 41; the open circles are obtained by dividing the structure factor by the appropriate thermal factor as discussed in eq. 41.

A disturbing feature of the data is that $S(q, \omega \to 0)$ is nonzero, even for $T \ll T_c$. In a superconductor one would expect (and NMR relaxation rate data confirm) that a gap in the spin excitation spectrum would open up below T_c; such a gap is apparently not observed in the neutron data.

Some neutron scattering data exists for the $YBa_2Cu_3O_{7-\delta}$ system.[39,42] As an account of it may be found elsewhere in this volume, it will be only briefly discussed here. A broad peak in $S(q,\omega)$ near $q = (\pi/a, \pi/a)$ was observed in a $T_c = 50K$ sample. The correlation length associated with this peak was claimed to be ~ 1 lattice constant at $T = 12K$. In contrast to the $La_{1.85}Sr_{.15}CuO_4$ data. The scattering intensity at $T = 12K$ is very small at $\omega = 3$ meV and is an increasing function of increasing ω, up to the highest frequency measured ($\omega = 12$ meV). The BCS gap for a weak-coupling superconductor with $T_c = 50K$ would be approximately $2\Delta = 15$ meV. However, early work[39] on a $YBa_2Cu_3O_7$ sample found little magnetic scattering for $\omega < 25$ meV.

We now consider Raman scattering data. In Raman scattering one irradiates a material with laser light of some frequency Ω and some polarization, and measures the emitted light at some other frequency $\Omega + \omega$ and some other polarization. The ω and polarization dependence provides information about excitations at frequency ω in the solid. In a magnetic insulator, spin excitations may be Raman active because the incident light causes atoms to move; changing the distance between neighboring magnetic ions modulates the local exchange coupling and therefore excites short wavelength, high energy spin excitations. In materials such as La_2CuO_4 and $YBa_2Cu_3O_{6.0}$ this process has a characteristic dependence on the polarization of the incident and outgoing light, and vanishes for some combination of polarizations. This allows the magnetic signal to be separated from other contributions. The background ought in any case to be small, because La_2CuO_4 and $YBa_2Cu_3O_{6.0}$ are insulators with a large gap (~ 1.5 eV) to charge excitations, so in principal the only excitations between ~ 0.1 eV (above the upper limit of the phonon spectrum) and 1.5 eV should be spin excitations.

As discussed in detail in Professor Chakravarty's article in this volume, in the insulating materials there is a clear peak as a function of energy transfer in the Raman intensity for the "allowed" polarizations. The peak is absent in the disallowed channel. The lineshape may be quantitatively fit in terms of the quantum Heisenberg model. However, to observe the magnetic excitations clearly, a "resonant enhancement" is required; in other words the frequency Ω of the incident radiation must be tuned to be in resonance with some intermediate state.

In the superconducting cuprates, the situation is less clear. The peak in the allowed polarization channel broadens rapidly with doping.[43,44] The relative magnitude of the Raman intensity in the disallowed channel increases. However the difference between the channels persists even in the $T_c = 90K$ materials. A naive interpretation of the data would be that the spin fluctuations dwindle as the doping is increased, but are nonzero even in the $YBa_2Cu_3O_7$ material. Issues which complicate the interpretation include the presence of charge degrees of freedom which may be Raman active in all polarization channels and which may affect the spin excitations (e.g., when a carrier hops from one site to another a spin is also destroyed at one site and recreated at another), and the apparent weakening of the resonance enhancement of the Raman signal.

In sum, the Raman experiments are consistent with the idea that some remnant of the spin fluctuations observed in the insulating phase persist in superconducting samples (although at present no theoretical framework exists by which this remnant may be more precisely characterized) and alternative, "non-magnetic" explanations are not ruled out.

The neutron scattering experiments strongly suggest (at least in reduced T_c materials) the presence of short range, finite lifetime, incommensurate antiferromagnetic fluctuations in at least some superconducting cuprates, but substantial further experimental effort is required before the details of the spin excitation spectrum are known.

V. NUCLEAR RESONANCE EXPERIMENTS

Nuclear magnetic resonance (NMR) and nuclear quadrupole resonance (NQR) experiments may be used to probe the low frequency limit of the dynamic spin correlation function $\chi''(q,\omega)$. NMR and NQR experiments have produced a number of surprising results, which have been interpreted as implying the presence of strong antiferromagnetic correlations in superconducting cuprates and as suggesting the opening of a small "spin gap" for $T \lesssim 140K$ in $T_c = 60K$ $YBa_2Cu_3O_{6.6}$ and $La_{1.85}Sr_{.15}CuO_4$. In what follows we explain the reasoning behind these conclusions. Specifically, we first give the motivation for and assumptions implicit in the Hamiltonian used to analyse NMR and NQR experiments on the cuprates, then we discuss some general implications of the experimental data, and finally we review specific model calculations. We have not given a survey of all relevant literature. In a partial attempt to remedy this defect, we list here some of the experimental[45–54] work bearing on this issue.

NMR experiments provide information on the electronic spin dynamics of a solid because the nuclear moments are coupled to the magnetic degrees of freedom the electrons in the solid. Low frequency fluctuations of these magnetic moments cause the nuclear spin to change its quantum state. The energy splitting of the nuclear spin states is negligible, thus the nuclear relaxation experiments probe the $\omega \to 0$ limit of the magnetic susceptibility of the conduction electrons. The magnetic degrees of freedom of the conduction electrons may include the orbital angular momentum of an electron about a given nucleus and, of course, the spin degrees of freedom of the electrons. It has been argued[45] that the orbital angular momentum is quenched in the cuprate superconductors. Thus, e.g., a transition of an electron from a $3d_{x^2-y^2}$ state to a d_{z^2} state would involve a finite and large compared with room temperature excitation energy. Such transitions then would give rise to the van Vleck contribution to the magnetic susceptibility, but could not cause nuclear spin relaxation at temperatures of order room temperature or lower. This line of reasoning suggests that the nuclear resonance experiments probe the low-lying *spin* dynamics of the electrons in cuprate superconductors. This assumption is made in all analyses of NMR and NQR experiments in the cuprates of which I am aware, and it will be assume henceforth.

Before proceeding with a detailed analysis, we outline the basic experimental anomaly. In a conventional Fermi liquid, the spin-lattice relaxation rate, W, has the Korringa temperature dependence $W \sim T$. This temperature dependence may be understood as follows: for a Fermi liquid the density of states for spin excitations is constant at low energies. The rate is given by the number of thermally excited spin excitations, which in this case scales as T. In the high T_c cuprate superconductors, the Cu relaxation rate ^{63}W varies more slowly than the Korringa rate for $T > T^{*}$.[45,46,48] The crossover temperature T^* varies from material to material, but is of order 100-150K. Therefore, for $T > T^*$, the density of states for spin excitations must increase as T decreases. This increase in the density of states may occur roughly uniformly throughout the Brillouin zone, or it may be concentrated near a particular q value, as would be the case for a system near a magnetic instability. Now in the $T_c = 90K$ $YBa_2Cu_3O_7$ material the oxygen relaxation rate ^{17}W is observed to be accurately Korringa (i.e. $^{17}W \sim T$) for all experimentally accessible $T > T_c$.[47] We have already argued that spin excitations provide the relaxation mechanism and that a one band model of these is appropriate. Therefore the same degrees of freedom relax both the Cu and O nuclear moments, and so some symmetry must cause the vanishing of the matrix element coupling the O nucleus to the enhanced part of the spin density of states seen by the Cu. Such a symmetry cannot be relevant if the Cu spin density of states is enhanced uniformly over the zone. Therefore, we conclude that the Cu spin density of states must be enhanced most strongly at some particular point of the Brillouin zone. If one makes the plausible assumption that a given O nucleus is coupled predominantly to the spins on its two nearest neighbor copper sites, the spin density of the states must be enhanced most strongly near the zone corner, $(\pi/a, \pi/a)$, i.e. the enhanced spin density of states must be associated with the development of nearly commensurate antiferromagnetic correlations. In antiferromagnetically correlated spin fluctuation, each O would be between two essentially oppositely directed Cu moments, so the transferred hyperfine field from these moments would cancel at the O site. From this argument, advanced by Hammel et al.,[49] Shastry[16] and Mila and Rice,[55] we conclude that the relaxation experiments provide strong evidence of temperature dependent antiferromagnetic correlations in $YBa_2Cu_3O_7$. However, for $T < T_c$

the copper and oxygen relaxation rates have the same temperature dependence,[47] suggesting that the correlations are T-independent below T_c.

In reduced T_c $YBa_2Cu_3O_{7-\delta}$ materials the oxygen,[54] yttrium[49] and (for $T < T_x \approx 130K$) copper[53] relaxation rates decrease more rapidly than linearly with decreasing temperature. This departure from Korringa behavior indicates that for all q the low ω density of spin fluctuations decreases as T is lowered. This behavior is not understood even phenomenologically at present, but it is tempting to associate it with the "spin-gap" behavior mentioned in section II.

We now turn to explaining the Hamiltonian used to analyse NMR and NQR experiments in the cuprates. We have previously argued that a one band model is appropriate for the interesting low ω spin dynamics in the CuO_2 planes. We therefore consider a model of the electronic spin degrees of freedom in which there is at most one spin 1/2 degree of freedom per unit cell. If an electronic spin is present in unit cell n in a CuO_2 plane, we may regard it as arising from a Wannier state centered on cell n. To obtain the coupling of the Cu nuclear spin in cell n to the electron spin centered in cell m one may in principal determine the projection of the Wannier state onto all Cu atomic orbitals in cell n, and obtain the coupling of each Cu atomic orbital to the nuclear moment from an atomic physics calculation or experiment. Because the Wannier function will extend over more than one cell, the nuclear spin in cell n may be coupled to electron spins in several adjacent cells, as well as to the electron spin in cell n.

It is generally accepted that the Wannier function relevant to the low-lying spin degree of freedom in unit cell n has $d_{x^2-y^2}$ symmetry about the Cu nucleus in unit cell n. In unit cell n Cu atomic orbitals of this symmetry may participate in the Wannier function. These orbitals have vanishing amplitude at the Cu nucleus because of the angular momentum barrier. Therefore, the Fermi contact interaction between the Cu nuclear spin and the electronic spin associated with the Wannier function at site n vanishes. The electron-nuclear coupling is via dipole and core polarization terms which are small in comparison to the contact interaction for a s-orbital. The dipolar coupling is anisotropic, because the Cu $d_{x^2-y^2}$ orbital is anisotropic.

Because the Wannier function centered on a given site has a finite range, it has a non-vanishing projection onto orbitals on adjacent sites. In particular, the Wannier function on one site may have a non-vanishing projection onto Cu 3s or 4s orbitals on adjacent sites. These orbitals are coupled to the appropriate Cu nuclear spin via the Fermi contact interaction. In cuprate superconductors the large magnitude of the contact interaction relative to the dipole and core polarization terms apparently compensates for the small admixture of the adjacent-site 3s or 4s Cu orbitals in the Wannier function relative to the large admixture of the same-site Cu $d_{x^2-y^2}$ orbital, so that the coupling of the Cu nuclear moment to the electron spin on the same site is comparable in magnitude to the coupling to the electron spin on adjacent sites. This observation is due to Mila and Rice.[55] In sum, the Hamiltonian describing the coupling of the Cu nuclear moment to the electronic degrees of freedom of the solid may be written

Spin Dynamics of Superconducting Cuprates

$$^{63}H_{e-n} = {}^{63}\vec{I}_n \mathbf{A} \vec{S}_n + B \sum_\delta {}^{63}\vec{I}_n \cdot \vec{S}_{n+\delta} \qquad (5.1a)$$

Here $^{63}\vec{I}_n$ is the operator for the Cu nuclear moment in unit cell n (^{63}Cu is the Cu isotope with a nuclear moment). $\vec{S}_n$ is the electronic spin operator in cell n. δ labels the four near neighbor unit cells of unit cell n. The onsite coupling $\mathbf{A}$ is expected to be a tensor with two independent components $A_\parallel$ and $A_\perp$: $A_\parallel$ pertains to spins perpendicular to the CuO_2 plane and $A_\perp$ pertains to spins lying in the CuO_2 plane. (In principal further anisotropy could exist because the high-T_c materials are orthorhombic. In practice, the tetragonal symmetry of the form given here for $\mathbf{A}$ suffices). The "transferred hyperfine" coupling B has been assumed isotropic, on the basis of the physical argument that it is due to the Fermi contact interaction. Estimates of the values of the hyperfine coupling consistents $A_\parallel$, $A_\perp$ and B are given by Mila and Rice[55] and Monien, Pines and Slichter.[56]

One may obtain the Hamiltonian describing coupling of an oxygen nuclear moment to the low-lying electronic spin degrees of freedom similarly. Note that each oxygen is shared between two unit cells in the CuO_2 plane; thus we have

$$^{17}H_{e-n} = \sum_{\delta'} {}^{17}\vec{I}_n \mathbf{C} \cdot \vec{S}_{n+\delta'} \qquad (5.1b)$$

Here $^{17}I_n$ denotes the oxygen nuclear moment (^{17}O is the oxygen isotope with a nuclear moment), δ' labels the two unit cells sharing oxygen site n, and $\mathbf{C}$ is the hyperfine coupling constant. $\mathbf{C}$ is a tensor with three independent components. Eq. (5.1b) was derived from a model for the insulating state of the CuO_2 planes by Shastry,[16] and was later discussed by other authors[22,57,58] from various points of view. The anisotropy of $\mathbf{C}$ has been interpreted as implying that the important oxygen orbital is the $O_{p\sigma}$ orbital, in agreement with the discussion in section II. In presently available analyses of relaxation data, C has been taken to be isotropic.[22,57,58,59]

One may obtain the hyperfine Hamiltonian for yttrium in the $YBa_2Cu_3O_{7-\delta}$ system similarly. The yttrium atom sits between two CuO_2 planes and couples to four unit cells in each one has

$$^{89}H_{en} = D \sum_{\delta'',\eta} {}^{89}\vec{I}_n \cdot \vec{S}_{n+\delta'',\eta} \qquad (5.1c)$$

Here $^{89}\vec{I}_n$ is the nuclear moment operator for an yttrium atom, δ' labels the four CuO_2 cells in one plane nearest the yttrium atom, and η labels the two CuO_2 planes nearest the yttrium atom. The hyperfine coupling D is generally regarded as isotropic, and no data inconsistent with this assumption is known to the author. This term was derived by Mila and Rice.[55]

It is worth emphasizing that the form of the hyperfine Hamiltonians, (5.1a-c) is of rather general validity. The essential assumptions are the the only low-lying electronic degrees of freedom coupling to the nuclear moments are spin degrees of freedom and that the relevant electronic spin degrees of freedom are described by a one-band model. The Hamiltonians could be derived in the context of band theory or from a model for a Mott insulating state. Further, it is not necessary that the spin degree of freedom reside wholly or ever mostly on the Cu sites. Of course, the details of which orbitals which make up the spin degrees of freedom will determine the magnitudes of the various hyperfine coupling constants. Indeed, the experimental results (explained below) that $A_\parallel \cong -4B$ and $A_\parallel$ strongly support microscopic models in which the Cu $d_{x^2-y^2}$ is the most important Cu orbital, because the anisotropy of the on-site term is most plausibly explained by a dipole electron-nuclear coupling, and if an on-site Fermi contact term were appreciable, the transferred term would not be comparable in magnitude to the on-site term.

Having discussed the Hamiltonian, we turn to the experimental data. We begin with the knight shifts. The experiment is simple in principle. The nuclear moment precesses at the nuclear larmor frequency in a magnetic field, however, the effective field at the nuclear site differs from the externally applied field because the applied field $\vec{H}_{ext}$ polarizes the electronic spins, so that $\langle \vec{S}_n \rangle = \chi_s \vec{H}_{ext}$ (χ_S is the spin susceptibility, which is expected to be isotropic in the cuprate superconductors). The finite expectation value of $\langle \vec{S}_n \rangle$ leads via Eqs. 5.1 to an extra effective field at the nuclear site, proportional to the hyperfine coupling, the spin susceptibility, and the external field. The extra effective field causes a shift in the nuclear Larmor frequency from the value expected in a field equal to the externally applied field. This shift is the spin Knight shift; from it the product of the hyperfine coupling and the spin susceptibility may be deduced. Thus the Knight shift experiments fix $(A_\parallel + 4B)$, $(A_\parallel + 4B)$, C and D if the spin susceptibility is known. As discussed in Ref. 58, the anisotropy of the Cu relaxation rate fixes the ratio $A_\perp/4B$. Thus all of the coupling constants in the model specified be Eq. 5.1 may be determined from experiment.

There is, however, a complication. The electrons have magnetic degrees of freedom other than the low-lying spin degrees of freedom considered in eqs. 5.1. These orbital degrees of freedom may couple to the nuclear moments, giving additional contributions to the Knight shift. If the arguments leading to Eqs. 5.1 are correct, these other degrees of freedom are separated from the ground state by a large energy gap so that their contribution to the susceptibility will be to a good approximation temperature independent for temperatures of order room temperature and below. The spin susceptibility, however, must vanish at $T = 0$ in a singlet superconductor. Thus to determine the spin knight shift, $K_s(T)$ at temperature T one measures the total knight shift K as a function of temperature and defines

$$K_s(T) = K(T) - K(T = 0) \tag{5.2}$$

The spin Knight shift for the planar Cu site has been measured in $YBa_2Cu_3O_7$ ($T_c = 90K$)[50] and $YBa_2Cu_3O_{6.6}$ ($T_c = 60K$).[53,54] In both materials it is anisotropic; in both materials the shift for $\vec{H}$ parallel to the c axis, (ie. perpendicular to the CuO_2 planes) vanishes while for $\vec{H}$ perpendicular to C the shift is non-zero. This plus the isotropy of χ_s implies (using Eq. 5.1) that $A_\parallel + 4B = 0$ in both systems. The anisotropy of the nuclear relaxation rate, to be discussed below, suggests that neither $A_\parallel$ nor B vanishes. The result, that $A_\parallel = -4B$, despite the very different microscopic origin of the two terms is apparently a coincidence. That the coincidence occurs at two different doping levels of the CuO_2 planes suggests that the local electronic structure is insensitive to doping. Further information concerning the doping dependence of the hyperfine couplings may be obtained from antiferromagnetic resonance experiments on insulating $YBa_2Cu_3O_{6.0}$ as pointed out by Monien, Pines and Slichter.[56] This material is an insulating antiferromagnet. The ordered moments lie in the CuO_2 plane and their magnitude is known. Because there are local moments, the Cu nuclear spins precess in zero applied field. From Eq. 5.1 and the value of the ordered moment one may deduce the value of $A_\parallel - 4B$ from the antiferromagnetic resonance frequency. The value so obtained is in agreement with the result of the knight shift experiments in $YBa_2Cu_3O_7$, if the value for χ_s quoted in section III is used. This agreement is only suggestive, as the value of χ_s is not directly measurable so is not known with certainty. However, it seems clear from the doping independence of $A_\parallel + 4B$ that the Cu hyperfine couplings change by less than 10% between $YBa_2Cu_3O_7$ and $YBa_2Cu_3O_{6.6}$, and it is plausible that the doping independence extends to the $YBa_2Cu_3O_{6.0}$ material. This is surprising. The value of the transferred hyperfine coupling B in particular depends on the overlap of wavefunctions in one cell with those in another. One would naively expect this overlap to vary with doping.

Knight shift measurements for planar Cu with $\vec{H}$ in the CuO_2 plane indicate that the spin susceptibility in $YBa_2Cu_3O_7$ is T independent for $T > T_c$, consistent with direct measurement of χ and with conventional Fermi liquid behavior. Such measurements in the $YBa_2Cu_3O_{6.6}$ material indicate a substantial T-dependence of χ_s in the normal state, with $K_s(T_c = 60K) \cong 0.3\ K_s(T = 300K)$.[53] This T-dependence is in at least qualitative agreement with that of the measured static susceptibility, but the two have not been compared in detail.

Knight shift measurements have also been performed for planar oxygen[47,51,18] and yttrium nuclei in the $T_c = 90K$ $YBa_2Cu_3O_7$ material, and for yttrium[49] reduced T_c $YBa_2Cu_3O_{7-\delta}$ samples. The magnitude of the yttrium hyperfine coupling constant is claimed to be independent of doping, and the T-dependence of the yttrium knight shift is apparent consistent with the T-dependence of the independently measured static susceptibility. The yttrium knight shift data is subject to some uncertainty because its value is very small, and because data were only taken for $T \gtrsim 100K$, so Eq. 5.2 could not be used and other arguments were made to estimate the orbital contribution to the knight shift.

In sum, the knight shift measurements permit the determination of the value of the product of the hyperfine coupling constants defined in Eqs. 5.1 and the spin susceptibility. Presently available data are consistent with the assumption that the coupling constants (most notably the transferred coupling B) are doping independent.

We turn now to the nuclear relaxation rate. A nuclear moment initially polarized in a particular direction will decay because of its coupling to conduction electron spin degrees of freedom. By applying the golden rule to Eqs. 4.1 one sees that the relaxation rate, W, of a nuclear moment is proportional to the square of a hyperfine coupling and to various on site and nearest neighbor spin-spin correlation functions, and may be written

$$W \sim \frac{kT}{\hbar \omega_0} \sum_q F(q) \chi''(q, \omega_0) \qquad (5.3)$$

Here $\omega_0 \to 0$ is a nuclear level splitting or Larmor precession frequency, $\chi''(q,\omega)$ is the dynamic susceptibility of the electronic spin system and F(q) is a combination of a form factor and hyperfine coupling constants. F(q) is different for different nuclei and, because the hyperfine couplings for the Cu nucleus are anisotropic, F(q) in this case depends upon the direction in which the nuclear moment is polarized. The form factors may easily be computed from eqs. 5.1.[16,22,57,58] The oxygen and yttrium form factors are found to vanish for $q = (\pi/a, \pi/a)$, while the copper form factor does not. The relaxation of the oxygen and yttrium nuclei is therefore not sensitive to commensurate (or nearly commensurate) antiferromagnetic fluctuations while that of the copper nuclei is. Thus within the model for nuclear spin relaxation developed in this chapter, differences in the magnitude and T dependence between the copper and, e.g., oxygen relaxation rates must be due to temperature dependent antiferromagnetic correlations.

Because the hyperfine coupling constants and form factors are known, one may extract quantitative information concerning $\chi''(q, \omega \to 0)$ by assuming a model form for χ'', and fitting it to experiment. Several groups have pursued this program. In the remainder of this section we summarize their results. We consider first the $YBa_2Cu_3O_7$, $T_c = 90K$ material, which have been extensively experimentally studied and where fairly consistent story has emerged, and then the reduced T_c $YBa_2Cu_3O_{7-\delta}$ and $La_{1.85}Sr_{.15}CuO_4$ systems, where the situation is much less well understood.

A general form for the low frequency limit of the dynamic susceptibility in a paramagnetic spin system is

$$\chi''(q, \omega \to 0) = \frac{\chi'(q, T)\omega}{\Gamma(q, T)} \qquad (5.4)$$

Here $\chi'(q,T)$ is the (possibly temperature dependent) static susceptibility at wave-vector q, and $\Gamma(q,T)$ is a spin relaxation rate. In a Fermi liquid, one would have $\Gamma(q) \sim v_F q$ while in a system with a gap ϵ_o to spin excitations at wave-vector q one would expect $\Gamma_q \sim e^{-\epsilon_o/T}$.

In the $YBa_2Cu_3O_7$ material a consistent picture has emerged.[22,58] A $\chi''(q,\omega)$ appropriate to a Fermi liquid with commensurate or nearly commensurate antiferromagnetic correlations quantitatively accounts for all available data. In particular, there is no evidence for a "spin gap": $\chi''(q, \omega \to 0)$ is either T independent or increases as T decreases. The antiferromagnetic correlations must be nearly commensurate (i.e. peaked near $q = (\pi/a, \pi/a)$) (although how large a deviation from commensurability is allowed by the data has not been established), and must be characterized by a correlation length which varies approximately as $T^{-1/2}$ at least for $T > 100K$, and is of order 3 lattice constants at $T = 100K$.[58] The T dependence of the correlation length has clearly ceased for temperatures less than the superconducting transition temperature T_c.[47,58] Different data lead to different conclusions as to whether the T-dependence of the correlation length ceases at T_c or at another temperature scale $T^* > T_c$. Regardless, the energy scale governing the T-dependence of the correlation length is very low when compared with the electronic energy scales discussed in section III. The origin of this small energy scale is not understood. In Fermi liquid based models, such a small energy scale can only be obtained by "fine-tuning" microscopic parameters.[22,58] Such fine tuning seems unreasonable, given the qualitatively similar behavior of the planar Cu relaxation rate in various superconducting cuprates.

In terms of Eq. 5.4, the oxygen and yttrium relaxation rates are well fit by a $\chi'(q,T) \approx \chi_s$, where χ_s is the $q = 0$ spin susceptibility discussed in section III and a $\Gamma(q,T) \approx 0.4\ eV$[58] (for a typical q not near $q = (0,0)$ or $q = (\pi/a,\ \pi/a)$), these values are changed from band theory values[58] ($\chi_s \sim 3\chi_s^{band}$ and $\Gamma \sim \Gamma^{band}/4$) but the renormalizations are not unreasonably large.

The copper relaxation rates are dominated by the zone corner contributions, with q near $Q^* = (\pi/a,\ \pi/a)$. One finds[58] $\Gamma(Q^*,T) \sim \chi'(Q^*,T) \sim 1/T$ and $\chi'(Q^*,T = 100K) \sim 20\chi_s$, and $\Gamma(Q^*,T = 100K) \sim 20\ meV \sim \Gamma/20$. Thus the antiferromagnetic enhancements are large, although the correlation length is not. Note that at e.g. $T = 100K$ χ'' is found to vary by a factor of ~ 400 between zone center and zone corner. The magnitude of this variation comes directly from the enhancements of the relaxation rates over their Fermi liquid value, and is not dependent on a specific model for χ''. However the values for χ' and Γ^{-1} and the correlation length depend on the model for the antiferromagnetic enhancements. The values quoted come from a mean field theory treatment; a model with different exponents would lead to different values. It is at present not clear whether the large antiferromagnetic enhancements required to explain the NMR and NQR experiments are consistent with the results of neutron scattering experiments discussed in section IV.

We now consider relaxation rate experiments on reduced T_c $YBa_2Cu_3O_{7-\delta}$ and on the $La_{1.85}Sr_{.15}CuO_4$ materials. There is less experimental data available, and it is not clear how to model the data which are available. As discussed previously, the static susceptibility in the normal state shows substantial T dependence. This T dependence is not understood. In reduced T_c $YBa_2Cu_3O_{7-\delta}$ the oxygen[54] and yttrium[49] relaxation rates decrease more rapidly than T as T is lowered, suggesting that spin spectral weight at typical q values is pushed gradually away from $\omega = 0$ as T is lowered. In yttrium it is claimed that the ratio of the relaxation rate to the square of the Knight shift has the Korringa temperature dependence ($^{89}W/K_s^2 \sim T$).[49] For oxygen, it is claimed that the ratio of the relaxation rate to one power of the Knight shift has the Korringa temperature dependence[54] ($^{17}W/K_s \sim T$). Neither of these T dependences is understood. In contrast to the $YBa_2Cu_3O_7$ material, in the reduced T_c $YBa_2Cu_3O_{7-\delta}$ system the ratio of the copper relaxation rate to the temperature is non-monotonic, increasing as T is lowered for $T > 150K$ and decreasing as T is lowered for $T < 120K$.[53,54] However, the ratio of the copper to the oxygen relaxation rates is a monotonic function of temperature, increasing as T is lowered as in the $YBa_2Cu_3O_7$ material.[54] The behavior of the Cu relaxation rate in the $La_{1.85}Sr_{.15}CuO$ system is similar to that found in the $T_c = 60K$ $YBa_2Cu_3O_{6.6}$ material.[48] Other relaxation rates have not been measured.

The T-dependence of the copper relaxation rate in $YBa_2Cu_3O_{6.6}$ and $La_{1.85}Sr_{.15}CuO_4$ is not understood. It has been interpreted as a precursor of the superconducting transition[53] and, in purely magnetic terms, as due to the opening of a spin gap.[57] The spin gap must be small because the temperature variation of all relaxation rates is not exponential. The apparently monotonic T dependence of the ratio of the copper to the oxygen relaxation rates suggests that antiferromagnetic correlations continue to grow as T is reduced, even as spin fluctuation spectral weight is pushed away from $\omega = 0$. It is clear, however that more experimental, phenomenological and theoretical work is required.

The discussion so far ahs been based on a one component model. Monien, Pines and Slichter,[56] Monien and Pines,[59] and Cox and Tree[57] have considered two component models. It is the opinion of this author that the discussion in section II and the quantitative success of one-component models[22,58] make it unlikely that these are relevant.

A phenomenological model which accounts for the T dependence of the Cu relaxation rate (at least at $T \gtrsim 120K$ in $YBa_2Cu_3O_7$) has been proposed by Varma, Abrahams, Littlewood, Ruckenstein and Schmitt-Rink.[60] In their model, which is a one-component model, susceptibilities such as $\chi''(q,\omega)$ are given by the sum of two terms: a quasiparticle contribution which is of approximately Fermi liquid form and another contribution arising from a polarization part which is roughly q independent and is assumed to scale as ω/T for $\omega < T$ and to be constant for $\omega > T$. Using their ansatz they can account for the T-dependence of the Cu relaxation rate for $T \gtrsim 120K$ in the $YBa_2Cu_3O_7$ material. The attractive feature of their proposal is that several experiments, including NMR, optical conductivity, tunneling and Raman scattering may be explained within the same framework. However, it will be difficult to account for the oxygen and yttrium data in the $T_c = 90K$ material in this model.

To summarize, the magnetic relaxation data suggest the presence of T-dependent antiferromagnetic correlations in superconducting cuprates and, in materials with $T_c < 90K$, an anomalous reduction in the low ω spin density of states as T is lowered. The essential assumptions on which these results depend are (i) that the relaxation experiments measure the coupling of the nuclear moments to the electronic spin degrees of freedom and (ii) that a one band model provides an accurate description of the low ω spin degrees of freedom. Given these two assumptions, the difference in the magnitude and T dependence of the Cu and O relaxation rates can only be due to the buildup of T-dependent antiferromagnetic correlations. However, it is not clear whether the correlations required to explain the magnetic relaxation data are consistent with what is seen in the neutron scattering experiments. As yet no microscopic model has been constructed which can explain the data without fine-tuning the parameters of the theory.

CONCLUSION

Very little is known with certainty about spin fluctuations in superconducting cuprates. On the theoretical side, it seems likely that the correct Hamiltonian with which to describe spin fluctuation phenomena is an effective one-band model such as the Hubbard or t-J model. Although these models have been intensively studied, there are few results which may be quantitatively compared with experiment at the carrier concentrations of interest. The crossovers from weak to strong coupling and from low to high doping are not yet understood. Further, it is not known whether these models produce high temperature superconductivity. On the experimental side, the situation is also unclear. The quasiparticle plasma frequency and bulk static magnetic susceptibility scale differently with doping. This behavior is consistent with that expected of holes doped into a system of strongly antiferromagnetically correlated spins, and therefore provides indirect evident for interesting spin dynamics in superconducting cuprates. More direct evidence comes from neutron scattering studies, in which finite range, finite lifetime, nearly commensurate antiferromagnetic correlations are observed. The detailed frequency, momentum and temperature dependence of these correlations has yet to be unravelled. Nuclear magnetic and quadrupole resonance experiments have been interpreted as providing a much more detailed picture of the very low frequency spin dynamics. The interpretation depends crucially upon two assumptions: that a one band model suffices to describe the spin dynamics and that spin fluctuations provide the dominant mechanism for nuclear moment relaxation. If these assumptions are correct, then the relaxation experiments imply that in the $YBa_2Cu_3O_7$ superconductor the dynamic susceptibility is that of a Fermi liquid with strong, temperature dependent, nearly commensurate antiferromagnetic correlations, and that in cuprate superconductors with lower transition temperatures more exotic behavior occurs.

Thus, there is experimental evidence of interesting magnetic behavior in superconducting cuprates, but as yet neither a comprehensive interpretation of the various data not an understanding of the relation between the observed magnetic behavior and high temperature superconductivity exists.

ACKNOWLEDGEMENTS

This article is one person's view of a large and rapidly growing area of research. It has proven impossible to survey all of the relevant experimental and theoretical work or to consider in detail all of the works I have cited, and I apologize to those whose work has been omitted or insufficiently discussed. My understanding of this subject owes much to conversations with P. C. Hammel, B. G. Kotliar, P. A. Lee, P. B. Littlewood, D. J. Scalapino, B. S. Shastry, M. Takigawa, C. M. Varma, R. E. Walstedt, W. W. Warren, Jr and, especially, to my two collaborators in analysis of magentic resonance experiments, Dr. Hartmut Monien and Professor David Pines.

REFERENCES

1. W. E Pickett, Rev. Mod. Phys **61**, 433,(1989).
2. See, e.g., P. B. Littlewood, C. M. Varma, S Schmitt-Rink and E. Abrahams, Phys Rev. **B39**, 12371 (1989), and P. B. Littlewood, unpublished.
3. S. Uchida, H. Takagi, Y. Tokura, N. Koshihara and T. Arima, in Strong Correlation and Electron Superconductivity, eds. H. Fukuyama, S. Maekawa and A. Malozemoff, Springer-Verlag, 1989, p. 194.
4. F. C. Zhang and T. M. Rice, Phys Rev **B37**, 3759 (1988).
5. M. Inui, S. Doniach, P. J. Hirschfeld and A. E. Ruckenstein, Phys Rev **B37**, 2320, (1988).
6. B. I. Shraiman and E. D. Siggia, Phys.Rev. Lett. **62**, 1564, (1989).
7. see, e.g. P. A. Lee, T. M. Rice, J. W. Serene, L. J. Sham and J. W. Wilkins, Comments on Condensed Matter Physics **12B**, 99 (1986).
8. A. K. McMahan, R. M. Martin and S. Satpathy, Phys. Rev. **B38**, 6650 (1988) and M. S. Hybertsen, M. Schluter and N. E. Christiansen, Phys. Rev. **B39**, 9028 (1989).
9. See, e.g., C. G. Olson, R. Liu, A.-B Yang, D. W. Lynch, A. J. Arko, R. S. List, B. W. Veal, Y. C. Chang, P. Z. Jiang and A. P. Paulikas, Science 245, 731 (1989) and T. Takahashi, H. Matsuyama, H. Katayama-Yoshida, Y. Okabe, S. Hosoya, K. Seki, H. Fugimoto, M. Sato and H. Inokuchi, Phys. Rev. **B39**, 6636 (1989). These and other angle-resolved photoemission studies find a Fermi edge consistent with bandstructure calculations. The bandstructure involves hybridized Cu-d and $O - p_\sigma$ orbitals. If the holes went into other orbitals weakly coupled to the Cu, the observed Fermi edge would be different.
10. Y. Guo, J. M. Langlois and W. A. Goddard III, Science **239**, 896 (1988).
11. M. S. Hybertsen, private communication.
12. D. M. Newns, M. Rasolt and P. C. Pattnaik, Phys. Rev. **B38**, 6513 (1988).
13. Ju. H. Kim, K. Levin and A. Auerbach, Phys. Rev. **B39** 11633 (1989).
14. M. Grilli, B. G. Kotliar and A. J. Millis, unpublished.
15. C. Lacroix, Solid State Comun. **54**, 991 (1985).
16. B. S. Shastry, Phys. Rev. Lett. **63**, 1288 (1989).

17. V. J. Emery and G. Reiter,Phys. Rev. **B38**, 11938 (1988).
18. S. R. White, D. J. Scalapino, R. L. Sugar, N. E. Bickers and R. T. Scalettar, Phys. Rev. **B39**, 839 (1989).
19. H. J. Schulz, unpublished
20. J. Zaanen and O. Andersson, unpublished.
21. D. J. Scalapino, this volume.
22. N. Bulut, D. W. Hone,D. J. Scalapino and N. E. Bickers, unpublished.
23. J. E. Hirsch and H. Q. Lin, Phys. Rev. **B37**, 5070 (1988).
24. J. R. Schreiffer, X. G. Wen and S. C. Zhang, Phys. Rev. Lett. **60**, 944 (1988) and A. Kampf and J. R. Schreiffer, unpublished.
25. D. Baerswyl, C. Gros and T. M. Rice, Phys. Rev. **B35**, 8391 (1986).
26. P. A. Lee, this volume.
27. V. J. Emery, this volume.
28. C. L. Kane, P. A. Lee, T. K. Ng, B. Charkaborty and N. Read, Phys. Rev. B, in press.
29. D. R. Harshman, G. Aeppli, B. Batlogg, G. P. Espinosa, R. J. Cava, A. S. Cooper and L. W. Rupp, Phys. Rev. Lett. **63**, 1187 (1989).
30. D. R. Harshman,L. F. Schneemeyer, J. V. Waszczak, G. Aeppli, R. J. Cava, B. Batlogg, L. W. Rupp, E. J. Ansaldo and D. L. Williams, Phys. Rev. **B39**, 851 (1989).
31. See, e.g. W.W. Warren, Jr. and R. E. Walstedt, Proceedings of the 10th International Symposium on Quadrupole Resonance Spectroscopy, Takayama, Japan, August 1989, to be published.
32. F. Mila and T. M. Rice, Physica **C157**, 561 (1989).
33. A. Junod, A. Bezinge, D. Cattani, J. Cors, M. Decrous, O. Fischer, P. Genoud, L. Hoffman, J. L. Jorda, J. Muller and E. Walker, Jpn J. Appl. Phys **26**, 1119 (1987); A. Junod, A. Bezinge and J. Muller, Physica **C152**, 50 (1988).
34. Y. Yamaguchi, M. Tokumoto, S. Waki, Y. Nakagawa and Y. Kimura, Proceedings of the Tsukuba Seminar on High Temperature Superconductivity, eds K. Masuda, T. Arai, I. Iguchi and R. Yoshizaki, (University of Tsukuba Press, Tsukuba: 1989) p. 31.
35. J. Rossat-Mignod, L. P. Regnault, M. J. Jurgens, C. Vettier, P. Burlet, J. Y. Henry and G. Lapertot, Proceedings of the 1989 International Conference on the Physics of Highly Correlated Electron Systems, Santa Fe, in press.
36. C. H. Pennington, D. J. Durand, C. P. Slichter, J. P. Rice, E. D. Bukowski and D. M. Ginsberg, Phys. Rev. **B39**, 2902 (1989).
37. D. C. Johnston, Phys. Rev. Lett. **62**, 957 (1989).
38. J. Orenstein, G. A. Thomas, A. J. Millis, S. L. Cooper, D. H. Rapkine, T. Timusk, L. F. Schneemeyer and J. V. Waszczak, unpublished.
39. T. Bruckel, H. Capellmann, W. Just, O Scharpf, S. Kemmler-Sach, R. Kiemel and W. Schaefer, Europhysics Letters 4, 1189 (1987).
40. R. J. Birgeneau, Y. Endoh, Y. Hidaka, K. Kakurai, M. A. Kastenr, T. Murakami, G. Shirane,T. R. Thurston and K. Yamada, Mechanisms of High Temperature Superconductivity, eds H. Kamimura and A. Oshiyama, Springer-Verlag (Berlin: 1989), p. 120.

41. G. Shirane, R. J. Birgeneau, Y. Endoh, P. Gehring, M. A. Kastenr, K. Kitazawa, H. Kojima, I. Tanaka, T. R. Thurston and K. Yamada, Phys. Rev. Lett. **63**, 330 (1989).
42. J. M. Tranquada, W. J. L. Buyers, H. Chou, T. E. Mason, M. Sato, S. Shamoto and G. Shirane, unpublished and J. M. Tranquada, this volume.
43. K. B. Lyons, P. A. Fleury, L. F. Schneemeyer and J. V. Wasczcak, Phys. Rev. Lett. **60**, 732 (1989).
44. S. Sugai, S. Shamoto and M. Sato, Phys. Rev. **B38**, 6436 (1988).
45. R. E. Walstedt, W. W. Warren, Jr., R. F. Bell, G. F. Brennert, G. P Espinosa, R. J. Cava, L. F. Schneemeyer and J. V Waszczak, Phys. Rev. **B38**, 9299 (1988).
46. C. H. Pennington, D. J. Durand, C. P. Slichter, J. P. Rice, E. D. Bukowski and D. M. Ginsberg, Phys. Rev. **B39**, 2902 (1989).
47. M. Takigawa, P. C. Hammel, R. H. Heffner, Z. Fisk, K. C. Ott and J. D. Thompson, Phys. Rev. Lett. **63**, 1865 (1989).
48. T. Imai, T. Shimizu, H. Yasuoka, Y. Ueda, K. Yoshimura and K. Kosuge, unpublished.
49. H. Alloul, T. Ohno and D. Mendels, Phys. Rev. Lett. **63**, 1700 (1989).
50. S. E. Barrett, D. J. Durand, C. H. Pennington, C. P. Slichter, T. A. Friedman, J. P. Rice and D. M. Ginsberg, unpublished.
51. E. Oldfield, C. Coretsopolous, S. Yang, L. Reven, H.C. Lee, J. Shore, O. H. Han, E. Ramli and D. Hinks, Phys. Rev. **B40**, 6832 (1989).
52. M. Horvatic, Y. Berthier, P. Butaub, Y. Kitaoka, P. Segransan, C. Berthier, H. Katayama-Yoshida, Y. Okabe and T. Takahasi, unpublished.
53. W. W. Warren, Jr., R. E. Walstedt, G. F. Brennert, R. J. Cava, R. Tycko, R. F. Bell and G. Dabbagh, Phys. Rev. Lett. **62**, 1193 (1989) and R. E. Walstedt, this volume.
54. M. Takigawa, this volume.
55. F. Mila and T. M. Rice, Physica C **157**, 561 (1989) and unpublished.
56. H. Monien, D. Pines and C. P. Slichter, unpublished.
57. D. L. Cox and B. R. Trees, unpublished.
58. A. J. Millis, H. Monien and D. Pines, unpublished.
59. H. Monien and D. Pines, unpublished.
60. C. M. Varma, P. B. Littlewood, S. Schmitt-Rink, E. Abrahams and A. E. Ruckenstein, Phys. Rev. Lett. **63**, 1996 (1989).

DISCUSSION

S. Chakravarty: Andy, I think it's worth emphasizing, even though you've said it that in this analysis, however you do the analysis for the (π, π) contribution, one would lead to the conclusion that the antiferromagnetic correlation length would change by say, a factor of two between 300 degrees and 100 degrees. On the other hand, the neutron scatterers, and we're talking about the energy integrated $S(q)$, made a statement in their paper that there's no change in correlation length between 300 degrees and 5 degrees.

J. M. Tranquada: I don't think you can trust that statement any more. The way the data has been reinterpreted in terms of incommensurate peaks, the temperature dependence hasn't been taken into account.

A. Millis: May I just add one comment on that. It is clearly explicitly stated in the paper on lanthanum-strontium that Professor Chakravarty is referring to that the correlation length has no temperature dependence but I have been informed by Professor Birgenau (private communication) – I asked him explicitly about that – that at the moment all he is willing to say is that the temperature dependence is not nearly as strong as in the insulating material. But it is still not clear whether the neutron and NMR agree.

S. Chakravarty: I'm not taking an issue with it, what I'm saying is that one needs a theory where somehow or the other neutron could look at $(\pi - \pi)$, and whatever could happen could happen, but because NMR is looking over all q's and adding them up together one could see something different. Now, for example, Phil's theory that he was talking about yesterday, the orbital scattering. That could get you out of that problem altogether. The neutron correlation length could do anything at (π, π), and it is not so intimately tied with NMR.

D. Pines: The nice thing about the theory that Andy has described is that it makes very specific quantitative predictions, predictions for the neutron scattering, so that when the neutron scattering experiments are good enough to be compared with the theory, then one will see whether or not these predictions are correct.

A. Millis: In our calculations we use mean field exponents. If you like some other theory which has different exponents, the numbers are going to change by a lot; this may affect the comparison with neutron experiments.

C. M. Varma: Isn't it true that after all the calculations in the antiferromagnetic fluctuation model, what's basically being done is of the same phenomenological form that the five friends suggested. You have this extra parameter, the Tx, which is effectively zero for us, and you put something like 20 degrees, but the point simply is that without this new parameter you can get in the 90 degree material, just as good as this, if you take into account the log corrections for the Fermi liquid term.

The second point is that I completely agree with David, that this theory makes a very specific statement about what has to happen to $\chi''(q)$ for the thing to at all make sense. And what it requires is so implausible – it requires that at (π, π), the susceptibility be about a hundred times larger than the Pauli susceptibility. That's so implausible that I can't...well, I personally don't have to wait for experiments, for that.

Then the question comes about the 60 degree material. In looking through, over the last three years, enormous amount of data in these materials, I've become very wary of looking at properties which are qualitatively different from one kind of copper oxide material to another kind of copper oxide material. The 60 degree T_c material in NMR is qualitatively different than the 90 degree material. I also have seen on the 60 degree materials, two specific heat curves, one from Illinois, and one from Geneva. And these show essentially nothing in the specific heat in the superconducting transition. So I would be very wary of trying to make theories on the basis of samples whose quality is very much in doubt.

D. Pines: Chandra, I think we should leave that to the experimentalists. We will hear both from Russ Walstedt and from Masashi Takigawa about the experimental situation...

C. M. Varma: It is too important a matter to be left only to the experimentalists!

A. Millis: Before we go on may I make one more remark. The antiferromagnetic correlations we require are very strong. The issue that I would raise for Dr. Varma and collaborators is that as far as I know, if you believe that NMR for oxygen and copper is coming from the same spin degrees of freedom, then the different magnitudes and T dependences of the relaxation rates for Cu and O directly imply the existence of strong antiferromagnetic correlations. These correlations are directly implied by the NMR data.

C. M. Varma: The basic point is that we could have put in a q-dependence and got this kind of theory. I simply do not believe there is a such a q dependence. I think there is something more subtle going on. I don't quite know what that is, but it is perfectly all right not to know, then to propose things which are just not going to be true.

D. Pines: This begins to be a discussion involving matters of taste. The basic question is do you want to try to explain simultaneously, at this point in time (so to speak) the results for both copper and oxygen. At the moment the theory which Milis has described is the only candidate phenomenological theory which can do that quantitatively...

P. W. Anderson: That's not true.

D. Pines: It remains to be seen if there are others which...

P. W. Anderson: Why did you say something which is not true?

D. Pines: Because I have yet to see a preprint which does the other. That's why. There is no preprint, to my knowledge, that provides a quantitative...

P. W. Anderson: It's been sent to *Nature* magazine.

D. Pines: Great.

D. Pines: Quantitative explanation, simultaneously, of Knight shifts, and of the relaxation rates...

P. W. Anderson: I haven't done the Knight shifts.

D. Pines: ...and relaxation rates for both copper and oxygen.

P. Lee: If you attempt to analyze the 60 degree $1/T_1$ data, and the strontium-doped material, which shows very distinctive deviation from this linear behavior, would you be forced into introducing a spin gap?

A. Millis: Yes, in the sense that because $(T_1 T)^{-1}$ decreases as T decreases the low ω spin density of states must decrease as T decreases. It is interesting that $(T_1 T)^{-1}$ for yttrium has the T dependence of the spin Knight shift squared (according to Alloul et al. PRL **63** 1700 (1988)) and $(T_1 T)^{-1}$ for oxygen has the T-dependence of the spin Knight shift (Takigawa, this conference). Explaining these T dependences in a spin gap picture may be difficult.

P. W. Anderson: I have several comments. One of them is that I was very happy with the statement Andy made about my theory, and it's quite correct. Second one was that the temperature fits that I made contained roughly one and a half parameters. There is one parameter of course nobody knows, which is the orbital contribution to the coupling constant, only one parameter that you have no estimate of. The other two parameters are approximately the right values.

A. Millis: Yes.

P. W. Anderson: These fits, although I've not actually been willing to take the various data, I believe they fit all the data simultaneously with the same curve, basically. Those are universal to all the data, do not require kind of a separate long argument for each individual substance. Those are the two comments about the T_1 fits.

One comment: I want to say that the anisotropy of the susceptibility and its temperature dependence is something that no one ever mentions, but I think it is important. The susceptibility itself is much less temperature dependent than is the anisotropy, and the anisotropy of the susceptibility has temperature dependences that are in the same ballpark as the kinds of temperature dependences we're talking about here. And this is true even for bismuth which has an isotropic spin susceptibility from absolutely flat, absolutely Fermi, and yet there is an anisotropy that seems to be somewhat temperature dependent. I'm with Chandra that one should really look for things that are common to most of these materials.

V. Emery: You didn't say much about the incommensurabilities I believe the best data on that is still in doped lanthanum copper oxide, and whether or not you call it incommensurability, there's a lot of weight away from (π, π). How much of a problem that is for you?

A. Millis: I don't have a precise answer. I have been informed by Dr. Hartmut Monien that using the correlation length which we have stated for $YBa_2Cu_3O_7$, we can tolerate the sort of incommensurability observed in $La_{1.85}Sr_{.15}CuO_4$, which is of the order of ten or fifteen per cent. However, doing the calculation is tricky, because there are two length scales: the correlation length, which is temperature dependent, and then the incommensurability, and you get funny temperature dependence, basically, if the ratio of these length scales changes from greater than unity to less than unity. And clearly, as you make the correlation length longer, you can tolerate less incommensurability. So the short answer is that I don't know in detail.

D. Pines: It will be extremely interesting to have the same quality NMR data on the 2,1,4 material for both the oxygen and the copper, and indeed the lanthanum, because again, once one has that, then I think the issue will become much clearer.

Antiferromagnetic Fluctuation Dynamics in the YBCO Superconductors

R. E. Walstedt
AT& T Bell Laboratories, Murry Hill, New Jersey 07974

It has long been suspected that antiferromagnetic fluctuations[1,2] enhance the nuclear spin-lattice relaxation rate T_1^{-1} at CuG sites in $YBa_2Cu_3O_{6+x}$. Recent calculations based on this assumption using mean-field models[3,4] for a single dynamic susceptibility $\chi''(\vec{q},\omega)$ have yielded simultaneous fits to experimental relaxation rates for ^{63}Cu, ^{17}O, and ^{89}Y nuclei in the conducting planes of YBCO. In this note, we use these ideas to discuss recent shift and relaxation data for the aforementioned nuclei in the 60K phase of YBCO,[5,7] and to compare their behavior with similar data for the 90K phase.

Data for the Cu(2) spin-paramagnetic shift in $YBCO_{6.64}$ (T_c - 60K)[8] are shown in the inset to Fig. 1, where it is seen to decline sharply with temperatures as T− > T_c. At the same time, Cu(2) relaxation data for this and comparable samples,[5,6] when plotted as $(T_1T)^{-1}$ vs. T (not shown), exhibit a peculiar maximum just above T = 100K. We discuss these results in terms of a mean-field model susceptibility[4] $\chi''(\vec{q},\omega) \propto \omega\chi_0^s F(\vec{q})$, where χ_0^s (T) is the uniform spin susceptibility and $F(\vec{q})$ models a fluctuation peak near $\vec{q} = (\pi,\pi)$. Using $(T_1T)^{-1} = (2\gamma^2 k_B T/g^2\mu_B^2) \sum_{\vec{q}} A(\vec{q})^2 \chi''(\vec{q},\omega/\omega)$, where $A(\vec{q})$ is the q-dependent hyperfine coupling parameter, we can extract the behavior of $\sum_{\vec{q}} A(\vec{q})^2 F(\vec{q})$ by plotting the modified Korringa ratio (MKR) $(T_1(Korr.)/T_1(T))(K_{2ab}^s(T)/K_{2ab}^s(300K))$ vs. T, where $T_1(Korr.)^{-1} = 4\pi k_B T\gamma_n^2 K_{2ab}^2/(N_s\gamma_e^2\hbar)$ is the relaxation rate corresponding to shift K_{2ab}^s in a system of non-interacting fermions. The factor N_s is placed in the denominator to account for multiple sources of hyperfine fluctuations.[9] The MKR for $^{63}Cu(2)$ obtained with a sample of $YBCO_{6.64}$ (T_c = 60K)[5] is plotted in Fig. 1 (dots). For comparison we also plot $T_1(Korr.)/T_1(T)$ for $^{63}Cu(2)$ in 90K material[10] (open circles), assuming K_{2ab}^s = 0.30%,[11] independent of temperature. Both sets of Cu(2) data show a temperature variation $\sim T^{-1/3}$, with the enhancement for the 60K sample nearly an order of magnitude greater. By this analysis, the maxima in $(T_1T)^{-1}$ reported for the 60K phase are simply an artifact of the declining density of states at low temperature.

As a control for the foregoing procedure, we also plot the MKR for ^{89}Y NMR data on $YBCO_{6.63}$[7] in Fig. 1 (filled squares). The spin-paramagnetic shift component K_s is obtained from the data of Ref. 7 as follows. By plotting measured ^{89}Y

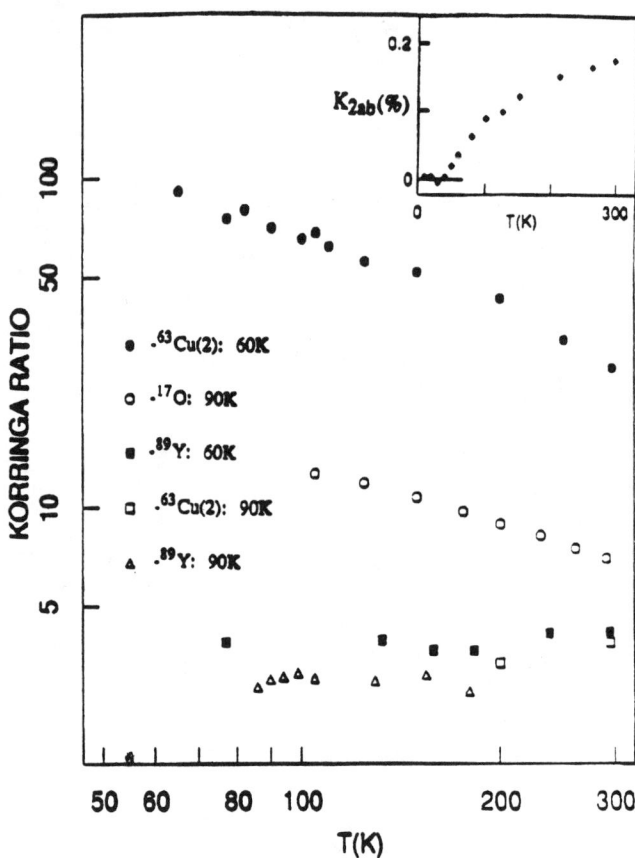

FIGURE 1 The modified Korringa ratio (MKR) and the Korringa ratio, is plotted vs. T for ^{63}Cu(2) and ^{89}Y in $T_c = 60K$ material (solid points). For comparison, the simple Korringa ratio (=MKR for T-independent shift) is plotted for ^{63}Cu(2), ^{89}Y, and ^{17}O(2,3) nuclei in $T_c = 90K$ material (open points). The inset shows the temperature variation of the spin-paramagnetic shift for ^{63}Cu(2) in a (60K) sample of YBCO$_{6.64}$.

shifts against the Cu(2) shift data (Fig. 1, inset) and extrapolating to the zero of the copper shift, we find $K_0 \sim 175$ ppm for the zero of spin paramagnetic ^{89}Y shift. Using this procedure, we see that the MKR is independent of temperature, in accord with the antiferromagnetic fluctuation model result.[4,5] A similar result has been reported for ^{17}O nuclei in 60K phase material.[12] The point here is that the hyperfine form factors for ^{89}Y and ^{17}O, i.e., $A(\vec{q})$, vanish in the vicinity of

$\vec{q} = (\pi, \pi)^{4,13}$ rendering these nuclei impervious to the antiferromagnetic fluctuation peak. We compare the latter results with the Korringa ratio $T_1(\text{Korr.})/T_1$ for ^{89}Y in 90K material[14] (open squares), using the zero of shift as determined above, and with the same quantity for ^{17}O, using new T_1 data with the isotropic spin-paramagnetic shift value given by Takigawa et al.[16] (triangles). These results show that the relaxation of ^{89}Y and ^{17}O nuclei in either phase is unaffected by the antiferromagnetic fluctuation peak, for the reason noted above.

To conclude, the Cu(2) shift results (Fig. 1, inset) appear to be successfully interpreted as a temperature-varying density of states. The value of the spin susceptibility at T_c is estimated to be lower than that of the 90K phase by a factor -10, suggesting a similar drop in the density of states. Indeed, specific heat results for the 60K phase give little or no disturbance at T_c,[16] in comparison with a very substantial specific heat peak for the 90K phase.[17] In terms of mean-field model susceptibility models[3,4] both 60K and 90K phase data give a monotonically increasing antiferromagnetic correlation length as temperature is lowered, with much stronger correlations in the 60K phase. It is interesting to note that these results are in qualitative accord with a recently calculated phase diagram for the three-band Hubbard model.[18]

The author wishes to acknowledge the collaboration of W. W. Warren, Jr., and many illuminating discussions with A. Millis.

REFERENCES

1. T. Imai, T. Shimizu, T. Tsuba, H. Yasuoka, T. Takabatake, Y. Nakazawa, and M. Ishikawa, J. Phys. Soc. Japan **57**, 1771 (1988).
2. R. E. Walstedt, W. W. Warren, Jr., R. F. Bell, G. F. Brennert, G. P. Espinosa, R. J. Cava, L. F. Schneemeyer, and J. V. Waszczak, Phys. Rev. B**38**, 9299 (1988).
3. N. Bulut, D. Hone, D. J. Scalapino, and N. E. Bickers, unpublished.
4. A. Millis, H. Monien and D. Pines, unpublished.
5. W. W. Warren, Jr., R. E. Walstedt, G. F. Brennert, R. F. Cava, R. Tycko, R. F. Bell, and G. Dabbagh, Phys. Rev. Lett. **62**, 1193 (1989); W. W. Warren, Jr., R. E. Walstedt, G. F. Brennert, R. F. Bell, G. P. Espinosa, and R. J. Cava, Proc. Int. Conf. on High-Temperature Superconductivity, Stanford, CA (1989).
6. T. Imai, T. Shimizu, H. Yasuoka, Y. Ueda, K. Yoshimura, and K. Kosuge, unpublished.
7. H. Alloul, T. Ohno, and P. Mendels, Phys. Rev. Lett. **63**, 1700 (1989).
8. R. E. Walstedt, W. W. Warren, Jr., R. F. Bell, R. J. Cava, G. P. Espinosa, L. F. Schneemeyer, and J. V. Waszczak, unpublished.
9. In the present instance, $N_s = 4$ for Cu(2), 2 for O(2,3), and 8 for Y. This correction is only accurate if one neglects dynamical correlations between Cu(2) spins and is thus only approximately correct here.

10. R. E. Walstedt, W. W. Warren, Jr., R. F. Bell, and G. P. Espinosa, Phys. Rev. B**40**, 2572 (1989).
11. S. E. Barrett, D. J. Durand, C. H. Pennington, C. P. Slichter, T. A. Friedmann, J. P. Rice, and D. M. Ginsberg, unpublished.
12. M. Takigawa *et al.*, see paper presented at this conference.
13. B. S. Shastry, Phys. Rev. Lett. **63**, 0000 (1989).
14. G. Balakrishnan, R. Dupree, I. Farnan, D. McK. Paul and M. E. Smith, J. Phys. C**21**, L847 (1988).
15. M. Takigawa, P. C. Hammel, R. H. Heffner, Z. Fisk, K. C. Ott, and J. D. Thompson, Phys. Rev. Lett **63**, 1865 (1989).
16. J. E. Crow, private communication.
17. A. Junod, A. Bezinge, and J. Muller, Physics C**152**, 50 (1988).
18. P. B. Littlewood, C. M. Varma, and E. Abrahams, Phys. Rev. Lett. **63**, 2602 (1989).

Discussion

E. Abrahams: The use of the Korringa product is very misleading. The Korringa product has the square of the Knight shift in it and the reason for this is (Andy Millis had it on his viewgraph) that for a free electron gas, the characteristic energy scale and q=0 spin susceptibility are inversely proportional. So if you use the square of the Knight shift in the Korringa Product, that means you're using the square of the susceptibility. However the physics is, that only one susceptibility should go in there, and the other susceptibility really was the density of states, or some inverse energy scale. Then, if you have an anomalously temperature-dependent susceptibility, which you do have in the oxygen-depleted compounds using the Korringa products as given for a free electron gas is completely misleading.

R. Walstedt: Well I think Masashi is going to show you...in fact, Andy already showed you what happens when you just take out one factor...if we take out one factor of χ, instead of two, it's not quite as bad, but it still goes up quite high. Suppose you just normalize here, and just take out a factor of five that occurs by the time you get to 60 degrees, you still end up at a very large number.

D. Pines: There is perhaps a more illuminating way to look at this, which relates to examining the ratio of the copper/oxygen relaxation rates as a function of temperature. What one sees, is that this ratio goes, say, at 100 degrees, from being nineteen in the case of the 90 degree Kelvin material, to being around forty-eight, in the case of the 60 K material. That ratio gives a rather direct measurement of what the antiferromagnetic enhancement is doing as a function of temperature.

R. Walstedt: Let me speak up for the yttrium; the yttrium is not consistent with this.

D. Pines: ...It picks up the behavior of $\chi_o(T)$, but does not see, in our calculations, the temperature dependence of antiferromagnetic correlation length.

R. Walstedt: The yttrium is really behaving somewhat differently, and don't throw out the yttrium...it is in a more highly symmetric location in this lattice. It has eight oxygens around this, which all contribute to the fluctuations, and so forth. So I think there's more enhancement going on at the oxygen than the yttrium, and more at the copper than at the oxygen. The yttrium may be the kind of neutral probe that everything needs to be referred to.

A. Millis: Within this one-component picture, oxygen and yttrium relaxation rates should have essentially the same T-dependence, and they're behaving quite differently, as far as I can see. Because yttrium really does have two powers of the susceptibility – if you believe Russ's statement that the Korringa product is just one overall at all temperatures – and oxygen only seems to have one. So there is something funny going on there.

D. Pines: Whereas yttrium and oxygen seem to behave the same way in the 90 degree material.

T. M. Rice: You show that there is a substantial Curie term in the susceptibility. Could it be that there are some relatively free copper moments which could give rise to fast relaxation process. So, in other words, could it be that this relaxation rate is only for part of your sites?

R. Walstedt: By now the relaxation rate has been observed by us at Murray Hill, by Imai *et al.* in Tokyo, and by Takigawa here in Los Alamos. We all have different samples with different amounts of the background Curie term, and we all get the same T_1 results. I would have to say that there is very good correspondence between different workers on all of these results at this point.

M. Takigawa: I have a question about the yttrium data. Even with an improved estimate of the yttrium chemical shift, do you still have a nice Korringa relation, i.e. $T_1 T K^2$ = constant when the temperature and concentration are changed?

R. Walstedt: You mean changing the zero of the shift?

M. Takigawa: That's right.

R. Walstedt: I haven't really worked that out. We now have to go back and look at Alloul's analysis in a new light, because his analysis implied a zero for the shift which is different from what we now find and so there's an inconsistency there which bothers me a little bit, because he claims they had also a Korringa product which was near unity, for the yttrium. We use a different Shift zero, we find unity, I couldn't believe it so I had to do it twice. We still find unity. I don't know what's going on there.

D. J. Scalapino: We have done some calculations of χ on a copper-oxygen lattice. The reason we did this was that we held the impression that the coupling to the yttrium nuclei was more naturally coming from the oxygen sites rather than the copper sites. Frank Adrian did some work on this. We found within an RPA calculation that if we put a Coulomb interaction U on the copper, it acted via the CuO transfer to produce an enhanced response on the O.

R. Walstedt: Frank Adrian.

D. J. Scalapino: Frank Adrian argued that the yttrium nuclei had a transferred hyperfine interaction from the four oxygen above and the four oxygen below it.

A. Millis: How many independent spin degrees of freedom do you have, I mean is this simply a matrix element?

D. J. Scalapino:: At the Stanford meeting, where we first presented our work, we discussed both a single-band Hubbard model and a three-band CuO_2 Hubbard model. In general χ is a 3 × 3 matrix for the CuO_2 three-band model. However, if the nuclear spins only couple to the Cu spins, then just the d-d part of χ is needed. With U only on the copper site, we found similar results for the Cu and O spin relaxation times for the CuO_2 three-band and the one-band Hubbard models.

A. Millis: I think the bottom line must be that $\chi''(q,\omega)$, as ω goes to zero, is a one-band thing. And so, all of this stuff that you talked about, or that Adrian talked about, I think can only be matrix elements, and I don't see how it can give a different temperature dependence.

D. Pines: Or turn it around. If there is a genuine different temperature dependence then this argues against the one-band model.

D. J. Scalapino: Our picture has been from the beginning that we were dealing with a one-component situation, as opposed to the early two-component models, but how those carriers are spread out on the lattice could make a difference if the yttrium is coupled to the oxygens.

A. Millis: But that's what I mentioned: how could it make a difference to the temperature dependence as opposed to the values of the matrix elements?

D. J. Scalapino: We found, within an RPA calculation for the three-band model, that there are two terms which contribute to χ on the oxygen site. One is the $U = 0$ noninteracting band contribution and the second comes from the copper through the Cu-O one-electron transfer and reflects the strong U induced antiferromagnetic correlations on the copper. Thus if the Cu and O nuclei are dominantly coupled to the Cu spins and Y is coupled to the O spins, there could be a different response for the Y nuclei.

D. Pines: Are you saying perhaps, Doug, that the matrix element is temperature dependent, therefore Andy's point of view and yours can be reconciled, if you argue this matrix element is temperature dependent. Are you saying that?

D. J. Scalapino: The transfer matrix element is just the hopping t. It's the RPA denominator for the copper susceptibility that is coupled to χ_{oxygen} that is quite temperature sensitive and has a different temperature dependence from the $U = 0$ band contribution which also enters χ_{oxygen}. What I'm trying to say is that within RPA we find two pieces in χ_{oxygen} and they have different temperature dependencies. Only one has the same temperature dependence as χ on the copper site.

V. Emery: I wonder if you have any evidence from different oxygen concentrations, but within the plateau region, of any mixed-phases ? They might occur particularly at the lower oxygen concentration.

R. Walstedt: Not that I know of. Different workers seem to determine their composition in different ways, so it's hard to know exactly, you know, how you compare. But, for example, the Tokyo data claims that they're at composition 6.52. We're at 6.64, and 6.7, and their relaxation data map onto ours perfectly. So somehow, within that plateau, it seems that things are fairly constant, and we really see no evidence of mixed-phase behavior.

V. Emery: It's not really a matter of going from one place to the other in the plateau, but stick to a particular oxygen concentration and ask if you have two different hole concentrations – that kind of inhomogeneity – within one sample. Perhaps you would find two relaxation rates in your data.

R. Walstedt: Well, all I can say is we really have no evidence for that. The relaxation was done with NQR. All of these samples give NQR peaks at the same frequencies, and we do the relaxation on those peaks. It is very hard to say. If there are small inhomogeneities, we may not pick them up, but we find no evidence for problems.

Copper and Oxygen Nuclear Magnetic Resonance in $YBa_2Cu_3O_{6.63}$ (T_C = 62K)

Masashi Takigawa

Los Alamos National Laboratory, Los Alamos, NM 87545

In this short note, I will discuss the results of the Knight shift and the nuclear relaxation rate at the planar Cu and O sites in $YBa_2Cu_3O_{6.63}$ (T_c=62K) in comparison with the previous results in $YBa_2Cu_3O_7$ (T_c=90K).[1,2]

There has been a controversy about whether the magnetic properties of the CuO_2 planes are described by a single component spin which mainly resides on the Cu sites or two distinct spin degrees of freedom are required, associated with Cu d-spins and O holes. An experimental clue to this question has been given by the Cu and O Knight shift in the T_c=60K material. Unlike the T_c=90K material, various principal components of the Cu and O Knight shift are strongly T-dependent in the normal state. A remarkable fact is that the principal components of the Cu and O Knight shift follow the same T-dependence. More precisely, all these components are coupled to a common T-dependent spin susceptibility.

$$K_i(T) = A_i \chi_s(T) + K_i(0) , \qquad (1)$$

with $K_i(0)$ being consistent with the orbital contribution to K_i determined from the results in the T_c=90K material.[1] It should be noted that the axial part of the Cu and O spin Knight shift are coupled exclusively to the spin density on the Cu-3d and O-2p states, respectively.[1] Therefore, the above result strongly supports the single component model in the sense that the spin density on the Cu and O sites behave as parts of the same spin system.

This result allowed us to extract the temperature dependence of the spin susceptibility χ_s associated with the CuO_2 plane as shown by the line in Fig. 1a. χ_s shows a strong reduction with decreasing temperature in a wide temperature range above T_c. This apparent suppression of the spin excitations in the reduced T_c material is in strong contrast to the Pauli-like T-independent susceptibility in the T_c=90K material.

The nuclear relaxation behavior in the reduced T_c material is also quite different from that in the T_c=90K material. $1/(T_1T)$ at the O sites shows a significant reduction with decreasing temperature in the normal state. It is found that the T-dependence of $1/(T_1T)_0$ is very similar to that of the spin Knight shift, i.e. $1/(T_1T)_0 \propto \chi_s(T)$, as shown in Fig. 1a. This is in contradiction to the argument by Alloul et al. that $1/(T_1T)$ at the Y sites is proportional to the square of K_{spin} (Y).[3] It is found, however, that the contradiction is resolved simply by taking the chemical shift at the Y sites to be 170 ppm in agreement with Walstedt[4] instead of 300 ppm given by Alloul et al.[3] $1/(T_1T)$ at the Cu sites shows a broad maximum at 160 K in agreement with the published data (Warren et al.[5] and Yasuoka et al.[6]). The reduction of $1/(T_1T)$ below 160 K in the normal state has been discussed as

Spin Dynamics of Superconducting Cuprates

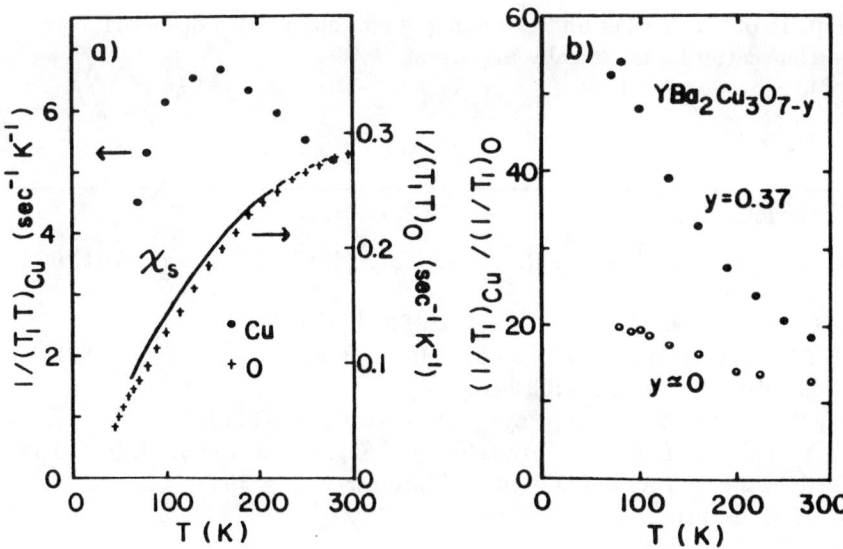

FIGURE 1a Temperature dependences of $1/(T_1T)$ at the planar Cu and O sites and the spin susceptibility of the CuO_2 planes (solid line, arbitrary units) in $YBa_2Cu_3O_{6.63}$. b). The ratio of $(1/T_1)$ at Cu to $(1/T_1)$ at O is plotted as a function of temperature for the T_c=90K (from ref. 2) and the T_c=60K materials.

an evidence of a gap opening in the spin excitation spectrum. These behaviors are far less understood than the T_c=90K material. It is interesting, however, to see to what extent the model for the antiferromagnetic (AF) correlation among Cu spins extracted from the nuclear relaxation data in the T_c=90K material[2,7] can be extended to the T_c=60K material.

Because of the q-dependent hyperfine coupling, $1/(T_1T)$ at O sites is determined by spin fluctuations in a rather broad q-space around the zone center.[2,7] Therefore, $1/(T_1T)_O \propto \sum_{q\sim 0} Im\chi(q,\omega_o)/\omega_o \propto \chi_s(T)/\Gamma$, where Γ is the energy width at a typical q not close to $Q=(\pi,\pi)$. The fact that $1/(T_1T)_O \propto \chi_s(T)$ implies that Γ is T-independent. $1/(T_1T)$ at the Cu sites, on the other hand, is determined by spin fluctuations near $Q=(\pi,\pi)$. If we assume that the staggered susceptibility χ_Q is the product of the uncorrelated susceptibility χ_{Q_o} and the enhancement factor due to AF correlation, $\chi_Q = \chi_{Q_o}f(\xi)$ (ξ: AF correlation length), and χ_{Q_o} has the same T-dependence as χ_s, the ratio of $1/(T_1T)_{Cu}$ to $\chi_s(T)$ or equivalently $(1/T_1)_{Cu}/(1/T_1)_O$ can be a measure of the AF enhancement factor $f(\xi)$ ($f(\xi) \propto \xi^2$ in the analysis in ref. 7). The ratio $(1/T_1)_{Cu}/(1/T_1)_O$ is shown in Fig. 1b for both the T_c=90K and the T_c=60K materials. This ratio is smoothly increasing with decreasing temperature in the T_c=60K material and is much larger than that in the T_c=90K material. Therefore, a peak in $1/(T_1T)_{Cu}$ may not necessarily require a spin gap. To obtain further understanding, direct observation of the AF correlation by neutron scattering is crucially important.

REFERENCES

1. M. Takigawa et al., Phys. Rev. B39, 7371 (1989), Phys. Rev. Lett. 63, 1865 (1989).
2. P. C. Hammel et al., Phys. Rev. Lett. 63, 1992 (1989).
3. H. Alloul, T. Ohno and D. Mendels, Phys. Rev. Lett. 63, 1700 (1989).
4. R. E. Waldstedt, to be published.
5. W. W. Warren, Jr., et al., Phys. Rev. Lett. 62, 1193 (1989).
6. H. Yasuoka et al., Strong Correlation and Superconductivity, ed. H. Fukuyama, S. Maekawa and A. P. Malozemoff (Springer-Verlag, 1989).
7. A. J. Millis, article in this volume.

Discussion

D. Pines: I'll just show the transparency that Masashi mentioned. It displays the preliminary fit to the 60 degree material we have obtained using exactly the same kind of phenomenology that was required for the 90 degree material. In other words, it is an indication that nothing mysterious is going on, that one needs no spin gap in order to explain the experimental results. The only thing that has changed is the correlation length, which must be longer, and one sees very clearly that the correlation length goes up. At 90K, in the 60K material the square of the correlation length is around 20 whereas in the case of $YBa_2Cu_3O_7$ it is about 8. So there's a difference there of a factor of two or so in the correlation length.

S. Doniach: I'd just like to make the suggestion that really we have a two-fluid situation here in the 60 degree material. Something is making the susceptibility go down. Apparently once you've done that you can explain everything using the antiferromagnetic correlations, but you've got to say why it went down in the first place. This could be a two-fluid picture, where you're getting some kind of changing of the carrier gas so that part of it, you know, is going away from the susceptibility, and the rest is magnetic, and I think some kind of pairing might be going on above T_c in these materials.

D. Pines: Seb, there is perhaps a more prosaic explanation. If one believes that the change in χ_o with doping is correlated with the change in the correlation length, ξ, the way you see doping most clearly is that , as you reduce the oxygen content, you see the correlation length go up, and that would explain why, at a fixed temperature, comparing the O_7 and the $O_{6.63}$, χ_o has gone down because the correlation length has gone up. So the physics is that because the spin liquid is occupied in being highly antiferromagnetic at large q, as a consequence the q=0 susceptiblity is reduced.

And that would explain two puzzles: one, why χ_o is temperature dependent in the reduced oxygen material; two, why the magnitude of χ_o at a fixed temperature is down for the reduced oxygen material compared to the O_7. A problem with that explanation is that it would also suggest you would get more of a temperature dependent χ_o in the O_7 material than seems to be seen. If, however, there were a threshold for that effect, this particular problem would go away.

M. Takigawa: If some second component has a significant contribution to the susceptibility, this will have a different weight on oxygen and copper from the underlying copper spin. Experimentally, the ratio of the spin density at copper to that at oxygen is independent of hole doping. This means the second spin component, if it exists, makes negligible contribution to the susceptibility.

H. Ott: How can you get this, what you labeled K_a, K_b, for the oxygen on a polycrystalline sample?

M. Takigawa: The NMR frequency shift is the sum of quadrupole shift and the Knight shift. I'm looking at the quadrupole satellite line, and there is good reason to expect that quadrupole shift is larger along the Cu-O bonding axis. So the Knight shift obtained from such satellite lines which have a larger quadrupole shift can be assigned to the Cu-O bond direction, i.e. K_a for $O(2)$ and K_b for $O(3)$ sites.

A. Millis: A comment and then a question. First of all, just in terms of the fit to the copper relaxation...one of the most attractive things, to my mind, about the phenomenology proposed by Chandra Varma and company is that there is no low-energy scale except the temperature. Now, it seems that when you compare the 90 and the 60 Kelvin $YBa_2Cu_3O_{7-\delta}$ superconductors our parameter T_x, which is a low-temperature scale, changes in proportion to T_c, and that leads one to wonder if you can get by without any T_x at all, and whether there really is nothing except T.

And a question: how do I reconcile this big change in the spin Knight shift with relatively small change in the observed susceptibility measured in bulk? Does the separation into Van Vleck and chain contributions change?

M. Takigawa: That's one thing, but actually the Van Vlech term cancels with the diamagnetic term so the measured susceptibility is almost equal to the spin term.

A. Millis: Is this the same as the measured, then?

M. Takigawa: No, the measured susceptibility still has some contribution from the second phase, which has a Curie term and there's a chain contribution which we don't know quite exactly.

A. Millis: I guess my question is, how big a contribution from chains, and what not, do you have to put in to make your measurement of the Knight shift consistent with the measured bulk susceptibility?

M. Takigawa: I haven't made any analysis of the chain susceptibility.

Spin Correlations and Pairing States in Superconducting $YBa_2Cu_3O_7$

David Pines

Department of Physics, University of Illinois at Urbana-Champaign, 1110 West Green Street, Urbana, IL 61801

In these remarks, I should like to emphasize the fact that just as is the case for the normal state properties, magnetic resonance experiments provide significant constraints on any description of the excitations in the superconducting states of the cuprate oxides. Measurements of the Knight shift[1] and relaxation rate[2] for ^{63}Cu nuclei and ^{17}O nuclei[3] in the superconducting state of $YBa_2Cu_3O_7$ reveal the following:

- Only a one-component description of the planar excitations responsible for the Knight shift and spin relaxation rate is viable. Thus while theories which invoke two components (spin and charge, say) for planar excitations to explain the different temperature-dependence of the ^{17}O and ^{63}Cu relaxation rates measured in the normal state cannot yet be ruled out, the constancy of the ratio, R, of these rates in the superconducting state between Tc and Tc/4[3] means there can be no spin/charge separation in the superconducting state.

- Both Knight shift[1] and relaxation rate[2,3] experiments tell us that spins participate fully in the superconducting pairing process; thus while a local moment description of the Cu(2) spins in the normal state is viable, in the superconducting state these spins display a behavior which cannot be distinguished from the charge degrees of freedom, and hence must be regarded as itinerant.

- The temperature independence of the relaxation rate ratio, R, requires as well that the anti-ferromagnetic correlations invoked to explain the normal state magnetic behavior[3,4] be temperature-independent below T_c.

- The temperature dependence of the $^{63}Cu(1)$ Knight shift[1] as well as the $^{63}Cu(1)$ relaxation rate[2] are markedly different from those measured for planar Cu(2) nuclei; hence the chain and plane excitations must be distinct in the superconducting state, (as they are in the normal state).

- Both the $^{63}Cu(2)$ and $^{17}O(2)$ relaxation rates display unconventional, i.e. non-BCS s-state pairing, behavior; there is no coherence peak, and the relaxation rate drops dramatically in the vicinity of T_c.

- The temperature dependence of the $^{63}Cu(2)$ Knight shift is inconsistent with a BCS weak-coupling description.[1,5]

Consider further the behavior of the planar excitations. Can these be described using an "unconventional" version of BCS theory? For example, can one invoke anisotropic s-wave pairing and attribute the absence of a coherence peak to quasiparticle-impurity scattering? The answer is "no" because the scattering rate required to reduce this peak significantly leads to a gradual reduction of the relaxation rate in the vicinity of T_c, in marked contrast to the measured rapid drop. A second possibility is singlet d-wave pairing. As Bertram Batlogg has told you, Monien and I find that if the excitations pair in a singlet d-state, with a line of nodes, an energy gap $\sim 5kT_c$, and strong coupling corrections which act to enhance the specific heat jump at T_c, we can obtain a quantitative fit to the ^{63}Cu(2) spin-lattice relaxation rate between T_c and $T_c/4$, and, within experimental error, fit as well the measured temperature dependence of the ^{63}Cu(2) Knight shift.[5] While such a fit is encouraging, it is not definitive, since we have not taken into account the inelastic scattering processes which lead quasiparticles in the normal state to possess comparatively short lifetimes ($\hbar/\tau \sim k_B T$), nor have we demonstrated how the experimental conditions for measuring the penetration depth might be such that one does not measure, in the temperature dependence of the latter,[6] any significant trace of the line of nodes posited to explain the resonance experiments.

An alternative possibility for reducing substantially the coherence peak might be anisotropic s-state pairing, combined with temperature-dependent feedback effects (vertex corrections). Monien and I found that one such mechanism, temperature-dependent antiferromagnetic enhancement in the superconducting phase, would have many of the desired consequences; however this possibility is ruled out by the measured temperature independent ratio of the ^{63}Cu and ^{17}O relaxation rates. A second feedback mechanism has been examined by Kuroda and Varma;[7] whether it can lead to quantiative agreement with experiment over a substantive range of temperatures below T_c remains to be determined.

REFERENCES

1. S. Barrett, *et al.*, Phys. Rev. B, in the press.
2. R. E. Walstedt, *et al.*, Phys. Rev. B **36**, 5727 (198; T. Imai *et al.*, J. Phys. Soc. Japan **57**, 2280 (1988).
3. P. C. Hammel *et al.*, Phys. Rev. Lett. **63**, 1392 (1989).
4. A. Millis, H. Monien, and D. Pines, preprint.
5. H. Monien and D. Pines, Phys. Rev. B, in press.
6. L. Krusin-Elbaum *et al.*, Phys. Rev. Lett. **62**, 217 (1989).
7. Y. Kuroda and C. Varma, preprint.

Discussion

D. J. Scalapino: I know I've asked this before, but maybe just for clarification...Could you say a bit more about how the calculation of $(1/T_1)$ for d-wave pairing is carried out?

D. Pines: We take strong coupling corrections into account by using a standard interpolation formula for the temperature dependence of the gap, which contains two fit parameters: the specific heat jump at T_c, which determines the slope of Δ^2, and the zero temperature gap.

D. J. Scalapino: I understand. Let me ask one more question.

The large antiferromagnetic fluctuations, which are being used in the normal state,..what are you doing with them in your calculation of $1/T_1$, and maybe more specifically, are they in that curve that you showed us as the solid line?

D. Pines: They are indeed. They are assumed to be temperature independent in the curve that I showed you. We had earlier, before the results of the Los Alamos group had been known, explored as a feedback mechanism the possibility that the antiferromagnetic correlations were temperature dependent, and we saw then that one could, even with an anisotropic s-state, begin to get rid of the coherence peak. But the expression I showed you has no temperature dependent antiferromagnetic correlations. Everything is measured with regard to the normal state. We, as you know, believe they are strong in the normal state. We are saying they are independent of temperature in the superconducting state and that's the result we get.

J. R. Schrieffer: Just to point out, that d-wave can mean very different things in different materials. I think this is now well known, but I just want to go on record as again stating that, if you take a Fermi surface that is up against a zone boundary – for example, in the insulator, due to antiferromagnetic order, or in the intermediate spin-range correlation-fluctuation model where you have a pseudo-gap (and it may well be that the carriers do not span an entire free Fermi surface, but are pockets against this nested pseudo-Fermi surface) – in that case, you may well have finite gaps all around the physical Fermi surface of each of the hole pockets, so that for all intents and purposes, from temperature dependence, you would call node-less. However, there can well be zeros of the gap where there are no holes, and hence it's like when a tree falls in a forest, is there a sound? I only raise that as a possibility. We have advanced this, I don't know if it's true, but I think one has to be very specific when you say d-wave. It may well be that d-wave is a symmetry unobservable by nature.

D. Pines: I fully agree, and your point is, I think, very well taken.

C. M. Varma: I want to make a couple of remarks. One is this question of having superconductivity with significant antiferromagnetic fluctuations or antiferromagnetic order. It's very straightforward to see, that in an antiferromagnet you cannot have singlet superconductivity without simultaneously having some triplet superconductivity. An invariant in the free energy is allowed in which there is a singlet times the triplet times the antiferromagnetic order parameter. And my guess would be that the coefficient of a term like this would be large if the antiferromagnetism is in fact promoting superconductivity. In which case, in NMR there should be some very remarkable signals, like seen in liquid ^{3}He. Or at least, there should be some things which should be looked for.

D. Pines: Chandra, is that the case for heavy electrons, where we believe that in some cases you have something which looks, more or less, d-wave-like, and yet we know there is antiferromagnetism?

C. M. Varma: Yes, but the antiferromagnetism has an ordered magnetic moment of 10^{-2} Bohr magnetons, and...

D. Pines: Here we have no long-range antiferromagnetic order...

C. M. Varma: Well, I say this in relation to some of the ideas that Bob has been expressing...

D. Pines: He also does not require long-range order. And I think you are making an argument based on the existence of long-range antiferromagnetic order. You're making a Ginsburg-Landau free energy argument, which neither of us would want you to do.

C. M. Varma: Well, to reiterate, some of the things that Bob maintains – this business of d-wave looking like s-wave because you have local antiferromagnetic gaps in the band structure already... If those are true, then my statement would also be of experimental concern...maybe not. I don't know.

The second thing I wanted to do was to thank you again for mentioning this mechanism of not having a coherance peak. In talking to people yesterday, there's one thing which I realized has not been clear. The particular signature of the pairing being due to our kind of excitation spectrum. This gives a signature rather different than the classic BCS theory with all its bells and whistles. Which is that, even though the gap divided by T_c is larger than BCS, the specific heat jump appropriately normalized, is smaller than BCS, which is very unusual.

J. W. Schrieffer: May I respond to that?

I would like to mention that if one considers pairing in the presence of a static SDW, the broken spin symmetry induces a strong mixing between singlet and triplet superconductivity, if the latter exists at all, as Wen, Zhang and I showed. Moreover, if only fluctuating short range antiferromagnetism exists, then this mixing is motionally averaged out if the mean spin fluctuation frequency is larger than the pairing gap.

H. Ott: In very pure, simple superconductors, the coherence peak is reduced and the argument is that it is reduced because of the anisotropy of the gap, which is very small. And you can correct that by adding some impurities, and then the coherence peak becomes much bigger. Could it not be that here it is not seen simply because the gap anisotropy is very large?

D. Pines: I think not, because that would lead to a relaxation rate which falls off rather gradually below T_c. It would not lead to something which really drops like a rock. I think that all of us who have looked at that are in agreement on that.

P. W. Anderson: A very brief comment. Any mechanism involving, or any state involving separation of charge and spin has no problem with this observation, and also, as I said yesterday, has a very steep dropoff below T_c.

R. Heffner: How do you deal with a d-wave picture in terms of the measured temperature dependence of the penetration depth?

D. Pines: I think we use the lovely poetry and prose that Bob Schrieffer did, namely that you're not really sampling, in the experiment, the places where the nodes are. And Hartmut Monien and I have given some general arguments for how this might be true. What is clearly the case is that when you want to work, say, with an anisotropic s-wave state, you must get a feedback mechanism of the kind that Chandra has calculated, in order to get the relaxation rate to drop rapidly. You have to work very hard with an anisotropic s-wave state to make it work.

M. Takigawa: I have a comment on the Los Alamos Knight shift data, that you have shown. The data is based on somewhat of an overestimate of the diamagnetic field. With an improved estimate of the diamagnetic field, K_c of Cu stays almost constant below T_c.

V. Emery: I have a comment on consistency with the neutron scattering data. In the data that Andy Millis showed, most of the emphasis has been on whether or not there is a gap below T_c, but actually if you look at it what it's showing is that there are spin fluctuations below the superconducting gap. They're still there down at 5 Kelvin. Whereas to explain $1/T_1$ data, you need spin fluctuations to be frozen out except in the neighborhood of zeros, so there is not clearly consistency between the neutron scattering and the NMR data.

D. Pines: There is not yet a consistent calculation – we have one underway in Urbana, there may be some underway elsewhere – of the dynamic form factor in the superconducting state. That's just something that no one seems to have done, at least we could find it nowhere in the literature, so we've been doing it. But at the moment none of us is prepared to compare with the superconducting state.

D. J. Scalapino: I have data with me from Bulat's calculation, and if you'd like to see it we can watch what happens in d- and s-wave calculated throughout the zone, but within again this very simple RPA idea. What you do is you come in and dress the RPA χ_0 and superconducting propagators, and make them either d- or s-wave. It still doesn't do what Chandra has done, what you'd like to see is what happens if you put in strong spin fluctuations. But you can see the dynamics of what happens, and one interesting old point, first pointed out by Anderson and Suhl, is that if you look at χ of q of 0 to ask what's happening, you find that it gets eaten out as you go superconducting, out to a distance in q-space of one over the coherence length. And indeed that part gets eaten out. If you come and look around (π, π), you find very

very little change in that regime, because you're well down inside the coherence length. And that, as David is saying, is I think linked deeply to this data that we hear about, of the ratio of $1/T_1$ (copper) to $1/T_1$ (oxygen), coming up and being more or less fixed, once you go below T_c. What we find very interesting, and we're anxious to see more data, is there are subtle variations in that data below T_c, and you get different results for s-wave and d-wave if you do the dynamic calculation within RPA adding the superconducting propagators.

C. M. Varma: Can I make a comment on what Doug said? The changes that occur below superconductivity occur both in the real part of χ, that you described, the part that ω near zero is influenced only up to $1/\xi$, and then they also occur in the imaginary part of χ. By far the most important effect is in the imaginary part of χ, because there the fluctuations are going all the way down to zero, and NMR is sensitive precisely to the slope at ω tending to zero. And that effect is there at essentially all momentum, not just long wave-lengths.

D. J. Scalapino: I completely agree with you, other than – within the framework that Andy has spoken of and we've looked at – to keep this ratio, somehow – $1/T_1$ (copper) to $1/T_1$ (oxygen) – somewhere near where it was. We at least need to keep the momentum form factors more or less OK. Because if you go mess with those you'd be in big trouble. I'm not sure about other theories, but...that's why I spoke of the real part.

D. Pines: To put it another way, what Doug is saying, is that the antiferromagnetic correlations – if they exist – are uninfluenced by superconductivity.

P. W. Anderson: I do think it's really worth saying that somehow, in going from one day to another, we have somehow shifted our psyches in a way which I find, you know, absolutely mind-boggling. And after listening, very carefully, to Bertram saying, and a number of people saying, that there's no evidence whatever that the gap is of any conventional kind, amidst all kinds of evidence there are many different gaps for many different phenomena; we have no evidence that there's anything remotely as simple as a d-wave pairing, or something like that – my feeling, basically, is that comparisons of this sort are extremely premature.

D. Pines: They may be premature; you know one tries to crawl or something before one walks...

P. W. Anderson: But maybe in the right direction.(Laughter)

D. Pines: ...One hopes one is going in the right direction.

5. QUANTUM SPIN LIQUIDS

Chair — E. Abrahams

A. Zee
Institute for Theoretical Physics University of California, Santa Barbara, California 93106

Semionics: a theory of high temperature superconductivity

A more-or-less self-contained review of semionics, but not an exhaustive survey of the literature, is given.

I. INTRODUCTION

I will describe a theory of high temperature superconductivity based on the possibility that the quasi-particles in high temperature superconductors may obey fractional quantum statistics. I would like to emphasize the underlying physics of this subject rather than detailed calculations. Hopefully, this review will prove to be more-or-less self-contained.

It is generally believed that the essential physics of high-temperature superconductors can be captured by treating them as two-dimensional systems. If we subscribe to this belief, then it would be natural for us to ask whether there is any physics that occurs only in two dimensional space but not in three dimensional space.

Indeed there is. Two dimensional space is significantly different from three dimensional space. It is not simply less of a good thing compared to higher dimensional spaces. As we descend from three dimensions to two dimensions, a new concept emerges, namely the concept of "going around." In two dimensional space, a particle can go around another. Clearly, no such concept of "going around" can be defined in three dimensional space. Thus, less may actually be more.

In two dimensional space, we can thus postulate that there exists an infinitely long ranged phase interaction between particles such that when one particle moves around another through an angle φ the wave function acquires a phase $e^{i(\theta/\pi)\varphi}$. Here θ is an arbitrary real parameter defined mod 2π that characterizes some interaction properties of the particles. We are certainly allowed to postulate such an interaction. Whether particles participating in such an interaction actually exist is of course another question.

Another way of underscoring the possibility of this phase interaction is to remark that the angle φ has an intrinsic geometric meaning in two dimensional space but not in three dimensional space. (In three dimensional space only the solid angle has intrinsic geometric meaning. This observation leads us to the monopole, but that's another story.)

We can effectively interchange two particles by moving one particle around the other through $\varphi = \pi$ and then translating the two particles. After this process, the wave function acquires the phase $e^{i\theta}$. Thus, the particles behave like bosons for $\theta = 0$ and like fermions for $\theta = \pi$. The infinite range phase interaction we postulated provides a natural generalization of the usual quantum statistics. In general, for θ equal to neither 0 nor π, the particles are said to obey fractional statistics[1-9] and they are known as fractional statistics particles or anyons.[2]

Notice that in the preceding paragraph, we could equally have moved one particle around the other through $\varphi = -\pi$, in which case the phase would have been $e^{-i\theta}$. Were we in three dimensional space, we could lift the path $P(\varphi = \pi)$ "out of the plane" and deform it smoothly into the path $P(\varphi = -\pi)$, thus requiring $e^{i\theta} = e^{-i\theta}$ and hence $e^{i\theta} = \pm 1$. Bose-Einstein and Fermi-Dirac statistics are the only possibilities in three dimensional space. In contrast, in two dimensions, the paths $P(\varphi = \pi)$ and $P(\varphi = -\pi)$ are homotopically distinct, that is, they cannot be smoothly deformed into each other. Thus, different phases may be associated with the two paths.

There are at present three arguments or scenarios (but no proof, of course) suggesting that the quasi-particles in high temperature superconductors obey fractional statistics. Furthermore, these arguments are mutually consistent in that they all determine the statistics angle θ to be $\pi/2n$ for n an integer. In particular, for the simplest case $n = 1$, the quasi-particles are semions, particles half-way between bosons and fermions in their statistical properties. When we interchange two fermions, the wave function acquires a phase factor (-1); two bosons, a phase factor $(+1)$; two semions, a phase factor i. The word "semion", being half Latin and half Greek, may be particularly appropriate in describing this hybrid between fermion and boson. In the literature, the term "half fermion" is also used but may be potentially misleading as a semion is no more a half fermion than a half boson.

The physics underlying the theory of superconductivity to be described here may be stated heuristically as follows. We all know that superfluidity and superconductivity are intimately connected with quantum statistics and Bose condensation.

The tendency towards superconductivity is represented traditionally by the tendency of electrons to pair into bosons.

A novel possibility is for an electron (or hole) to "turn itself into a boson." The assertion here is that the electron can't quite transmute itself into a boson, but can only get there half-way. The naively fermionic degrees of freedom in the system are actually semionic. The argument is that the semions have a strong tendency to pair. Note that a pair of semions makes a boson: upon interchange of two such pairs, the wave function acquires a phase $(e^{i\frac{\pi}{2}})^4 = 1$. The phase $e^{i\frac{\pi}{2}}$ occurs four times. This again indicates the potential for confusion associated with the term "half fermion." Two half fermions do not a fermion make.

Why would semions want to pair? Heuristically, we may say that a collection of semions is phase frustrated: the wave function has to change its phase whenever one particle moves around another. One way of relieving this frustration is for each semion to find another partner so as to form a boson and to move through life as a couple, without having to worry about phases. The hope is that the condensation of these bosonic modes may lead to superconductivity. We must caution the reader however that this picture is strictly heuristic. The sizes of these bosonic bound states are larger or comparable to the average separation between the semions. It is more accurate to think of the collection of semions as a quantum liquid with phase coherence between them.

While some of what I will say in this review will apply to anyons in general, some specific results, particularly those on superfluidity, may hold only for special values of θ. Since the semion case appears to be the most likely, I will use the term semion rather than anyon when I want to indicate that arbitrary values of θ won't do.

Clearly, fractional statistics violates time reversal T and parity P invariances: the phase upon the interchange of two particles changes sign if we reverse or reflect the path. (In two dimensions, parity is of course defined as reflection in one spatial axis.) Since the underlying dynamics of electrons in solids satisfies T and P, the dynamics of the collective excitations can violate T and P only if T and P are violated spontaneously,[10] that is, if the ground state violates T and P.

I like to summarize the subject of anyon superconductivity by listing the four big questions we have to answer.

(1) Under what circumstances does the ground state of a reasonable model Hamiltonian violate T and P? Is it a chiral spin state?
(2) Assuming that the answer to question (1) is yes, we would like to know the quantum numbers of the excitations about this ground state? What are their statistics? What is the long distance effective theory that describes their low energy dynamics?
(3) Is a liquid of anyons a superfluid, and if the anyons are charged, a superconductor?
(4) What are some experimental tests of the scenario presented here?

Wen and I have published a series of papers,[11-19] and also a paper in collaboration with Wilczek[20] addressing these questions. In this article,[21] I will focus largely on this body of work, and will not make any attempt to survey the literature exhaustively and to recount the history. The subject is growing so rapidly at present that any effort at an exhaustive survey can only be foolhardy as well as futile. I apologize to those authors whose work is not represented here. I must mention, however, that the first suggestion that the quasi-particles in high temperature superconductors obey fractional statistics was made by Kalmeyer and Laughlin,[22] and the groundwork and general direction of the subject were outlined by Laughlin[23] in a profound and original article. This work was in turn based on earlier works by Anderson[24] and his collaborators, Baskaran, Zou, and others,[25] and on the suggestion by Kivelson, Rokshar, and Sethna[26] and Dzyloshinski, Polyakov, and Wiegmann[27] that statistics transmutation and fractional statistics may be relevant to superconductivity. I would also like to emphasize that our work overlaps with that of Wiegmann[28] on a number of points.

II. FRACTIONAL STATISTICS

By now fractional statistics is a rather well-known and ancient subject. I will give a brief review of a few salient points, partly to establish notation. It is easy to write down a Lagrangian describing a collection of anyons:

$$L = \sum_{i=1}^{N} \frac{1}{2} m \left(\frac{d\vec{x}_i}{dt} \right)^2 + \hbar \frac{\theta}{\pi} \sum_{i \neq j}^{N} \frac{d}{dt} \varphi_{ij}. \qquad (2.1)$$

Here $\vec{x}_i$ is a two-dimensional vector specifying the location of the ith particle and φ_{ij} is the azimuthal angle between particles i and j (measured relative to some fixed and irrelevant reference axis). The second term in (1) is a total time derivative and does not contribute classically. Quantum mechanically, however, a phase

$$e^{\frac{i}{\hbar} \int dt L} = e^{i \frac{\theta}{\pi} \sum_{i \neq j} \Delta \varphi_{ij}} \times (\text{usual phase factor}) \qquad (2.2)$$

is induced in the wave function. In particular, if particle i goes half-way around particle j (thus effectively interchanging the two particles after a translation) a phase $e^{i\theta}$ is induced.

From (2.1) we find that the canonical momentum

$$\vec{p}_i = m\dot{\vec{x}}_i - \hbar \vec{a}_i(x_i) \qquad (2.3)$$

where

$$a_{i\alpha}(x_i) = \frac{\theta}{\pi} \sum_{j \neq i} \frac{\epsilon_{\alpha\beta}(x_i - x_j)^\beta}{|x_i - x_j|^2} \qquad (2.4)$$

Here the spatial indices $\alpha, \beta = 1, 2$. Repeated α, β indices are summed. The Hamiltonian is then

$$H = \sum_{i=1}^{N} \frac{1}{2} m \left(\frac{d\vec{x}_i}{dt} \right)^2$$
$$= \sum_{i=1}^{N} \frac{1}{2m} (\vec{p}_i + \hbar \vec{a}_i(\vec{x}_i))^2 \qquad (2.5)$$

A gauge potential has been generated naturally by the phase interaction. Indeed, (2.5) tells us that we may regard the anyons as particles carrying statistical charge and statistical flux. We thus recognize the phase interaction as essentially a generalized Dirac-Aharonov-Bohm effect.

To distinguish between the charge and flux associated with the gauge potential a from the charge and flux associated with the electromagnetic gauge potential, the qualifier "statistical" is sometimes added. I will generally drop this qualifier, trusting the context to make the distinction clear.

Instead of working with (2.5), we can also perform a singular gauge transformation and have a free Hamiltonian

$$H' = \sum_{i=1}^{N} \frac{\vec{p}_i^{\;2}}{2m} . \qquad (2.6)$$

In this gauge, the wavefunction is constrained by the rather nasty condition that upon interchange of two anyons the wavefunction acquires the phase $e^{i\theta}$.

The difficulty of the problem may be appreciated by noting that in the singular gauge the N-body wave function cannot be constructed out of the one-body wave function, as is the case for bosons and fermions. Strictly speaking, even the notion of wave function is not defined. For instance, the two-body wave function $\psi(x_1, x_2)$ are reached from some reference positions (x_1^0, x_2^0). The wave function $\psi(x_1, x_2)$ has a phase $e^{i\frac{\theta}{\pi}\varphi}$ relative to $\psi(x_1^0, x_2^0)$, where φ is the angle measuring the number of times the particles have wound around each other in going from (x_1^0, x_2^0) to (x_1, x_2). Thus, if $\varphi = 17° + 2n\pi$, say, the phase depends on n unless $\theta = 0$ or π. In other words, history matters. In the N-body case, the phase depends on the positions of the other particles.

It is natural, and more fundamental, to use the path integral formalism of quantum physics. We are taught by Feynman to sum e^{iS} over all paths leading from some initial configuration to some configuration of interest, where S is the classical action associated with each path. When it happens that the set of all paths can be decomposed into homotopically distinct classes, we are allowed by general principles to associate a factor $\Phi(C)$ with the paths in class C in the sum. The factor $\Phi(C)$ must satisfy general principles such as that of composition: $\Phi(C_1)\Phi(C_2) = \Phi(C_1 C_2)$. In other words, $\Phi(C)$ furnishes a representation of the braid group. For the two-body case, C is just labelled by the number of times one particle winds around the other. For N bodies, the situation is more complicated to describe in words but easy enough to visualize.

In (2.5), we have implicitly treated the particles as bosons. Because of the identity $0+\theta = \pi - (\pi - \theta)$, anyons with statistics angle θ may be treated as bosons with a phase interaction characterized by θ or as fermions with a phase interaction characterized by $-(\pi - \theta)$.

What is the physical effect of fractional statistics? Most simply, consider two anyons. The Schrödinger equation describing the relative motion of two anyons has the form (which we can derive from (2.5) after separating out the center of mass motion)

$$\left[-\frac{1}{r}\frac{\partial}{\partial r}r\frac{\partial}{\partial r} - \frac{1}{r^2}\left(\frac{\partial}{\partial \varphi} - \frac{\theta}{\pi}\right)^2 + V(r)\right]\psi = E\psi. \qquad (2.7)$$

(We have included a potential $V(r)$ between the anyons.) For the lowest energy state the centrifugal term is equal to $\frac{1}{r^2}\min\left[m + \frac{\theta}{\pi}\right]^2$ where m is an integer. For bosons, we have the centrifugal term $= \frac{0}{r^2}$, while for fermions, $\frac{1}{r^2}$. Anyons may thus be regarded either as bosons with an additional "centrifugal repulsion" between them or as fermions with an additional "centripetal attraction" between them.

One of the first questions asked in the early days of fractional statistics was whether fractional statistics, which in this discussion we postulated as an infinite range phase interaction between point particles, could be accommodated in a local quantum field theory. The answer, as given in Ref. 3, may be stated in the form of a recipe. Given any $\mathcal{L}_0$ with a conserved current j_μ we can consider the Lagrangian

$$\mathcal{L} = \mathcal{L}_0 + a_\mu j^\mu - \alpha\epsilon_{\mu\nu\lambda}a^\mu f^{\nu\lambda}. \qquad (2.8)$$

The last term in (2.1) is known as a Chern-Simons term. Greek letters $\mu, \nu, \lambda, \ldots = 0, 1, 2$ denote spacetime indices while Latin letters $i, j, \ldots = 1, 2$ will denote spatial indices. The 3-indexed antisymmetric symbol $\epsilon_{\mu\nu\lambda}$ (and hence the Chern-Simons term) is of course possible only in (2+1) dimensional spacetime. Varying with respect to a_μ, we obtain

$$2\alpha\epsilon_{\mu\nu\lambda}f^{\nu\lambda} = j_\mu \qquad (2.9)$$

Let there be an object which carries q_0 units of the charge $q_0 = \int d^2x j_0$. Take the object as sitting at rest. Far away from this object $f_{12} = 0$ according to (2.9) and so the gauge potential a_μ is a pure gauge. However, it must be topologically nontrivial since

$$\oint_C dx^i a_i = \int dS f_{12} = \frac{1}{4\alpha}\int d^2x j_0 = q_0/4\alpha \qquad (2.10)$$

where C is a contour encircling the object. Thus, the gauge potential is

$$a_i = \frac{q_0}{8\pi\alpha}\partial_i\varphi \qquad (2.11)$$

where φ is the azimuthal angle around the object. If we now move another object of charge q'_0 slowly around this object, a phase proportional to $\oint_C dx^i a_i$ is induced in the wave function. This phase is the origin of fractional statistics. An object carrying charge $\int d^2x j^0 = 1$ would then have statistics given by

$$\theta = \frac{1}{8\alpha}. \quad (2.12)$$

(A brief explanation of this formula, including that of a notorious factor of two, may be found in Ref. 11.) The only requirement is that $\mathcal{L}_0$ has a conserved current. It may be a Lagrangian of non-relativistic point particles or relativistic fields. Or, as in Ref. 3, it may be a Lagrangian containing solitons, in which case the solitons would acquire fractional statistics. Of course, the effect represented by θ is to be added to whatever statistics the particle already has in $\mathcal{L}_0$. For example, if $j_\mu = \bar\psi \gamma_\mu \psi$ is a fermionic current, the particle associated with the field ψ would have a total statistics angle $\theta_t = \left(\pi + \frac{1}{8\alpha}\right)$.

From (2.8) it is clear that under time reversal T or parity P the statistics parameter θ is taken to $-\theta$. Thus, we may restrict our attention to θ between 0 to π Incidentally, we see also that $\pi + \epsilon$ is equivalent to $\pi - \epsilon$.

That two semions pair to form a boson is also clear from (2.8). If we double the charge, we can absorb the factor of 2 into a_μ so that according to (2.12) the angle θ scales by a factor of 4.

We see from (2.9) that the flux is concentrated entirely on the particle. Thus, a point particle would carry an infinitely thin flux tube. We refer to such particles as ideal anyons. In real materials, anyons may be expected to represent a collective excitation of finite size. The gauge potential is itself an effective representation of the collective dynamics of the electrons in the system. Thus, Lagrangians such as (2.8) are to be regarded as effective long distance Lagrangians describing the low frequency long wavelength dynamics in an expansion in powers of derivatives and fields. Thus, in general we would expect the Lagrangian $\mathcal{L}$ to contain the Maxwell term $-\frac{1}{4g^2} f_{\mu\nu}^2$, not to mention other higher order terms.

By comparing powers of derivatives in the Maxwell term and the Chern-Simons term $\alpha \epsilon_{\mu\nu\lambda} a^\mu f^{\nu\lambda}$ we see that the gauge potential acquires a gauge invariant mass of order αg^2. Thus, the flux carried by each particle is spread out over a length $\sim \theta/g^2$. Since the Maxwell term has more derivatives than the Chern-Simons term, the long distance effect of the latter, namely fractional statistics, is not affected.

More explicitly, (2.9) is now modified to be

$$\frac{1}{g^2} \partial^\nu f_{\mu\nu} + 2\alpha \epsilon_{\mu\nu\lambda} f^{\nu\lambda} = j_\mu. \quad (2.13)$$

Equation (2.10) continues to hold provided we take the contour C to be at infinity where we expect the gauge invariant field strength $f_{\mu\nu}$ to vanish, a fact we can easily verify by solving (2.13). Thus, far away from the particle, we still have the topologically non-trivial gauge potential in (2.11) and hence fractional statistics. As two such non-ideal anyons get closer together, the phase they would acquire from going around each other would be corrected appropriately. We can picture a non-ideal anyon as a particle surrounded by a bag of statistical electric and magnetic fields, a bag of characteristic size $1/(\alpha g^2) \sim \theta/g^2$. Note that the characteristic response time of the bag is also of order $1/(\alpha g^2)$. As the particle moves around, the bag, with its finite response time, may not be able to keep up. In the theory of high temperature superconductivity discussed here, the anyon is also electrically charged. While part of the energy-momentum density of the anyon resides in the bag, the electric charge density is concentrated on the particle. Thus, in high frequency phenomena, the center of mass and the center of charge of an anyon would not coincide in general.

It is important to recognize that the mass discussed here is gauge invariant. Thus, a purely gauge degree of freedom as in (2.11) may propagate to infinity and generate fractional statistics. In contrast, were we to add a term $\beta a_\mu a^\mu$ we would have obtained a gauge variant mass of order β/α and thus all effects of the gauge field would have short range. There will be no fractional statistics. Thus, the appearance of an $a_\mu a^\mu$ term signals a statistics-changing phase transition.[13] Another way of saying this is to note that the $a_\mu a^\mu$ term has lower mass dimension than the Chern-Simons term while the Maxwell term has higher dimension.

III. THE ANYON FLUID

With this brief review of fractional statistics, we are now ready to explain anyon superconductivity. A logical order of presentation would follow the order of the "four big questions" I listed in Section I. Instead, however, I will start with question 3 on my list: does a collection of anyons behave as a superfluid, and for charged anyons, as a superconductor?

I want to discuss question 3 first because I would like to start by emphasizing the essential physics underlying anyon superconductivity. Also, of my four questions, this question has been answered most definitively. I believe that most of the leading researchers on this subject now agree that the anyon fluid is indeed a superfluid. (At least, this has been demonstrated within the extended mean field approximation used in this field.) Finally, the problem posed by this question is a wonderfully elegant challenge to theoretical physicists. It can be simply stated: somebody gives you a bunch of particle such that when two of them are interchanged their wave function acquires a factor of i, do they form a superfluid?

Let us begin by noting that the problem for ideal anyons is so simple that besides the dimensionless parameter θ it involves only the mass of the anyon m. By dimensional analysis, the only quantity with the dimension of energy or temperature we can construct is $\hbar^2 n/m$ where n is the density of anyons. Thus, if the anyon fluid is superfluid, the transition temperature T_c is given by

$$T_c = f(\theta)\hbar^2 n/m \qquad (3.1)$$

where $f(\theta)$ is some unknown function, presumably of order one. For La-Cu-O with 10% doping, this comes out to be

$$T_c \simeq (200°K)(m_e/m)f(\theta) \tag{3.2}$$

and thus the anyon superfluid may well be relevant to high temperature superconductivity provided that the effective mass of the anyon is not too large compared to the electron mass.

The low density limit or equivalently the high temperature limit (in which the thermal wavelength is small compared to the interparticle spacing) was studied some years ago by Arovas, Schrieffer, Wilczek and Zee.[6] The first virial correction b to the low density gas law $P = nT(1 + bn + ...)$ can be computed. As may be expected, $b(\theta)$ is positive for $\theta = \pi$ and negative for $\theta = 0$. Fermi statistics is effectively repulsive, while Bose statistics is attractive. The virial correction crosses zero precisely at the semionic value of $\theta = \pi/2$. In this leading order, the problem reduces essentially to a two-body problem, but already $b(\theta)$ shows an interesting feature: a cusp at $\theta = 0$. It is tempting to conjecture, in analogy to the quantum Hall effect, that the next virial coefficient would contain interesting structures according to whether or not θ/π is a rational number.

Much more interesting is the $T = 0$ or high density limit, in which case we are confronted with a formidable strongly correlated quantum many-body problem. The one simplification is that in this limit the kinetic energy which is of order $\hbar^2 n/m$, is much larger than the Coulomb energy $e^2/n^{-\frac{1}{2}}$, in other words, for $n \gg \left(\frac{me^2}{\hbar^2}\right)^2$, we can neglect the Coulomb interaction between the anyons. An analogous approximation was made in the Laughlin theory of the quantum Hall effect.

Let us begin by making an elementary but general remark. We would like to show that the charged anyon fluid exhibits the Meissner effect and thus superconducts. Since we are interested in the linear response of the system to an external magnetic field, we can in fact turn off the external electromagnetic field. In other words, it suffices to show that the anyon fluid is a superfluid. More precisely, we show that the low energy excitation spectrum contains a linearly dispersing phonon mode, namely a Goldstone boson with the right quantum numbers so that it can be eaten by the photon via the standard Anderson-Higgs mechanism. The photon acquires a mass and thus we have the Meissner effect.

With these preliminary remarks out of the way, we are now ready to try to understand the basic physics of why an anyon fluid may be a superfluid. As explained in Section II, semions may be regarded either as fermions with a gauge attraction between them or as bosons with a gauge repulsion between them. Thus, theorists may choose between two possible treatments.

The literature on the subject bifurcates according to the choice made. Laughlin[29] used the mean field approximation to be described below to make a first attempt in understanding the statistical mechanics of the anyon fluid, treating the particles as fermions. Later, Fetter, Hanna and Laughlin[30] did a random phase calculation in which they went beyond the mean field approximation. They concluded that the excitation is gapless and that the system behaves as a superfluid.

The calculation of Fetter, *et al.*, is technically involved and the physical origin of the gapless excitation is far from clear. Wen and I were motivated in Ref. 17 by a desire to understand the appearance of this gapless excitation in more physical terms. We chose to treat the particles as bosons as we felt that it would be valuable to look at the underlying physics from a different starting point than that of Fetter, *et al.*

The work of Fetter, *et al.*, was later clarified and extended by Chen, Wilczek, Witten and Halperin.[31] They were able to prove a number of rigorous statements[32] for ideal anyons and to provide some additional insights on the origin of superfluidity. Related work was also carried out by Hosotani and Chakravarty.[33]

In this review we will describe superfluidity largely from the boson point of view, but first let me give an overview of how the same physics may be described in two different ways.

Suppose we start with a free fermi gas, which is certainly not a superfluid. The theorist's task is to show that the gauge attraction due to semionic statistics would lead to pairing and to superfluidity. This may be quite plausible since we know from BCS theory that an arbitrarily small attraction will lead to pairing between fermions.

Alternatively we may start with a free bose gas at zero temperature. As Bogoliubov[34] has shown, any short ranged repulsion between the bosons leads to superfluidity.[35] For non-ideal semions as may actually occur in real materials we would expect such short ranged repulsion. The task here is then to ascertain that the gauge repulsion due to semionics does not destroy the superfluidity.

Viewed in this light, both the fermi gas and the bose gas are poised on the brink of instability. The fermi gas has a continuum of low frequency excitations, while the bose gas has a single quadratically dispersing excitation $\omega \propto k^2$ at low frequencies. The claim is that a gauge attraction, from the fermi end, or a gauge repulsion, from the bose end, would change the low frequency excitation spectrum drastically to a linearly dispersing spectrum. From this perspective, it may well appear that superfluidity is generic, and that the behaviors of the fermi and bose gases are in some sense exceptional.

Two interesting questions arise in this context. Is there a critical value for θ as we move away from either the fermi or bose end, or does superfluidity set in immediately? Does superfluidity occur only for special values of θ? The discussion below suggests that the answer to the second question may be yes: superfluidity may occur only for $\theta/\pi = 1/q$ for q an integer. Thus, the terminology semion superfluid may be preferable to anyon superfluid since the latter may suggest erroneously that superfluidity occurs for any value of θ. A lawyer would not use the term "anyon superfluid."

The slope of the linearly dispersing excitation, namely the speed of the phonon, is of course directly connected to the compressibility of the quantum fluid. A quantum fluid is incompressible if it does not have any low frequency, that is, gapless excitations. A compressible quantum fluid is a superfluid. Our task is thus to show that the semion fluid has a gapless excitation and hence is not incompressible. Note that it is not enough to show that the semion fluid is compressible. After all, a free fermi gas is compressible. We also have to show that it has only one gapless excitation and not a continuum of them.

At least naively, we may think it unlikely for a quantum fluid to be incompressible: some special physics must be involved. Indeed, we know of one example of an incompressible quantum fluid, namely the quantum Hall fluid. It thus behooves us to understand the difference between the quantum Hall fluid and the semion fluid. The quantum Hall fluid consists of charged fermions moving in an external magnetic field. The fermions are forced by the magnetic field to move in Landau orbits. The quantum correlations due to Fermi statistics make the fluid incompressible for special values of the filling factor, that is, for special values of the density in a given magnetic field. In the semion fluid, in contrast, the semions do not move in an external magnetic field. Rather, as I explained earlier, they may be regarded as fermions or bosons carrying with them a bag of statistical magnetic (and electric) fields. The crucial difference is that in the semion fluid, the particles carry the magnetic field with them. They are not forced to move in circles, but in straight lines to first approximation. Thus, heuristically, we may expect the semion fluid to be compressible.

This discussion, while underscoring the difference between the Hall fluid and the semion fluid, also hints that the two fluids are intimately connected. The discussion below will make precise this connection.

IV. SEMION SUPERFLUIDITY

Now that we have outlined some heuristic physical considerations, how can we study the anyon fluid in more detail? Arovas, et al.,[6] suggested that in the high density limit, with the particles close together, we can adopt a mean field approximation by treating the particles as moving in a uniform magnetic field b generated by the other particles. If we move a particle around a loop enclosing an area A and hence nA particles, the wavefunction acquires a phase $e^{i(\frac{\theta}{\pi})(nA)(2\pi)}$ which we interpret as the phase e^{ibA} due to a mean field b. Thus

$$b = 2n\theta. \qquad (4.1)$$

Incidentally, this is just the time component of (2.9)

$$n = j_0 = 2\alpha\epsilon_{ij} f_{ij} = 4\alpha b \qquad (4.2)$$

which is just (4.1) with the relation between α and θ given in (2.12).

The validity of the mean field approximation is controlled by the ratio of the magnetic length $l \sim b^{-\frac{1}{2}} \sim (n\theta)^{-\frac{1}{2}}$ to the average interparticle separation $n^{-\frac{1}{2}}$. Thus, if the anyons are strictly point particles, the mean field approximation is expected to be valid only for small θ. The mean field treatment of the semion case, with $\theta = \pi/2$, involves an extrapolation.

Semionics: a theory of high temperature superconductivity

However, as was emphasized earlier, in all likelihood the anyons in real materials will have its flux spread over some finite size $1/\mu$. For such non-ideal anyons, the mean field approximation may be valid even for $\theta = \pi/2$, as long as $1/\mu \gtrsim n^{-\frac{1}{2}}$. In this case there may be a critical density for superfluidity to set in.

We have a gas moving around in a mean field b. The single-body energy states are given by the Landau levels, each with a degeneracy $bA/2\pi$. The ground state of our system thus appears to be highly degenerate: we can put each of the N bosons into any of the $bA/2\pi$ states in the lowest Landau level.

We can define, as usual, the ratio between the number of particles and the number of states in each Landau level:

$$\nu \equiv \frac{N}{bA/2\pi} = \frac{2\pi n}{b}. \tag{4.3}$$

(If we were treating the particles as fermions, then because of Pauli exclusion ν would measure the number of filled Landau levels. Here, all the bosons can be put into the lowest Landau level. Nevertheless, we will still refer to ν as the filling factor.) We see that with the mean field b the filling factor is equal to

$$\nu = \frac{\pi}{\theta}. \tag{4.4}$$

It would appear that by adopting the mean field approximation we have actually gone backward: in the mean field approximation the semion fluid looks like a Hall fluid which we know to be incompressible for certain values of the filling factor. However, there is a crucial difference. Here the magnetic field is not a fixed external field, but reacts dynamically to the density according to (4.1). It is this dynamical character of the gauge field that will produce the gapless linearly dispersing excitation. Indeed, as we will see, this phonon mode may be represented as an oscillation in the gauge field $f_{\mu\nu}$. While for both the anyon fluid and the Hall fluid we have the equation in (4.1) we interpret it quite differently. For the anyon fluid, we think of b and n as varying with θ held fixed, while for the Hall fluid, we think of n and the local filling factor π/θ as varying with b held fixed.

While we do not know how to write down the N-body wave function, we can calculate its energy easily. The energy of the lowest Landau level is given by

$$\frac{1}{2}\hbar\omega_c \equiv \frac{\hbar^2 b}{2m} = \frac{\hbar^2 \theta n}{m} \tag{4.5}$$

and thus the energy density is given by

$$\epsilon = \frac{\hbar^2 \theta n^2}{m}. \tag{4.6}$$

This represents an energy due to quantum statistics. The phase acquired by the wave function upon the exchange of two anyons implies that the anyon system is frustrated and the anyons have non-zero kinetic energies even in the ground state.

For comparison, the exchange energy of a free fermion gas in two dimensional space is $\epsilon = \frac{\hbar^2 \pi n^2}{m}$ which coincidentally happens to be reproduced by (4.6) for $\theta = \pi$.

Incidentally, if we had treated semions as fermions, the energy would be exactly twice as high. We can see this easily by noting that for $\theta = \pi/2$, $\nu = 2$ according to (4.4). Since the energies of the Landau levels go as $\frac{1}{2}\hbar\omega_c$, $\frac{3}{2}\hbar\omega_c$, $\frac{5}{2}\hbar\omega_c$, and so on, we have $\left(\frac{3}{2} + \frac{1}{2}\right) = 2\left(\frac{1}{2}\right)$. The true energy density of a gas of semions is presumably $\xi \left(\frac{\hbar^2 \pi n^2}{2m}\right)$ with ξ between 1 and 2.

To proceed rigorously, we should write down the N-body ground state. While we do not know to write down this state, we can describe how it must be constructed. Since each boson can be put in any of the states in the lowest order Landau level, the ground state is extremely degenerate. This extreme degeneracy is mathematical, rather than physical however. Classically, the bosons move in Larmor orbits. The quantum degeneracy corresponds to the fact that for non-interacting bosons the Larmor orbit of each boson can be centered anywhere in space. Physically, a repulsion between the bosons should remove the degeneracy. We expect the ground state to be one of uniform density. Classically, we can picture space being uniformly filled with Larmore orbits. For $\theta = \pi/q$, the filling factor is $\nu = q$ and we put q bosons into the quantum state corresponding to each orbit. (Strictly speaking, the states corresponding to different orbits are not orthogonal but the overlap is exponentially small.) The radius of the orbit is of the order of the magnetic length which for θ of order 1 is comparable to the interparticle separation $l \sim b^{-\frac{1}{2}} \sim (n\theta)^{-\frac{1}{2}}$.

This semi-classical picture suggests that the values $\theta = \pi/q$ with q an integer are special. For irrational values of θ, we would have to put more bosons into some orbits than others and it is not clear how to achieve a uniform density state. In the alternative fermion treatment, $\theta = \pi/q$ is special because then q Landau levels are filled.

For $\theta = \pi/q$, we would thus expect a non-degenerate N-body ground state separated from the higher lying states by a gap that depends on the repulsion. To prove this expectation would require solving the problem of bosons with mutual short ranged repulsion moving in a magnetic field. This problem is highly nontrivial as neither the repulsion nor the magnetic field can be treated as a small perturbation. Suppose we try to treat the repulsion as a perturbation. We are faced with a highly degenerate perturbation calculation, in contrast to the situation in Bogoliubov's problem.

The quadratic dependence of ϵ on n insures that the density will be uniform. We can gain energy by moving particles from regions of high density to low density. Note that to obtain this characteristic quadratic dependence it is important that the bosons move in a magnetic field which can react to the bosons rather than in a fixed and static magnetic field. Thus, the approximation used here may be called an "extended mean field approximation.'. Note also that this quadratic dependence is effectively the same as a short ranged repulsion between the bosons.

Semionics: a theory of high temperature superconductivity

In the ground state, the magnetic field b is constant. The low lying long wavelength excitation corresponds to a "breathing mode" in which b and the particle density vary slowly over space and time.

Let us thus consider a hydrodynamic treatment. Including the kinetic energy we write down the energy functional

$$E = \int d^2x \left\{ \frac{1}{2}mnv^2 + \gamma(n-n_0)^2 - \gamma n_0^2 \right\} \tag{4.7}$$

with $\gamma = \hbar^2 \theta/m$. We have added a chemical potential term to fix the total number of particles in the system.

The pressure is easily calculated

$$P = -\frac{\partial}{\partial A}\gamma An^2 = \gamma n^2 \tag{4.8}$$

The hydrodynamics equation is thus

$$\frac{\partial \vec{v}}{\partial t} + (\vec{v}\cdot\vec{\partial})\vec{v} = -\frac{1}{mn}\vec{\partial}P = -\frac{2\gamma}{m}\vec{\partial}n \tag{4.9}$$

Linearizing (4.9) around the ground state with uniform density n_0 and the equation of continuity

$$\frac{\partial n}{\partial t} + \vec{\partial}(n\vec{v}) = 0 \tag{4.10}$$

we find the wave equation

$$\frac{\partial^2 n}{\partial t^2} + \frac{2\gamma n_0}{m}\nabla^2 n = 0 \tag{4.11}$$

so that the speed of sound is given by

$$c_s^2 = 2\gamma n_0/m = 2\hbar^2\theta\, n_0/m^2 \tag{4.12}$$

Thus, the low lying elementary excitation of the system is expected to be a gapless phonon. The form of this result could have been determined by dimensional analysis. The θ dependence is consistent with the fact that $c_s = 0$ for an ideal Bose gas. It was crucial here that the potential energy is proportional to n^2 rather than to n, (in which case c_s would clearly vanish). This is a direct consequence of the "magnetic" interaction between the bosons, or in other words, the exchange pressure of the fractional statistics as discussed above.

We may calculate the coherence length if we can regard (4.7) as the Ginzburg-Landau theory. The potential term in the Ginzburg-Landau theory can be identified as

$$V(n) = \mu n + \gamma n^2 \tag{4.13}$$

where μ is the chemical potential to make $V(n)$ be minimized at $n = n_0$. Using a standard formula, we find the coherence length ξ_0 to be

$$\xi_0^2 = \frac{\hbar^2}{2m|\mu|} = \frac{\hbar^2}{4m\gamma n_0} = \frac{1}{4\theta n_0} \tag{4.14}$$

It is instructive to express the dynamics of this long wavelength mode in terms of the gauge potential a_μ by using (4.2) and its spatial-component counterpart as contained in (2.9)

$$nv_i = 4\alpha\epsilon_{ij}f_{0j}. \tag{4.15}$$

Substituting into (4.7) and dropping some irrelevant higher dimensional terms we see that the Lagrangian of the system is none other than the Maxwell Lagrangian

$$\mathcal{L} = \frac{-1}{16\pi g^2}f_{\mu\nu}^2 \tag{4.16}$$

with $g^2 = \theta m/2\pi\hbar^2$. The phonon is represented by the "photon" of this Lagrangian. (We have chosen length and time units so that the speed of the phonon is equal to one.) In a propagating electromagnetic wave of wave number $\vec{k}$ as described by (4.16) the electric field is oscillating perpendicular to the direction of propagation. According to (4.15) the particles are oscillating back and forth along $\vec{k}$. Notice that what we are describing here is the "guiding-center motion" of the particles. The much faster (at the cyclotron frequency ω_c) cyclotron motion around the guiding center has been averaged out, so to speak.

As Wen and I emphasized in Ref. 18, it is crucial that the Chern-Simons term $\epsilon_{\mu\nu\lambda}a^\mu f^{\nu\lambda}$ has disappeared.[36] If not, a gauge invariant mass for a_μ would be generated and we would obtain an incompressible fluid rather than a superfluid.

In effect, what we have done here is to integrate out the "matter fields", that is the particles, and represent the low frequency long wavelength dynamics solely in terms of the gauge fields. This may be described more formally as follows. The Lagrangian for a system of non-interacting anyons by

$$\mathcal{L} = \mathcal{L}_m - \alpha\epsilon_{\mu\nu\lambda}a^\mu f^{\nu\lambda} - \frac{1}{4g^2}f_{\mu\nu}^2 \tag{4.17}$$

with the "matter Lagrangian"

$$\mathcal{L}_m = \psi^\dagger i(\partial_0 + ia_0)\psi + \frac{1}{2m}\psi^\dagger(\partial_i + ia_i)^2\psi. \tag{4.18}$$

Integrating out the matter fields ψ we obtain the effective action

$$S = S_m(a) - \int d^3x \left[\alpha\epsilon_{\mu\nu\lambda}a^\mu f^{\nu\lambda} + \frac{1}{4g^2}f_{\mu\nu}^2\right]. \tag{4.19}$$

Semionics: a theory of high temperature superconductivity

In mean field approximation, we expand $a_\mu = a_\mu^0 + a'_\mu$ where a_μ^0 is the solution of the mean field equation

$$\left.\frac{\delta S_m}{\delta a_\mu}\right|_{a^0} = 2\alpha \epsilon_{\mu\nu\lambda} f^{0\nu\lambda} - \frac{1}{g^2} \partial^\nu f^0_{\mu\nu}. \tag{4.20}$$

This is, of course, essentially (2.9) or (2.13) again. To second order in a'_μ, we have the action

$$S = S\big|_{a^0} + \int\int \pi_{\mu\nu}(a^0) a'^\mu a'^\nu \tag{4.21}$$
$$- \int d^3x \left[\alpha \epsilon_{\mu\nu\lambda} a'^\mu f'^{\nu\lambda} + \frac{1}{4g^2} f'^2_{\mu\nu}\right]$$

where $\pi_{\mu\nu}(a^0) = \left.(\delta^2 S_m/\delta a_\mu \delta a_\nu)\right|_{a^0}$. Expanding $\pi_{\mu\nu}(a^0)$ in powers of derivatives, we finally have the effective action

$$S = S\big|_{a^0} + \int d^3x \left[\beta \epsilon_{\mu\nu\lambda} a'^\mu f'^{\nu\lambda} + \frac{1}{4g'^2} f'^2_{\mu\nu} + \ldots\right]$$
$$- \int d^3x \left[\alpha \epsilon_{\mu\nu\lambda} a'^\mu f'^{\nu\lambda} + \frac{1}{4g^2} f'^2_{\mu\nu}\right]. \tag{4.22}$$

But the mean field a^0 describes a constant "magnetic" field. Thus, the determination of $\pi_{\mu\nu}(a^0)$ is precisely the quantum Hall problem. The physics of the quantum Hall effect and the physics of anyon superfluidity are intimately related.

The coefficient β may be determined from the theory of the Hall effect or simply by counting. We note that the β term first says that in a constant magnetic field there is a current

$$j_i = 4\beta \epsilon_{ij} f'_{0j} \tag{4.23}$$

in response to an external electric field $f'_{0j} \equiv e_j$. We can now determine β as follows. Balancing the Lorentz force against the electric force gives $v_j b = \epsilon_{ij} e_j$ and so the current $j_i = nv_i = (n/b)\epsilon_{ij} e_j$. Thus $\beta = n/4b = n/(8n\theta) = 1/8\theta = \alpha$ where we have used (4.1) and (2.12). The induced and bare Chern-Simons terms in (4.22) cancel each other exactly and we have

$$S = S\big|_{a^0} + \text{Maxwell action} + \ldots. \tag{4.24}$$

The low lying excitation is described by a Maxwell Lagrangian as in (4.16). Thus, the incompressible Hall fluid corresponds to the anyon superfluid.

This discussion is rigorous except for the implicit assumption that we can express S_m in a gradient expansion. This is justified if the quantum Hall state has a finite gap, which corresponds precisely to our assumption earlier that when a small repulsion is introduced between the bosons in a magnetic field a unique N-body ground state separates off with a gap between it and the next state. In other words, we have to prove the incompressibility of the boson Hall fluid. In contrast this point is easily proved were we to treat semions as fermions. In that case, the gap in question is just the Landau gap.

That the semion fluid is a superfluid thus appears to be fairly well established. There has also been numerical studies of a small number of anyons (typically eight) on small lattices.[37] These studies also indicate the onset of superfluidity.

V. SUPERCONDUCTIVITY, SUPERFLUIDITY, AND COMPRESSIBILITY

We can now turn on a background electromagnetic field and verify that the Meissner effect and hence superconductivity indeed occur. We recall that the semions are electrically charged and thus the current $j_\mu = (n, nv_i)$ is in fact the electromagnetic current. Let us couple the electromagnetic gauge potential A_μ to the current in (2.9) so that the Lagrangian in (4.16) becomes

$$\mathcal{L} = \frac{-1}{16\pi g^2} f_{\mu\nu}^2 + 2\alpha \epsilon_{\mu\nu\lambda} A^\mu f^{\nu\lambda}. \tag{5.1}$$

Integrating out a_μ, we see that a mass term $\sim A_\mu^2$ is induced and hence we obtain the Meissner effect. The photon A_μ has eaten the fake photon a_μ. This is of course just the Higgs mechanism whereby a charged superfluid becomes a superconductor.

More carefully, we note that the three spatial dimensional electromagnetic current density is given by $\frac{e}{d} j_\mu$, where d is the interplane distance. (Recall that high temperature superconducting materials consist of weakly coupled layers.) The current-current correlation function is easily calculated in momentum space:

$$\frac{e^2}{d^2} \langle j_\mu(k) j_\nu(-k) \rangle = \frac{e^2}{16\theta^2 d^2} \epsilon_{\mu\lambda\sigma} \epsilon_{\nu\rho\tau} \langle f_{\lambda\sigma}(k) f_{\rho\tau}(-k) \rangle \tag{5.2}$$

$$= \frac{e^2}{4\theta^2 d^2} \epsilon_{\mu\lambda\sigma} \epsilon_{\nu\rho\tau} k_\lambda k_\rho \langle a_\sigma(k) a_\tau(-k) \rangle$$

$$= \frac{\pi e^2}{\theta^2 d^2} g^2 \left(g_{\mu\nu} k^2 - k_\mu k_\nu \right) / k^2$$

$$= \frac{\pi e^2}{\theta^2 d^2} g^2 \left(g_{\mu\nu} + \text{gauge degree of freedom} \right).$$

We obtain the Meissner effect and thus superconductivity. In coordinate space (5.2) can be written as

$$\frac{e^2}{d^2}\langle j_\mu(x)j_\mu(0)\rangle = \frac{\pi e^2}{\theta^2 d^2}g^2\delta^{(3)}(x) = \frac{\pi e^2}{\theta^2 d}g^2\delta^{(4)}(x) \quad \text{(no sum over } \mu\text{)}. \quad (5.3)$$

The delta function in 1+3 dimension is given effectively by $\delta^{(4)}(x) = \delta^{(3)}(x)/d$. Putting back the speed of the sound c_s we get

$$\frac{e^2}{d^2}\langle j_i(x)j_i(0)\rangle = \frac{\pi e^2}{\theta^2 d}g^2 c_s^2 \delta^{(4)}(x) = \frac{e^2 n_0}{md}\delta^{(4)}(x), \quad \text{(no sum over } i\text{)} \quad (5.4)$$

which is the expected result. We see that the London penetration depth is

$$\lambda_L^2 = \frac{mc^2 d}{4\pi e^2 n_0} \quad (5.5)$$

where c is the speed of light. The ratio between the London penetration depth and the coherence length is given by κ:

$$\kappa^2 = \frac{\lambda_L^2}{\xi_0^2} = \frac{mc^2 d}{e^2}\frac{\theta}{\pi}. \quad (5.6)$$

The reader may have recognized that the gauge potentials a_μ and A_μ are dual to each other in the sense discussed in Ref. 14. Our effective theory (5.1) can be regarded as the dual form[38,39] of the conventional Ginzburg-Landau theory in 2+1 dimensional space-time.

Although the effective theory (5.1) give rise to the Meissner effect, the value of the flux quantum is not determined by the theory. This is because (5.1) only describes the local small fluctuations in the superfluid. In the following we would like to show that in any quantum fluid state of the charge e anyon system, hc/qe flux always has finite energy if the anyons have $\theta = p\pi/q$. First, according to the argument of Byers and Yang[40] hc/e flux has finite energy because the anyon wave function acquires a phase 2π after an anyon goes around the flux. We also know that the anyon wave function acquires a phase $2p\pi/q$ after an anyon goes around another anyon. Because p and q are incommensurable, there exists an integer n such that $np/q = -1/q$ mod 1. Thus the bound state of the n anyons and hc/qe flux has a finite energy because an anyon going around this bound state acquires a phase of 2π. We would like to stress that in the above we only show that hc/qe flux has a finite energy, this does not imply that hc/qe is the smallest flux quantum. When both p and q are odd integers, a $2q$-particle bound state is a boson, since the statistics angle for such a bound state is given by $(2q)^2\theta$. The condensation of these bosons lead to a superconducting state, and the smallest flux quantum in this superconducting state is $hc/2qe$. We do not know whether there exists a superconducting state for anyon with odd p and q such that the flux quantum is hc/qe. For fermions ($p = q = 1$), Yang[41] showed that the flux quantum is always less than or equal to $hc/2e$ in the superconducting state of charge e fermion system.

When one of p and q is an even integer, q anyon bound states are bosons. In the boson condensed state the flux quantum is hc/qe, which coincides with the value obtained from general consideration.

From our discussion we see that the compressibility and the superfluidity of quantum fluid are closely related. We demonstrate that a compressible quantum fluid with phonons as the only low lying excitations is actually a superfluid. Here by superfluid we mean a quantum fluid with Meissner effect, i.e., the current-current correlation is given by (5.2) for small momenta. Landau's argument[42] assures us that the quantum fluid under consideration supports dissipationless flow, but it does not implies the existence of off diagonal long range order[41] or Meissner effect which is the essence of superfluidity. Our dual relation (2.9) is needed.

Due to the importance of this result it is worthwhile to give another derivation of the relation between compressibility and superfluidity. We start with the effective low energy Hamiltonian of the phonon

$$H = \int \frac{1}{2} m n_0 \, v^2 \, d^2x + \int \frac{1}{2}\varphi(x-x')n(x)n(x')d^2x d^2x' \tag{5.7}$$

where $\varphi(x)$ is some short-ranged interaction satisfying

$$\int d^2x \frac{1}{2}\varphi(x) = \gamma. \tag{5.8}$$

In momentum space, we have the linearized version of the Hamiltonian

$$\begin{aligned} H &= \int \frac{d^2k}{(2\pi)^2} \left(\frac{m}{2n_0 k^2} \dot{n}_k \dot{n}_{-k} + \frac{1}{2}\varphi_k n_k n_{-k} \right) \\ &= \int \frac{d^2k}{(2\pi)^2} \left(\frac{n_0 k^2}{2m} \pi_k \pi_{-k} + \frac{1}{2}\varphi_k n_k n_{-k} \right) \end{aligned} \tag{5.9}$$

where π_{-k} is the canonical conjugate of n_k

$$[\pi_{-k}, n_{k'}] = -i\hbar (2\pi)^2 \delta^2(k - k'). \tag{5.10}$$

Introducing the phonon annihilation operator

$$a_k = \frac{1}{\sqrt{2\hbar}} \left(\sqrt{\frac{\omega_k}{\varphi_k}} \pi_k - i \sqrt{\frac{\varphi_k}{\omega_k}} n_k \right) \tag{5.11}$$

we may write H as

$$H = \int \frac{d^2k}{(2\pi)^2} \hbar \omega_k a^\dagger_k a_k + \text{const.} \tag{5.12}$$

where

$$\omega_k^2 = \frac{\varphi_k n_0}{m} k^2 \tag{5.13}$$

Semionics: a theory of high temperature superconductivity

$$[a_k, a^\dagger_{k'}] = (2\pi)^2 \delta(k - k'). \tag{5.14}$$

The current is given by

$$\vec{j}_k = -\frac{i\vec{k}}{k^2} \dot{n}_k = -i\vec{k}\frac{n_0}{m}\pi_{-k}. \tag{5.15}$$

The time-ordered product of the currents is found to be

$$\tilde{K}^{ij}(t;k) \equiv \left\langle T\left(j_k^i(t)j^j_{-k}(0)\right)\right\rangle$$
$$= \frac{1}{2}k^i k^j \frac{n_0^2 \varphi_k}{m^2 \omega_k} e^{-i\omega_k|t|} \tag{5.16}$$

In frequency space

$$\tilde{K}^{ij}(\omega, k) = i\frac{k^i k^j n_0^2 \varphi_k}{m^2} \frac{1}{\omega^2 - \omega_k^2} \tag{5.17a}$$

Similarly, we may obtain the time-ordered products between the current and the density

$$\tilde{K}^{0i} = -i\frac{k^i n_0}{m} \frac{\omega}{\omega^2 - \omega_k^2} \tag{5.17b}$$

$$\tilde{K}^{00} = i\frac{k^2 n_0}{m} \frac{1}{\omega^2 - \omega_k^2} \tag{5.17c}$$

We know that the Meissner effect corresponds to a term like $\sim A_i^2$ in the effective Lagrangian or equivalently a δ^{ij} term in the current-current correlation function at low frequency and long wavelength. But $\tilde{K}^{ij}$ as shown in (5.17) does not contain a δ^{ij} term. Indeed, that $\tilde{K}^{ij} \propto k^i k^j$ is precisely the statement that the phonon mode represents a density wave. The resolution is of course that $\tilde{K}^{ij}$ in (5.17) represents only the phonon pole and that in general the current-current correlation function at small ω and $\vec{k}$ contains another term, thus

$$K^{\mu\nu} = \tilde{K}^{\mu\nu} + \Delta K^{\mu\nu}. \tag{5.18}$$

At small ω and k, $\Delta K^{\mu\nu}$ can only be a polynomial ω and $\vec{k}$, thus $\Delta K^{00} = a$, $\Delta K_{0i} = bk_i$, and $\Delta K_{ij} = ck_i k_j + d\delta_{ij}$ where a, b, c, d are polynomials. That $\Delta K^{\mu\nu}$ is polynomial corresponds to the statement that all other excitations are gapful. These higher excitations contribute to $K^{\mu\nu}$ terms of the form $\int_M^\infty d\sigma \frac{1}{\omega - (\sigma + k^2 + ...)}$ with $M > 0$. These terms can be expanded as a power series in ω and $\vec{k}$.

The polynomial $\Delta K^{\mu\nu}$ is determined by the requirement that $K^{\mu\nu}$ satisfies the Ward identity

$$k^\mu K_{\mu\nu} = 0. \qquad (5.19)$$

We see immediately that

$$d = \frac{in_0}{m} + c\vec{k}^2 - a\omega^2 \rightarrow in_0/m. \qquad (5.20)$$

Thus, at zero frequency and momentum the correlation function

$$K_{ij} \rightarrow \tilde{K}_{ij} + i\frac{n_0}{m}\delta_{ij}. \qquad (5.21)$$

The δ_{ij} term describes the Meissner effect. (In the field theory literature, $\Delta K_{\mu\nu}$ is determined completely by the additional requirement that it behaves suitably at high frequency and momentum. Here, however, since we are dealing with an effective theory, we are not allowed to impose this additional constraint.)

This derivation of the relation between compressibility and superfluidity also clarifies an apparently puzzling point: the Wigner crystal (in two dimensions, say) has two gapless linearly dispersing modes but is most definitely not a superfluid. The point is that with two gapless modes we do not have enough information to fix $\Delta K^{\mu\nu}$. This underscores the difference between non-relativistic field theory and relativistic field theory, a difference that some particle theorists are apt to forget.

We remark that the one phonon state $a_k^\dagger |0\rangle$ is just $i\sqrt{\frac{2\varphi_k}{\hbar\omega_k}}\, n_k^\dagger |0\rangle$. It corresponds to the wave function $\psi(x) = \sum_{i=1}^{N} e^{i\vec{k}\vec{x}_i}\psi_0(x)$ which is just the variational wave function considered by Bijl and Feynman.[43]

From the above discussion, it is clear that the relation between compressibility and superfluidity should be valid in any dimension, even though the mathematical manipulations used at the beginning of this section are only valid in (2+1)-dimensional spacetime.

VI. VORTICES AND FINITE TEMPERATURES

The astute reader may notice that our low energy effective "Maxwell" Lagrangian (4.16) does not violate T and P. This makes sense: due to semion pairing we should see T and P violation only at short distances or high frequencies and we have not included any higher energy excitations. In a superfluid, the higher energy excitations are vortices. In our "fake Maxwell" description, the vortices correspond to "electric charges". This is easy to see. Around an electric charge there is an electric field $f_{0i} \propto x_i/r^2$. According to (4.15), this describes a vortex flow $v_i \propto \epsilon_{ij}x_j/r^2$. Thus, we can incorporate vortices by introducing a complex scalar field ϕ coupled to a_μ via the usual minimal coupling $(\partial_\mu - iqa_\mu)\phi$. The "charge" of ϕ, q, may be determined by appealing to the quantization of fluid flow around the vortex.

At the same time we show how to determine q we will also go through the instructive exercise of showing how our effective description of phonons and vortices may be derived directly from a Ginzburg-Landau Lagrangian. In this section we will find it convenient to absorb q into a_μ so that the vortex has unit charge. The coupling g^2 is suitably renormalized. This changes the normalization of the electromagnetic current of course. See (6.4) below.

In this so-called duality picture[38,39] we represent the phonon or the superfluid component by a gauge potential a_μ and the vortex or the normal fluid component by a complex scalar field ϕ carrying "electric" charge with respect to a_μ. Let us briefly review how this comes about. In the standard Ginzburg-Landau theory, the potential for the complex order parameter field Φ, has the form $V(\Phi) = a|\Phi|^2 + b|\Phi|^4$. In the superconducting phase, a is negative, while in the insulating phase a is positive.

In the superconducting phase, we have a Goldstone mode η corresponding to the phase of Φ : $\Phi = |\Phi|e^{i\eta}$. The dynamics of η is described by the Lagrangian

$$\mathcal{L} = \frac{1}{2h^2}(\partial_\mu \eta)^2 \qquad (6.1)$$

where we have chosen spacetime units so that the phonon velocity is unity. (Throughout this section we will treat the electromagnetic gauge field A_μ as a background classical field that we can switch on at the end of our calculation.) In (2+1)-dimensional spacetime, we can go to a dual representation by introducing a gauge potential a_μ and a gauge field $f_{\mu\nu} = \partial_\mu a_\nu - \partial_\nu a_\mu$ related to η by

$$\partial_\mu \eta = \epsilon_{\mu\nu\lambda} f^{\nu\lambda} \qquad (6.2)$$

We will now explain a choice of the arbitrary overall constant β, which sets the scale of $f_{\mu\nu}$, so that the electromagnetic current J_μ takes on a convenient form. Had we switched on electromagnetism through the covariant derivative $(\partial_\mu - iA_\mu)\Phi$ we would have gotten, instead of (6.1), $\mathcal{L} = \frac{1}{2h^2}(\partial_\mu \eta - A_\mu)^2$. Thus, we identify the electromagnetic current (or equivalently the velocity current in the superfluid) $J_\mu = \partial_\mu \eta / h^2$ (this is standard, of course) and hence $J_\mu = \beta \epsilon_{\mu\nu\lambda} f^{\nu\lambda}$ according to (6.2). (Note that we have suppressed the charge $(2e)$ carried by the order parameter Φ; this factor will be restored at the end. We also set $\hbar = 1$.)

In addition to the gapless excitation η, the superfluid also contains a gapful excitation, namely the vortex, one at each place where $|\Phi|$ vanishes. In the dual picture, the vortex is described by an "electric" charge coupled to a_μ. As was just mentioned we will now choose β so that the vortex carries unit "electric" charge. First, note that with (6.2), the Lagrangian (6.1) in the dual picture is just good old Maxwell

$$\mathcal{L} = -\beta^2 h^2 f_{\mu\nu} f^{\mu\nu} \qquad (6.1')$$

as in (4.16). Thus, if we couple a_μ to a current j_μ by adding $a_\mu j^\mu$ to (6.1') we obtain the Maxwell equation

$$4\beta^2 h^2 \partial^\mu f_{\mu\nu} = j_\nu. \tag{6.3}$$

Integrating the time component of this equation over space, normalizing the vortex charge $\int d^2x j_0 = 1$, and keeping in mind the quantization of vortex circulation, namely $\oint dx_i \partial_i \eta = 2\pi$, we determine that $\beta = 1/(4\pi)$ and thus

$$J_\mu = \frac{1}{4\pi}\epsilon_{\mu\nu\lambda}f^{\nu\lambda}. \tag{6.4}$$

In summary then, the following effective Lagrangian

$$\begin{aligned}\mathcal{L} =& \frac{1}{4g^2}(2f_{0i}^2 - c^2 f_{ij}^2) + |(\partial_0 - ia_0)\phi|^2 \\ &+ v^2|(\partial_i - ia_i)\phi|^2 - m^2\phi^\dagger\phi\end{aligned} \tag{6.5}$$

describes the long distance dynamics of the superfluid. This is to be supplemented by the equation for the electromagnetic current

$$J_i = \frac{1}{2\pi}\epsilon_{ij}f_{0j}. \tag{6.6}$$

Note that the Lagrangian in (6.5) is non-relativistic with the phonon velocity c and the vortex velocity v. We have taken the low momentum dispersion of the vortex to have the form $\omega = \sqrt{m^2 + v^2 k^2} \simeq m^2 + \frac{v^2}{2m^2}k^2$. Note that m^2 represents the gap for creating a vortex-antivortex pair and m/v the effective "mass" of the vortex.

Note that the Lagrangian does not yet violate T and P. To see that it does not violate P, for instance, we note that according to (6.4), the gauge potential a_μ should transform as an axial vector under parity. This leaves the Lagrangian (6.5) invariant provided we also interchange ϕ and $\phi^\dagger$, as is physically reasonable. Thus, this Lagrangian may be useful even outside the context of semion superconductivity. Wen and I have invoked this duality picture to calculate the universal conductance observed in thin films at the superconductor-insulator phase transition.[39]

In the field theory literature it is known that a gauge invariant T and P violating coupling[44] may be introduced as follows:

$$\begin{aligned}\mathcal{L} =& \frac{1}{4g^2}(2f_{0i}^2 - c^2 f_{ij}^2) \\ &+ |(\partial_0 - ia_0 - i\eta\epsilon_{ij}f_{ij})\phi|^2 \\ &+ v^2|(\partial_i - ia_i - i\xi\epsilon_{ij}f_{0j})\phi|^2\end{aligned} \tag{6.7}$$

where η and ξ are phenomenological parameters. The coupling of the vortex current $\sim \phi^\dagger(\partial_i - ia_i)\phi$ to the superfluid current $\sim \epsilon_{ij}f_{0j}$ implies that the superfluid density would be different near the core of a vortex and the core of an anti-vortex. Thus, a vortex and an anti-vortex have different energies.

We can integrate out ϕ to obtain an amended low energy effective Lagrangian. Clearly, it will have the form "Maxwell Lagrangian" plus a "higher Chern-Simons term" where the higher Chern-Simons term denotes the non-relativistic form[45] of $\epsilon^{\mu\nu\lambda} f_{\mu\sigma} \partial_\nu f_{\lambda\sigma}$. (We know from our construction that this term must be gauge invariant.) That it is a higher derivative term implies that T and P violation vanish at low frequency. Clearly, this conclusion will propagate itself into the effective Lagrangian for A_μ when we turn on electromagnetism.

VII. DIFFERENT STATES OF THE ANYON SUPERFLUID

At first sight, Laughlin's theory of the fractional Hall effect appears rather strange in that it hardly mentions the many-body Hamiltonian. We learn that it is more important to specify the state and to incorporate in it the correct quantum correlations. Recently, Wen and Niu[46] have pointed out that the filling factor ν is not sufficient to characterize the quantum Hall state. Different wave functions describing different quantum Hall states, all corresponding to the same filling factor, may be constructed. Which of these states is favored is a question of energetics and depends on the detailed interaction between the electrons in the quantum Hall system.

Here we have an analogous situation. There may exist many different anyon superfluids corresponding to the same value of θ. As I have already alluded in our discussion of the ideal anyon versus the non-ideal anyon, the statistics parameter θ specifies only the long distance phase interaction between anyons. In real materials, we would expect in general all sorts of complicated short distance interaction as well. Which anyon superfluid is actually realized can only be determined by studying the detailed Hamiltonian.

In Section IV we learned in effect a recipe for constructing an effective theory of anyon superfluidity once we are given an effective theory of the quantum Hall system. The recipe, as shown in the discussion from (4.17) to (4.24) amounts to adding a Chern-Simons term, and if we wish, a Maxwell term to the effective action for the quantum Hall system. As was shown by the work of Girvin and MacDonald[47] and others, the low energy properties of the quantum Hall system defined by $S_m(a^0)$ as in (8.17) may be described by a Lagrangian $\mathcal{L}_{GL}$ of the Ginzburg-Landau type. Different quantum Hall states are described by different G-L effective theories. Using the effective theories of the quantum Hall states we will construct the effective theories of the anyon superfluid states.

Let us consider a specific example. Assume the anyon has statistics $\theta = \pi - \frac{\pi}{q}$ and $\alpha = \frac{q}{8\pi}$. In this section we treat the anyon as fermions with an extra phase interaction. The filling factor of the associated quantum Hall problem is $\nu = q$. The simplest quantum Hall state with filling factor $\nu = q$ is given by q filled Landau levels. The G-L theory of such an integral quantum Hall state is given by

$$\mathcal{L}_{GL} = \sum_{I=1}^{q} \Big[\Phi_I^\dagger i(\partial_0 + ih_{I0} + ia_0 + ieA_0) \Phi_I$$
$$+ \frac{1}{2m} \Phi_I^\dagger (\partial_i + ih_{Ii} + ia_i + ieA_i)^2 \Phi_I \qquad (7.1)$$
$$+ \frac{1}{4\pi} h_{I\mu} \partial_\nu h_{I\lambda} \epsilon^{\mu\nu\lambda} + V_I(\Phi_I) \Big].$$

We introduce a separate set of fields Φ_I and $h_{I\mu}$ for each Landau level. The coupling of $h_{I\mu}$ to Φ_I turns the bosonic fields Φ_I into fermionic fields. (The coefficient $\frac{1}{4\pi}$ of the Chern-Simons term in (7.1) corresponds to $\alpha = \frac{1}{8\pi}$ in the notation of (2.8) and hence $\theta = \pi$.) Here $a_\mu + eA_\mu$ are treated as background potentials. This describes the quantum Hall problem in which fermions move in a fixed background $(a_\mu + eA_\mu)$. (This corresponds to the dynamics defined by $S_m(a)$ in (8.17) with a_μ^0 denoted by $a_\mu + eA_\mu$ here.) We now dualize $\mathcal{L}_{GL}$ according to the prescription given in Section V, representing the phase degree of freedom in Φ_I by gauge potentials $a_{I\mu}$ and the vortex degree of freedom by complex scalar fields ϕ_I. The current carried by Φ_I to which $(h_{I\mu} + a_\mu + eA_\mu)$ couples is given by $\frac{1}{4\pi} \epsilon^{\mu\nu\lambda} f_{I\nu\lambda}$. We thus have

$$\tilde{\mathcal{L}}_{GL} = \sum_{I=1}^{q} \frac{1}{4\pi}(h_{I\mu} + a_\mu + eA_\mu)\epsilon^{\mu\nu\lambda} f_{I\nu\lambda}$$
$$+ \frac{1}{4\pi} h_{I\mu} \partial_\nu h_{I\lambda} \epsilon^{\mu\nu\lambda} + \phi_I^\dagger i(\partial_0 + ia_{I0}) \phi_I \qquad (7.2)$$
$$+ \frac{1}{2m_I} \phi_I^\dagger (\partial_i + ia_{Ii})^2 \phi_I - \Delta_I \phi_I^\dagger \phi_I$$

We now integrate out $h_{I\mu}$ to obtain

$$\tilde{\mathcal{L}}'_{GL} = \sum_{I=1}^{q} \Big[\frac{1}{4\pi}(a_\mu + eA_\mu) f_{I\nu\lambda} \epsilon^{\mu\nu\lambda}$$
$$+ \frac{1}{4\pi} a_{I\mu} \partial_\nu a_{I\lambda} \epsilon^{\mu\nu\lambda} + \phi_I^\dagger i(\partial_0 + ia_{I0}) \phi_I \qquad (7.2')$$
$$+ \frac{1}{2m_I} \phi_I^\dagger (\partial_i + ia_{Ii})^2 \phi_I - \Delta_I \phi_I^\dagger \phi_I \Big].$$

Note that a Chern-Simons term for each of the gauge potentials $a_{I\mu}$ is induced. Thus, the quasi-particle created by ϕ_I carries fermionic statistics. This fermionic quasi-particle corresponds to a hole in the Ith Landau level in the original picture.

Now we follow our recipe and proceed to the effective theory for the anyon superfluid by adding a Chern-Simons term for a_μ with a coefficient $\alpha = \frac{1}{8(\pi/n)} = \frac{q}{8\pi}$

Semionics: a theory of high temperature superconductivity

(since we are considering anyons with $\theta = \pi - \frac{\pi}{q}$.) Thus, we have the effective Lagrangian of the anyon superfluid

$$\mathcal{L}_{SF} = \frac{q}{8\pi} a_\mu f_{\nu\lambda} \epsilon^{\mu\nu\lambda} - \frac{1}{4g^2}(f_{\mu\nu})^2$$
$$+ \sum_{I=1}^{q} \left[\frac{1}{4\pi}(a_\mu + eA_\mu)f_{I\nu\lambda}\epsilon^{\mu\nu\lambda} + \frac{1}{4\pi} a_{I\mu}\partial_\nu a_{I\lambda}\epsilon^{\mu\nu\lambda} \right.$$
$$\left. - \frac{1}{4g_I^2}(f_{I\mu\nu})^2 + a_{I\mu}j_I^\mu \right] \tag{7.3}$$

For good measure, we have added Maxwell terms for a_μ and $a_{I\mu}$; they and other higher derivative terms will be present in general.

A different quantum Hall state with filling factor $\nu = q$ can be constructed by assuming that q fermions form a bound state. The effective filling factor for such q-fermion bound states is $\nu^* = \frac{\nu}{q^2} = \frac{1}{q}$. (The filling factor is the ratio of number density to the magnetic field. By regarding an q-fermion bound state as a fundamental unit, the number density scales down by q while the magnetic field scales up by q.) We now have fractional filling. The simplest quantum Hall state for $\nu^* = \frac{1}{q}$ is the Laughlin state

$$\psi(z_i) = \prod_{i<j}(z_i - z_j)^n \prod_i e^{-\frac{1}{4}|z_i|^2} \tag{7.4}$$

where z_i is the center-of-mass coordinate of the q-fermion bound states. The effective theory of the (fractional) quantum Hall state (corresponding to the Laughlin state (7.4)) is given by (in the dual form)

$$\tilde{\mathcal{L}}_{GL}'' = + \frac{q}{4\pi}(a_\mu + eA_\mu)\tilde{f}_{\nu\lambda}\epsilon^{\mu\nu\lambda}$$
$$+ \frac{q}{4\pi}\tilde{a}_\mu\partial_\nu\tilde{a}_\lambda\epsilon^{\mu\nu\lambda} + \tilde{a}_\mu\tilde{j}^\mu \tag{7.5}$$

where $\tilde{j}^\mu$ is the current of the quasi-particles in the Laughlin state which carries integral $\tilde{a}_\mu$ charge. According to our recipe, the corresponding anyon superfluid state is described by

$$\mathcal{L}_{SF}' = \frac{q}{8\pi} a_\mu f_{\nu\lambda}\epsilon^{\mu\nu\lambda} - \frac{1}{4g^2}(f_{\mu\nu})^2 \tag{7.6}$$
$$+ \frac{q}{4\pi}(a_\mu + eA_\mu)\tilde{f}_{\nu\lambda}\epsilon^{\mu\nu\lambda} + \frac{q}{4\pi}\tilde{a}_\mu\partial_\nu\tilde{a}_\lambda\epsilon^{\mu\nu\lambda} + \tilde{a}_\mu\tilde{j}^\mu.$$

To study the two superfluid states (7.3) and (7.6), let us introduce a test particle which carries unit a_μ charge. The test particle can be included by adding $a_\mu j^\mu$ to (7.3) and (7.6). We would like to study the statistics of the test particle in the superfluid states. We know from the Chern-Simons term in (7.3) and (7.6) that the test particle has statistics $\theta = \pi - \frac{1}{8\alpha} = \pi - \frac{\pi}{n}$ in absence of anyons (whose effects are represented by the last four terms in (7.3)). (We have assumed that the test particle is generated by a fermionic field.) However, in the presence of anyons, the test particle is screened and its statistics may be changed.

The effective Lagrangian (7.3) can be put into a more compact form by defining $a_{0\mu} = a_\mu$ and $j_0^\mu = j^\mu$:

$$\mathcal{L}_{SF} = \frac{1}{4\pi} \sum_{I,J=0}^{q} a_{I\mu} \Lambda_{IJ} \partial_\nu a_{I\lambda} \epsilon^{\mu\nu\lambda} + \sum_{I=0}^{q} a_{I\mu} j_I^\mu + \ldots \tag{7.7}$$

where the $(q+1) \times (q+1)$ matrix Λ is given by

$$\Lambda = \begin{pmatrix} q & 1 & 1 & \ldots & 1 \\ 1 & 1 & 0 & \ldots & 0 \\ 1 & 0 & \ddots & & 0 \\ \vdots & \vdots & & \ddots & 0 \\ 1 & 0 & \ldots & 0 & 1 \end{pmatrix}. \tag{7.8}$$

Introducing $c_{I\mu}, I = 0, \ldots, q$ such that

$$\begin{pmatrix} a_{0\mu} \\ \vdots \\ a_{q\mu} \end{pmatrix} = c_{0\mu} \begin{pmatrix} 1 \\ -1 \\ \vdots \\ -1 \end{pmatrix} + c_{1\mu} \begin{pmatrix} 1 \\ 1 \\ 0 \\ \vdots \\ 0 \end{pmatrix}$$

$$+ c_{2\mu} \begin{pmatrix} 1 \\ 0 \\ 1 \\ \vdots \\ 0 \end{pmatrix} + \cdots + c_{q\mu} \begin{pmatrix} 1 \\ 0 \\ \vdots \\ 0 \\ 1 \end{pmatrix}. \tag{7.9}$$

We may rewrite (7.7) as

$$\mathcal{L}_{SF} = \frac{1}{4\pi} \sum_{I,J=1}^{q} c_{I\mu} \Lambda'_{IJ} \partial_\nu c_{J\lambda} \epsilon^{\mu\nu\lambda}$$
$$+ \left(\sum_{I=0}^{q} c_{I\mu} \right) j^\mu + \sum_{I=1}^{q} (c_{I\mu} - c_{0\mu}) j_I^\mu + \ldots \tag{7.10}$$

where the $q \times q$ matrix Λ' is given by

$$\Lambda' = \begin{pmatrix} 1 & & \\ & \vdots & \\ & & 1 \end{pmatrix} + (q+2) \begin{pmatrix} 1 & 1 & \cdots & 1 \\ 1 & 1 & \cdots & 1 \\ \vdots & \vdots & \ddots & \vdots \\ 1 & 1 & \cdots & 1 \end{pmatrix}. \tag{7.11}$$

Thus, the gauge potential $c_{0\mu}$ has no topological mass, while $c_{I\mu}$, $(I = 1, \ldots, q)$ all have non-zero topological mass since $\det \Lambda' \neq 0$. Therefore the low lying excitation (the phonon) in the superfluid state is described by

$$\mathcal{L}_{SF} = \frac{1}{4g^{*2}} (\partial_\mu c_{0\nu} - \partial_\nu c_{0\mu})^2 + \ldots \tag{7.12}$$

Now let us consider a bound state of the test particle with q_I particles generated by ϕ_I fields. Such a bound state carries a $c_{0\mu}$ charge of $Q_0 = 1 - \sum_{I=1}^{q} q_I$ and a $c_{I\mu}$ $(I = 1, \ldots, q)$ charge of $Q_I = q_I + 1$. If $Q_0 \neq 0$ the bound state generates a long range "electric" field of $c_{0\mu}$ and the energy of bound state has a logarithmic divergence. According to the duality picture, the bound state with $Q_0 \neq 0$ corresponds to a vortex in the superfluid state with a circulation proportional to Q_0. The bound state with $Q_0 = 0$ has finite energy corresponding to a screened test particle. Such a screened test particle has statistics

$$\theta = \pi - \pi (Q_1 \ldots Q_q) \Lambda'^{-1} \begin{pmatrix} Q_1 \\ \vdots \\ Q_q \end{pmatrix} + \pi = \left(1 - \sum_{I=1}^{q} q_I^2\right) \pi. \tag{7.13}$$

In particular, for $q = 2$, $\Lambda'^{-1} = \frac{1}{9} \begin{pmatrix} 5 & -4 \\ -4 & 5 \end{pmatrix}$ and $\theta = 0$ as long as $1 - q_1 - q_2 = 0$.

Similarly, one can show that the screened test particle in the superfluid state described by (7.6) has statistics

$$\theta = \frac{\pi}{q} + \pi. \tag{7.14}$$

Such a screened test particle carries a unit a_μ charge and a unit $\tilde{a}_\mu$ charge.

The statistics described by (7.13) and (7.14) are manifestly different. Because the test particle has different statistics, (7.3) and (7.6) definitely describe two different anyon superfluid states. Adding an anyon to the system, we may regard the added anyon itself as the test particle. Therefore (7.13) and (7.14) determine the statistics of the quasi-particle excitations in two different superfluid states.

VIII. CHIRAL SPIN STATE

Now that I have argued that a semion fluid is a superfluid, let us return to my questions 1 and 2, namely, how do semions emerge in real superconductors? The solid state physics of real superconductors has been described in detail elsewhere in this volume and in any case is beyond my domain of competence. In broad outline, high temperature superconducting materials are antiferromagnetic insulators before they are doped with holes. They become superconducting at a finite concentration of holes. The critical concentration is typically about $x_c \sim 0.05$ where x denotes the number of holes per site. The Néel order disappears for $x_c \sim 0.03$. The holes lead to frustration in the antiferromagnet. By using the slave-boson technique, it is possible to derive the effective spin Hamiltonian in the presence of holes by a systematic expansion.[48]

As the frustration increases with an increasing concentration of holes, the Néel order must give way and lead to some sort of disordered state. We can also see this clearly by thinking directly about holes hopping in a background of electrons with their spins specified according to Néel order. By hopping around, the holes bring electrons with the same spin together and thus tend to destroy the Néel order.[49] In fact, in a Néel ordered state a moving hole leaves a "string" of energy behind it. Thus, Néel ordering is not favored in the presence of holes.

The crucial question is then the following: What does the ground state of the frustrated Heisenberg Hamiltonian look like? For some range of parameters, is it a "quantum disordered spin liquid" state?

The exact solution of the frustrated Heisenberg Hamiltonian is of course a formidable problem. Instead, people have been trying to guess what the ground state may look like. In particular, the resonating valence bond approach[24,25] of Anderson and others represents an attempt to write down the ground state by an inspired guess. In general, it has proved rather difficult to describe this putative quantum disordered spin liquid state.

Fortunately, instead of tackling this difficult task, we can ask a clear cut and binary question about the ground state: Does it violate T and P invariances? If the answer is no, then the physics discussed in this article is largely irrelevant. But if the answer is yes, then, as we have emphasized already, the appearance of excitations with fractional statistics is a generic possibility.

In Ref. 20, Wen, Wilczek and I introduced the concept of chiral spin state and chiral spin liquid. In contrast to the resonating valence bond state, the chiral spin state can be characterized precisely. Recently, a complete classification has been given by Arovas and Haldane[50] and by Wen.[51]

Given the ground state $|\psi\rangle$, calculate the expectation value

$$E_{ijk} = \langle\psi|\vec{S}_i \cdot (\vec{S}_j \times \vec{S}_k)|\psi\rangle \tag{8.1}$$

We will take the sites of i, j, and k to lie on the vertices of an elementary square plaquette and to thus form a triangle, even though this is not necessary for some of the remarks to be made below. If the state $|\psi\rangle$ is invariant under time reversal T (so that $T|\psi\rangle = \eta|\psi\rangle$ for some phase η), then since $T^{-1}\vec{S}T = -\vec{S}$

$$\begin{aligned} E_{ijk} &= \langle T\psi|\vec{S}_i(\vec{S}_j \times \vec{S}_k)|T\psi\rangle \\ &= \langle\psi|(T^{-1}\vec{S}_i(\vec{S}_j \times \vec{S}_k)T)|\psi\rangle^* \\ &= -E_{ijk} \end{aligned} \qquad (8.2)$$

Thus, a non-zero value of E_{ijk} necessarily means that T is violated. Similarly, by considering reflection in the perpendicular bisector of the diagonal side of triangle formed by i, j, and k, we see that a non-zero E_{ijk} also violates P.

We call T and P violating states chiral spin states. E_{ijk} is clearly a chiral order parameter. It tells us that neighboring spin directions form a handed triad, either right or left handed. The quantum fluctuations of the spins may be such that the spin-spin correlation falls off exponentially, but the order parameters $\vec{S}_i \cdot (\vec{S}_j \times \vec{S}_k)$ may have long range correlations with each other.

Another way to characterize these states is to introduce electron creation operators $c_{i\alpha}^\dagger$ on site i, spin α, and the hopping operators

$$\hat{\chi}_{ij} \equiv c_{i\alpha}^\dagger c_{j\alpha}. \qquad (8.3)$$

($\hat{\chi}_{ij}$ is obviously not to be confused with the triple spin operator χ_{ijk} introduced above.) Under a local gauge transformation, whereby an electron at site j acquires the phase $e^{i\theta_j}$, we have $\hat{\chi}_{ij} \to e^{i(\theta_i - \theta_j)}\hat{\chi}_{ij}$. In the half-filled Hubbard model at infinite U, exactly one electron occupies each site, so the Hamiltonian is gauge invariant. Therefore, according to general principles, a gauge-variant object like $\hat{\chi}_{ij}$ cannot acquire a non-zero vacuum expectation value. Rather, the simplest gauge-invariant order parameters we can construct from $\hat{\chi}$ are of the general type

$$\mathcal{P}\ell_{123} = \langle \hat{\chi}_{12}\hat{\chi}_{23}\hat{\chi}_{31}\rangle \qquad (8.4)$$

or

$$\mathcal{P}\ell_{1234} = \langle \hat{\chi}_{12}\hat{\chi}_{23}\hat{\chi}_{34}\hat{\chi}_{41}\rangle \qquad (8.4')$$

where the $\hat{\chi}$'s circle a closed triangle (8.4) or plaquette (8.4'). Indeed, an expectation value of the latter type has been used to characterize the flux phase.[52,53,24]

A simple calculation shows that these plaquette order parameters are related to the spin expectations E. Specifically, we have

$$\mathcal{P}\ell_{123} - \mathcal{P}\ell_{132} = -\frac{i}{2} E_{123} \qquad (8.5)$$

and

$$\mathcal{P}\ell_{1234} - \mathcal{P}\ell_{1432} = \frac{i}{4}\left(-E_{123} - E_{134} - E_{124} + E_{234}\right). \qquad (8.5')$$

Thus the chiral spin states are alternatively characterized by their supporting a difference between the expectation values for plaquettes traversed in opposite directions. This of course emphasizes their P-violating nature. Unlike E_{123}, the advantage of the plaquette variables is that they can be defined even away from the one electron per site limit.

The chiral spin order parameter captures some, but not all, of the properties we would like to postulate of a quantum spin liquid. It has the desirable feature of leaving rotation and translation symmetry unbroken, but unfortunately it does not capture the long-range coherence we expect is necessary for an incompressible liquid. Inspired by analogy with the quantized Hall effect, however, we are led to the following definition. We say that we have a chiral spin liquid, when not only small triangles or plaquettes, but also large loops are ordered, in such a way that products around consecutive links enclosing a loop obey

$$\langle \prod_\gamma \hat{\chi}_{ij}\hat{\chi}_{jk}\ldots\hat{\chi}_{li}\rangle = e^{-f(\gamma)}e^{ibA(\gamma)}. \tag{8.6}$$

Here $f(\gamma)$ is a positive real function of the geometry of the loop γ (in our mean field models it will be proportional to the length of γ), but the crucial feature is the phase term proportional to the area $A(\gamma)$ enclosed by the loop. Identifying $\langle \prod_\gamma \hat{\chi}_{ij}\hat{\chi}_{jk}\ldots\hat{\chi}_{li}\rangle$ loosely as a sort of Wilson loop, we can think of $bA(\gamma)$ as the flux enclosed by the loop γ. We expect that the crucial properties of the spin liquid, and specifically the statistics of its quasiparticle excitations, are determined by the coefficient b. The mean field theories we construct below support ground states with order of this type.

The preceding discussion suggests that we perform a mean field analysis with the operator $\hat{\chi}_{ij}$ having some expectation value $\chi_{ij} = \langle \hat{\chi}_{ij} \rangle$ such that the product of χ_{ij} around closed loops corresponds to non-zero flux. Details may be found in Ref. 20. Here we sketch the development.

We represent the site coordinate $\vec{i} = (i_x, i_y)$ and denote the elementary lattice vectors $\tau_x = (1, 0)$ and $\tau_y = (0, 1)$. Let us also suppress the overall magnitude of χ_{ij} and set the lattice spacing to unity. We see that the product of χ_{ij} around an elementary plaquette is equal to $(-1) = e^{i\pi}$ and there is a flux π per plaquette. Note that since a flux of π is equivalent to a flux of $(-\pi)$. This state does not violate T and P.

The mean field Hamiltonian is then

$$H_{MF} = \frac{J}{2}\sum \mathrm{tr}(\Psi_i^\dagger \Psi_j \Gamma)\chi_{ji} = J\sum \chi_{ji}\hat{\chi}_{ij} \tag{8.14}$$
$$= J\sum \chi_{ji} c_{i\alpha}^\dagger c_{j\alpha}$$

where $\Gamma = \begin{pmatrix} 1 & \\ & -1 \end{pmatrix}$. A hop from j to i is accompanied by the quantum amplitude χ_{ji} and thus we have an electron hopping around a lattice in the presence of a magnetic field. This problem has been studied over the years by a number of authors. Recently, we have given an analysis[12] using elegant topological methods to conclude that for a flux of $2\pi p/q$ per plaquette the energy spectrum $E(k)$ exhibits q Dirac zeroes in the reduced Brillouin zone. More precisely, in k-space, there are q different locations around which $E(k)\,(k-k^*)$ vanishes like the first power of k^*'s. Thus low energy physics may be described effectively by q massless Dirac fields.

For the case at hand, $q=2$ and we can proceed explicitly. It is instructive to see how the Dirac modes arise. Passing to momentum space

$$\Psi_i = \sum_k e^{i\vec{k}\vec{i}}\Psi_k \quad , \quad c_i = \sum_k e^{i\vec{k}\vec{i}} c_k \qquad (8.15)$$

and introducing $\vec{w} = (0,\pi)$ so that $(-)^{i_y} = e^{i\vec{w}\vec{i}}$ we find

$$\begin{aligned} H &= J\sum_i \left(c^\dagger_{i+\tau_x} c_i (-)^{i_y} + c^\dagger_{i+\tau_y} c_i + h.c. \right) \\ &= J\sum_k \left(c^\dagger_{k+w} c_k e^{ik_x} + c^\dagger_k c_k e^{ik_y} + h.c. \right). \end{aligned} \qquad (8.16)$$

Thus, in the basis

$$\begin{pmatrix} c_k \\ c_{k+w} \end{pmatrix}$$

we have the two-by-two Hamiltonian

$$\mathcal{H} = 2J \begin{pmatrix} \cos k_y & \cos k_x \\ \cos k_x & -\cos k_y \end{pmatrix} \qquad (8.17)$$

and hence the energy eigenvalues

$$E = \pm 2J\sqrt{\cos^2 k_x + \cos^2 k_y} \qquad (8.18)$$

There are two isolated zeroes at $(k_x, k_y) = (\pi/2, \pm\pi/2)$ within the reduced Brillouin zone, corresponding to two massless Dirac modes in the continuum limit. The presence of $(-)^{i_y}$ in (8.13) and hence (8.16) is crucial for producing the Dirac modes. Were this factor absent, so that we just have an electron hopping freely on the lattice, the second line in (8.16) would just be

$$H = J\sum c^\dagger_k c_k (e^{ik_x} + e^{ik_y}) + h.c. \qquad (8.19)$$

and we will have the standard energy dispersion

$$E = -2J(\cos k_x + \cos k_y) \qquad (8.20)$$

and there would be no Dirac mode.

Suppose we now break T and P by allowing

$$\chi_{i,i+\tau_x+\tau_y} = (-)^{iy}\beta \tag{8.21}$$

and

$$\chi_{i,i-\tau_x+\tau_y} = (-)^{iy}\beta^*. \tag{8.22}$$

Note that this choice is necessary so that the flux around each triangle (when traversed in the same sense) is the same and equal to

$$\begin{aligned}H &= \beta \sum_i \text{tr}\left(\Psi^\dagger_{i+\tau_x+\tau_y}\Psi_i\Gamma\right)(-)^{iy} + \text{h.c.} \\ &+ \left(\beta \to \beta^*, \tau_x \to -\tau_x\right) \\ &= (-\beta)\sum_k \text{tr}\left(\Psi^\dagger_{k+w}\Psi_k\Gamma\right) e^{-ik(\tau_x+\tau_y)} + \text{h.c.} \\ &+ (\beta \to \beta^*, \tau_x \to -\tau_x).\end{aligned} \tag{8.23}$$

Carrying out the diagonalization, we see that the two Dirac modes acquire masses with the same sign. This is consistent with the fact, well-known by now, that Dirac fermion mass term violates T and P in (2+1)-dimensional spacetime.[54]

To study the quantum numbers of the excitations, we have to consider deviations from the mean field. Thus, let us write

$$\chi_{ij} = \chi^0_{ij} e^{i\int_i^j \vec{a}\cdot d\vec{x}} \tag{8.24}$$

where the phase fluctuation is described by a gauge degree of freedom. In the continuum limit we obtain the Lagrangian

$$\mathcal{L} = \sum [\bar{\psi}\gamma^\mu(i\partial_\mu + a_\mu + A_\mu)\psi - m\bar{\psi}\psi]. \tag{8.25}$$

We have also included the electromagnetic gauge potential A_μ. Since the Dirac fields are now massive, we can safely integrate them out, to obtain the effective Lagrangian

$$\mathcal{L}_{\text{eff}} = 4\,\frac{m}{|m|}\,\frac{1}{8\pi}\,(a_\mu + A_\mu)\partial_\nu(a_\lambda + A_\lambda)\,\epsilon^{\mu\nu\lambda}. \tag{8.26}$$

The factor 4 is because of the four fermions (counting the spin up and spin down modes).

Using the effective Lagrangian we may obtain the low energy properties of the chiral spin phase. First we would like to show that there is no zero magnetic field Hall effect in the chiral spin phase (*i.e.*, the conductance $\sigma_{xy} = 0$), even though it is no longer forbidden by the (broken) symmetries P and T. The electrical current is defined by

$$J_e^\mu = \frac{\partial \mathcal{L}_{\text{eff}}}{\partial A_\mu} = \frac{1}{\pi} \frac{m}{|m|} \epsilon^{\mu\nu\lambda} \partial_\nu (a_\lambda + A_\lambda). \tag{8.27}$$

The equation of motion for a_μ reads

$$\begin{aligned} 0 &= \frac{\partial \mathcal{L}_{\text{eff}}}{\partial a_\mu} = \frac{1}{\pi} \frac{m}{|m|} \epsilon^{\mu\nu\lambda} \partial_\nu (a_\lambda + A_\lambda) \\ &= J_e^\mu. \end{aligned} \tag{8.28}$$

This implies that the electrical current vanishes for any background electromagnetic field and the chiral spin phase at half filling is an insulator. This is hardly shocking, since our effective theory (8.26) is supposed to describe the low energy properties of the Heisenberg model (3.4), which contains no charge fluctuations. However it was not completely obvious *a priori* in our mean field approximation, which does allow charge fluctuations.

Now let us consider the excitations in the chiral spin phase. The simplest excitation to consider is an excited electron in the conduction band. We must emphasize that the appearance of an electron in the conduction band does not correspond to introducing an electron into our system, because integrating out a_0 still enforces the constraint $n_i = 1$. We will see that such an excited electron corresponds to a neutral spin 1/2 particle. At low energy the excited electron can be regarded as a test particle. The effective Lagrangian in presence of such a particle can be written as

$$\begin{aligned} \mathcal{L}_{\text{eff}} =& \frac{1}{2\pi} \frac{m}{|m|} \epsilon^{\mu\nu\lambda} (a_\mu + A_\mu) \partial_\nu (a_\lambda + A_\lambda) \\ &+ (a_\mu + A_\mu) j^\mu \end{aligned} \tag{8.29}$$

where j^μ is the current of the test particle. Now the electrical current and the equation of motion become

$$J_e^\mu = \frac{1}{\pi} \frac{m}{|m|} \epsilon^{\mu\nu\lambda} \partial_\nu (a_\lambda + A_\lambda) + j^\mu \tag{8.30}$$

and

$$0 = \frac{\partial \mathcal{L}_{\text{eff}}}{\partial a_\mu} = j^\mu + \frac{1}{\pi} \frac{m}{|m|} \epsilon^{\mu\nu\lambda} \partial_\nu (a_\lambda + A_\lambda). \tag{8.31}$$

The first term in (8.30) can be regarded as the contribution to electrical charge arising from the vacuum polarization. The equation of motion implies that the electrical charge of the excited electron is completely screened by vacuum polarization. The screened electron behaves like a neutral particle. Because the chiral spin vacuum is a spin singlet even when a_μ and A_μ are non-zero the vacuum polarization can not change the spin quantum number of the excited electron. Therefore the screened electron is really a spin 1/2 neutral particle. Due to the Chern-Simons term in (8.29) the statistics of the excited electron is also changed. Referring to (2.12) we find that the statistics are given by a phase factor $e^{i(\pi+\frac{\pi}{2}\frac{m}{|m|})}$. Thus the screened electron behaves like a half fermion. Properties of the chiral spin state have been studied by a number of authors.[55]

The following rather hand-waving argument may be helpful also. A defect in our featureless singlet spin liquid can be introduced by constraining the spin on one site to be, say, up. The density of the liquid is then reduced, because the site in question can only be reached by neighboring electrons already spinning up. In effect, one-half a site — and therefore, by incompressibility, one-half an electron — has been removed. Since the phase was $e^{i\pi}$ per encircled electron, it becomes $e^{i\pi/2}$ for encircling the defect. Now we can expect that upon our slowly delocalizing the constraint the system will relax to an energy eigenstate with spin 1/2. As long as there is a gap neither the total spin nor the phase accompanying transport around a loop far from the defect can be altered by the relaxation, which is a local process. Thus we expect that the defect relaxes into a spin 1/2, neutral, half-fermion quasiparticle.

We have characterized a class of states, but what about the Hamiltonian? In condensed matter physics, the detailed Hamiltonian at the atomic level is of course enormously complicated. Two strategies are followed. We may start with an idealized Hamiltonian that we may try to solve analytically, or failing that, numerically. An alternative philosophy is to emphasize the state rather than the Hamiltonian. By characterizing the symmetries of the state we may try to determine the quantum numbers of the excitations around that state. We then ask whether there is a universality class of Hamiltonians for which the state may plausibly be the ground state. This philosophy was particularly popularized by Laughlin's theory of the Hall effect. As I mentioned in Section VII, the idea is to take into account of the quantum correlation first and worry about the detailed interaction structure later. In the present context, at finite doping the effective spin Hamiltonian will presumably be exceedingly complicated, with all sorts of frustrating non-nearest neighbor couplings.

Let me summarize the present situation regarding my question 1 of whether the chiral spin state is the ground state of some reasonable Hamiltonian as follows.

(1) There exists[20] a class of Hamiltonians for which the chiral spin state is the ground state. (That state is however not a chiral spin liquid.) I should emphasize that this was shown exactly and not in the mean field approximation. Whether the Hamiltonian is "reasonable" is another story: it involves a six-spin interaction.

(2) In a mean field calculation, the chiral spin state is higher than a serious competitor, the dimer state, by six-tenth of a percent. Since fluctuations favor the chiral spin state, the mean field estimate may actually be prejudiced against the chiral spin state. In any case, we can easily add a term to the Hamiltonian that would favor the chiral spin state over the dimer state.

(3) Numerical computations on small lattices[56,57] with Hamiltonians with limited number of non-nearest neighbor spin couplings appear not to show the chiral spin state but these computations are typically sensitive to the boundary.

(4) Baskaran[58] has used a duality transformation to map frustrated spin models into models in the Ising universality class. Within a certain approximation, he claimed to have shown the appearance of chiral spin state.

(5) Laughlin and Zou[59] have undertaken a massive numerical program in which they fill the lower band of the single-electron spectrum in the chiral spin state found in ref. (20) and then Gutzwiller project. They use the resulting state as a variational state for the t-J model for example. The numerical results are encouraging.

(6) Finally, I would like to emphasize a crucial distinction between the chiral spin liquid and its various competitors, such as the dimer state and the staggered flux state, namely that the chiral spin liquid is topological in character. For instance, the coefficient of the Chern-Simons term in the effective Lagrangian describing physics around the state is an integer. As we wander about in the space of Hamiltonians, the integer can change only when the gap in our single particle spectrum closes. As we change the coefficients of various terms in the Hamiltonian, there is a finite region in parameter space in which the chiral spin state is at least a local minimum. More crudely, we can summarize by saying that the semion cannot turn into a boson smoothly.

IX. NON-LINEAR σ-MODEL

An important first step in elucidating the nature of a condensed matter system is the determination of the quantum numbers of the elementary excitations or quasiparticles. A long distance effective field theory approach is particularly suitable since the most essential quantum numbers, such as charge and fractional statistics, can be deduced by studying the physics at long distances. In this respect, field theory has an important role to play in condensed matter physics. Also, a field theory treatment often accommodates analytic calculation more readily than a lattice model.

I believe that the field theory treatment given here captures the essential long distance physics. By its very nature, a field theory discussion misses the details of the short distance physics which is simply absorbed into a cut-off. Of course, even in a lattice model, the interaction on the lattice scale already represents an approximation, often rather drastic, of the true underlying physics.

Using the coherent state formalism and evaluating the path integral representation of the partition function carefully,[60] one can show that Heisenberg spin Hamiltonians may be represented by the non-linear σ-model:[61,62]

$$\mathcal{L} = \frac{2}{g^2} \left(\partial_\mu \vec{n}\right)\left(\partial^\mu \vec{n}\right) \tag{9.1}$$

with the constraint

$$\vec{n}^2(x) = 1 \tag{9.2}$$

and the spacetime coordinates $x^\mu, \mu = 0, 1, 2$. The order parameter $\vec{n}$ represents an expectation value of the staggered spin operator $(-)^i \vec{S}_i$. In proceeding to this effective field theory description, one has implicitly assumed that the spin-spin correlation is larger than the lattice spacing. In this case, over the length scale of a few lattice spacing there is antiferromagnetic ordering and that an effective field or local spin order parameter $\vec{n}(x)$ can be defined. This implicit assumption appears as an assertion about smoothness in the path integral formalism.

Actually, the representation of the long distance physics of the Heisenberg spin Hamiltonians by the nonlinear σ-model is quite general. As soon as we recognize that the spin rotation symmetry is spontaneously broken by the ground state, it follows that there is a Nambu-Goldstone mode, namely the spin wave. On general grounds, we know that for small $\vec{k}$ the spin wave satisfies the linear dispersion relation $\omega^2 \simeq \vec{k}^2$ in an antiferromagnet and the quadratic dispersion relation $\omega \simeq \vec{k}^2$ in a ferromagnet. (This is consistent with the time reversal transformation T. Ferromagnetic ordering violates T and hence the spin wave dispersion should not be invariant under $\omega \rightarrow -\omega$. On the other hand, antiferromagnetic ordering is invariant under T followed by a translation through one lattice spacing.) Thus, with a suitable choice of units to set the spin wave velocity to unity, we may describe the long wavelength physics of an antiferromagnetic system by a "relativistic" field theory.

The nonlinear σ-model is not renormalizable in (2+1)-dimensions. The short distance physics we have neglected is represented by the cutoff Λ needed to define the theory and thus enters via higher order radiative corrections. The nature of the field theoretic vacuum may depend on short distance physics. The 2-dimensional quantum statistical mechanics of the spin system is described by the quantum field theory (9.1) in (2+1)-dimensional spacetime. At finite temperature T, the field theory is evaluated in Euclidean spacetime with time duration equal to $\beta = 1/T$. We will focus on the behavior at zero temperature, in which case we can study the field theory in Minkowskian spacetime.

Several comments are in order.

(1) It is clear from this discussion that the philosophy here is to start from the Néel ordered phase and to move towards the order-disorder transition. Because of the assumption of a spin correlation length over which $\vec{n}(x)$ can be defined, the field theory description may not be valid far into the disordered phase. In other words, we start with a theory with a Nambu-Goldstone boson and we show that it disappears as a coupling increases past some critical value.

(2) In imposing the constraint (9.2) one has supposed that the spin ordering varies spatially only in orientation. We have ignored fluctuations in the magnitude of $\vec{n}$.

(3) That there may be a disordered or spin liquid phase is suggested by elementary considerations. A fluctuation costs an amount in action of order $(1/g^2)$. Thus, at large coupling fluctuations should be so inexpensive that they occur freely and

where χ is a non-relativistic fermionic field describing the dressed holes. χ carries charge but not spin and may be thought of as a hole bound with a z quantum. Because of its gauge coupling to a_μ, it acquires a fractional statistics given by $\theta = \pi + \frac{\pi}{2n}$.

How trustworthy is the mean field approximation? Wen and I[17] have attempted to obtain some feel for this question by studying field theories similar to the ones discussed here in (1+1)-dimensional spacetime non-Abelian bosonization and other field theoretic techniques are then available to solve some of these theories more or less exactly, or at least without invoking the mean field approximation. Our general impression from these studies is that the results of the mean field approximation appears to be reliable as far as quantum numbers are concerned.

As mentioned earlier, there are at present three arguments (but certainly no proof) indicating that the excitations in antiferromagnetic systems in the disordered phase carry fractional statistics. The first argument was due to Kalmeyer and Laughlin,[22] who mapped the antiferromagnet onto an equivalent system of hard core bosons moving on a lattice in a background magnetic field. Later, Wen and I gave the field theoretic treatment described here. Finally, Wen, Wilczek and I showed that the effective low energy theory around the chiral spin state contains anyons. These arguments, while all based on mean field approximation in one way or another, are at least mutually consistent in that they all fix θ to be $\pi/2n$ for n some integer.

I conclude this section with the remark that in the field theory treatment described here we could also have added the Hopf term with a half-integral coefficient $(n + 1/2)$ instead of an integer coefficient n. The path integral would have been changed, in the sector where $\int d^3x \mathcal{H} = 2\pi k$, by the factor $e^{i2\pi(n+\frac{1}{2})k} = (-)^k$. According to the analysis of Ref. 3, this implies that the soliton in the non-linear σ-model is quantized as a fermion. The calculation of the fractional statistics carried by the z quantum goes through as before, with n replaced everywhere by $(n + \frac{1}{2})$. Thus, we find that $\theta = \pi/2(n + \frac{1}{2})$. In the simplest case, with $n = 0$, we have $\theta = \pi$ and the z quantum (or spinon in the terminology of resonating valence bond) is a fermion. By extension, the holon is a boson. This provides a concrete realization of the original suggestion of Kivelson, et al.,[26] that the fermionic and bosonic characters of the excitations may be interchanged. For $n \neq 0$, we have fractional statistics as before but now with $\theta = \pi/$(odd integer).

X. DUALITY AND VORTEX CONDENSATION

As is well known, the nonlinear σ-model contains solitons or skyrmions. In particular, it was shown in Ref. 3 that these skyrmions can be quantized in such a way that they have fractional statistics.

In Ref. 14 we showed that the statistics of the skyrmion is related to the statistics of the z quantum by the reciprocal relation

$$\frac{\theta_{\text{skyrmion}}}{\pi} = \frac{\pi}{\theta_z}. \tag{10.1}$$

This result may also be obtained by direct computation.

With this duality relation, it is easy to understand the value of the fractional statistics derived in the approaches mentioned in the last section. The fractional statistics $\theta_z = \frac{\pi}{2n}$ carried by the z-quantum just corresponds to the soliton in the dual picture being quantized with the statistics of the boson. (According to (10.1), the value $\theta_z = \frac{\pi}{2n}$ corresponds to $\theta_{\text{skyrmion}} = 2\pi n$.) Similarly, if the soliton is quantized as a fermion we obtain $\theta_z = \pi/(\text{odd integer})$, a possibility mentioned in the previous section.

This duality relation indicates that fractional statistics occurs generically, either in the particle or in the soliton sector. The inverse duality relation discussed here is mathematically reminiscent of the duality between the electric charge of the electron and the magnetic charge of the magnetic monopole.

I believe that the duality derivation for $\theta = \pi/2n$ given here may be more general than the three "classic" derivations mentioned in the last section and which all rely on the mean field approximation. The duality structure may be made rigorous by formulating it on a lattice[64] and appears to have a deep mathematical origin.

In the preceeding section we see that the nonlinear σ-model exhibits an ordered phase at small coupling (in which $\langle z \rangle \neq 0$) and a disordered phase at large coupling (in which $\langle z \rangle = 0$). Wen and I have suggested in Ref. 14 a duality picture in which we claimed that the disordered phase corresponds to a state in which skyrmions condense into the vacuum. This is rather plausible since around each skyrmion the z-field is twisting around. If the quantum ground state consists of a condensate of skyrmions, the z-field should be totally disordered.

We imagine then that we start with the ordered phase $\langle z \rangle \neq 0$ and construct the skyrmion living in this phase. In the field theory literature, an effective quantum operator creating a soliton can be constructed. In the limit in which the skyrmion size goes to zero, or equivalently if we are only interested in long distance physics, the skyrmion may be described by some effective complex scalar field φ. The field has to be complex because the skyrmion carries a conserved change. At long distances, the skyrmions are described by an effective Lagrangian with the terms

$$\mathcal{L}_{\text{eff}} = \ldots - (M^2 \varphi^\dagger \varphi) + \lambda(\varphi^\dagger \varphi)^2 + \ldots \tag{10.3}$$

where the mass squared of the vortex M^2 is some function of g^2 that can be calculated perturbatively. We conjecture that when g^2 exceed g_c^2, the parameter M^2 in (10.3) becomes negative and the vacuum develops an instability towards skyrmion condensation.

What about the ordered phase? If we are correct in describing the disordered phase as one in which a complex scalar field φ condenses and acquires a non-zero vacuum expectation value, then in this phase we can construct a vortex out of the field φ. According to the duality relation (10.1), the statistics of this vortex is equal to precisely that of the z-field. We believe that this vortex constructed out of the skyrmion field φ represents precisely the original z field.

In the ordered phase, z has a non-zero vacuum expectation value by definition, and the gauge potential a_μ acquires a mass according to the usual Higgs mechanism. On the other hand, in the disordered phase, z has zero vacuum expectation value and a_μ remains massless. In contrast, the effective skyrmion field φ has a non-zero vacuum expectation value, and it is A_μ that becomes massive. We summarize the situation in a table.

	order	disorder
a_μ	massive	massless
A_μ	massless	massive

Using this dual language, we have a very simple picture for the T and P broken states discussed earlier. The solitons or skyrmions may condense into the many possible Laughlin states described by the wave function of the solitons

$$\Phi(w_1, w_2, \ldots) \propto \prod_{i,j}(w_i - w_j)^{2n} e^{-\sum_k |w_k|^2/\xi^2} \tag{10.4}$$

where $w_j = x_j + iy_j$ is the coordinates of the jth skyrmion. When we exchange w_1 and w_2, say, we obtain a phase $(-1)^{2n}$. Here n is an integer when the skyrmion is boson as in the model. This is also consistent with the analysis of Ref. 15 where the fractional statistics of the skyrmion was determined to be characterized by the phase $e^{i 8\pi^2 (n/4\pi)} = e^{i 2\pi n} = (-1)^{2n}$. The Laughlin state (10.4) obviously break T and P and may correspond to the quantum disordered states discussed here.

But unfortunately the duality picture described here is not complete and the above argument is only heuristic. It would be very interesting to find a complete duality picture for the CP^1 (or CP^{N-1}) model. Perhaps by going to a lattice formulation we can formulate this duality picture more rigorously.[64]

XI. EXPERIMENTAL TESTS

Finally we come to the last question on my list. At this conference, we have heard a great deal of discussion about whether certain theories of high temperature superconductivity are falsifiable. I want to state for the record that semionics is certainly falsifiable.

To make contact with experiments, we may write down a Lagrangian of the Ginzburg-Landau type. The details are given in Ref 16. Here we would like to emphasize the generic features that such an effective phenomenological Lagrangian should have. First, two independent $U(1)$ gauge invariances, one associated with the statistical gauge potential a_μ and the other with the electromagnetic gauge potential A_μ, must be broken, thus requiring two separate Higgs fields or order parameters. (Borrowing the terminology of resonating valence bonds, we may think loosely of the two order parameters as corresponding to the spinon condensate and the holon pair condensate.) The second generic feature is the presence of a Chern-Simons term for a_μ so as to give the spinon fractional statistics. In the end, when we integrate out all but the electromagnetic field to obtain the effective electromagnetic Lagrangian, the T and P violation inherent in this term will show up as a higher derivative Chern-Simons term in the electromagnetic interaction, in accordance with our earlier discussion.

The existence of two $U(1)$ gauge invariances and two Higgs fields indicates another feature that may well be generic. For the Meissner effect and superconductivity to occur, both gauge invariances must be broken in general. This suggests that there may be three normal phases, one in which neither gauge invariances is broken, and two in which one or the other, but not both, of the gauge invariances is broken. Which of these is actually realized will depend on energetics and hence on the detailed values of the parameters in the theory.

Obviously, the most striking characteristic of the theory presented here that may be tested experimentally is T and P violation. Several tests have been proposed in the literature: (1) the polarization of light reflected from a sample may be rotated;[16] (2) vortex lines may have different energies according to whether they are parallel or anti-parallel to the magnetic field;[16] (3) flux quantization may be shifted[65] from its standard value of $2\pi n$; and (4) the effect on muon precession due to the magnetic field generated by the semions.[66]

I am somewhat skeptical of test (3), proposed by Zhang from general considerations of T and P violation without reference to any detailed microscopic theory. For what it is worth, the effect does not occur in the specific Ginzburg-Landau'ish theory proposed in Ref. 16. The reason is easy to understand. The value of the quantized flux is determined by long distance consideration. As I mentioned, the gauge invariance associated with a_μ is broken and thus long distance physics is controlled by the lowest dimension, the $a_\mu a^\mu$ term, rather than the Chern-Simons term. Test (1) may be impractical for a variety of real-world reasons. Also, there are some indications that the statistics parameter θ may alternate between different layers[59] in which case the effect in question would average out. In contrast, test (4) is a local probe. It requires, however, some understanding of where the muon prefers to sit.

Recently, there have been some ominous rumblings that preliminary experimental data on muon precession may already ruled out the more optimistic estimates of the effect. For a review, see the talk by Rice in this volume. In this context, Huxley's remark about beautiful theories confronting ugly facts naturally comes to mind. The case, however, is far from closed.

In particular, the discussion given in Section VII may be relevant here. In an experimental test, what are the particles whose fractional statistics are we effectively measuring? After all, as discussed in Section VII, fractional statistics may be screened by the collective dynamics of the fluid. Consider the "holons" associated with the holes doped into the anti-ferromagnetic Mott insulator. Since superconductivity sets in only after the anti-ferromagnetic order disappears, it is plausible to think of the "sphere of influence" of each holon, as measured by its effect on the anti-ferromagnetic order, as overlapping. If the holons are anyons at all, they are very fat non-ideal anyons. In such a dense fluid of overlapping anyons, screening effects may be of primary importance, as indicated by the discussion in Section VII. The elementary excitations in the fluid, on the other hand, may carry well-defined statistics.

I would like to conclude with a reminder about elementary logic. Even if the presently known high temperature superconductors prove not to be semionic superconductors, it does not follow that semionic superconductors do not exist. I am obviously biased but it is fair to say that semionics as a theory has an exceptional elegance. It would be a shame for Nature not to make use of it.

ACKNOWLEDGEMENTS

Much of the work discussed here was done in collaboration with X.G. Wen, to whom I am grateful for many stimulating and instructive discussions. I am also grateful to F. Wilczek for introducing me to fractional statistics at its creation and for many collaborative efforts. I would like to thank J.R. Schrieffer for inviting me to this conference, for telling me about fractional quantum numbers years before I got into the subject, and for giving me a copy of his book, which I plan to study in detail eventually.

REFERENCES

1. J. M. Leinaas and J. Myrheim, Il Nuovo Cimento 37B (1977) 1.
2. F. Wilczek, Phys. Rev. Lett. **49** (1982) 957.
3. F. Wilczek and A. Zee, Phys. Rev. Lett. **51** (1983) 2250.
4. F. Wilczek and A. Zee, unpublished, SBITP report.
5. Y. S. Wu and A. Zee, Phys. Lett. **147B** (1984) 325.
6. D. Arovas, R. Schrieffer, F. Wilczek and A. Zee, Nucl. Phys. **B251** (1985) 117.
7. D. Arovas, R. Schrieffer and F. Wilczek, Phys. Rev. Lett. **53** (1984) 722.
8. R. I. Nepomechie and A. Zee, in *Quantum Field Theory and Statistics*, ed. I.A. Batalin, *et al.*, (Adam Hilger, Bristol, 1987), p. 467.
9. R. McKenzie and F. Wilczek, Int. J. Mod. Phys. (to be published).

10. J. March-Russell and F. Wilczek, Phys. Rev. Lett. **61** (1988) 2066.
11. X. G. Wen and A. Zee, Phys. Rev. Lett. **61** (1988) 1025.
12. X. G. Wen and A. Zee, Nucl. Phys. **B316** (1989) 641.
13. X. G. Wen and A. Zee, J. de Physique **50** (1989) 1623.
14. X. G. Wen and A. Zee, Phys. Rev. Lett. **62** (1989) 1937.
15. X. G. Wen and A. Zee, Phys. Rev. Lett. **63** (1989) 461.
16. X. G. Wen and A. Zee, Phys. Rev. Lett. **62** (1989) 2873.
17. X. G. Wen and A. Zee, Nucl. Phys. **B326** (1989) 619.
18. X. G. Wen and A. Zee, "Compressiblity and superfluidity in the fractional statistics liquid," Phys. Rev. **B41** (1990).
19. X. G. Wen and A. Zee, "Quantum statistics and superconductivity," SBITP preprint, to be published in *Modern Trends in Particle Physics*, edited by N. Zovko.
20. X. G. Wen, F. Wilczek and A. Zee, Phys. Rev. **B39** (1989) 11413.
21. This article overlaps considerably with an earlier review (Ref. 19).
22. V. Kalmeyer and R. Laughlin, Phys. Rev. Lett. **59** (1987) 2095.
23. R. Laughlin, Science **242** (1988) 525.
24. P. W. Anderson, Science **235** (1987) 1196.
25. G. Baskaran and P. W. Anderson, Phys. Rev. **B37** (1988) 580; G. Baskaran, Z. Zou and P.W. Anderson, Solid State Comm. **63** (1987) 973; I. Affleck, Z. Zou, T. Hsu and P.W. Anderson, Phys. Rev. **B38** (1988) 745; E. Dagotto, E. Fradkin and A. Moreo, Phys. Rev. **B38** (1988) 2926.
26. S. A. Kivelson, D. S. Rokshar and J. P. Sethna, Phys. Rev. Lett. **60** (1988) 821; Phys. Rev. **B35** (1987) 8865.
27. I. Dzyaloshinski, A. Polyakov and P. Wiegmann, Phys. Lett. **127A** (1988) 112.
28. P. Wiegmann, Physica Scripta **T17** (1988); Physica **153C** (1988) 103; Phys. Rev. Lett. **60** (1988) 821; D. Khveshchenko and P. Wiegmann, SBITP preprint.
29. R. Laughlin, Phys. Rev. Lett. **60** (1988) 2677.
30. A. L. Fetter, C. B. Hanna and R. B. Laughlin, Phys. Rev. **B39** (1989) 9679.
31. Y. Chen, F. Wilczek, E. Witten and B. Halperin, IAS preprint, 1989.
32. I have stressed the distinction between ideal and non-ideal anyon because rigorous statements made by Chen *et al.*, for a system of ideal anyons are based on the fact that in such a system the centers of mass and charge coincide. (In other words, for $g^2 = \infty$, the bag has no extension and responds instantaneously.) The rigorous statements are valuable in that they impose certain requirements on the results of detailed calculation in the ideal limit $g^2 = \infty$. However, the uninitiated have sometimes gotten the erroneous impression that these rigorous statements hold in general.
33. Y. Hosotani and S. Chakravarty, IAS preprint.
34. N. N. Bogloiubov, J. Phys. (USSR) **11** (1947) 23.
35. The mathematical theorem that in two dimensional space an ideal bose gas does not condense at any non-zero temperature is something of a red herring in the present discussion. At zero temperature, the bosons can all be put into the ground state. This is however not a superfluid state, as evidenced by the fact that for a superfluid we expect the density to reach a constant within some

characteristic healing or coherence length from the wall of the container. The ground state wave function in a square box however has the form $\sin \frac{\pi x}{L} \sin \frac{\pi y}{L}$. A repulsion between the bosons is needed to smooth out the density profile. This represents a heuristic physical understanding of Bogoliubov's argument.

36. X. G. Wen and A. Zee, Ref. 18; P. Wiegmann, private communication; E. Fradkin, private communication, T. Banks and J. D. Lykken, SLAC preprint.
37. G. S. Canright, S. M. Girvin and A. Brass, Phys. Rev. Lett. **63** (1989) 2295.
38. M. Fisher and D. H. Lee, Phys. Rev. Lett. **63** (1989) 903.
39. X. G. Wen and A. Zee, to be published.
40. N. Byers and C. N. Yang, Phys. Rev. Lett. **7** (1961) 46.
41. C. N. Yang, Rev. Mod. Phys. **34** (1962) 694.
42. L. Landau, J. of Physics XI (1947) 91.
43. G. D. Mahan, *Many Particle Physics* (Plenum, New York, 1987).
44. S. K. Paul and A. Khare, Phys. Lett. **B174** (1986) 420; **B193** (1987) 253.
45. In Ref. 18, for the sake of simplicity, we carried out this computation only in a relativistic form.
46. X. G. Wen and Q. Niu, ITP preprint.
47. S. M. Girvin and A. H. MacDonald, Phys. Rev. Lett. **58** (1987) 1262; S. C. Zhang, T. H. Hansson and S. Kivelson, Phys. Rev. Lett. **62** (1989) 82; N. Read, Phys. Rev. Lett. **62** (1989) 86.
48. M. Inui, S. Doniach and M. Gabay, Phys. Rev. **B38** (1988) 6631.
49. W. F. Brinkman and T. M. Rice, Phys. Rev. **B2**(1970) 1324.
50. D. Arovas and D. H. M. Haldane, to be published.
51. X. G. Wen, to be published.
52. G. Kotliar, Phys. Rev. **B37** (1988) 3664.
53. I. Affeck and B. J. Marston, Phys. Rev. **B37** (1988) 3774.
54. W. Siegel, Nucl. Phys. **B156** (1979) 135; J. Schonfeld, Nucl. Phys. **B185** (1981) 157; S. Deser, R. Jackiw and S. Templeton, Phys. Rev. Lett. **48** (1983) 975.
55. P. Wiegmann, Physica, C153-155 (1988) 103; D. Khveshchenko and P. Wiegmann, ITP preprint; G. Baskaran, ITP preprint; R. Laughlin and Z. Zou, Stanford preprint.
56. E. Dagotto and A. Moreo, Phys. Rev. Lett. **63** (1989) 2148.
57. S. Kivelson, *et al.*, to be published.
58. G. Baskaran, Phys. Rev. Lett. **63** (1989) 2524.
59. R. Laughlin and Z. Zou, Stanford preprint, 1989.
60. D. Haldane, UCSD-report; Phys. Rev. Lett. **57** (1968) 1488.
61. A. A. Belavin and A. M. Polyakov, JETP Lett. **22** (1975) 245.
62. S. Chakravarty, B. I. Halperin and D. R. Nelson, Phys. Rev. Lett. **60** (1988) 1057; Phys. Rev. **B39** (1989) 2344.
63. X. G. Wen, Phys. Rev. **B39** (1989) 7227.
64. A. Shapere and F. Wilczek, Nucl. Phys. B, to appear; S. J. Rey and A. Zee, UCSB preprint TH 89/14.
65. S. J. Zhang, to be published.
66. B. Halperin, J. March-Russell and F. Wilczek, preprint 1989.

DISCUSSION

P. Lee: I have a question for Tony. What is the distinction between the ideal anyon and the non-ideal, are there any physical distinctions?

A. Zee: Yes, In particular the paper of Wilczek, Witten, Halperin and Chen, they prove a number of theorems, which they state as rigorous theorems I think some people have been confused by these theorems. These theorems are proved strictly for ideal anyons, because these authors use the fact that the stress-energy tensor, and the momentum density, and electromagnetic current density, are exactly equal, and that's because for an ideal anyon the electric charge and the energy-momentum follow each other instantaneously. But, in fact, with these non-ideal anyons you would think that this statistical gauge field is forming a bag. The bag as a characteristic response time. So at high frequencies, the bag cannot follow the shaking particle, around. So in general, the energy-momentum density and charge density, the electromagnetic charge density, will be out of phase, and it's precisely this effect which gives rise to some of these T and P violating effects, at high frequencies. If you like, a theorem says that at zero frequencies, these T and P violations are suppressed, and that's a rigorous statement.

Also perhaps I can say that it seems to us that since the effect of holons is to destroy the Néel order, presumably the holons would have to be pretty big; at least their spheres of influence must be almost overlapping. So if the holons are in fact the semions, then we should be talking about non-ideal semions, rather than mathematical, point semions.

P. Lee: I guess my real question is, would these non-ideal anions allow you to have a smaller value of these predictions of magnetization, and so on, in the presence of a charge?

A. Zee: That's a very interesting question. I don't know the answer. You mean to squirm out of the experimental...

P. Lee: Right. (Laughter)

A. Zee: I see that condensed matter physicists are just as good as particle physicists at getting out of ...

Z. Tesanovic: ...I'm glad to ask either Tony, or Duncan, or Paul: What is the current thinking of how anions are going to work in quasi-two-dimensional systems when there is some coupling between the layers?

A. Zee: My knowledge of the subject is derived from discussion with you, so...

Z. Tesanovic: If that's the current status...(Laughter) No but really is there some kind of understanding of this?

S. Doniach: I have a question for Tony. Can you get....this is partly clarification...the crucial thing is the cancelling of the Chern-Simons term, and in fact Paul Wiegmann just told us that in one case it doesn't cancel. So, under what conditions do you feel that it will cancel, and under what conditions won't it cancel?

A. Zee: I think that sounds like a question for Paul.

S. Doniach: Well, you said that it cancels, and I'm trying to ask you if you have a criterion that makes it cancel exactly, not an accidental cancellation.

A. Zee: In our discussion, it seems to me that as long as the ground state has a gap up to the next highest state, then you can make this gradient expansion, and then it seems to me that this cancellation is quite general. So I don't know if Paul's example is such that the ground state has no gap.

C. M. Varma: Can I ask a very dumb question? As I understand it (I learned it from Dong Hai Lee, actually) if you apply an external magnetic field, then at the right value of the magnetic field, this theory goes over into the quantized Hall effect, which is experimentally observed, in two-dimensional electron gases. Now given that, it seemed to me that that that would be the place to look for anionic superconductivity.

A. Zee: But then anions would be moving in a fixed, magnetic...

C. M. Varma: ...no, excuse me, I do not now apply a magnetic field – why does everybody get excited in relation to copper oxide and not look for anionic superconductivity in heterostructure junctions.

A. Zee: OK, let me rephrase you question, just to show that I understood you question. (Laughter) Your question is, you can look at the Hall system – for example, one-third filling – then the quasiparticles in that Hall system would have fractional statistics given by something like $\pi/3$, or whatever. Why isn't this gas not a superconductor?

C. M. Varma: No that is not my question. I have a much dumber question.

E. Abrahams: There are a lot better two-dimensional electron gases available than copper-oxide superconductors. That's the statement, with which (even with your limited knowledge) you probably agree. So the question was, why don't people or why don't theorists (to rephrase the question again) make suggestions for anyon behavior in such systems.

P. W. Anderson: In particular there's some very pretty organic two-dimensional systems which might be quite interesting.

C. M. Varma: Well, what I know is I can cool my heterostructure junction to one millidegree, and without a magnetic field and nothing ever happens.

P. Lee: No, but you have fermions here, you have to produce your anyons.

C. M. Varma: No but these anions are supposed to be produced by these two-dimensional fermions attaching to themselves their own fictitious field. Nobody legislated against the Coulomb interaction in two dimensions in heterostructure junctions.

T. M. Rice: Heterostructures are in the high-density limit, because the dielectric constant is very large...

C. M. Varma: I can go to the low-density limit, I can make electrons on the surface of helium, and they're being made at pretty low density...why are people looking at copper oxide, what's so special?

J. R. Schrieffer: Very simple question, which I think I've asked Tony before. Admittedly, there is a calculation which, for infinitely weak magnetic fields, shows that the Meissner kernel shows superconductivity. To what extent is there a proof, or even a plausibility argument, that the critical field is finite, and if so, how large is it?

A. Zee: I see. What you are saying is that you may have a non-analytic structure so that the effective Lagrangian cannot be expanded in powers of A.

J. R. Schrieffer: That's my question.

A. Zee: Right. That's the question. So we make that imlicit assumption.

J. R. Schrieffer: Is there anyone who knows the answer to it, or even have a good guess?

A. Zee: That I have to go ask the condensed matter expert.

P. Wiegmann: ...I think that any magnetic field destroys this cancellation very softly, and that H_{c1} in this order equals zero.

J. R. Schrieffer: Thank you.

A. Zee: So is there a critical magnetic field, are you saying that there's a critical magnetic field?

P. Wiegmann: This process is very soft, not like in BCS.

J. R. Schrieffer: That means there is no superconductivity in the earth's field, for example.

P. Wiegmann: Right. It is. There's a small violation, but a very small violation of the superconductive state in the Earth's field.

A. Millis: This is true in any two dimensional conventional superconductor without disorder. You put on a little field perpendicular to the plane and you get vortices and it's not a superconductor, at least at any finite temperature.

G. Kotliar: His statement is a zero-temperature statement.

P. Wiegmann: The anyon liquid will be immediately incompressible, if you add magnetic fields...but this may be very small.

Does the "background" in ARPES belong to $G_1(k,\omega)$?

G. A. Sawatzky

Laboratory for Applied and Solid State Physics, Materials Science Centre, Nijenborgh 18, 9747 AG Groningen

Angular resolved photoelectron spectroscopy (ARPES) is potentially one of the most important methods to distinguish between Fermi liquid and other behaviour of solids. This is really quite obvious since for high enough photon energy the spectral distribution of photoelectrons with a given momentum hk is given by the so-called sudden approximation and is equal to

$$Im\ G_1(k,\omega) = \sum_n \frac{A^2(k,\varepsilon_n^{N-1})}{\omega - (E_F - \varepsilon_n^{N-1}) - i\eta}$$

where $A^2(k,\varepsilon_n^{N-1}) = |<\phi_n^{N-1}|c_k\phi^N>|^2$

This is simply the probability of reaching an eigenstate ϕ^{N-1} with energy ε_n^{N-1} of the N-1 electron system upon the sudden removal of an electron from the N electron system. We don't really know what the lower limit of the photon energy is for the sudden approximation to break down however it is important to note that its breakdown would enhance the intensity for the structure closest to E_F (the quasi particle pole) relative to that of the incoherent part. The reason for this is that for low kinetic energies the photo electron can reabsorb the excitations left behind. It is worthwhile noting that the photon energies used in ARPES are rather low so there is reason to be concerned about the use of the above relation.

Recent ARPES experiments[1-3] have shown for $Bi_2Sr_2CuCu_2O_8$ that there is a clear Fermi cut off at k along TX close to that given by band theory. Aside from this, however, the spectral shape is quite distinct from that observed in normal metals in that it changes drastically with angle (k) and contains a large background extending to about 1 eV below E_F. As remarked by Anderson[4] this spectral shape looks more like an edge singularity $1/E^{1-\alpha}$ with a k dependent cut off than that expected from a quasi particle picture. The point here is that ARPES contains all the information of $G_1(k,\omega)$ and the fact that one sees such a large "background" extending to about 1 eV below E_F could be important as there is a Fermi cut off.

A major question then is what is the "background" due to? Is it misoriented parts of the crystal so the angle is not well defined? I don't think so because the "background" extends to too high energies. Is it due to electron energy loss structure? I don't think so because there is much too much intensity for an energy loss structure. In fact an estimate of the energy loss intensity can be made by looking at the "background" at about 7 eV just above the dominant valence band structure between 1-6 eV. If, as is usual, we assume an energy loss which is of a step function like character for each intrinsic peak the total loss at a given energy is proportional to the integral of the intrinsic intensity above it. With this assumption and the data of Olson *et al.* and Manzke *et al.*, we can estimate that at most about 10-20% of the background close to E_F is due to energy loss processes. In fact I think this is an over estimate since usually the loss intensity even increases with increasing loss energy. The thing remaining is that the "background" is not at all a background but a *major* intrinsic part of $G_1(k,\omega)$. If this is so then the background is telling us that we are dealing with strongly dressed electrons and that most of the spectral weight is in that part of the spectrum corresponding to the dressing left behind upon the sudden removal of an electron. Perhaps it is even so that all the spectral weight is in this incoherent part at least at E_F as suggested by Anderson [4] and more recently by Littlewood *et al.*[5] In this case the quasi particle pole strength goes to zero (Z=0) and the Fermi liquid picture breaks down.

I do not think however that the energy and angular resolution in the reported ARPES data is sufficient to distinguish between a quasi particle picture consisting of a highly dressed electron and a breakdown of Fermi liquid theory as described above. In this regard, it is interesting to demonstrate what one might see in ARPES for a highly dressed electron with a quasi particle pole strength of about 0.1 corresponding to a mass of 10 m_e.

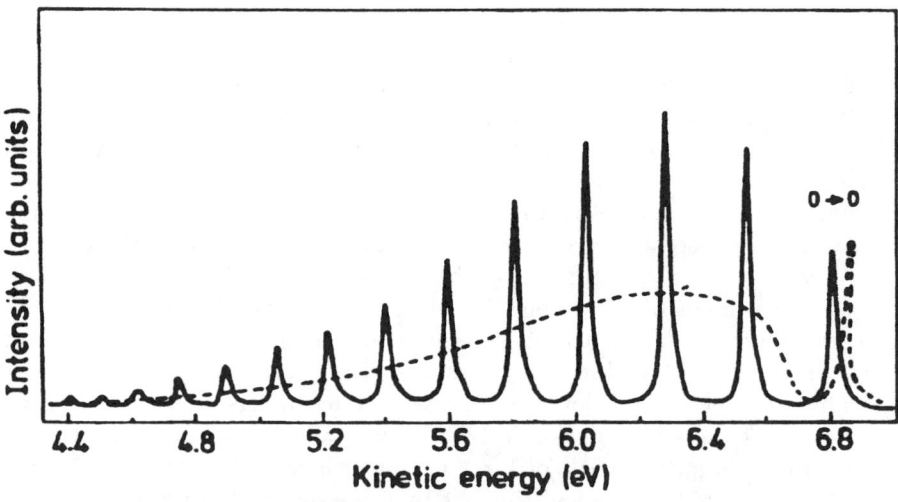

FIGURE 1 The solid line shows an idealized gas phase photoelectron spectrum of H_2 molecules using a photon energy of 21.2 eV. The intensities of the n'th H^+_2 vibrational state seen is $A^2(n) = |<_n \phi^{H_2^+}|_0\phi^{H_2}>|^2$. The 0=0 transition carries about 10% of the total spectral weight. The dashed line is a conceptual solid H_2 spectrum at a particular angular displaying the quasi particle peak followed by a broad incoherent "background." The quasi particle mass is about 10 m_e in this example.

In Fig. 1 I show an idealized photoelectron spectrum of H_2 molecules in the gas phase.[6] The vibrational progression seen corresponds to the fact that the equilibrium interatomic distance in H^+_2 is considerably larger than for H_2. The relative intensities is given by A^2 above but now with ϕ_n referring to a vibrational state of H^+_2 and ϕ_0 to the ground vibrational state of H_2. In a molecular solid the vibrational states would broaden into phonon bands (dashed line) and the quasi particle pole would correspond to the 0-0 transition and would be called a small polaron. If we replace the phonons by magnons the expected spectral weight distribution in this example is similar to what is expected for the high T_c's as calculated for the tJ model.[7] Including the experimental resolution of 30 meV could result in spectra similar to what is seen in refs. 1-3.

In order to distinguish between the quasi particle picture and a non Fermi liquid behavior will require even better resolution and a detailed study of the "background."

REFERENCES

1. T. Takahashi, H. Matsuyama, H. Katayama-Yoshida, Y. Okade, S. Hosoya, K. Seki, H. Fujimoto, M. Sato and H. Inokuchi, Nature **334**, 691 (1988).
2. T. Manzke, T. Bushlaps, R. Claessen and J. Fink, J. Europhys. Lett. **9**, 477, 1989.
3. C. G. Olson, R. Liu, A.-B. Yang, D. W. Lynch, A. J. Arko, R. S. List, B. W. Veal, Y. C. Chang, P. Z. Jiang and A. P. Paulikas, Science **245**, 731 (1989).
4. P. W. Anderson in "Strong correlations and superconductivity" (Edited by H. Fukuyama and S. Maekawa, Springer 1989) p. 2, and private communication.
5. P. B. Littlewood, C. M. Varma, S. Schmitt-Rink, and E. Abrahams, Phys. Rev. **39**, 12371 (1989), and Phys. Rev. Letters in press, Phys. Rev. B., in press.
6. L. Asbrink, Chem. Phys. Lett. **7**, 549 (1970).
7. W. Stephan, K. J. van Szczcpanski, M. Ziegler and P. Horsch (to be published).

DISCUSSION

R. S. List: I think the effects you showed are very interesting, and I agree if you have a signature that's good I would identify it definitely with some real intrinsic feature. I think however if you look at high T_c in general and the quality of the crystals that one deals with, you know, if you look at a featureless background and say immediately that that's intrinsic I think that's a source for a large possibility for errors. In particular I should comment on the fact that in several of our cleaves, we look at the section that we see the band is dispersing through in some crystals, in other crystals all we see is a "featureless background with no peak dispersing through it. In other words we have only dressing with no particle. The variability in the size of the background should perhaps suggest to us that not too much significance should be attached to it."

G. Sawatzky: That's why I essentially went through the whole argument at the beginning to tell you that it probably was not energy loss. In order to establish this one should in the same crystal, measure what the energy loss part is, and perhaps you would find that in those badly cleaved crystals, you would indeed have a very high energy loss part. Could be.

R. S. List: Well, Okay, but I think you might just have inhomogeneities of the crystal, you might have other...

G. Sawatzky: Could be also. Sure.

R. S. List: ...faces. And I think it is important to consider the angular dependence of the background along different directions in the Brillouin zone. Along Γ - Y, the Fermi surface is nearest to Γ and the background intensity persists furthest past the crossing while along the Γ - M, the crossing is furthest from Γ and the background intensity is lost just past the crossing. This suggests that sample inhomogeneities are contributing to the background. The mere fact that there is a background which is not entirely derived from electron-electron scattering does not by any means imply that the background is intrinsic. I'd be very happy if it were intrinsic, but such an interpretation is premature and certainly not unique.

P. Lee: Would you expect a very large mass enhancement if you interpret all the background as intrinsic?

G. Sawatzky: Well, as a matter of fact, if you estimate that very crudely, you come up with about a factor of ten, indeed. A mass of about ten.

P. Lee: But we don't see a mass of ten from the band dispersion.

G. Sawatzky: Oh, but then you have to be very careful, because once you have done your background subtraction, assuming that you have a peak, and energy loss, then of course you do get a peak out. That's basically what you get. However, if you had started without making that assumption, that the whole structure, in fact, was the one-electron Green's function, then it's difficult to say what exactly the dispersion is. Besides that, it's rather interesting I think to see that in small-polaron-type models, that the effective mass that you get out of the band narrowing...you know, in some cases goes as the square root of m/m^*, rather than as m/m^*. So, if m/m^* were...one-tenth, the square root of that is one-third, so we could still be Okay.

A. Malozemoff: Could you just clarify the energy scales of the region near the Fermi surface – of course it's very small compared to the region from which you were trying to deduce the ratio of the electron energy loss background to the overall integrated intensity. Is it plausible to use that same ratio in this other energy regime?

G. Sawatzky: Well that's the main point I was making, that energy loss – the energy loss spectrum, the energy loss part – is dependent only on the kinetic energy of the outgoing electron. The kinetic energy of the outgoing electron depends on the binding energy of the structure here, and the photon energy. So for any particular structure, I can choose my photon energy so that the final state kinetic energy is the same. And then I can look at the energy loss structure for that, Okay? So I simply wanted to do it for a completely different region of the spectrum, which does not suffer from the intrinsic type of structures, and see what kind of intensities that would have there. Now the real experiment is to do the energy loss experiment at high resolution, with an incoming electron beam of the corresponding kinetic energy and look at the energy loss in a specularly scattered beam, with high resolution, and exactly on the same crystal, under the same conditions, and that will tell you then really what the extrinsic part is, shall we say, the energy loss part. That's the best way to do it. And I would strongly suggest that somebody really do that.

Orbital Magnetic Moments in Anyon and Flux Models

T. M. Rice

Theoretische Physik, ETH-Hönggerberg, 8093 Zürich, Switzerland

Earlier this year, Halperin, March-Russell and Wilczek,[1] estimated the size of the orbital magnetic moment in an anyon gas with a density of order that of holes in the CuO_2 planes in high-T_c superconductors. They found that the induced magnetic field was of order 30 G. While the average value of the magnetic field will be reduced to zero through a surface current consistent with the superconducting properties of the anyon gas, there will be variations in density induced by the lattice or by a probe such as a positive muon which will cause a net field at an arbitrary muon site. Therefore they proposed[1] μSR as a sensitive test for a local magnetic field and therefore for the existence of an anyon gas.

There are close parallels between the anyon gas models and the flux states with uniform chirality. Example of such states are the commensurate flux phases (CFP) examined by Lederer, Poilblanc and myself (LPR).[2] These in turn fall in the class of generalized flux phases put forward by Anderson, Hristopoulos and Shastry[3] as interesting generalizations away from half-filling of the Affleck-Marston[4] state.

These states take the form

$$|\Psi\rangle = P_{d=0} \prod_l^{oce} c_{l\uparrow}^\dagger c_{l\downarrow}^\dagger |vac\rangle$$

where $c_{l\sigma}^\dagger$ creates an electron in the lth-Hofstadter eigenstate,[5] LPR[2] showed that if the flux in the Hofstadter problem is chosen to be commensurate with the density, that the Heisenberg energy is minimized. The kinetic energy term in the t-J model is reduced so that the CFP are favored only if $|J/t| \gtrsim 1$. The kinetic energy term however depends also on the gauge in which the flux is repesented i.e. it depends directly on the phase $\varphi_{i,i+\tau}$ on each bond in the supercell. The expectation value $\langle \Psi | c_{i\sigma}^\dagger c_{i+\tau,\sigma} |\Psi\rangle$ is directly proportional to the phase factor $(exp(i\varphi_{i,i+\tau}))^{2,6,7}$. The current operator on this bond ($\hbar = a = 1$)

$$j_{i,i+\tau} = i\,e\,t \sum_\sigma (c_{i,\sigma}^\dagger c_{i+\tau,\sigma} - h.c.)$$

has a non-zero expectation value in the presence of a finite density of holes, when $\sin(\varphi_{i,i+\tau}) \neq 0$. Lederer, Poilblanc and I[8] have made estimates of the orbital current patterns and the resulting spatially varying magnetic fields. As an example for hole doping (1/3) per Cu-site the pattern is simple, namely a circulating current on (1/3) of the plaquettes. This causes a spatially varying magnetic field in the supercell. The difference to the anyon gas comes from the inclusion of the lattice periodicity in this approach. The magnitude of the field modulations is again of a similar order of magnitude to that quoted earlier.

Very recently, a careful examination was made by the UBC-Columbia group[9] for such magnetic fields in the sample interior and they placed an upper limit $\lesssim 0.8\ G$ on such fields, considerably under the estimates quoted above.

Very recently some suggestions and proposals have been made for modifications of the CFP which would have a drastic effect on the orbital currents. Rodriguez and Doucot[10] and Nori, Abrahams and Zimanyi[11] have used slave boson techniques to look for more general mean field theories of the t-J model. In these approaches the flux Φ is associated to spinon motion and additional flux is introduced for the holons so that in the end there is no net current $j_{i,i+\tau}$ on any bond. The non-zero spinon flux generates a chiral spin order of the form discussed by Wen, Wilczek and Zee.[12] So these states also break time reversal symmetry, but the expectation value $\langle\Psi|c^{\dagger}_{i\sigma}c_{i+\tau,\sigma}|\Psi\rangle$ is now real. Nori $et\ al.$[11] argue that their state should be energetically favored at special dopings where the hole density is the inverse of an odd integer and that these states are not electrically conducting.

Another possible case is that of unconventional even parity BCS-states which break time-reversal symmetry. Such states have been considered by several groups. These states unlike the CFP do not break the translational invariance of the crystal. Even in the case of a larger crystal unit cell the superconductivity will not cause circulating orbital currents in the unit cell. However Volovik and Gor'kov[13] pointed out that fields due to circulating currents will in general occur at domain walls and at surfaces. These have been investigated in more detail by Sigrist, Rice and Ueda.[14] Kiefl $et\ al.$[9] report evidence of a small magnetic field in powder samples with small diameter grains. Whether this effect is a sign of such T-breaking states or has a more prosaic origin such as a dilute density of frozen Cu^{2+}-moments needs to be clarified.

In conclusion the failure to observe local magnetic fields in the bulk of various high-T_c superconductors rules out or at least places severe restrictions on models which predict that orbital magnetic currents accompany the high-T_c superconducting state.

ACKNOWLEDGEMENT

This work was made in collaboration with Pascal Lederer and Didier Poilblanc. We would like to thank B. I. Halperin, I. Affleck, F. C. Zhang and R. Heffner for interesting discussions on this topic.

REFERENCES

1. B. I. Halperin, J. March-Russell and F. Wilczek, Harvard Univ. preprint HUTP-89/A010.
2. P. Lederer, D. Poilblanc and T. M. Rice, Phys. Rev. Lett. **63**, 1519 (1989).
3. P. W. Anderson, B. S. Shastry and D. Hristopoulos, Phys. Rev. B **40**, 8939 (1989).
4. I. Affleck and J. B. Marston, Phys. Rev. B **37**, 3774 (1988); **39**, 11538 (1989).
5. D. R. Hofstadter, Phys. Rev. B **14**, 2239 (1976).
6. D. Poilblanc, Y. Hasegawa and T. M. Rice, ETH-preprint TH/89-28.
7. S. Liang and N. Trivedi, Univ. Ill., preprint.
8. P. Lederer, D. Poilblanc and T. M. Rice, ETH-preprint TH/89-53.
9. R. E. Kiefl *et al.*, preprint.
10. J. P. Rodriguez and B. Doucot, Grenoble preprint.
11. F. Nori, F. Abrahams and G. T. Zimanyi, preprint UCD89-25.
12. X. G. Wen, F. Wilczek and A. Zee, Phys. Rev. B **39**, 11413 (1989).
13. G. E. Volovik and L. P. Gor'kov, Pisma Zh. Eksp. Teor. Fiz. **39**, 550 (1984); [JETP Lett. **39**, 674 (1984)]; Zh. Eksp. Teor. Fiz. **88**, 1412 (1985); [Sov. Phys. JETP **61**, 843 (1985)].
14. M. Sigrist, T. M. Rice and K. Ueda, Phys. Rev. Lett. **63**, 1727 (1989).

DISCUSSION

A. Zee: Did you say that your conclusion depends on where the muon is?

T. M. Rice: Well, the muon is a local probe, so it only measures the magnetic field at that site. Now if the muon sits on a very high symmetry site – and you can have for example canceling fields between two planes if there were oppositely oriented – so if the muons sit only on very high symmetry sites, then there can be just cancellations at that site. But I think the experts claim that the muons sit on a low symmetry site, for example, one of the sites that they can sit on is a bridging oxygen, which is not a very high symmetry site.

R. M. Martin: If you have these possible plaquettes, like you gave for particular fillings, wouldn't in general you expect a lot of zero-point motion, possibly melting in general is a consideration for the problem.

T. M. Rice: Well, we haven't really looked into this question of whether this would melt. I think it's a bit more like an Ising problem in the sense that one has three, say distinct, arrangements, for putting these down inside the supercell. So it's not so obvious to me that it would necessarily melt.

R. M. Martin: And the other one that you have has the...

T. M. Rice: This is a one-fifth, so this has a bigger supercell, so that there is a more complex pattern of currents.

R. M. Martin: And the net circulation is zero, in that cell, right?

T. M. Rice: Yes. In this one, the net one is zero. But there still would be local fields. You know, if you look at a site somewhere and then you feel a field. If you calculate the field from this current-current pattern at a particular site in the crystal, it's not zero.

R. Walstedt: You're going to have to distinguish this distribution of fields from what you would get from a flux lattice, right?

T. M. Rice: No, this is without an external magnetic field. This is spontaneous without an external magnetic field.

R. Walstedt: Well how about with an external magnetic field?

T. M. Rice: I don't know the answer to that.

R. Walstedt: Because there is some peculiar broadening of NMR lines at low temperatures in the 1,2,3 that I think is beyond what you would expect from a flux lattice.

T. M. Rice: Yes, there is a peculiar broadening. You should ask Takigawa about that. I'm not proposing this as an explanation for his experiments, if that's the question.

E. Abrahams: I want to make a comment on these local internal fields that someone from Chen *et al.* might make if she/he were here. You might say, "Well, why don't you see these local fields in an NQR experiment? I believe their answer is that in order to have any real local magnetic fields – at least this is what happens in the continuum approximation that they use – you have to have a charge inhomogeneity, so the muon is a good candidate, because you introduce the muon, it has a charge, and that makes a charge inhomogeneity, and it moves some of the electrons – which are the underlying objects out of which you make the anyons – it moves them away. Changes the density; the statistical flux is tied to the density, therefore that makes an inhomogeneity in the statistical flux. That wants to be screened, and is screened by charges, and is screened by some orbital motion of charges locally around the muon, and that provides a real magnetic field. So I think their argument has to do with the fact that you introduce an actual charge inhomogeneity in order to see this.

R. Walstedt: You know, NQR lines are very strongly broadened by internal magnetic fields, and perhaps the difficulty in resolving that is because the NQR lines are frequently quite broad to start with. But that's also a candidate.

R. Heffner: Yes, but it's mostly the copper NMR line which is broadened. The oxygen is not.

R. Walstedt: Yeah, but you can't see the oxygen NQR in zero field.

E. Abrahams: Well certainly the anyon game is not in a position to distinguish between oxygen and copper, I think, at the moment. (Laughter) So with that clever remark let's go on to the next talk...

Comment on μSR Experiments to Search for Magnetic Fields from Flux Phases in High Temperature Superconductors

R. H. Heffner

Los Alamos National Laboratory, Los Alamos, NM 87545

I have been asked by the conference organizers and by Maurice Rice if I would report on results of recent muon spin relaxation (μSR) studies to search for anomalous internal magnetic fields which might arise from flux phases or "anyon" states[1] in the copper oxide superconductors. My focus is a recent draft preprint[2] from the University of British Columbia, Columbia University and G. E. Corporate Research and Development groups in which investigations on $YBa_2Cu_3O_7$ (YBCO) and $Bi_2Sr_2CaCu_2O_8$ (BSCCO) are reported. I should state at the outset that I have spoken with Y. J. Uemura, Columbia U., about the results and he has given me his permission to discuss their work.

The μSR experiments were carried out in nominally "zero" (≤ 0.1 Oe) magnetic field. It is useful to review a few salient features of zero-field μSR to fully understand the experimental results. If the μ^+ spin makes an angle Θ with the local internal magnetic field H at the μ^+ site, then the function $G_z(t,\Theta,|H|)$ describing the time evolution of the component of the μ^+ spin along its initial z direction is given simply by

$$G_z(t,\Theta,|H|) = \cos^2\Theta + \sin^2\Theta\cos(\omega_\mu t) . \tag{1}$$

Here $\omega_\mu = \gamma_\mu |H|$ and $\gamma_\mu = 8.5 \times 10^4$ S^{-1} Oe^{-1} is the muon gyromagnetic ratio. In a typical experiment this function must be averaged over both the angle Θ and the distribution of internal magnetic fields $P(|H|)$. Thus $G_z(t)$ describes the averaged quantity

$$G_z(t) = \int d\Omega h |H| P(|H|) G_z(t,\Theta,|H|) . \tag{2}$$

The second term in Equation (1) means that the time dependence of $G_z(t)$ in Equation 2 is the Fourier transform of the field distribution. Therefore, a Gaussian field distribution gives a Gaussian lineshape while a Lorentzian field distribution gives an exponential lineshape. In general a Gaussian field distribution arises from a concentrated system of internal spins (as from the copper nuclear dipole moments in YBCO), whereas a Lorentzian field distribution comes from a dilute concentration of internal spins. It also indicates fields much larger than the "RMS" value.

The experiments of Kiefl et al.[2] were performed on c-axis oriented samples of sintered YBCO, on thick films of BSCCO and on fine-grain powders (2,000-10,000 Å) of YBCO. The powdered YBCO was found to be aligned to ± 9° from neutron scattering rocking curves and the films were found to be 95% single phase (by X-ray scattering) with the c-axis perpendicular to the face of the film. The fine grain powders were ball milled after oxygenation.

Internal magnetic fields from flux phases are thought to arise either through orbital angular momentum produced by gradients in the anyon density (which could be caused by the perturbative presence of the charged μ^+) or from intrinsic spin magnetism arising from the spin statistics.[3] Current estimates[3] of the magnitude of the local fields give 30-50 Oe. These fields are less than H_{c1}, so that below the superconducting critical temperature T_c one expects Meissner currents to exclude the field from the superconducting domains. Hence below T_c one would expect a relatively large change in the field distribution sensed by the muon. (Here "large" is with respect to the nuclear dipole fields of the order 1-2 Oe.)

The experimental results of Kiefl et al. for both the sintered, large-grain YBCO and the BSCCO film below 150-200 K show that the μSR linewidths correspond to an RMS field distribution of order 1.5 Oe, in good agreement with that expected from nuclear dipole fields. Furthermore the line shape is Gaussian. It is known that the μ^+ does not occupy a site of high symmetry in the lattice, so that an accidental cancellation of higher field values (by symmetry) is very unlikely. Also no significant change in linewidth below T_c is observed. A reduction of the linewidth above 150-200 K due to thermally induced μ^+ hopping is seen, however. A weak orientation dependence of the linewidth ($\leq 10\%$) relative to the direction of the μ^+ spin is found, but is not likely to be relevant to the anyon problem.

In the abstract of their paper, the authors state[2] that fields larger than expected from nucleus dipoles are seen in the fine-grained YBCO powder sample. What they mean by this is that the observed lineshape at 6 K is *Lorentzian*, even though the 1/e point of the exponential relaxation function still corresponds to a field distribution of order 1-2 Oe, as seen for the other samples. One possible explanation for the change of lineshape from Gaussian to exponential is that the process of milling may have introduced local defects or oxygen deficiencies into the lattice resulting in small magnetic regions randomly distributed throughout the grains. If, for example, 1-2% of the copper atoms developed moments $\simeq 0.6\mu_B$ (seen in $YBa_2Cu_3O_6$), the observed μSR lineshape and relaxation rate would be produced.[4]

Maurice Rice has suggested at this conference that the exponential linewidths might be due to supercurrents flowing around the outside of the superconducting domains. Such a situation can arise theoretically[5] for time-reversal violating superconducting states. The magnetic field produced by the surface currents would be screened by the usual Meissner mechanism with a penetration depth of the order of the grain size mentioned above. Sigrist et al.[5] have calculated the expected spatial field distribution from which $P(|\mathbf{H}|)$ could be derived. It is very roughly Lorentzian. However, the magnitude of the field is again of order 30 Oe, which is far too large to explain the observed μSR linewidth.

In conclusion the μSR experiments of Kiefl et al. set an upper limit to flux-phase induced magnetic fields in high temperature superconductors of the order of 1 Oe. This is 1 to 2 orders of magnitude smaller than current theoretical estimates, and is in agreement with very early (May 1987) μSR experiments on $La_{1.85}Sr_{0.15}CuO_4$ by Kossler et al.[6] I also note that these μSR experiments are close to the limits of precision (~ 1 Oe) obtainable with μSR because of the "background" magnetic fields produced by the nuclear dipole moments.

REFERENCES

1. V. Kalmeyer, and R. B. Laughlin, Phys. Rev. Lett. 59, 2095 (1987); R. B. Laughlin, Phys. Rev. Lett. 60, 2677 (1988); J. March-Russell and F. Wilczek, Phys. Rev. Lett. 61, 2066 (1988); see also Ref. 3.
2. R. F. Kiefl, J. H. Brewer, I. Affleck, J. Carolan, P. Dosanjh, W. N. Hardy, R. Kadono, J. R. Kempton, S. R. Kreitzman, Q. Lee, A. H. O'Reilly, T. M. Riseman, P. Schieger, P. Stamp, T. Hsu, H. Zhou, L. P. Le, G. M. Luke, B. Sternlieb, Y. J. Uemura, H. R. Hart, and K. W. Lay, preprint, TRIUMF and Department of Physics, University of British Columbia, Vancouver, Canada V6T2A3.
3. Bertrand I. Halperin, John March-Russell and Frank Wilczek, Harvard University, preprint No. HUTP-89-A010, 1989; also T. M. Rice, private communication, this conference.
4. Cf. discussion in Robert H. Heffner and D. L. Cox, Phys. Rev. Lett. 63, 2538 (1989).
5. M. Sigrist, T. M. Rice, and K. Ueda, Phys. Rev. Lett. 63, 1727 (1989).
6. W. J. Kossler, J. R. Kempton, X. H. Yu, H. E. Schone, Y. J. Uemura, A. R. Moodenbaugh, M. Suenagar and C. E. Stronach, Phys. Rev. B. 35, 7133 (1987).

DISCUSSION

D. Pines: Would the inhomogeneities in the sample give rise to essentially enough in the way of imperfections that again, if the anyons were there, you might see their influence in NQR? My question relates to Elihu's comment a minute ago, that you need something else in order to see them in NQR, but I would think the imperfections, the normal regions, would provide that something else.

R. Heffner: Yes, but of course you know, you can't see a broadening of only a gauss in NQR.

C. M. Varma: Since one can be sure that there'll be calculations produced in which anyons produce only a gauss...

R. Heffner: I figured that.

C. M. Varma: ...may I ask you, how much can this technique be pushed? Can you go to a milligauss?

R. Heffner: No.

R. Heffner: It's relatively easy to see a gauss, but because the nuclear dipolar background, which is of order a gauss, must be added in quadrature with any additional field, anion fields less than a gauss are difficult to see with standard μSR techniques.

C. M. Varma: You mean in this kind of experiment the earths field has not been taken out, mu metal shielding or anything like that.

R. Heffner: I believe the earth's field has been reduced. They should have been able to reduce that field to a few tens of milligauss, quite easily, at the sample.

C. M. Varma: So what do you think should be the ultimate limit? What limits do you think can ultimately be set in an experiment like this?

R. Heffner: I would say a fraction of a gauss, easily. If you really think that the field is of the order of a gauss, one could try much harder.

H. Ott: Was it said how that powder was made?

R. Heffner: What do you mean, the fine grain powder? They just said they ground it up. And it has quite a range of sizes, as you noticed, about a factor of five range in the size. If you mean the actual annealing history, I don't remember that. They didn't go into great detail on it, though.

M. Takigawa: Does the upper limit you quoted apply also to uniform, or non-random fields?

R. Heffner: If it is not random, then it would in principle give rise to coherent precession of the muon spin. You can't really distinguish between coherent precession or inhomogeneous broadening at the level of one gauss.

M. Takigawa: But if that field is smaller than, say, a nuclear dipole field it will not give rise to precession.

A. Malozemof: I'd like to show another experiment which bears on the question of the compatibility of, possibly, anyon states, or other kinds of anomalous order parameters, with more conventional s-wave superconductivity, and these are persistent current experiments. This is the work of Rudolf Gross, Pradeep Chaudury and Gupta at IBM Yorktown. They made loops made out of epitaxial films of yttrium barium copper oxide, interrupted by...you see a gap here, a 50 micron gap...they put down a layer of silver, which they diffuse in a layer with an annealing treatment, to make a very low resistance contact to the lead. So they have a lead section of this loop, and they do a persistent current measurement, and magnetic studies, to look for the compatibility of the persistent currents. And what they observe here...these are experiments in which they cooled in a field, down to 4.2, turn off that field, and then they measure some tracts, flux, some moment here with a certain magnitude (these are the squares). And then as they heat up the transition temperature of the lead, the flux seeps out, and they are left with a small residual signal which presumably represents the flux exclusion, the flux trapping in the superconductor – a high T_c superconductor – so the difference in effect is a measure of the persistent current, and they confirm that the relaxation is very slow in this regime, so it looks like a true persistent current. And working through the numbers, although they're continually improving this experiment, but they find that the moments imply currents here which imply – you take the area of this overlap between the lead and the high T_c superconductor current density, which are now up to 600 amps per square centimeter, quite a substantial current density, and it's clearly limited by the lead on the basis of the field dependence. So I think as this kind of number grows, we'll see where it goes in the future. They're trying to push the areas smaller and smaller, but, of course, this indicates a compatibility of the lead order parameter with the high T_c superconducting order parameter.

6. NUMERICAL STUDIES

Chair — T. M. Rice

D. J. Scalapino
Department of Physics
University of California, Santa Barbara, CA 93106

Results from Numerical Simulations of the 2D Hubbard Model

Quantum Monte Carlo simulations have been used to study the physical properties of the two-dimensional Hubbard model. Here we review what has been learned about the magnetic, superconducting, and single-particle properties from these simulations.

I. INTRODUCTION

Numerical Monte Carlo simulations of interacting many-electron systems offer the possibility of obtaining detailed information on a variety of physical properties. Here we will review what has been learned from such studies about the one-band 2D Hubbard model. In this model the electrons move on a square lattice with a near neighbor one-electron transfer t and have an onsite Coulomb interaction U. There are three parameters which determine the state of the system, the ratio U/t, the average electron occupation per site $\langle n \rangle = \langle n_{i\uparrow} + n_{i\downarrow} \rangle$ and the temperature T. Within this parameter space, the one-band Hubbard model may exhibit various phases, ranging from an insulating antiferromagnetic phase to a metallic-like phase with strong antiferromagnetic fluctuations. Here we will examine what can be said about the electronic correlations and in particular the possibility of superconductivity near the metal-insulator transition.

In order to appreciate both the usefulness and the limitations of the Monte Carlo simulations, we begin in Section II with an overview of this technique. Recently there has been progress in stabilizing the calculations of the fermion matrices at low temperature[1-3] and in extracting dynamic spectral information from the imaginary-time Green's functions.[4] However, the problem associated with the fluctuations in sign of the fermion determinant remains a serious obstacle limiting the temperature and lattice sizes that can be simulated in the important region just off of half-filling.[5]

Section III contains a discussion of a number of quantities that have been calculated for the 2D Hubbard model and the resulting physical picture that emerges. For the half-filled band $\langle n \rangle = 1$, a finite size scaling analysis[6,3] shows that the ground state has long-range antiferromagnetic order. In addition, information obtained from the one-electron Green's function implies that there is a gap in the one-electron spectrum.[7] Well off of half-filling, $\langle n \rangle = 0.5$, the antiferromagnetic correlations die off in several lattice spacings, and there is no indication of a gap in the single-particle spectrum. Near half-filling there appear to be incommensurate antiferromagnetic correlations,[6-8] and there is an attractive pairing interaction in the d-wave particle-particle channel.[9] However, over the temperature range and lattice size presently accessible, there is no indication of either a long-range antiferromagnetic[6-8] or a superconducting[8,10] phase in the region near half-filling.

Section IV contains a summary and some remarks on the implication of these results.

II. MONTE CARLO SIMULATIONS

THE BASIC IDEA

For the purpose of discussing the results obtained from the Monte Carlo simulations and their limitations, it is useful to keep the following general picture of the procedure in mind. The basic idea is to replace the interacting many-electron system by a noninteracting many-electron system which moves in a Hubbard-Stratonovich field.[11] Then functionally integrating over this field puts back in the many-body interactions. A variety of such formulations, in which the functional integration over the Hubbard-Stratonovich field was approximately carried out (e.g., a static approximation with Gaussian fluctuations), were actively pursued in the 1970's.[12-14] The advances during the past decade are based upon stochastic sampling techniques[15,3] which in principle provide a controlled evaluation of the functional integral.

Consider the Hubbard model[16]

$$H = -\sum_{\substack{\langle ij \rangle \\ \sigma}} t \left(c_{i\sigma}^{\dagger} c_{j\sigma} + c_{j\sigma}^{\dagger} c_{i\sigma} \right) + U \sum_i n_{i\uparrow} n_{i\downarrow}, \tag{1}$$

where $c_{i\sigma}^{\dagger}$ is the creation operator for an electron on site i with spin σ and $n_{i\sigma} = c_{i\sigma}^{\dagger} c_{i\sigma}$. The size of the one-electron overlap t sets the width of the band structure $8t$, and U is the strength of the onsite Coulomb interaction. In order to separate out the two-body interaction, the imaginary time interval $(0, \beta)$ is divided into L segments of width $\Delta\tau$ so that

$$e^{-\beta H} = \left[e^{-\Delta\tau H} \right]^L \cong \left[e^{-\Delta\tau T} e^{-\Delta\tau V} \right]^L. \tag{2}$$

Here T and V are the kinetic and potential energy parts of H, Eq. (1). In the last term we have used a Trotter breakup.[17] This introduces errors of order $tU\Delta\tau^2$ which can be systematically dealt with by reducing $\Delta\tau$. The purpose of the Trotter breakup is to allow us to reduce the interaction to a quadratic form in the fermion operators by introducing a Hubbard-Stratonovich field.

For the Hubbard model we can use a discrete version of the Hubbard-Stratonovich transformation introduced by Hirsch,[18]

$$e^{-\Delta\tau U n_{i\uparrow} n_{i\downarrow}} = \frac{e^{-\frac{\Delta\tau U}{2}(n_{i\uparrow} + n_{i\downarrow})}}{2} \sum_{S_{i\ell} = \pm 1} e^{-\Delta\tau S_{i\ell} \lambda (n_{i\uparrow} - n_{i\downarrow})}. \tag{3}$$

Here, a discrete Hubbard-Stratonovich field $S_{i\ell} = \pm 1$ is introduced at each site of the space-imaginary-time lattice, and λ is set by $\cosh(\Delta\tau\lambda) = \exp(\Delta\tau U/2)$. In this way, the two-body interaction at the ith site on the ℓth $\Delta\tau$-time slice, is replaced by an onsite one-body spin-dependent potential $\lambda S_{i\ell}\sigma$ plus a constant shift in the chemical potential

$$U n_{i\uparrow} n_{i\downarrow} \to S_{i\ell} \lambda (n_{i\uparrow} - n_{i\downarrow}). \tag{4}$$

Of course one must sum over all possible field configurations $\{S_{i\ell}\}$ on the space-time lattice. Thus, for example, the partition function becomes

$$Z = \text{Tr}\, e^{\beta(H - \mu N)} = \sum_{\{S\}} \text{Tr} \prod_{\ell, \sigma} e^{-\Delta\tau \sum_{i,j} (c_{i\sigma}^{\dagger} h(\ell, \sigma) c_{j\sigma} - \mu c_{i\sigma}^{\dagger} c_{i\sigma})}. \tag{5}$$

Here $h(\ell, \sigma)$ is a one-body Hamiltonian for the motion of an electron with spin σ propagating through the Hubbard-Stratonovich field on the ℓth time slice. It consists of the one-electron transfer kinetic energy and the one-electron spin dependent site energies, Eq. (4).

Carrying out the fermion trace in Eq. (5) gives

$$Z = \sum_{\{s\}} \det M_\uparrow(S) \det M_\downarrow(S), \qquad (6)$$

with

$$M_\sigma(\{S\}) = \left[I + e^{\beta\mu} \prod_\ell e^{-\Delta\tau h(\ell,\sigma)}\right]. \qquad (7)$$

If the Hubbard-Stratonovich field were independent of τ_ℓ, the det $M_\sigma(\{S\})$ would reduce to the usual expression

$$\det M_\sigma = \prod_i \left(1 + e^{-\beta(\varepsilon_{i\sigma}-\mu)}\right), \qquad (8)$$

with $\varepsilon_{i\sigma}$ the eigenvalues of the one-electron Hamiltonian $h(\sigma)$. As discussed in Ref. 15, Eqs. (6) and (7) generalized Eq. (8) to the case in which $h(\sigma)$ is τ-dependent. In this way, the interacting problem is changed to solving a noninteracting problem for a given Hubbard-Stratonovich field configuration $\{S_{i\ell}\}$ and then *summing* over all possible configurations.

We have of course traded one difficult problem for another, but they are computationally quite different. The original problem involved traces containing quartic fermion operators or alternatively functional integrals over Grassman variables, while the Hubbard-Stratonovich formulation involves sums over configurations specified by sets of real numbers $\{S_{i\ell}\}$. In the latter case we can seek a numerical solution using stochastic sampling procedures.[19] If the product of the spin-up and spin-down fermion determinants in Eq. (6) is a positive semi-definite function of the Hubbard-Stratonovich $\{S_{i\ell}\}$ field, it can serve as the weight function for a Monte Carlo importance sampling procedure just as $e^{-\beta V(x_i)}$ does in classical statistical mechanics. Then a Monte Carlo algorithm can be implemented[15] which generates configurations distributed according to

$$P(\{S\}) = \frac{\det M_\uparrow\{S\} \det M_\downarrow\{S\}}{\sum_{\{S'\}} \det M_\uparrow\{S'\} \det M_\downarrow\{S'\}}. \qquad (9)$$

These configurations are then used to calculate the physical quantities of interest which may range from thermodynamic Green's functions to equal time expectation values.

For example, the one-particle Green's function

$$G_{ij}(\tau) = -\text{Tr}\, e^{-\beta H} T c_{i\sigma}(\tau) c_{j\sigma}^\dagger(0)/Z \qquad (10)$$

is given by summing

$$G_{ij}(\tau;\{S\}) = \left(1 + e^{\beta\mu} e^{-\Delta\tau h(\ell-1,\sigma)} \ldots e^{-\Delta\tau h(1,\sigma)} \ldots e^{-\Delta\tau h(\ell,\sigma)}\right)^{-1} \qquad (11)$$

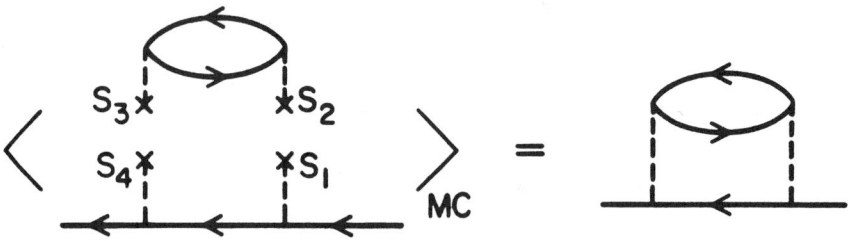

FIGURE 1 Schematic illustration of the relationship between the Monte Carlo averages and Feynman graphs.

over configurations with the weighing factor $P(\{S\})$ given in Eq. (9). Here $G_{ij}(\tau;\{S\})$ is the one-particle Green's function in the presence of the external space-time field $\{S\}$. If the configurations $\{S\}$ are selected according to the probability distribution $P(\{S\})$, then

$$G_{ij}(\tau) = \langle G_{ij}(\tau;\{S\}) \rangle_{MC} = \frac{1}{M} \sum_{\{s\}} G_{ij}(\tau;\{S\}), \qquad (12)$$

where M is the number of Monte Carlo configurations. This is schematically illustrated in Fig. 1, which shows how typical Feynman graph contributions arise. Note that the background fluctuations arise from $P(\{S\})$. It is important to realize that all Feynman graphs are of course included and the simulation is non-perturbative. Naturally one must examine the $(\Delta \tau)^2$ errors, extrapolating to $\Delta \tau \to 0$ if necessary. Furthermore, if one averages over, say, 10^4 independent configurations, one expects of order 1% statistical errors. As with all Monte Carlo procedures, one needs to ascertain that the configurations are independent, and one must have access to all regions of phase space. These can be nontrivial requirements, and detailed checks obtained from exact diagonalization of small systems as well as various perturbation theory calculations in the small U limit must be made.

Returning to the calculation of physical quantities, once $G_{ij}(\tau)$ is evaluated, we can take its spatial Fourier transform to obtain $G_p(\tau)$. Then setting $\tau \to 0^-$, we have $\langle n_p \rangle$, and, for example, the grand canonical ensemble average of the kinetic energy is just

$$\langle T \rangle = \sum_{p\sigma} \varepsilon_p G_\sigma(p, \tau \to 0^-). \qquad (13)$$

One can also evaluate the τ-Fourier transform of $G(p,\tau)$ at the Matsubara frequencies $\omega_n = (2n+1)\pi T$.

$$G(p, i\omega_n) = \int_0^\beta e^{i\omega_n \tau} G(p, \tau)\, d\tau = \frac{1}{i\omega_n - (\varepsilon_p - \mu) - \Sigma(p, i\omega_n)}, \quad (14)$$

and in this way calculate $\Sigma(p, i\omega_n)$.

Two-particle Green's functions are also directly obtained. The procedure is simple to understand if one keeps in mind the calculational scheme we have described. Consider a two-particle Green's function such as the s-wave pair field propagator

$$P(\tau) = -\langle T\Delta(\tau)\Delta^\dagger(0)\rangle \quad (15)$$

with

$$\Delta = \frac{1}{\sqrt{N}} \sum_\ell c_{\ell\uparrow} c_{\ell\downarrow}. \quad (16)$$

For a given Hubbard-Stratonovich configuration, it is straightforward to calculate $P(\tau)$, by contracting the spin-up and spin-down operators appearing in Eq. (15). All that can happen is that the spin-up and spin-down electrons interact with their respective Hubbard-Stratonovich fields. This gives

$$\frac{1}{N} \sum_{ij} \int_0^\beta G_{ij\uparrow}(\tau; \{S\})\, G_{ij\downarrow}(\tau; \{S\})\, d\tau. \quad (17)$$

Now, averaging this expression over the set of Monte Carlo generated Hubbard-Stratonovich field configurations $\{S\}$ gives the full interacting result shown in Fig. 2.

$$P = \frac{1}{N} \sum_{ij} \int_0^\beta \langle G_{ij\uparrow}(\tau; \{S\})\, G_{ij\downarrow}(\tau; \{S\})\rangle_{\text{MC}}\, d\tau. \quad (18)$$

Note that if we had taken the product of the average one-particle Green's functions

$$\overline{P} = \frac{1}{N} \sum_{ij} \int_0^\beta \langle G_{ij\uparrow}(\tau; \{S\})\rangle_{\text{MC}} \langle G_{ij\downarrow}(\tau; \{S\})\rangle_{\text{MC}}\, d\tau, \quad (19)$$

rather than the average of the product, we would have obtained the sum of the set of graphs corresponding to two dressed but noninteracting propagators shown as the left graph in Fig. 2. As we will discuss, by comparing P and $\overline{P}$, one can investigate the nature of the pairing interaction.

As this discussion demonstrates, one of the most important features of this technique is its close relationship to the usual finite temperature Green's function approach. This means that one's intuition, built up from standard many-body Feynman diagram techniques, can be directly applied. The Monte Carlo simulation is simply a way of implementing the calculations to all orders.

FIGURE 2 The pair field susceptibility P, Eq. (18), expressed in terms of dressed single particle Green's functions and a particle-particle interaction vertex. The graph on the left represents two dressed but not-interacting Green's functions and corresponds to $\overline{P}$, Eq. (19).

PROBLEMS AND LIMITATIONS

With this background, we turn to a discussion of the problems and limitations of present fermion simulations. It is instructive to compare the quantum problem with the classical one. Clearly they both share in common the fact that simulations are carried out on *finite spatial lattices*. One would like to *extrapolate* to the bulk limit and, for the quantum problem in two dimensions, to zero temperature ($\beta = L \cdot \Delta\tau \to \infty$). The problems of generating independent configurations and accessing the phase space, especially near phase transitions, are common to both the quantum and classical simulations. These difficulties may well limit what one can conclude.

However, beyond these common problems, the quantum simulations have their own unique set of difficulties. For a classical problem with short-range interactions, the number of arithmetic operations for a Monte Carlo sweep which allows a change at each lattice site varies as the number of spatial sites N. For a many-electron problem with a short-range interaction, such as the Hubbard model, the non-local nature of the fermion determinants leads to algorithms which scale more rapidly than the space-time volume NL. In fact, a straightforward evaluation of the fermion determinant would require $(NL)^3$ operations per site, leading to $(NL)^4$ operations for a sweep of the space-time lattice. However, using the specific structure of the fermion matrix and the local nature of the interaction, a Slater-Koster-like updating scheme has been developed which reduces this to $N^3 L$ operations for a sweep through the space-time lattice.[15] Nevertheless, the additional factor of N^2 for a 10 × 10 spatial lattice corresponds to a factor of 10^4. In addition, to reach a temperature of, say, $0.1 t$ requires that $L = 10^2$ if $\Delta\tau = 0.1$. Thus the quantum simulations of many-electron systems are inherently more computationally intense than classical problems with short-range interactions.

In addition, the fermion matrices

$$B = \prod_\ell e^{-\Delta\tau h(\ell,\sigma)} \qquad (20)$$

which must be calculated become ill-conditioned as the temperature is lowered. For the non-interacting system at a temperature $T = 0.1\,t$, the eigenvalues of B range from e^{-40} to e^{40}. This spread becomes even larger when $U \neq 0$. Unless the contributions from the large and small eigenvalues can be separated, the latter will be completely swamped by round-off errors. Until recently, this problem limited most simulations to temperatures $T \gtrsim 0.2\,t$. However, as discussed by Sugiyama and Koonin[1] and independently by Sorella et al.,[2] an orthogonalization procedure can be used to separate the diverse energy scales. Using a Gram-Schmidt orthogonalization scheme, we have developed a stable calculational procedure[3] for the grand canonical ensemble which allows the fermion matrices to be evaluated at temperatures as low as 10^{-3} of the bandwidth $8t$.

Thus there has been progress in the development of fermion algorithms. However, there remains one central difficulty, the so-called "sign" problem. Unfortunately, even though the partition function Z is of course positive, this is not necessarily the case for the contributions from individual configurations, and $\det M_\uparrow(\{S\}) \det M_\downarrow(\{S\})$ can fluctuate in sign. Only when a specific symmetry is present, such as particle-hole symmetry for the half-filled Hubbard model, will the products of the spin-up and spin-down determinants be positive. When $\det M_\uparrow \cdot \det M_\uparrow$ can change sign, the probability distribution is taken proportional to the absolute value of the product of the determinants

$$\tilde{P}(\{S\}) = \frac{|\det M_\uparrow(\{S\}) \det M_\downarrow(\{S\})|}{\sum_{S'} |\det M_\uparrow(\{S'\}) \det M_\downarrow(\{S'\})|}. \tag{21}$$

Then the expectation value of an observable O is given by

$$\langle O \rangle = \frac{\langle O \operatorname{sgn}(\tilde{P})\rangle_{\tilde{P}}}{\langle \operatorname{sgn}(\tilde{P})\rangle_{\tilde{P}}}, \tag{22}$$

where the subscript $\tilde{P}$ indicates an average with respect to the probability distribution given in Eq. (21). If the average sign, $\langle \operatorname{sgn}(\tilde{P})\rangle_{\tilde{P}}$, becomes small, then there will be large statistical fluctuations in the Monte Carlo results.

In this regard, it is unfortunate that an interesting suggestion by the Trieste group,[2] that at certain fillings the average sign approaches a constant as $\beta \to \infty$, has proved unfounded. Specifically, the Trieste group used a ground-state projector Monte Carlo technique in which $e^{-\beta H}$ acts on an initial state to project out the ground state. The basic procedure is similar to that for the grand canonical ensemble, and again fermion determinants which depend upon the Hubbard-Stratonovich field configuration enter. They noted that if the average fermion determinantal sign approached a constant as $\beta \to \infty$, it would be possible to extract the ground state energy E_0 by simply *ignoring* the sign. Furthermore, since many quantities of physical interest can be expressed as derivatives of E_0, they could also be evaluated. In particular, the Trieste group argued that with a Hartree-Fock trial function, the average sign $\langle S \rangle$ approached a constant for a 4×4 lattice with $U = 8$ and a filling of $\langle n \rangle = 0.625$. However, we have repeated their calculations[5] and find the results shown in Fig. 3. At large β, we find that $\langle S \rangle$ vanishes to within statistical error. Here, when no error bars are shown, they lie within the plotting symbols. In fact, one expects on general grounds that the average sign *decays exponentially* with increasing β. This is clearly born out by the semilog plot of this data, shown in Fig. 4.

Thus we conclude that ignoring the sign of the determinant is an *uncontrolled* approximation. Nevertheless, it can in some cases be quite good. In particular, this has been found to be the case for the ground state energy and the antiferromagnetic structure factor $S(\pi,\pi)$ on 4×4 lattices where there are exact Lanczos results with which to compare.[20] However, ignoring signs can also lead to qualitatively incorrect results, as is shown for the d-wave pair-field susceptibility in Fig. 5. Here, when the signs are kept, P_d increases as the temperature is lowered, while if the signs are ignored, P_d has a spurious maximum and then *decreases* with decreasing temperature. This neglect of signs led to early erroneous[10] results for the low-temperature dependence of P_d.

We have extended this study to a variety of band fillings and interaction strengths U and in all cases, except half-filling, find that the average sign decreases exponentially with β. We have also done this for the grand canonical ensemble and find that the average sign decreases exponentially with increasing β. However, at a given value of β, the average sign depends upon the average filling. Figure 6 shows the average sign for the grand canonical ensemble as a function of band filling for $U = 4$ and various lattice sizes and β values. Unfortunately, the rapid decrease in the average sign just below half-filling makes it difficult to study this extremely interesting regime. However, as one can see, the sign for the one-quarter-filled case is relatively benign.

Another limitation on numerical simulations is the difficulty of extracting dynamic, real frequency, spectral information from the imaginary-time results. Formally one simply continues the Matsubara frequencies by the replacement $i\omega_n \to \omega + i\delta$ (after explicitly evaluating terms in which $i\omega_n$ appears in fermi or bose factors). However, the continuation of numerical data is known to be an ill-conditioned problem which is exponentially sensitive to statistical fluctuations. Recently we have had some success with a least-square procedure[4] which fits $G(p,\tau)$ to

$$\overline{G}(p,\tau) = -\sum_i \frac{a_i e^{-\tau\omega_i}}{1 + e^{-\beta\omega_i}}, \qquad (23)$$

using a smoothness constraint. Then this form is continued, giving an estimate of the spectral weight

$$\overline{A}(p,\tau) = \sum_i a_i \delta(\omega - \omega_i). \qquad (24)$$

Various approaches to this problem are presently being studied, and we will show in Section III some recent results which have been obtained.

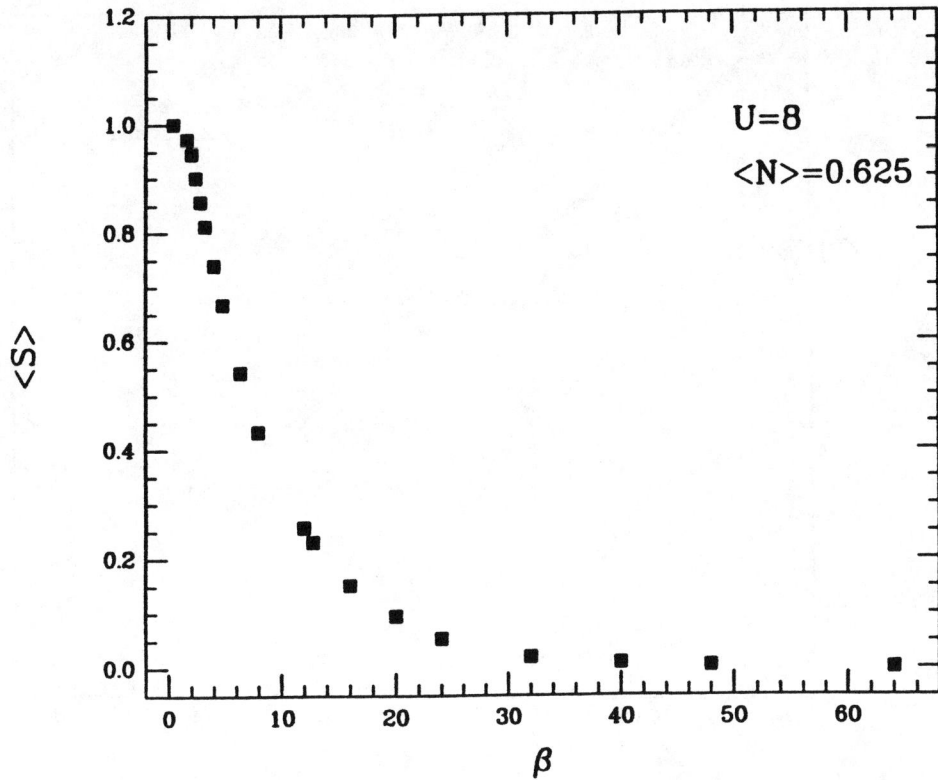

FIGURE 3 The average value of the sign $\langle S \rangle$ as a function of β on a 4×4 lattice with $U = 8$ and a filling $\langle n \rangle = 0.625$. These results were obtained with a ground-state projector Monte Carlo technique. (From Ref. 5.)

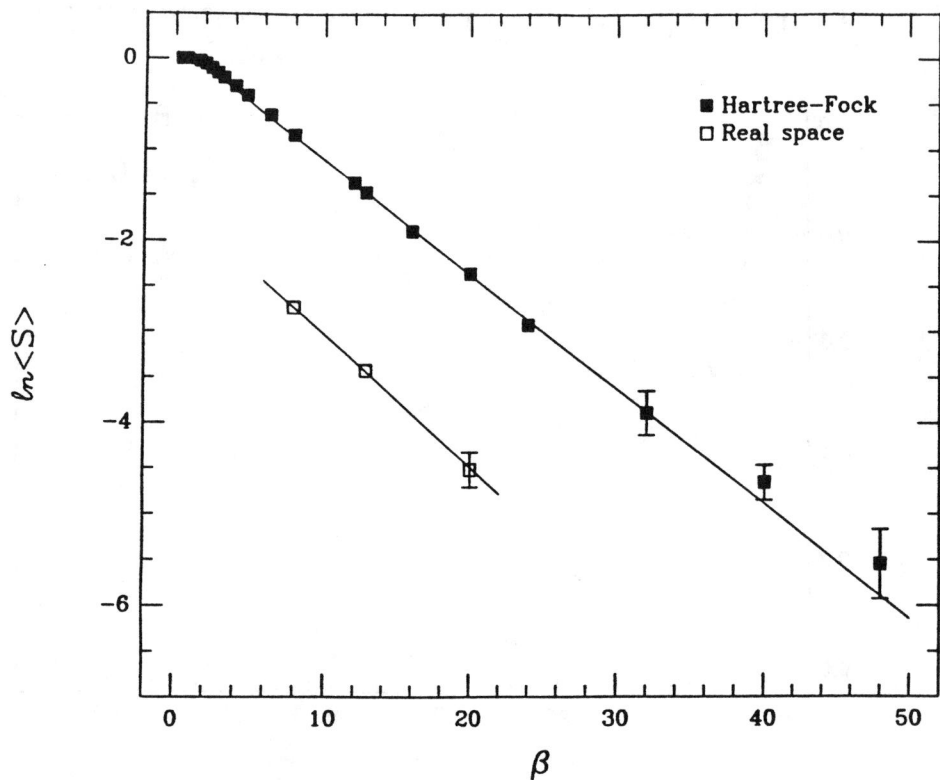

FIGURE 4 The same results as the previous figure showing $\ln\langle S\rangle$ versus β. The filled squares are for a Hartree-Fock trial function and the open squares for a real-space trial function. The straight lines are least-squares fits to the large β portion of the data.

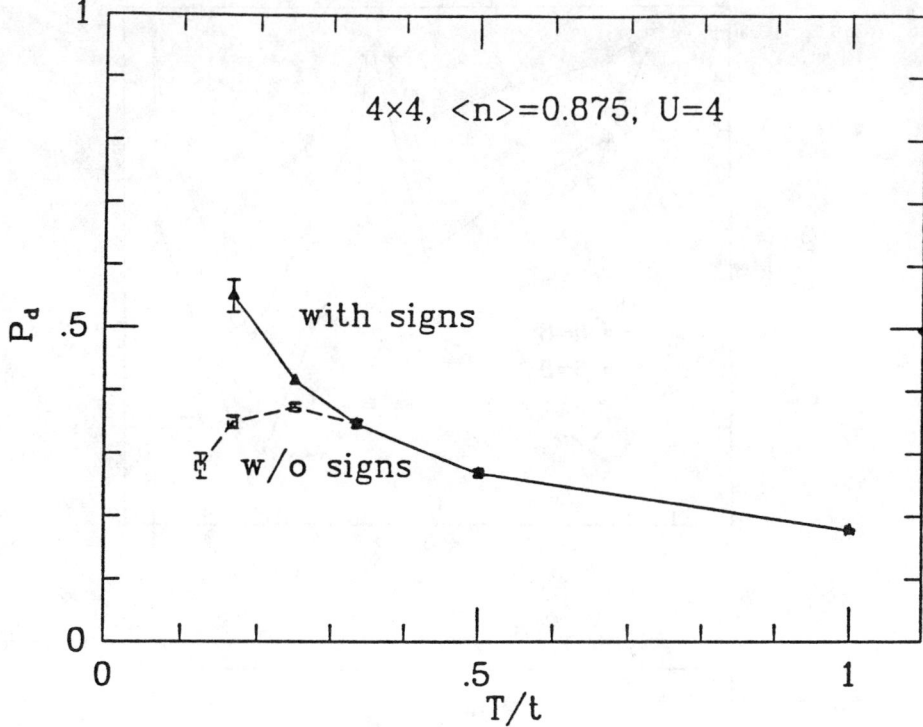

FIGURE 5 The d-wave pair field susceptibility P_d versus T evaluated with and without the sign of the fermion determinant. (From Ref. 5.)

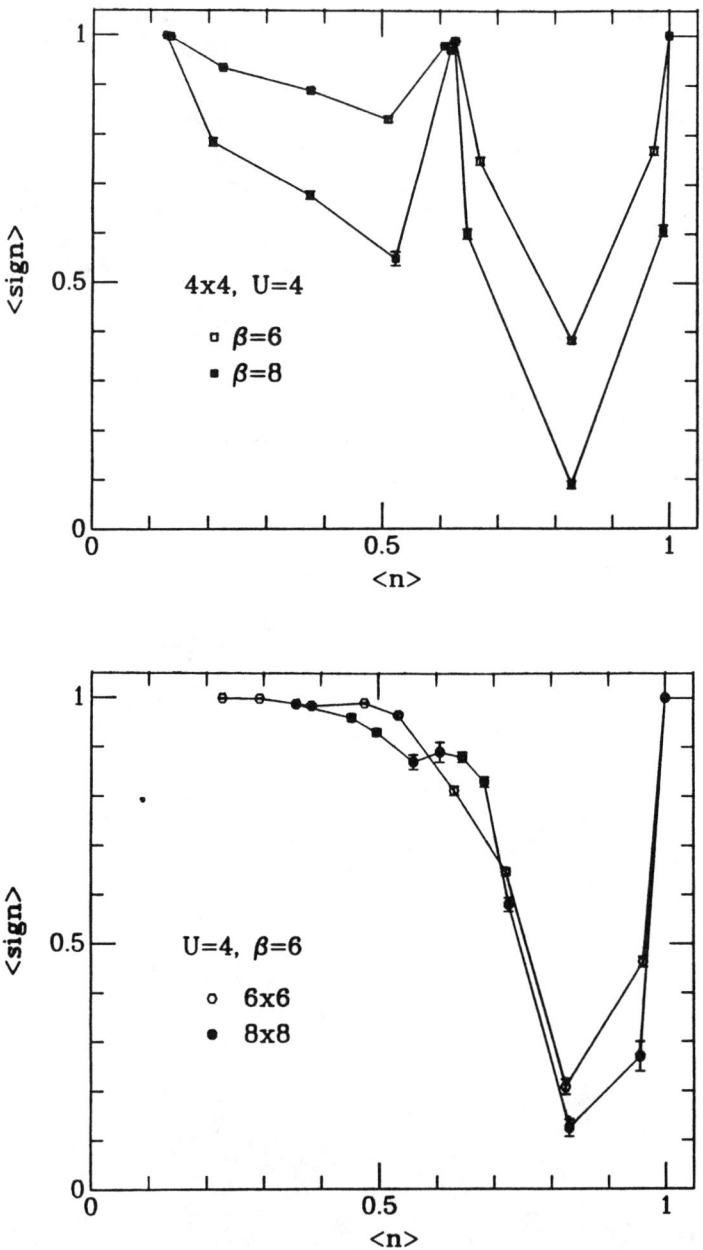

FIGURE 6 (a) and (b). Average sign $\langle S \rangle$ as a function of the band-filling $\langle n \rangle$ for various lattice sizes and temperatures. (From Ref. 3.)

LATTICE SIZES AND TEMPERATURES

In the following section we will describe data obtained on various sized lattices at different temperatures. A typical lattice is 8 × 8 with periodic boundary conditions. If each site is thought of as a Cu atom, then this would correspond roughly to a 30 A × 30 A 2D film. The largest lattice for which we will show results is 16 × 16. The temperature, as well as other energies, will be measured in units of t, and we will often give the value of $\beta = T^{-1}$. Thus $\beta = 10$ implies $T = t/10$ or, since the bandwidth is $8t$, a temperature of order 1/80 of the bandwidth. If the bandwidth $8t = 1\,\text{eV}$, then $\beta = 10$ would correspond to a temperature of 150° K. As previously noted, with our most recent algorithm[3] we have run at temperatures as low as 10^{-3} of the bandwidth for $\langle n \rangle = 1$ when there are no determinantal sign problems.

III. RESULTS

In reviewing[21] what has been learned from simulations of the 2D Hubbard model, it is important to keep in mind the limitations we have just discussed. For example, local quantities set by high-energy scales such as the square of the onsite moment or the effective one-electron hopping can be determined from simulations on relatively small lattices at moderate temperatures. Thus useful information for all band fillings can be obtained. On the other hand, the nature of the long-range correlations poses a more difficult problem. Here one requires large lattices and low temperatures. Thus, in this case, while real progress has been made at half-filling where there is no sign problem, the behavior just off of half-filling is not yet completely understood because of the difficulty in going to sufficiently large lattices and low temperatures. Similar lattice size constraints limit the grid spacing on which we can examine the electron momentum distribution and explore the Fermi surface (or lack of Fermi surface). Finally, it is difficult to obtain real frequency data unless one has very low-noise Monte Carlo data.

Thus in the following review we will begin with some results for local quantities and then turn to a discussion of the half-filled band. For important early work described below, see Hirsch.[21] We will comment on the effects of doping and contrast some of the half-filled band results with those obtained from the quarter-filled band (which as we have seen has a relatively benign sign problem). We will then turn to the most interesting and unfortunately the most technically difficult question of what happens in the region near half-filling.

LOCAL PROPERTIES

To begin at the most local level, consider the average of the square of the site local moment $m_{zi} = n_{i\uparrow} - n_{i\downarrow}$. This is equal to the difference between the density of electrons per site and twice the probability that a site is doubly occupied,

$$\langle m_{zi}^2 \rangle = \langle n_{i\uparrow} + n_{i\downarrow} \rangle - 2\langle n_{i\uparrow} n_{i\downarrow} \rangle. \tag{25}$$

At half-filling $\langle n_{i\uparrow} + n_{i\downarrow} \rangle = 1$, and as U/t increases the probability of double occupation decreases, leading in the large U limit to a full moment per site. Figure 7a shows $\langle m_{zi}^2 \rangle$ versus U for the case of a half-filled band. Figure 8 shows $\langle m_{zi}^2 \rangle$ versus $\langle n \rangle$ for various values of U. As expected, increasing U at all fillings leads to a local moment, but U is most effective in the region near half-filling.

As noted, a directly related quantity to $\langle m_{zi}^2 \rangle$ is the probability of double occupancy $\langle n_{i\uparrow} n_{i\downarrow} \rangle$ which is shown for the case of a half-filled band in Fig. 7b. Note that when U is equal to the bandwidth $8t$, the probability of double occupancy has decreased to approximately 20% of its value with $U = 0$. As pictured in various Gutzwiller schemes,[22] the avoidance of double occupancy leads to a reduction in the effective hopping rate narrowing the band. Using the Monte Carlo simulation, we have evaluated the effective one-electron transfer matrix element

$$\frac{t_{\text{eff}}}{t} = \frac{\langle c_{i\sigma}^\dagger c_{j\sigma} + c_{j\sigma}^\dagger c_{i\sigma} \rangle_U}{\langle c_{i\sigma}^\dagger c_{j\sigma} + c_{j\sigma}^\dagger c_{i\sigma} \rangle_0}. \tag{26}$$

The reduction in the effective hopping as a function of U for the case of a half-filled band is shown in Fig. 9. This matrix element can also be calculated in the weak and strong coupling limits. For weak coupling

$$\frac{t_{\text{eff}}}{t} = 1 - \left(\frac{U}{t}\right)^2 \frac{\Delta \varepsilon_2}{\Delta \varepsilon_0} \tag{27}$$

with

$$\Delta \varepsilon_0 = \frac{2}{N} \sum_k \varepsilon_k$$

$$\Delta \varepsilon_2 = \frac{1}{N^3} \sum_{kk'q} \frac{f(\varepsilon_k) f(\varepsilon_{k'}) (1 - f(\varepsilon_{k+q})) (1 - f(\varepsilon_{k'-q}))}{\varepsilon_k + \varepsilon_{k'} - \varepsilon_{k+q} - \varepsilon_{k'-q}}. \tag{28}$$

Here $\varepsilon_k = -2t(\cos k_x + \cos k_y)$ and f is the Fermi factor, and we consider the limit $T \to 0$. The weak coupling result, Eq. (27), is plotted as the dashed curve in Fig. 9. In strong coupling

$$\frac{t_{\text{eff}}}{t} = \frac{4 \left[\frac{4t}{U}\right] \left(|\langle \vec{S}_i \cdot \vec{S}_j \rangle| + \frac{1}{4}\right)}{\Delta \varepsilon_0}. \tag{29}$$

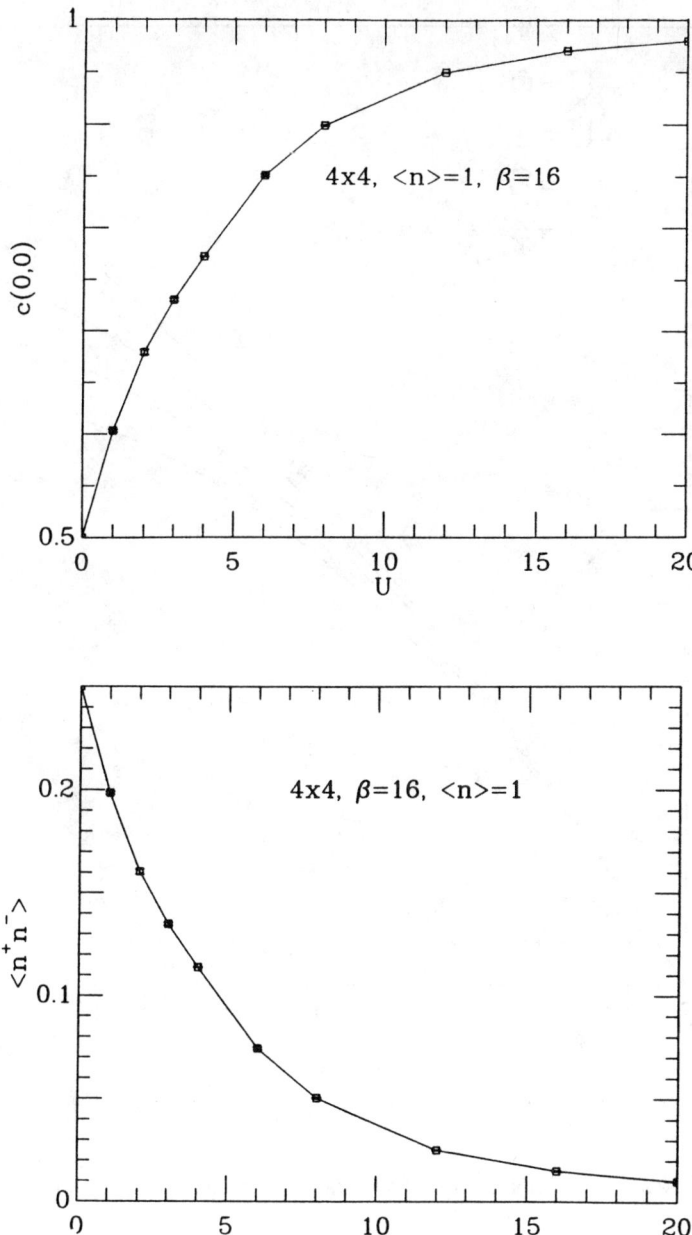

FIGURE 7 (a) The squared local moment $\langle m_{zi}^2 \rangle$, and (b) the double occupancy $\langle n_{i\uparrow} n_{i\downarrow} \rangle$ versus U on a 4×4 lattice with $\langle n \rangle = 1$ and $\beta = 16$. (From Ref. 3.)

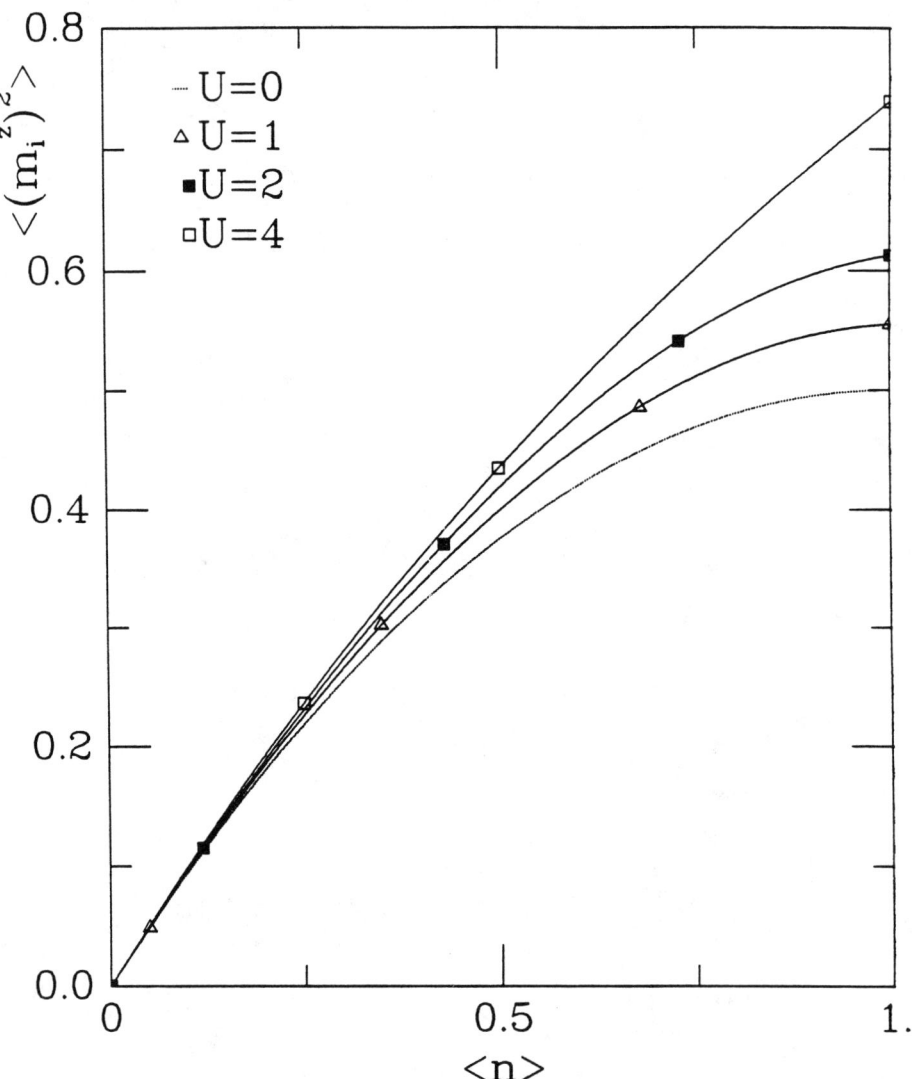

FIGURE 8 The squared local moment versus $\langle n \rangle$ for various values of U on a 12×12 lattice with $\beta = 6$. (From Ref. 7.)

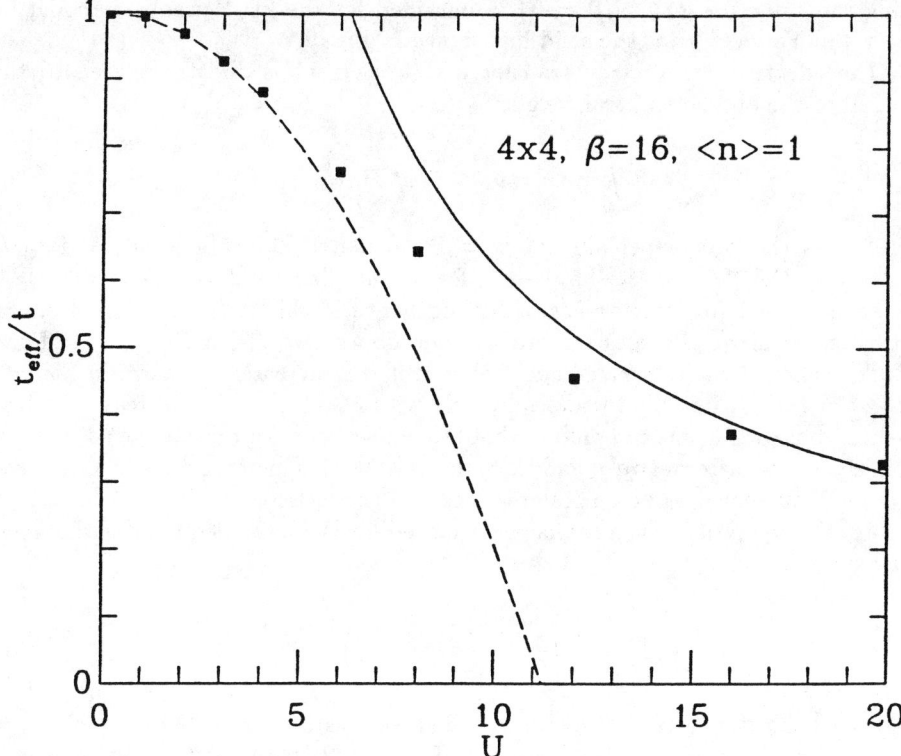

FIGURE 9 The effective hopping versus U on a 4×4 lattice with $\beta = 16$. The solid line is a strong coupling result and the dashed line is obtained from perturbation theory. (From Ref. 3.)

Using the value for $|\langle \vec{S}_i \cdot \vec{S}_j \rangle| = 0.335$ obtained for the 2D Heisenberg model by Reger and Young[23] gives the solid line shown in Fig. 9.

The effective one-electron transfer matrix element is directly related to the integral of the optical spectral weight[24] $\sigma(\omega)$,

$$8 \int_0^\infty d\omega\, \sigma(\omega) = 4\pi e^2 E_0(t_{\text{eff}}/t). \tag{30}$$

Here E_0 is the energy per site for $U = 0$ calculated at a filling $\langle n \rangle$, and t_{eff}/t is given by Eq. (26). As we have seen, for a half-filled band with $\langle n \rangle = 1$, t_{eff}/t decreases as t/U for large values of U, and hence in this case the integral of the optical spectral weight goes to zero. As one dopes away from $\langle n \rangle = 1$, "adding mobile carriers," the integrated spectral weight will increase. As shown in Fig. 10, for $U/t = 4$, t_{eff}/t initially increases linearly with $(1 - \langle n \rangle)$ as $\langle n \rangle$ decreases from 1. Phenomenologically one can picture that this linear behavior reflects the addition of mobile carriers. Alternatively, and more accurately, this increase reflects a decrease in the effective mass as vacancies are placed on the lattice.

Another quantity which we have calculated for the half-filled band is the magnetic susceptibility

$$\chi(T) = \left\langle \left[\frac{1}{N} \sum_\ell \left(n_{\ell\uparrow} - n_{\ell\downarrow} \right) \right]^2 \right\rangle \Big/ T. \tag{31}$$

Figure 11 shows χ versus T for an 8×8 lattice with $U = 4$. This temperature dependence of χ is similar to that for a 2D $s = 1/2$ Heisenberg antiferromagnet. However, as shown in Fig. 7, the local moment of the Hubbard model is not fully developed for $U = 4$.

LONG-RANGE ORDER

These results have shown how a repulsive onsite Coulomb interaction gives rise to local moments and at the same time reduces the one-electron transfer. These are both local effects. To explore the longer range electronic correlations, we begin by calculating the equal-time magnetic moment correlation function

$$C(\ell) = \langle m_{i+\ell}^z m_i^z \rangle. \tag{32}$$

This correlation function is shown for two different band fillings in Figs. 12a and 12b for a 10×10 lattice with $U = 4$ and $\beta = 10$. The half-filled case $\langle n \rangle = 1$, Fig. 12a, shows antiferromagnetic correlations extending throughout the lattice, while in the quarter-filled case, Fig. 12b, these correlations decay after one lattice spacing.

The Fourier transform of $C(\ell)$ gives the magnetic structure factor

$$S(q) = \sum_\ell e^{i\vec{q}\cdot\vec{\ell}} \langle m_{i+\ell}^z m_i \rangle. \tag{33}$$

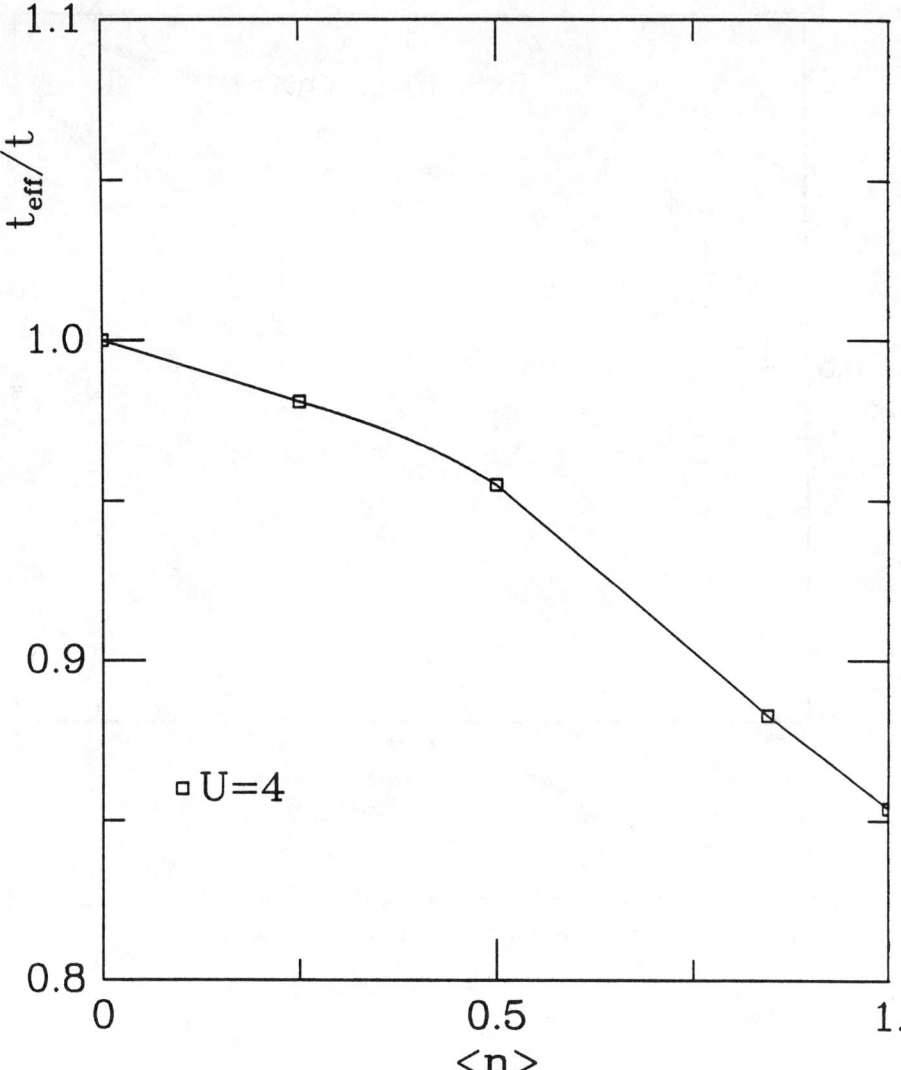

FIGURE 7 (a) The squared local moment $\langle m_{zi}^2 \rangle$, and (b) the double occupancy $\langle n_{i\uparrow} n_{i\downarrow} \rangle$ versus U on a 4 × 4 lattice with $\langle n \rangle = 1$ and $\beta = 16$. (From Ref. 3.)

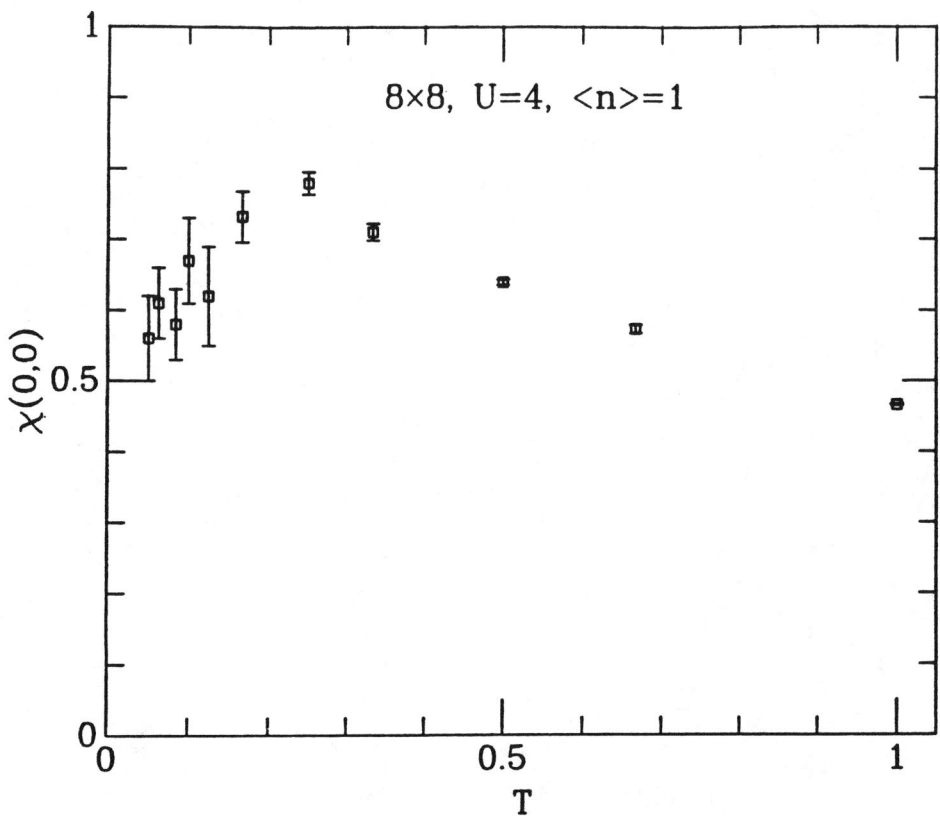

FIGURE 11 The magnetic susceptibility χ versus T for the half-filled Hubbard model. These results are from an 8×8 lattice with $U = 4$. (From Ref. 3.)

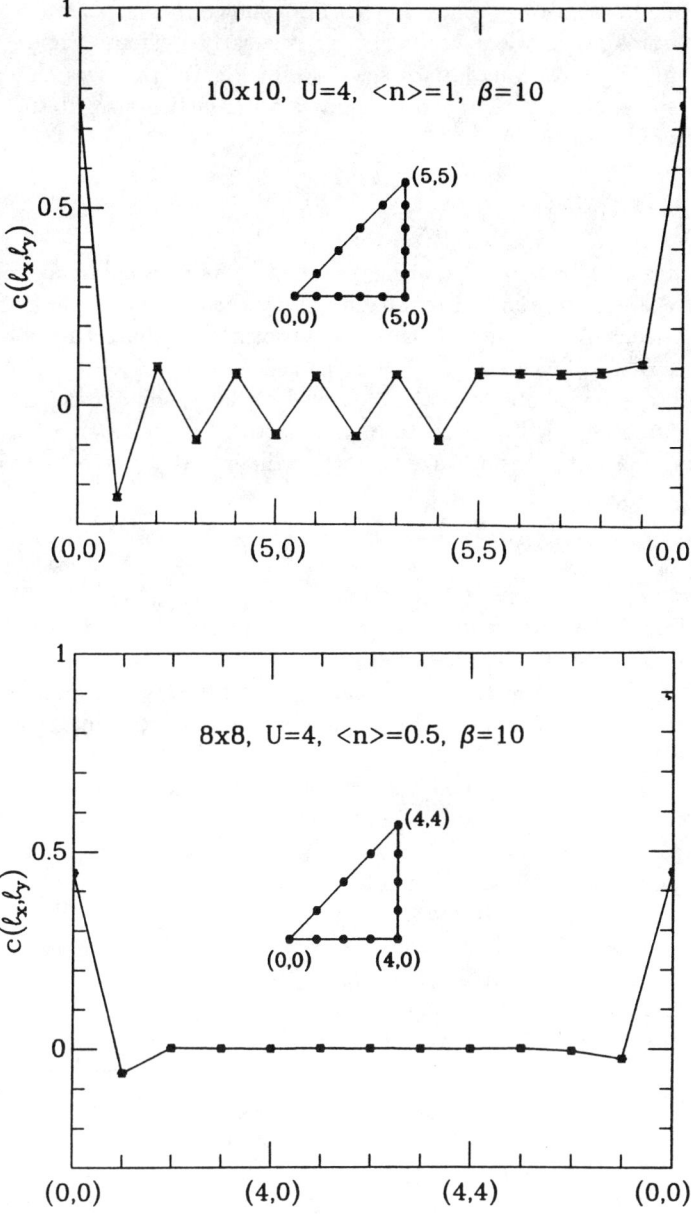

FIGURE 12 Magnetic moment correlation function $C(\ell_x, \ell_y)$. The horizontal axis traces out the triangular path shown in the center of the figure. Strong antiferromagnetic correlations are visible in (a) which is for a half-filled band, but are nearly absent in (b) which is at quarter-filling. (From Ref. 3.)

For the half-filled case, some results for $S(q)$ are shown in Fig. 13, where the peak at the antiferromagnetic wave vector (π, π) is clearly evident. Figure 14 shows this peak versus β for various lattice sizes. From Fig. 14 one sees that the zero-temperature extrapolation of $S(\pi, \pi)$ increases with lattice size. If there is long-range antiferromagnetic order,

$$\lim_{N \to \infty} \frac{S(\pi, \pi)}{N} = \frac{m^2}{3}, \tag{34}$$

where m is the antiferromagnetic order parameter. Within spin-wave theory the leading order correction is known to vary inversely with the linear dimension $N_x = \sqrt{N}$ of the lattice.[25] Assuming that this is appropriate for the 2D Hubbard model, we extrapolate $S(\pi, \pi)/N$ versus $1/N_x$ for $\langle n \rangle = 1$ and $U = 4$ as the dashed line in Fig. 15. The order parameter can also be obtained by extrapolating the spin-spin correlation function $C(\ell)$, Eq. (32), to infinite lattice spacing. Using the two most distant points on a lattice, spin-wave theory predicts that

$$C\left(\tfrac{N_x}{2}, \tfrac{N_x}{2}\right) = \frac{m^2}{3} + O\left(\tfrac{1}{N_x}\right). \tag{35}$$

For $\langle n \rangle = 1$ and $U = 4$, the extrapolation using $C(N_x/2, N_x/2)$ is shown as the solid line in Fig. 15. A non-vanishing value for m implies that there is long-range antiferromagnetic order in the ground state.

Hirsch and Tang[6] were the first to carry out such extrapolations for various values of U and found that the order parameter increased gradually with U, approaching the value obtained by Reger and Young[23] for the $s = 1/2$ Heisenberg antiferromagnet when U exceeded the bandwidth $8t$. At small values of U, the variation of m was consistent with weak coupling predictions, suggesting that the ground state of the half-filled 2D Hubbard model has antiferromagnetic long-range order for all values of U greater than zero.

As discussed, when the Hubbard model is doped off of half-filling, the simulations become more difficult because of sign problems. Unfortunately, as shown in Fig. 6, this problem is particularly acute just in the most interesting region, $0.7 < \langle n \rangle < 0.95$. For this reason, we are at present unable to carry out a definitive zero temperature-infinite lattice extrapolation near half-filling. Nevertheless, it is clear that the antiferromagnetic correlations are rapidly suppressed as the system is doped away from half-filling.[6–8] Figures 16a and 16b show $S(q)$ on an 8×8 lattice at $\beta = 6$ for band-fillings of $\langle n \rangle = .83$ and $.72$, respectively. Compared with Fig. 13, one sees that the antiferromagnetic peak is significantly smaller. An approximate fit of this data, shown by the solid lines, suggests that the peak has split, shifting out along the $(\pi, \pi - \Delta q)$ and $(\pi - \Delta q, \pi)$ edges rather than moving down along the diagonal $(\pi - \Delta q, \pi - \Delta q)$. This is consistent with both RPA[26] and conserving approximation[27] results. Over the accessible temperature range $\beta \lesssim 6$, one finds no evidence indicating the growth of long-range incommensurate order. We will return at the end of this section to a further discussion of this regime and the nature of the pairing correlations for the nearly half-filled band.

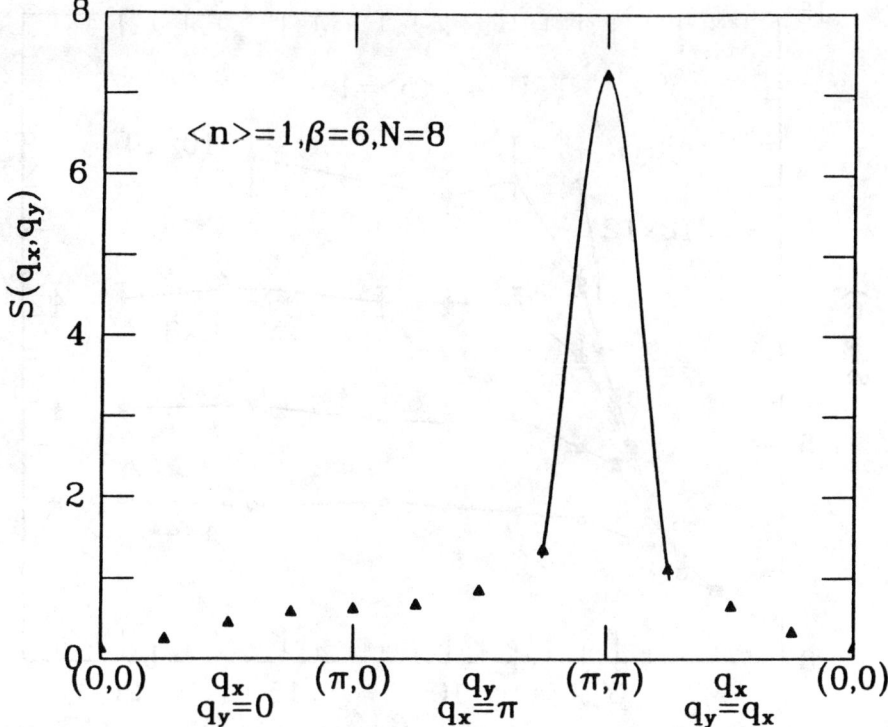

FIGURE 13 $S(q_x, q_y)$ versus (q_x, q_y) on an 8×8 lattice with $\langle n \rangle = 1$, $U = 4$ and $\beta = 6$. The solid line is a fit to guide the eye. (From Ref. 7.)

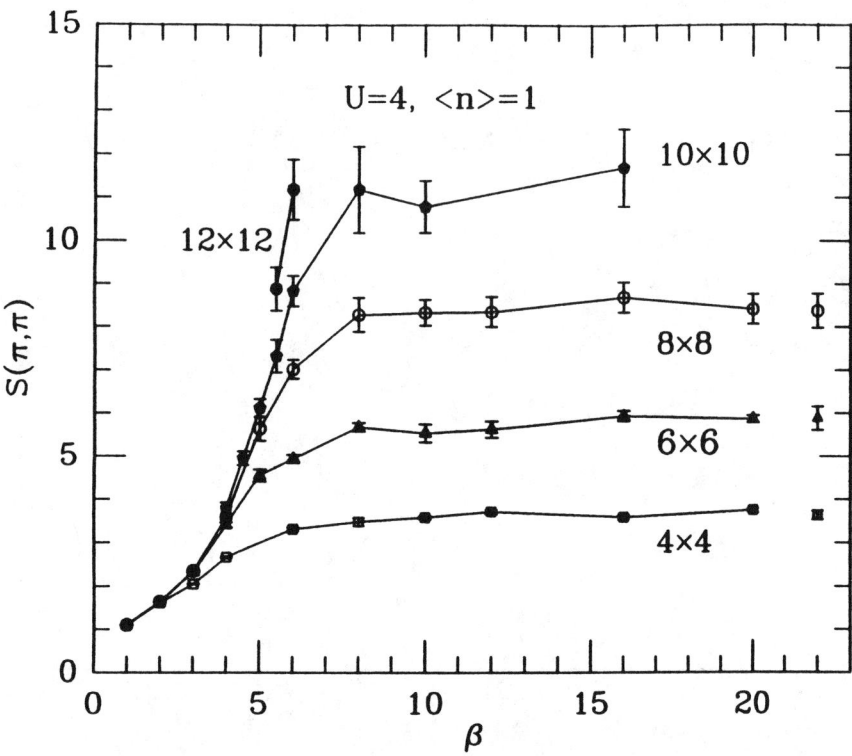

FIGURE 14 The antiferromagnetic structure factor $S(\pi,\pi)$ as a function of inverse temperature for a variety of lattice sizes. The points at $\beta = 22$ were done with a projector Monte Carlo method with fixed particle number, while the rest were obtained with the grand-canonical algorithm. (From Ref. 3.)

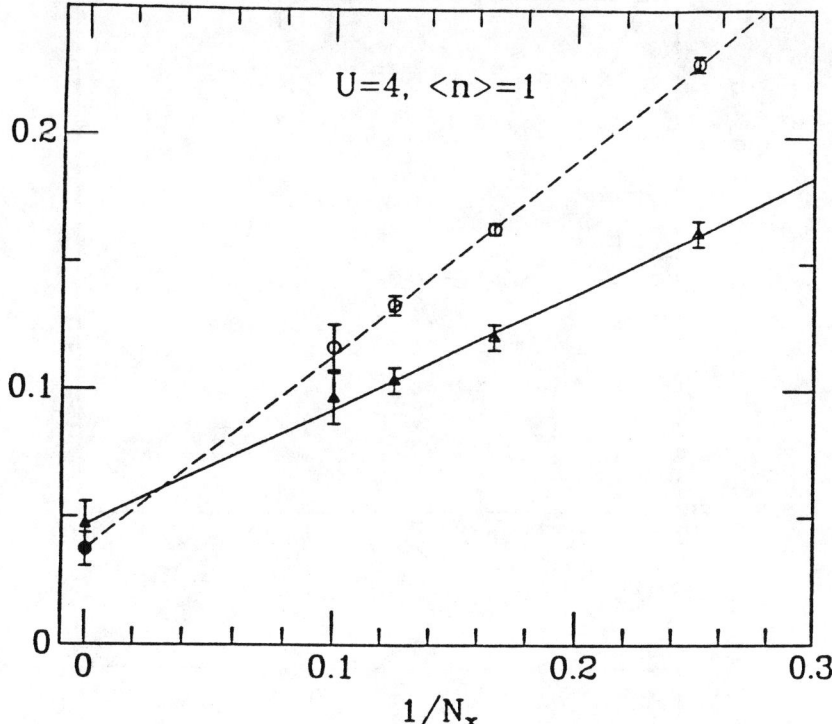

FIGURE 15 Plots of the zero-temperature limit of $S(\pi,\pi)/N$ (circles) and $C(N_x/2, N_x/2)$ (triangles) vs. $1/N_x$. The lines are least-squares fits to the data which extrapolate to the solid symbols for the infinite system. This extrapolation to a non-zero value implies that the ground state has long-range order. (From Ref. 3.)

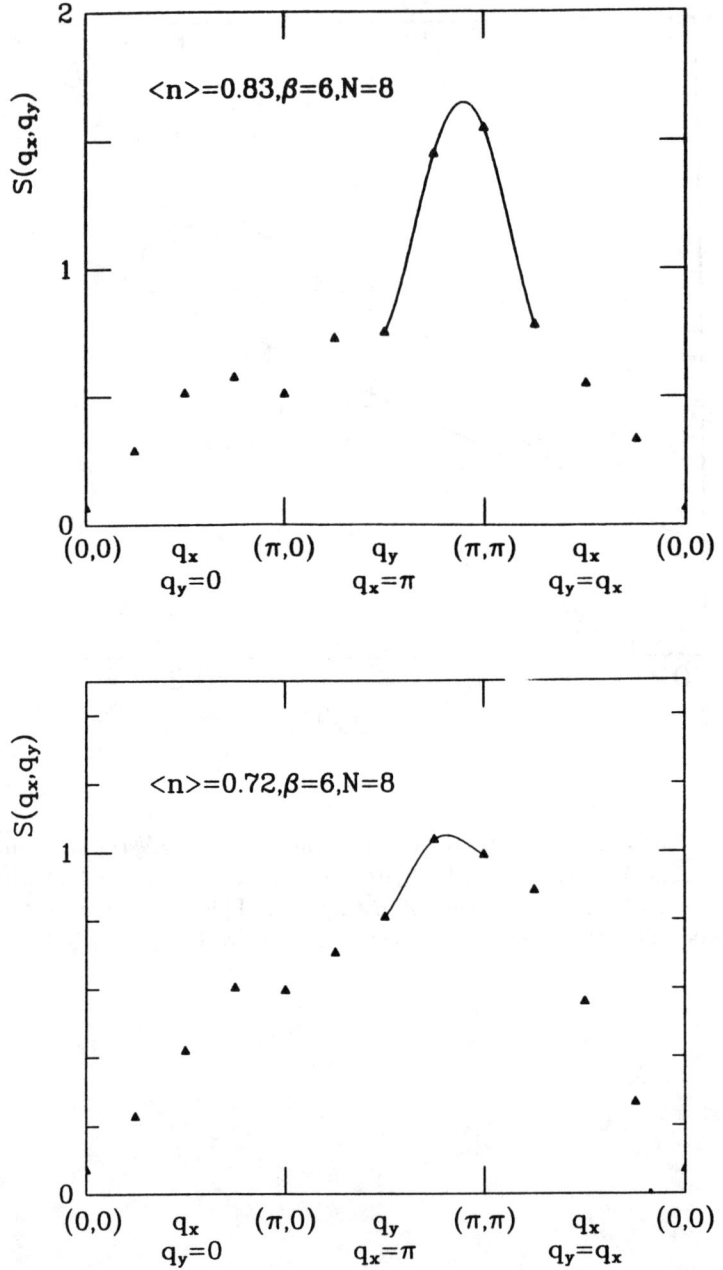

FIGURE 16 $S(q_x, q_y)$ versus (q_x, q_y) on an 8 × 8 lattice with $U = 4$, $\beta = 6$ and (a) $\langle n \rangle = .83$ and (b) $\langle n \rangle = .72$. The solid line is a fit to guide the eye. (From Ref. 7.)

SINGLE PARTICLE PROPERTIES

It is interesting to examine other correlations such as those contained in the single-particle Green's function

$$G_{ij}^\sigma(\tau) = \langle T c_{i\sigma}(\tau) c_{j\sigma}^\dagger(0) \rangle. \tag{36}$$

For example, the single-particle momentum distribution is given by

$$\langle n_{k\sigma} \rangle = \sum_\ell e^{i\vec{k}\cdot\vec{\ell}} G_{i+\ell,i}^\sigma(0^-). \tag{37}$$

The relationship between the band filling $\langle n \rangle$ and μ is set by

$$\langle n \rangle = \frac{1}{N} \sum_{i\sigma} G_{ii}^\sigma(0^-) \tag{38}$$

so that the compressibility

$$K = \frac{1}{\langle n \rangle^2} \frac{d\langle n \rangle}{d\mu} \tag{39}$$

can be determined from G_{ij}^σ. Finally, by its definition, the one-electron self-energy can be obtained from the Fourier tranform of G at the Matsubara frequencies $\omega_n = (2n+1)\pi T$,

$$\Sigma(k, i\omega_n) = i\omega_n - (\varepsilon_k - \mu) - G^{-1}(k, i\omega_n). \tag{40}$$

Calculationally, it turns out that $G_{ij}^\sigma(\tau)$ has the smallest statistical fluctuations of the quantities evaluated in the Monte Carlo simulations. For this reason, one can obtain useful information about these one-electron properties on fairly large lattices. This is of course important for quantities such as $\langle n_k \rangle$ where the spacing of the allowed momenta depends on the lattice size.

Figures 17 and 18 show the single-particle momentum distribution $\langle n_k \rangle$ with $\vec{k}$ along the (1,1) direction for the half-filled and quarter-filled bands with $U = 4$ and $\beta = 6$. The open triangles are for a 16×16 lattice. The remaining triangles are from different sized lattices ranging from 6×6 to 14×14. The solid line in Figs. 17a and 18a is the noninteracting Fermi function $f(\varepsilon_k) = (\exp(-\beta(\varepsilon_k - \mu)) + 1)^{-1}$, with $\varepsilon_k = -2t(\cos k_x + \cos k_y)$ and μ adjusted to give $\langle n \rangle = 1$ and 0.5, respectively. The solid line in Fig. 17b corresponds to the mean field spin-density-wave result

$$\langle n_k \rangle_{\mathrm{mf}} = \frac{1}{2}\left(1 - \frac{\varepsilon_k}{E_k}\right), \tag{41}$$

with $E_k = \sqrt{\varepsilon_k^2 + \Delta^2}$ and Δ determined on a 16×16 lattice from

$$\frac{1}{U} = \frac{1}{N} \sum_k \frac{1}{E_k}. \tag{42}$$

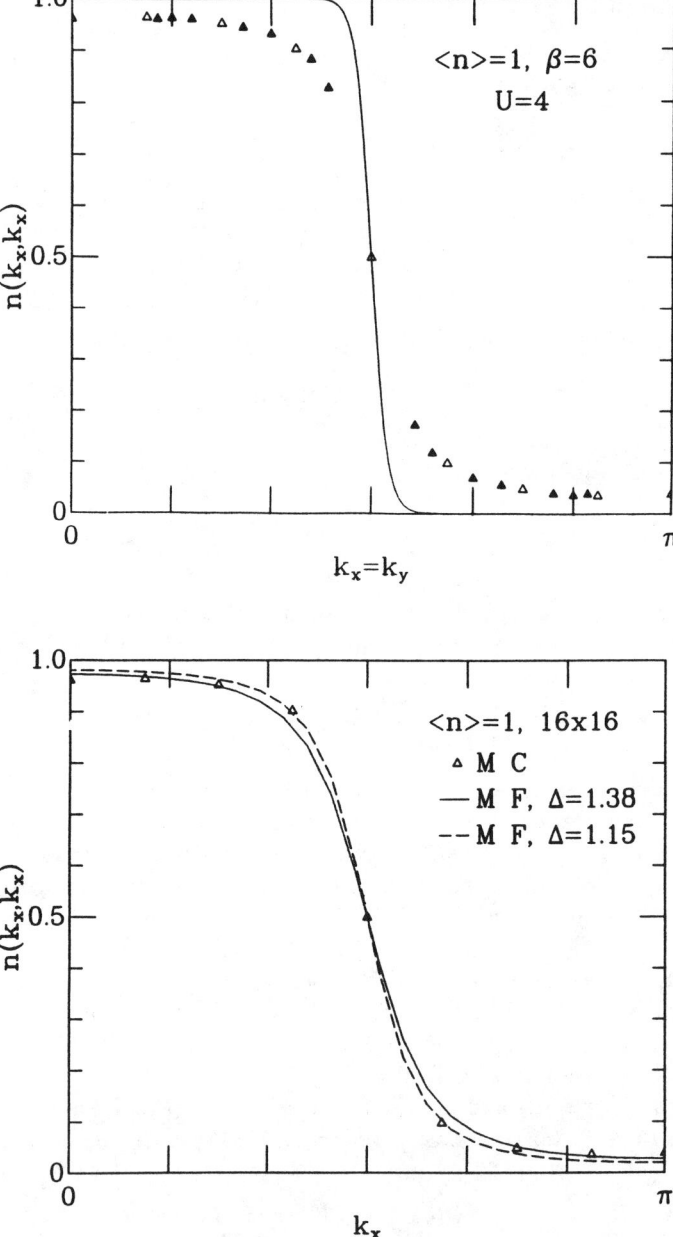

FIGURE 17 The momentum distribution $\langle n_k \rangle$ versus $\vec{k} = (k_x, k_x)$ for $\langle n \rangle = 1$. Here $U/t = 4$ and $\beta = 6/t$. (a) Results for 6×6, ..., 16×16 lattices are shown. The open triangles are for the 16×16 lattice. The solid line is the fermi function for a half-filled, non-interacting system at a temperature $\beta = 6/t$. (b) Monte Carlo results for a 16×16 lattice, shown as open triangles, are compared with the predictions of a mean-field spin-density-wave approximation. (From Ref. 7.)

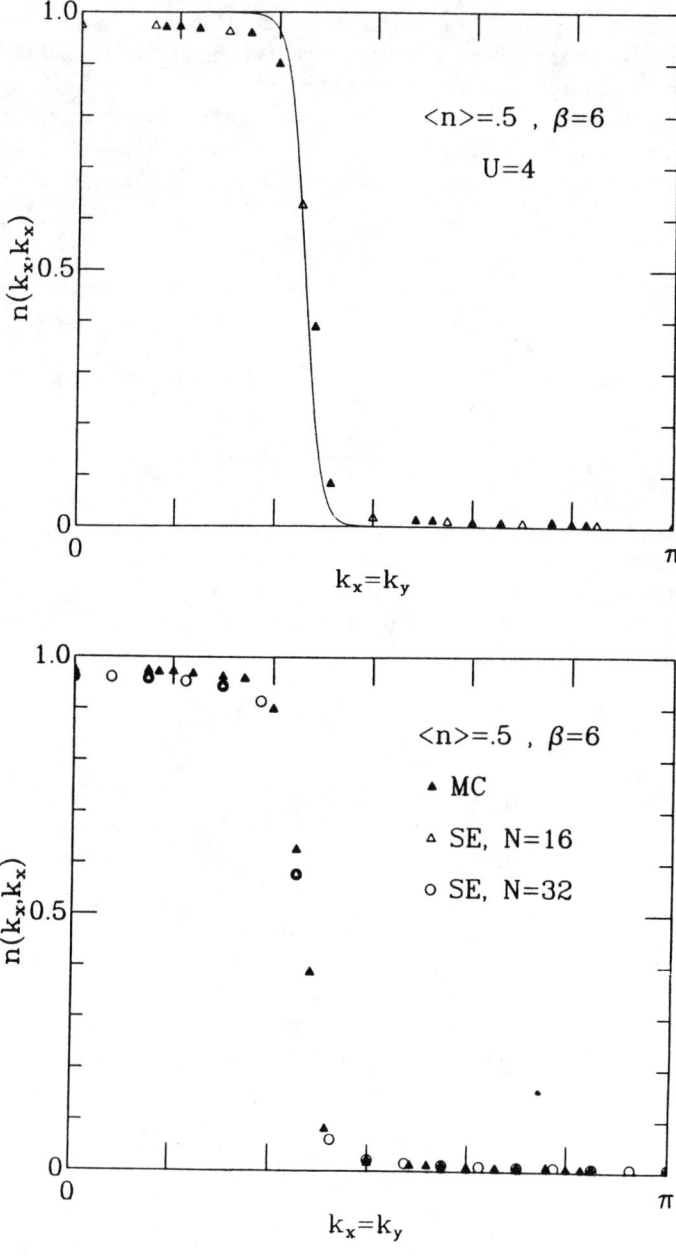

FIGURE 18 The momentum distribution $\langle n_k \rangle$ versus $\vec{k} = (k_x, k_x)$ for $\langle n \rangle = 0.5$. As in Fig. 17, $U/t = 4$ and $\beta = 6/t$. (a) Results for 6×6, ..., 16×16 lattices are shown with the open triangles corresponding to the 16×16 lattice. The solid line is the Fermi function for a quarter-filled band at $\beta = 6/t$. (b) Comparison of the Monte Carlo results solid triangles with the results for $\langle n_k \rangle$ obtained by keeping the lowest-order self-energy graph shown in Fig. 1. (From Ref. 7.)

The mean-field result gives a reasonable fit to the Monte Carlo data. However, the mean-field gap is too large, and a better fit near $k = (\pi/2, \pi/2)$, the dashed line in Fig. 17b, can be obtained using a reduced gap.[28]

For the quarter-filled band, the momentum distribution, shown in Fig. 18, is much closer to that of the noninteracting Fermi function. The leading correction to the noninteracting result can be obtained from the second-order self-energy

$$\Sigma_2(k, i\omega_n) = \frac{U^2}{N^2} \sum_{p,q} \frac{f(\varepsilon_p) - f(\varepsilon_{k-q})f(\varepsilon_p) - f(\varepsilon_{p+q})f(\varepsilon_p) + f(\varepsilon_{k-q})f(\varepsilon_{p+q})}{i\omega_n - (\varepsilon_{k-q} + \varepsilon_{p+q} - \varepsilon_p)}. \tag{43}$$

The open triangles and circles in Fig. 18b were obtained by evaluating

$$\langle n_k \rangle = T \sum_n \frac{e^{-i\omega_n 0^-}}{i\omega_n - (\varepsilon_k - \mu) - \Sigma_2(k, i\omega_n)} \tag{44}$$

on a 16×16 and 32×32 lattice respectively. They are in reasonable agreement with the Monte Carlo data.

The compressibility K, Eq. (39), provides another probe of the electronic correlations and the metallic or insulating character of the 2D Hubbard model. In the noninteracting, $U = 0$, limit,

$$K = \frac{2}{NT} \sum_k f(\varepsilon_k)(1 - f(\varepsilon_k)), \tag{45}$$

and at low temperatures $K = N(\mu)$ with $N(\mu)$ the single-particle density of states. As is well known, near half-filling as $\mu \to 0$, the saddle points at the corners of the Fermi surface lead to a logarithmic Van Hove singularity in $N(\mu)$ and

$$K \simeq \frac{\ln(16t/\mu)}{\pi^2 t}. \tag{46}$$

Thus in the noninteracting limit, $K \sim d\langle n \rangle/d\mu$ diverges as $\mu \to 0$. However, when $U \neq 0$ the long-range antiferromagnetic order is expected to lead to a gap in the single-particle spectrum. In this case $\langle n \rangle$ versus μ exhibits a flat portion over the region in which the gap has opened. The compressibility is determined by the slope of $\langle n \rangle$ versus μ at the edge of the gap (private communication from A. Millis). Figures 19a and 19b show Monte Carlo results for $\langle n \rangle$ versus μ for $U = 4$ and 2 respectively. The dashed line shows that $U = 0$ noninteracting result. There appears to be a plateau around $\mu \to 0$, as expected, if a gap in the one-electron spectrum is present. Further calculations on larger lattices at lower temperatures are required to determine the slope at the gap edge.

It is also interesting to examine the behavior of $\Sigma(k, i\omega_n)$. For the half-filled case, mean-field theory gives

$$\Sigma(k, i\omega_n) = \frac{\Delta^2}{i\omega_n + \varepsilon_k}, \tag{47}$$

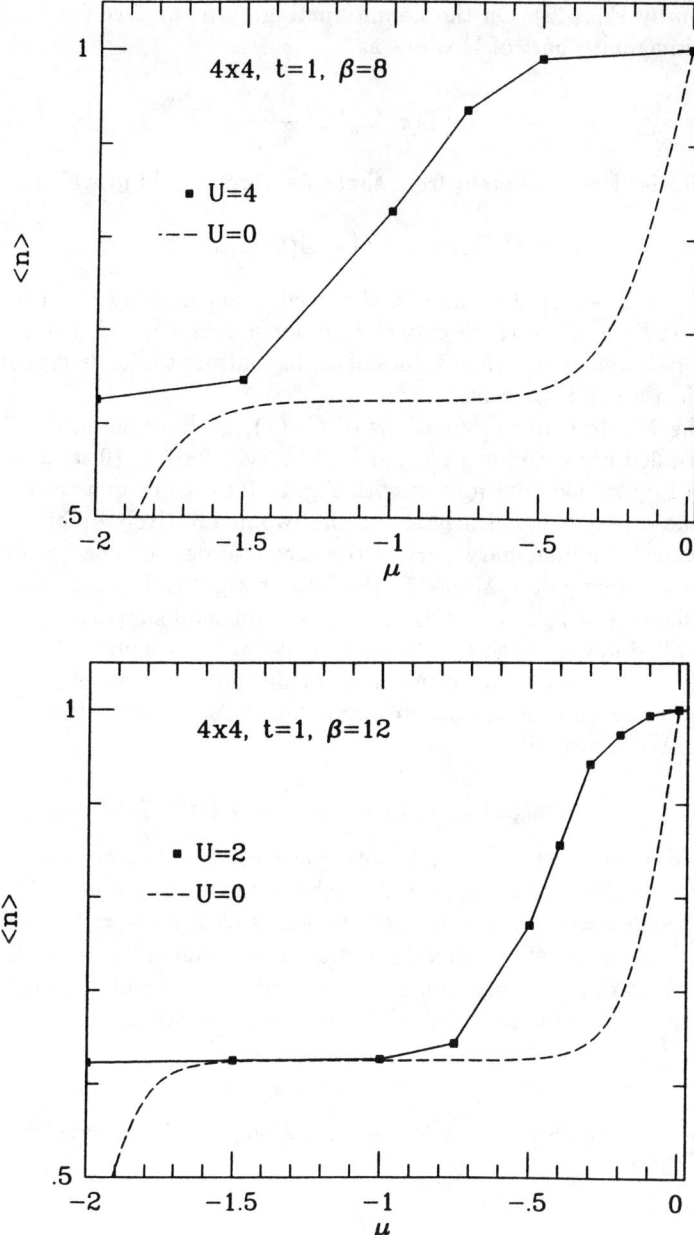

FIGURE 19 The average electron occupation per site $\langle n \rangle$ versus μ on a 4×4 lattice. The dashed line is for $U = 0$ and the points are Monte Carlo results for (a) $U/t = 4$ and $\beta = 8/t$, and (b) $U/t = 2$ and $\beta = 12/t$. (From Ref. 7.)

with Δ given by Eq. (42). On the noninteracting Fermi surface for $\langle n \rangle = 1$ where $\varepsilon_k = 0$, the imaginary part of Σ varies as

$$\text{Im}\,\Sigma(k, i\omega_n) \simeq \frac{-\Delta^2}{\omega_n}. \tag{48}$$

This behavior is clearly different from that of a Fermi liquid in which

$$\text{Im}\,\Sigma(k, i\omega_n) = (1 - Z(k, i\omega_n))\,\omega_n, \tag{49}$$

with $Z^{-1}(k_F, i\pi T \to 0)$ the jump in the occupation defining the Fermi surface. Since $Z(k_F, i\pi T \to 0) > 1$, we expect that for a Fermi liquid, $\text{Im}\,\Sigma(k, i\omega_n)$ will exhibit a negative slope for small values of ω_n in contrast to the divergent behavior, Eq. (48), when a gap is present.

From the Monte Carlo calculations of $G_{ij}(\tau)$, we have determined $\Sigma(k, i\omega_n)$. For the half-filled case, we have plotted $\text{Im}\,\Sigma(k, i\omega_n)$ for $k = (0, \pi)$ and $(\pi/2, \pi/2)$ versus ω_n in Fig. 20. Comparing this with Fig. 21 for the one-quarter-filled band at a k value where $\langle n_k \rangle \cong 0.5$ clearly shows the two qualitatively different behaviors. Figure 22 shows the imaginary part of the second-order self-energy, Eq. (43), for the same parameters as the Monte Carlo data of Fig. 21. The agreement between the results shown in Figs. 21 and 22 is quite sensible and supports the notion that the quarter-filled system with $U = 4$ behaves like a Fermi liquid.

Recently, as discussed above, we have made some progress in extracting real frequency information from the Monte Carlo data.[4] For example, the single-particle density of states is formally

$$N(\omega) = -\tfrac{1}{\pi}\text{Im}\,G_{ii}(i\omega_n \to \omega + i\delta). \tag{50}$$

Using a least-square fitting routine with positive definite and smoothness constraints, we have obtained results for $N(\omega)$ shown in Fig. 23 for a half-filled band on an 8×8 lattice at $\beta = 12$. The existence of a gap is clear for $U = 4$ and 8. Although a gap is expected to exist for all values of U at half-filling, the finite resolution of the numerical continuation procedure makes it difficult to resolve it for $U \leq 2$. This approach has also been used to calculate the single-particle spectral weight

$$A(p, \omega) = -\tfrac{1}{\pi}\text{Im}\,G(p, i\omega_n \to \omega + i\delta), \tag{51}$$

and results from this are shown in Fig. 24. A gap in the spectral weight which increases with U is seen.

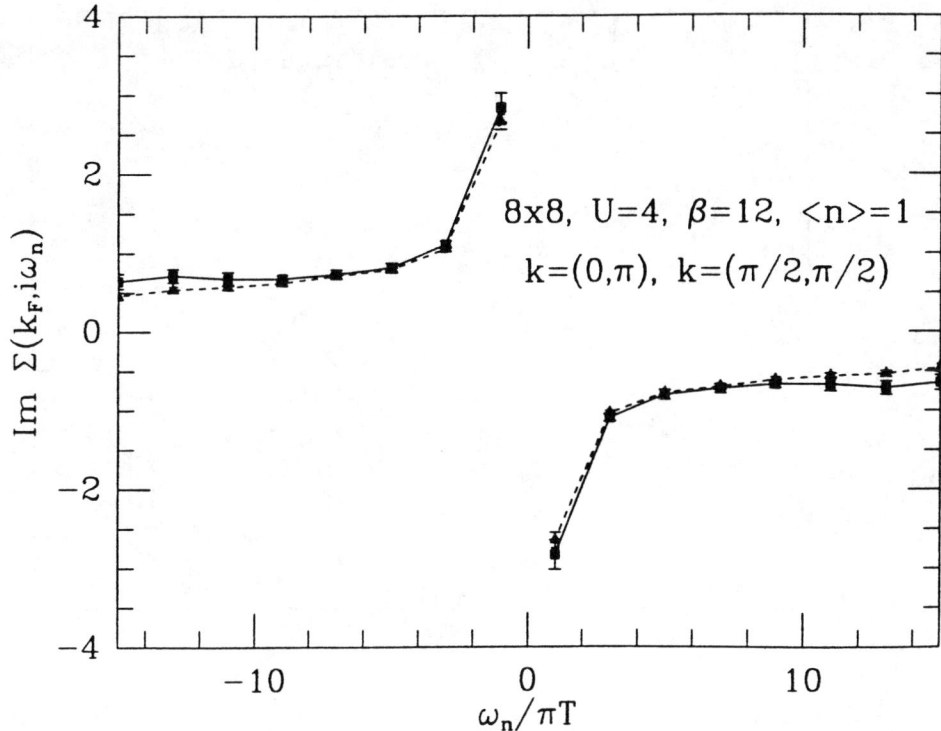

FIGURE 20 The imaginary part of the self-energy $\Sigma(k_F, i\omega_n)$ versus $\omega_n/\pi T$ for a half-filled band ($\langle n \rangle = 1$) and $k_F = (0, \pi)$ (squares) and $k_F = (\pi/2, \pi/2)$ (triangles). These results were obtained from Monte Carlo runs on an 8×8 lattice with $U/t = 4$ and $\beta = 12/t$. (From Ref. 7.)

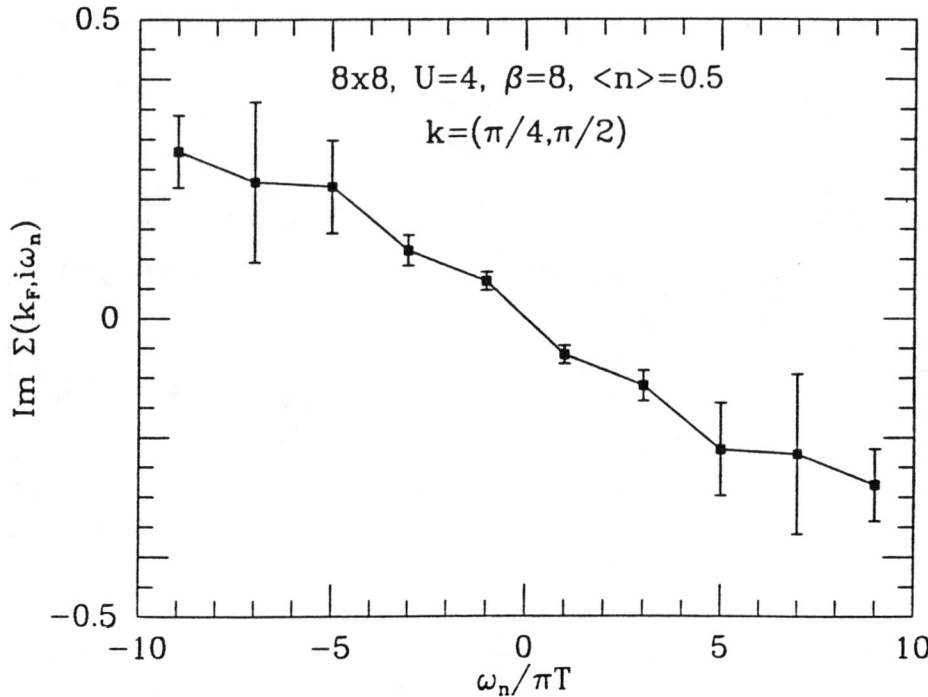

FIGURE 21 The imaginary part of the self-energy $\Sigma(k_F, i\omega_n)$ versus $\omega_n/\pi T$ for a quarter-filled band ($\langle n \rangle = 0.5$) with $k_F = (\pi/4, \pi/2)$. These results were obtained from Monte Carlo runs on an 8×8 lattice with $U/t = 4$ and $\beta = 8/t$. (From Ref. 7.)

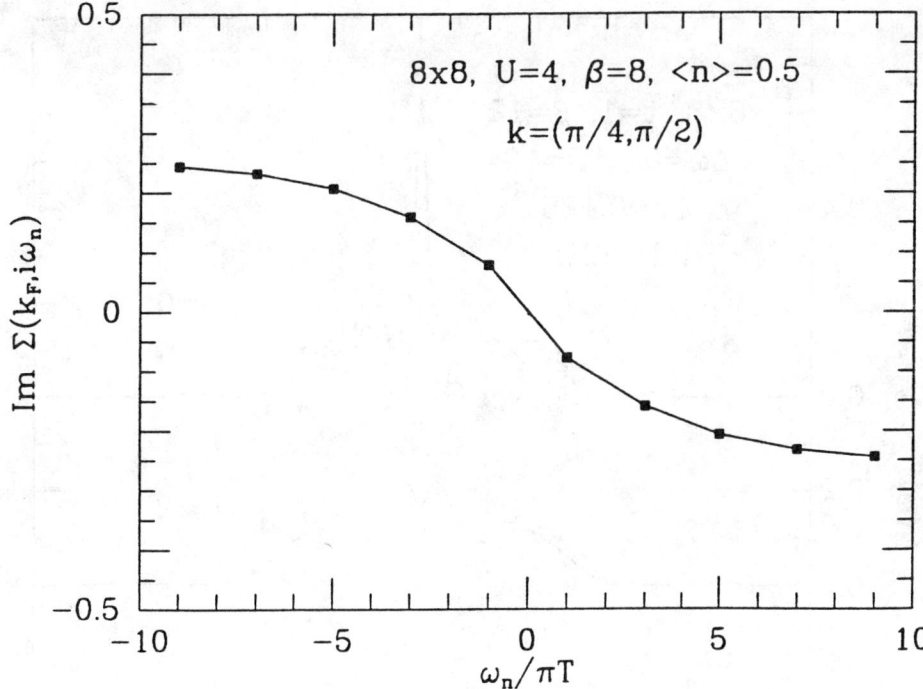

FIGURE 22 The imaginary part of the lowest-order self-energy graph (Fig. 1) versus ω_n for $k_F = (\pi/4, \pi/2)$ for $\langle n \rangle = 0.5$. These results were calculated from Eq. (43) for an 8×8 lattice with $U/t = 4$ and $\beta = 8/t$ in order to compare with the Monte Carlo results shown in Fig. 21. (From Ref. 7.)

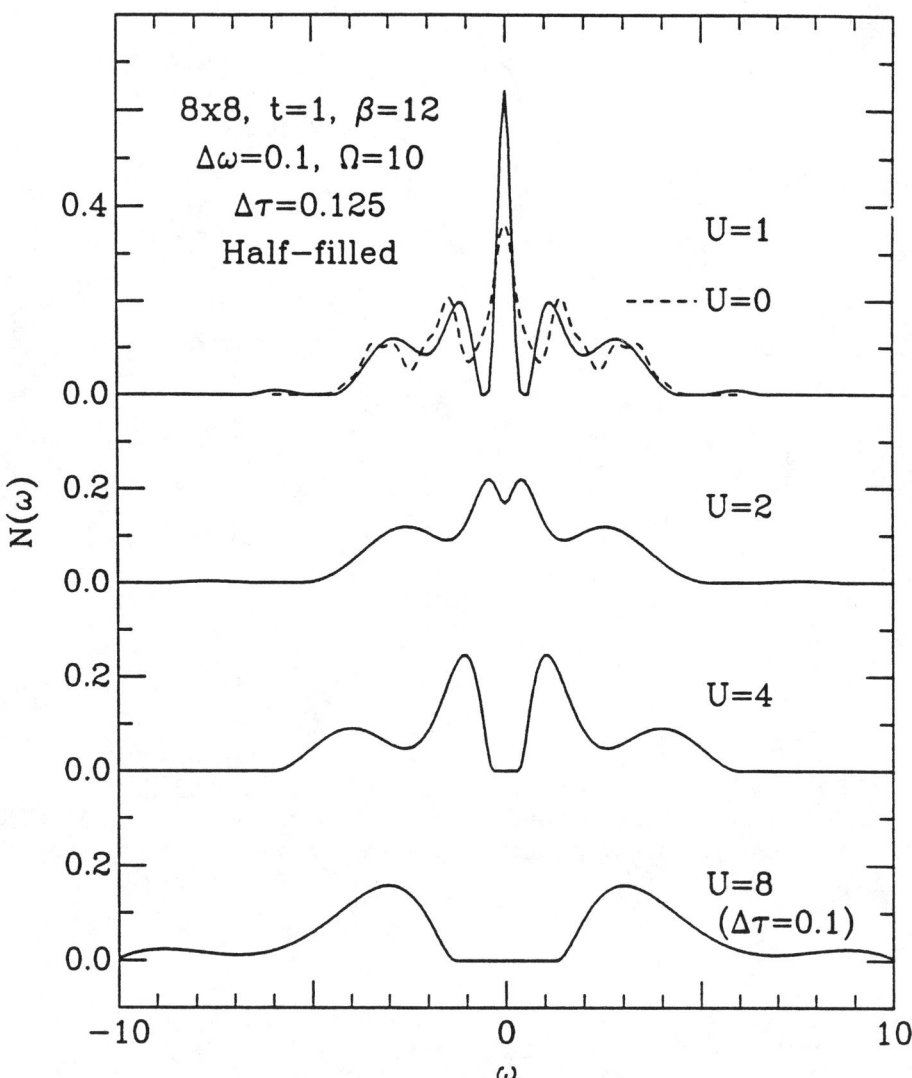

FIGURE 23 Density of states of the half-filled 2D Hubbard model for an 8×8 lattice at $\beta = 12$ for several values of U. The exact results for $U = 0$ (broadened to a resolution of 0.6) are shown by the dashed line. (From Ref. 4.)

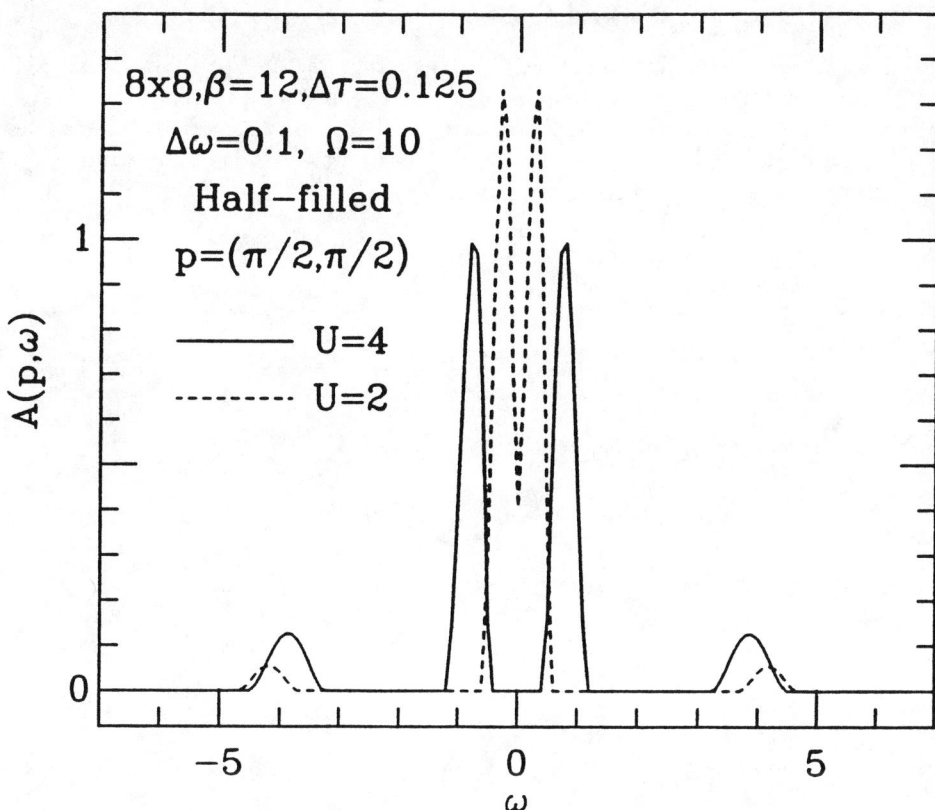

FIGURE 24 Spectral weight function $A(p,\omega)$ of the half-filled 2D Hubbard model for an 8×8 lattice at $\beta = 12$ for a wave vector $p = (\pi/2, \pi/2)$ on the Fermi surface.

SUPERCONDUCTING CORRELATIONS

Just as the correlation function $C(\ell)$, structure factor $S(q)$ and the $\omega = 0$ magnetic susceptibility $\chi(q)$ provide insight into the magnetic correlations, various pair-field functions provide information on the superconducting correlations in the 2D Hubbard model. In close analogy to the magnetic case, one can examine the equal-time spatial correlations of the pair-field

$$\bar{D}(\ell) = \langle \Delta_{\ell+i,i} \Delta_\ell^\dagger \rangle, \tag{52}$$

its corresponding structure factor

$$D(q) = \sum_\ell e^{iq\cdot\ell} \bar{D}(\ell) \tag{53}$$

or the static $q = 0$ pair-field susceptibility

$$P = \int_0^\beta d\tau \langle \Delta(\tau) \Delta^\dagger(0) \rangle. \tag{54}$$

Here, the simplest pair-field s-wave operator is

$$\Delta_\ell = c_{\ell\uparrow} c_{\ell\downarrow} \tag{55}$$

and

$$\Delta = \frac{1}{\sqrt{N}} \sum_\ell \Delta_\ell. \tag{56}$$

More generally, we will be interested in various pair fields having different rotational symmetries. For example, the s^* and d-wave pair fields are given by

$$\Delta_{s^*\ell} = \tfrac{1}{2}\left(c_{\ell\uparrow} c_{\ell+\hat{x}\downarrow} - c_{\ell\downarrow} c_{\ell+\hat{x}\uparrow}\right) + \tfrac{1}{2}\left(c_{\ell\uparrow} c_{\ell+\hat{y}\downarrow} - c_{\ell\downarrow} c_{\ell+\hat{y}\uparrow}\right) \tag{57}$$

and

$$\Delta_{d\ell} = \tfrac{1}{2}\left(c_{\ell\uparrow} c_{\ell+\hat{x}\downarrow} - c_{\ell\downarrow} c_{\ell+\hat{x}\uparrow}\right) - \tfrac{1}{2}\left(c_{\ell\uparrow} c_{\ell+\hat{y}\downarrow} - c_{\ell\downarrow} c_{\ell+\hat{y}\uparrow}\right). \tag{58}$$

In momentum space these correspond to the familiar forms

$$\Delta_{s^*} = \frac{1}{\sqrt{N}} \sum_\ell \Delta_{s^*\ell} = \frac{1}{\sqrt{N}} \sum_p (\cos p_x + \cos p_y) c_{p\uparrow} c_{-p\downarrow} \tag{59}$$

and

$$\Delta_d = \frac{1}{\sqrt{N}} \sum_p (\cos p_x - \cos p_y) c_{p\uparrow} c_{-p\downarrow}, \tag{60}$$

respectively.

Before looking for superconductivity in the repulsive Hubbard model, it is worthwhile to see how superconductivity appears in the *attractive* "negative-U" Hubbard model.[29] At half-filling the repulsive Hubbard model is mapped to the negative-U Hubbard model with a canonical transformation in which $c_{\ell\uparrow} \to c_{\ell\uparrow}$ and $c_{\ell\downarrow} \to (-1)^{\ell_x+\ell_y} c_{\ell\downarrow}^\dagger$. Under this transformation, the zz component of the $q = (\pi,\pi)$ magnetic structure factor goes over to the $q = (\pi,\pi)$ charge-density-wave structure factor, and the transverse (x,y) structure factors go over to the s-wave, $q = 0$, pair-field structure factor. Thus the existence of long-range antiferromagnetic order in the ground state of the half-filled repulsive Hubbard model implies that the ground state of the negative-U Hubbard model has both charge-density-wave and superconducting long-range order. Away from half-filling, superconductivity is favored and the 2D system is expected to undergo a Kosterlitz-Thouless transition at a finite temperture. Figure 25 shows a schematic phase diagram for the 2D negative-U Hubbard model.

We have carried out Monte Carlo simulations of the CDW and pair-field structure factors for $U = -4$ and band fillings of $\langle n \rangle = 1$ and $\langle n \rangle = 0.5$. Results for different sized lattices versus β for $\langle n \rangle = 0.5$ are shown in Fig. 26. For $\langle n \rangle = 1.0$, both $D(0)$ and CDW(π,π) behave as $S(\pi,\pi)$ shown in Fig. 15. The infinite lattice extrapolations are shown in Fig. 27. Here one clearly sees that at half-filling, $\langle n \rangle = 1.0$, both the CDW and pair-field order parameters are finite. However, at one-quarter filling, $\langle n \rangle = 0.5$, only the pair field extrapolates to give a finite order parameter. At quarter-filling, the CDW correlations are short range, and while the peak in the CDW structure factor is displaced from (π,π), the scaling behavior of the peak is the same as that of CDW(π,π).

Returning to the repulsive U Hubbard model, we have calculated the $\vec{q} = (0,0)$ equal-time pair-field structure factor for $\langle n \rangle = 0.875$ and $U = 4$ using s^* and d-wave pair-field operators. The results are plotted in Fig. 28. Comparing these results with Fig. 26, we conclude that the d-wave and extended s-wave correlations for the repulsive U Hubbard model do not grow with the lattice size at the temperatures that can be reached.[8]

A means of further testing the nature of the interaction between quasi-particles in the 2D Hubbard model is to compare the correlated and uncorrelated pair-field susceptibilities.[9] By the correlated susceptibility we mean the full pair-field susceptibility

$$P_\alpha = \int_0^\beta d\tau \langle \Delta_\alpha(\tau) \Delta_\alpha^\dagger(0) \rangle \tag{61}$$

determined by averaging the product of the Green's functions as in Eq. (18). The uncorrelated susceptibility $\overline{P}_\alpha$ is obtained in the same Monte Carlo run by taking the product of the average one-particle Green's functions, Eq. (19). If P_α is smaller than $\overline{P}_\alpha$, then the interaction between the dressed quasi-particles suppresses the pairing, while if P_α is larger than $\overline{P}_\alpha$, there is an attractive interaction in the

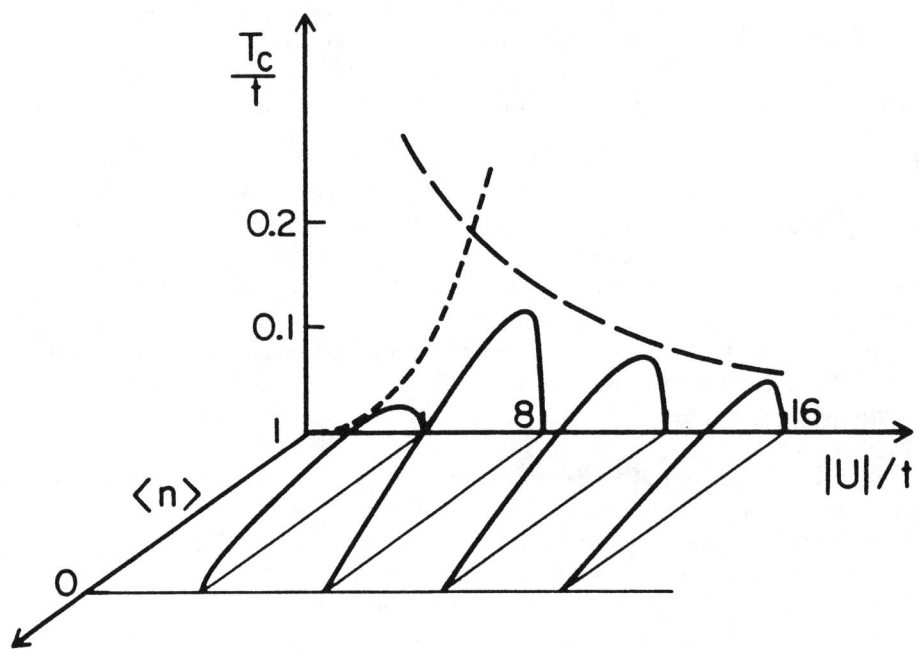

FIGURE 25 Schematic T_c versus $|U|/t$ and $\langle n \rangle$ for the 2D negative-U Hubbard model. The short dashed line is the weak coupling BCS preduction and the long dashed curve is the strong-coupling energy $t^2/|U|$ divided by t. (From Ref. 29.)

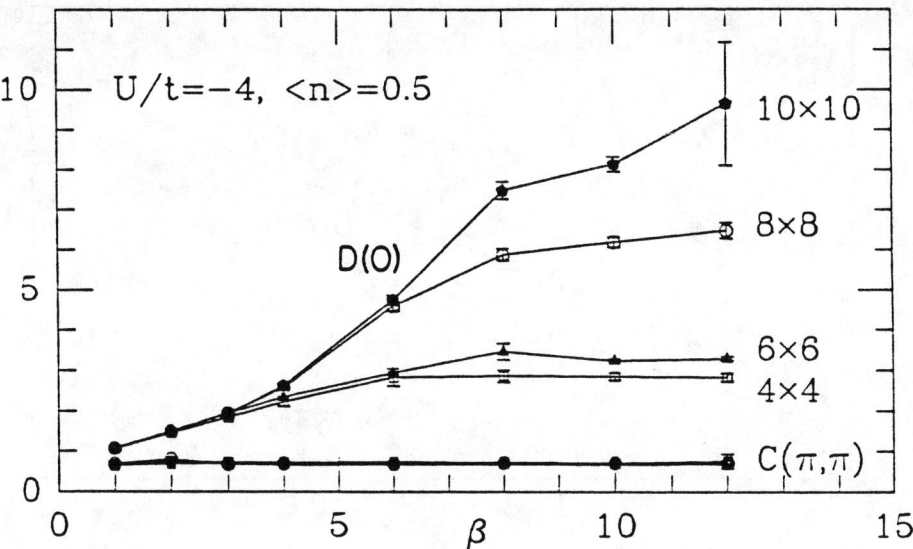

FIGURE 26 The structure factors $D(0)$ and $\text{CDW}(\pi,\pi)$ versus β for different sized lattices with $U = -4$ and $\langle n \rangle = 0.5$. (From Ref. 29.)

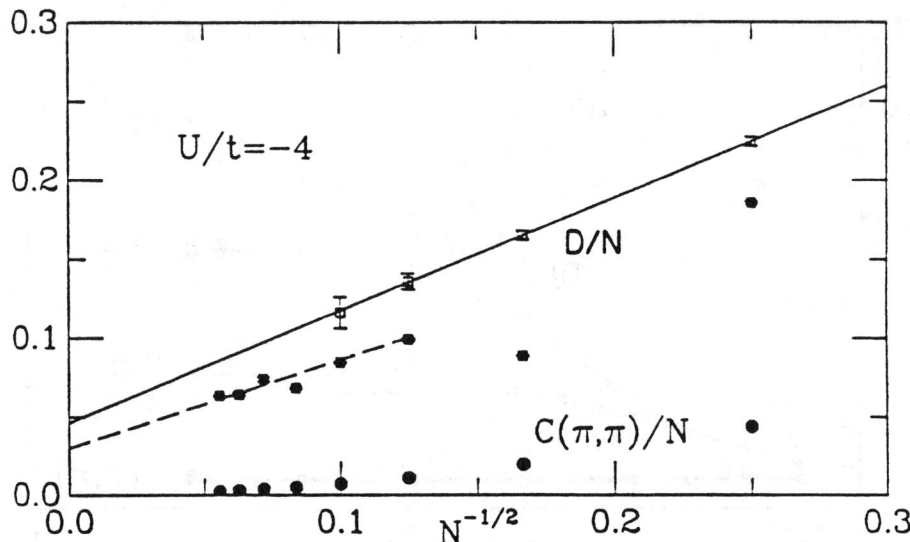

FIGURE 27 Zero-temperature extrapolations for $D(0)/N$ and $CDW(\pi,\pi)/N$ versus $N^{-1/2}$ for $\langle n \rangle = 1.0$ (□), where D and CDW are equal; and $\langle n \rangle = 0.5$ with (■) for D and (●) for C. Here $U = -4$. (From Ref. 29.)

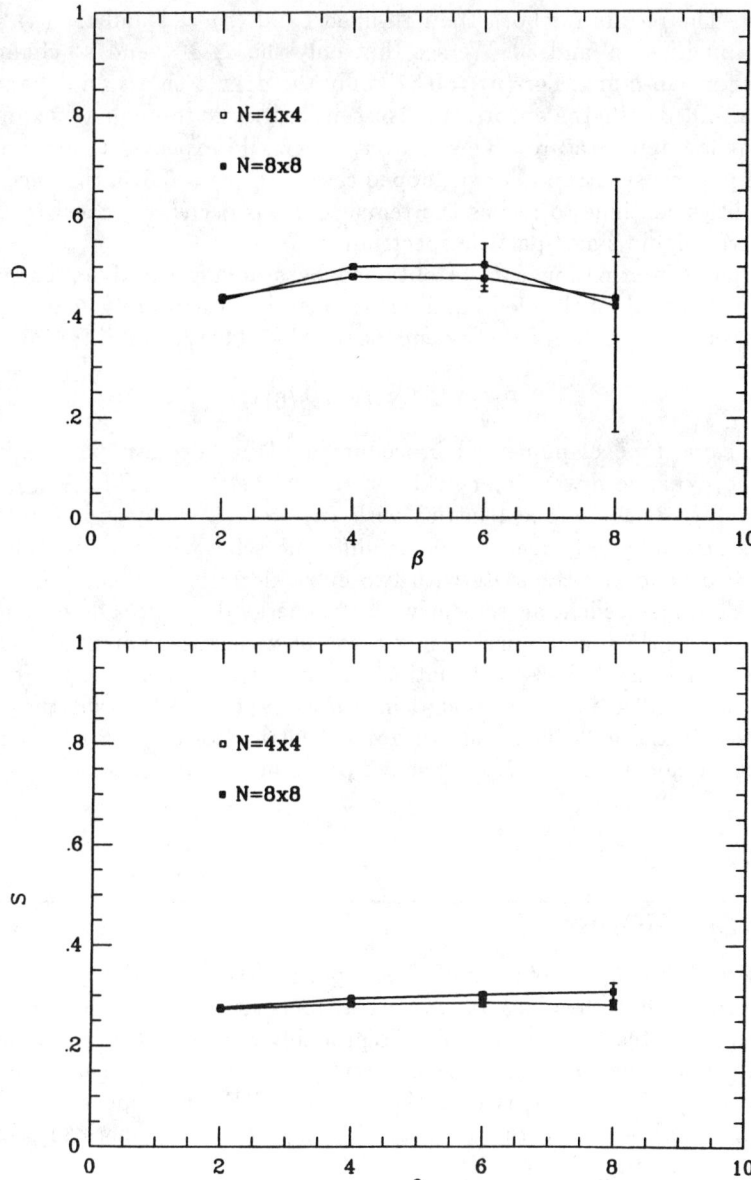

FIGURE 28 (a) The d-wave equal-time correlation function $D_d(0) = \langle \Delta_d \Delta_d^+ \rangle$, and (b) the s-wave equal-time correlation function $D_s(0) = \langle \Delta_s \Delta_s^+ \rangle$ versus β for 4×4 and 8×8 lattices. Note the absence of growth with lattice size.

α channel. The results for both the half-filled band $\langle n \rangle = 1$ and for $\langle n \rangle = 0.875$ are shown in Figs. 29 and 30. We see that only the $d_{x^2-y^2}$ and s^* channels are attractive for half-filling. For $\langle n \rangle = 0.875$ only the $d_{x^2-y^2}$ shows an enhancement. For the half-filled case, the uncorrelated susceptibilities go through a maximum and saturate at low temperatures. This is what one would expect if there is a gap in the single-particle spectrum. For the doped case with $\langle n \rangle = 0.875$, the uncorrelated susceptibilities continue to rise as the temperature is decreased, consistent with a gapless Fermi liquid single-particle spectrum.

Additional information can be obtained by studying the dynamic pair-field susceptibility. For the half-filled band where there is no sign problem it is possible to obtain very good data for the τ-dependent pair-field susceptibility

$$P_\alpha(\tau) = \langle \Delta_\alpha(\tau) \Delta_\alpha^\dagger(0) \rangle. \tag{62}$$

With the same type of numerical procedure used in calculating $N(\omega)$, one can extract the dynamic d-wave pair-field spectral weight[30] $\frac{1}{\pi} \text{Im} P_d(\omega + i\delta)$. This is plotted in Fig. 31 for a 4×4 lattice with $\langle n \rangle = 1$, $U = 4$, and $\beta = 12$. The two peaks near $\omega = \pm 2$ give the energy differences between the ground state of the half-filled band and the state with two extra electrons or holes. These energy differences are in excellent agreement with the energy differences determined from recent Monte Carlo simulations[31] using a complex chemical potential and recent Lanczos calculations.[20] These calculations show that on a 4×4 lattice, two holes added to a half-filled band are bound in a d-wave state. Note that the addition of two holes to the half-filled band of a 4×4 lattice corresponds to a doping of $1/8 = .125$. A calculation for $P_{s^*}(\omega)$ shows peaks near $\omega = \pm 6$.

IV. CONCLUSIONS

Numerical simulations of the half-filled repulsive 2D Hubbard model have shown that the ground state has long-range antiferromagnetic order for a wide range of interaction strengths U. The critical U is probably zero. The electron momentum distribution and compressibility are consistent with there being a gap in the single-particle density of states. This is further supported by the appearance of a pole in $\Sigma(k, i\omega_n)$ as $\omega_n \to 0$ and the observation of a gap in the numerically continued single-particle density of states $N(\omega)$.

As the system is doped away from half-filling, antiferromagnetism is rapidly suppressed. For moderate coupling $U = 4$, the peak in the magnetic structure factor which is at (π, π) for the half-filled band appears to split into two peaks which shift out along the $(\pi, \pi-\Delta q)$ and $(\pi-\Delta q, \pi)$ edges and become significantly weaker. Unfortunately, in this interesting range of doping $.7 < \langle n \rangle < 0.9$, fluctuations in the sign of the fermion determinant limit the temperature and lattice sizes that can be simulated. Over the temperature range presently accesssible, of order $1/50$

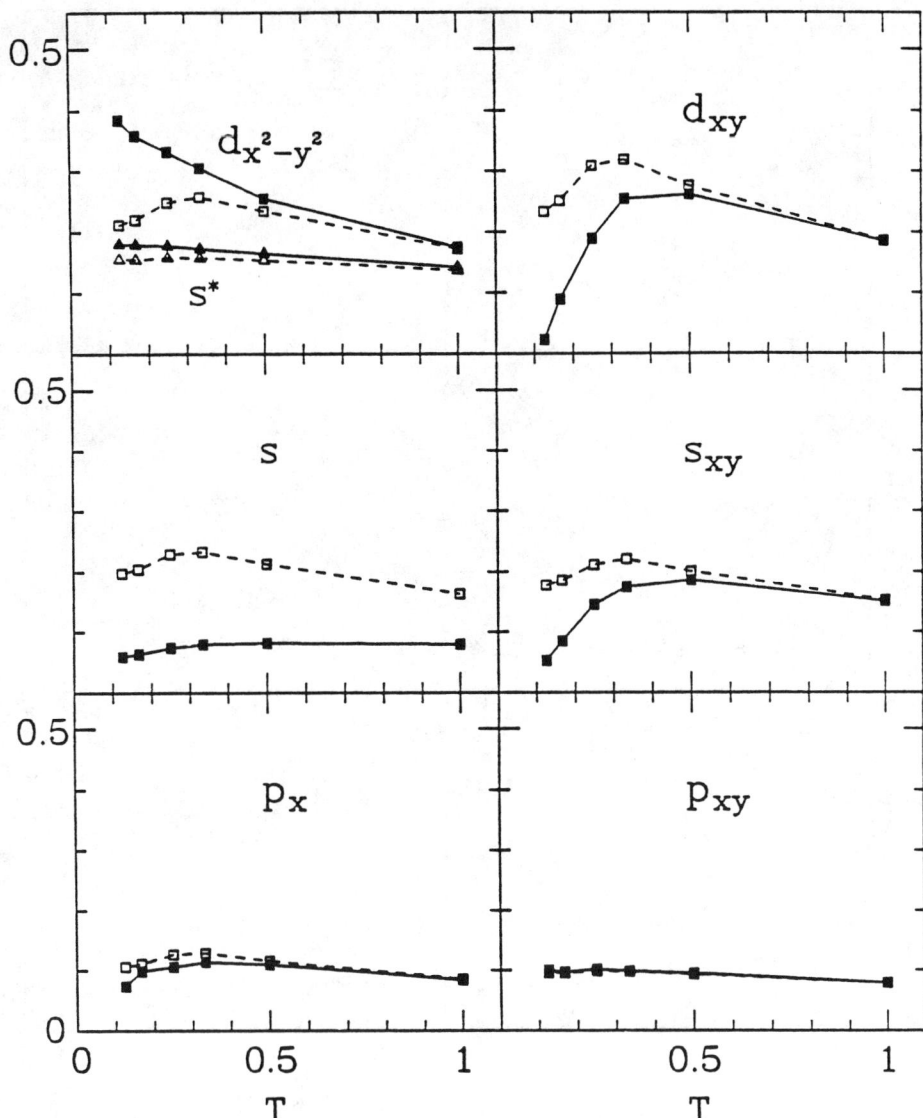

FIGURE 29 The pair-field susceptibility P and the uncorrelated pair-field susceptibility $\overline{P}$ at half-filling ($\langle n \rangle = 1$) are plotted versus T for various pair-field modes. Only the $d_{x^2-y^2}$ and s^* show an enhancement due to the interaction. The other

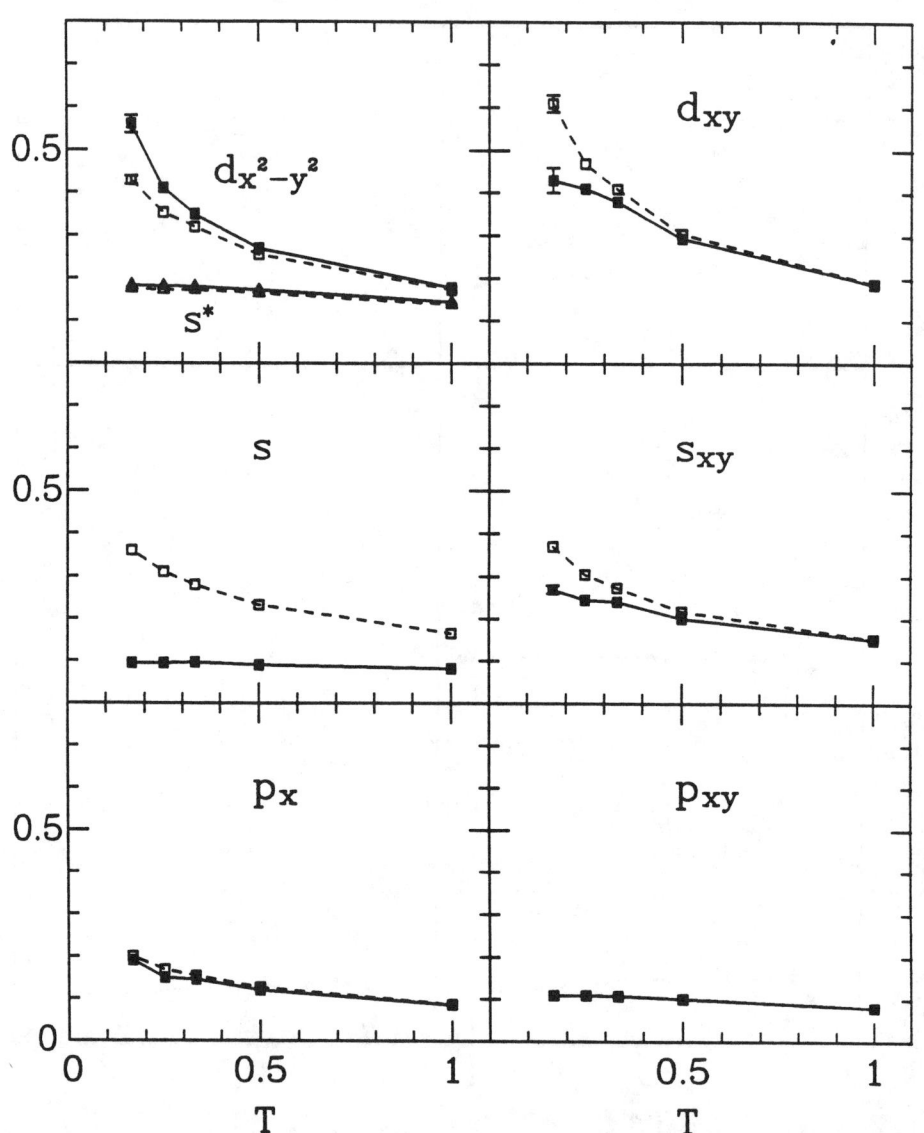

FIGURE 30 The pair-field susceptibilities P and $\overline{P}$ versus T for $\langle n \rangle = 0.875$. (From Ref. 9.)

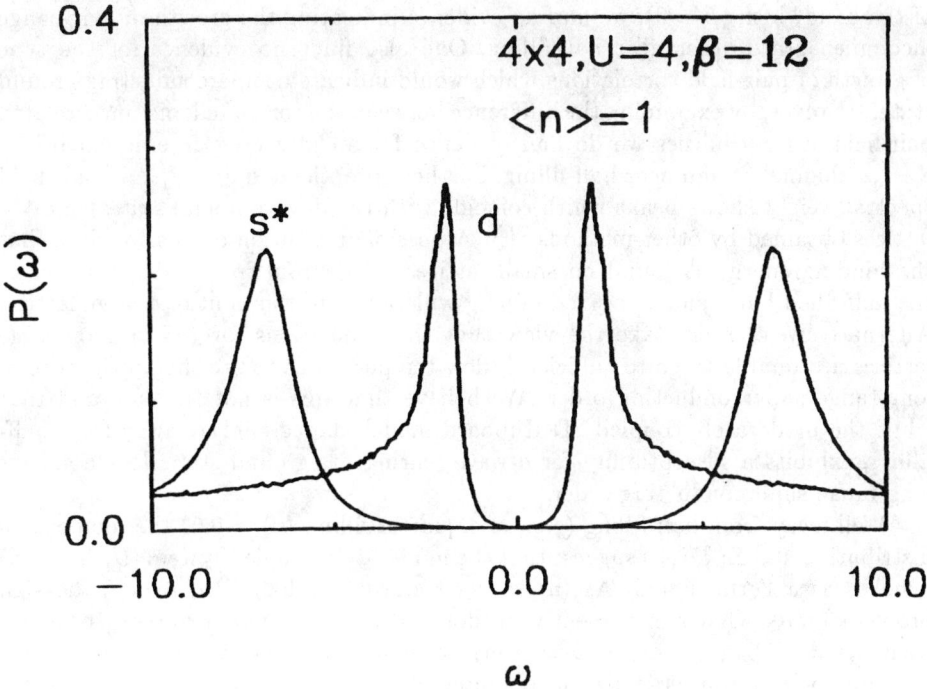

FIGURE 31 The d-wave and s^*-wave pair-field spectral weight $\mathrm{Im}\, P_\alpha(\omega+i\delta)/\pi$ versus ω. (From Ref. 30.)

of the bandwidth ($\beta = 6$), we find no evidence indicating the growth of long-range incommensurate order near half-filling. One also finds no evidence for the type of growth of pair-field correlations which would indicate a superconducting ground state. However, by examining the difference between the correlated and uncorrelated pair-field susceptibilities we do find evidence for an attractive interaction in the $d_{x^2-y^2}$ channel at and near half-filling. Furthermore, the dynamic $d_{x^2-y^2}$ pair-field spectral weight shows peaks which coincide with the d-wave bound states on 4×4 lattices obtained by other methods.[20,31] A possible resolution of this could be that the binding energy Δ found on small lattices arises from special degeneracies of the half-filled band and at a fixed doping will vanish in the limit of a large lattice. Alternatively, one can take the view that the simulations for $\langle n \rangle \neq 1$ on large lattices are unable to go to sufficiently low temperatures to see the appearance of long-range superconducting order. We believe that this is not the case and that while the moderately coupled 2D Hubbard model, doped slightly away from half-filling, exhibits a susceptibility for d-wave pairing, its ground state does not have long-range superconducting order.

Well away from half-filling (e.g., one-quarter-filling $\langle n \rangle = 0.5$) the momentum distribution and $\Sigma(k, i\omega_n)$ suggest that the moderately coupled system ($U/8t = 0.5$) behaves as a Fermi liquid. As $\langle n \rangle$ moves towards the half-filled region, the sign problem limits what can presently be done. However, runs on a 16×16 lattice with $\langle n \rangle = .875$, $U = 4$ and $\beta = 6$ give a locus of points for $\langle n_k \rangle = 0.5$ which appears to be quite close to the noninteracting Fermi surface, Fig. 32. We are presently continuing to analyze $\Sigma(i, i\omega_n)$ on this locus of points to see if there is any indication of a pseudo gap[28] or anomalous behavior of $Z(k, i\omega_n)$.

Thus, while the magnetic behavior of the 2D Hubbard model is in qualitative agreement with the observed properties of the cuprate materials, the relationship of the superconducting properties remain uncertain. The dominant pairing response of the 2D Hubbard model near half-filling appears in the $d_{x^2-y^2}$ channel where there is an attractive pairing interaction. Nevertheless, over the present accessible temperature range there is no indication of long-range superconducting correlations. Perhaps what has been found means that the one-band 2D Hubbard model is close to having a superconducting phase but that some additional feature is required to tip the balance. For example, an analysis[32] of the La_2CuO_4 bandstructure suggests that if the electronic structure near the Fermi level is mapped to a one-band Hubbard model, it is important to include a next-near-neighbor one-electron transfer t'. Prior to high T_c, such a term was invoked to move the Van Hove singularity away from the perfect nesting filling and allow the pairing interactions to take advantage of the increased density of states without the competing nesting instability.[33] Recently, both RPA[34] and numerical simulations of a 2D Hubbard model[35] with t' included have shown that it can lead to an enhancement of the d-wave pairing. Naturally one can also look to CuO_2 three-band Hubbard models in which charge-transfer processes[36] play a role or to models in which the ratio of the exchange interaction J to the bandwidth can be larger than in the single-band Hubbard model.[37]

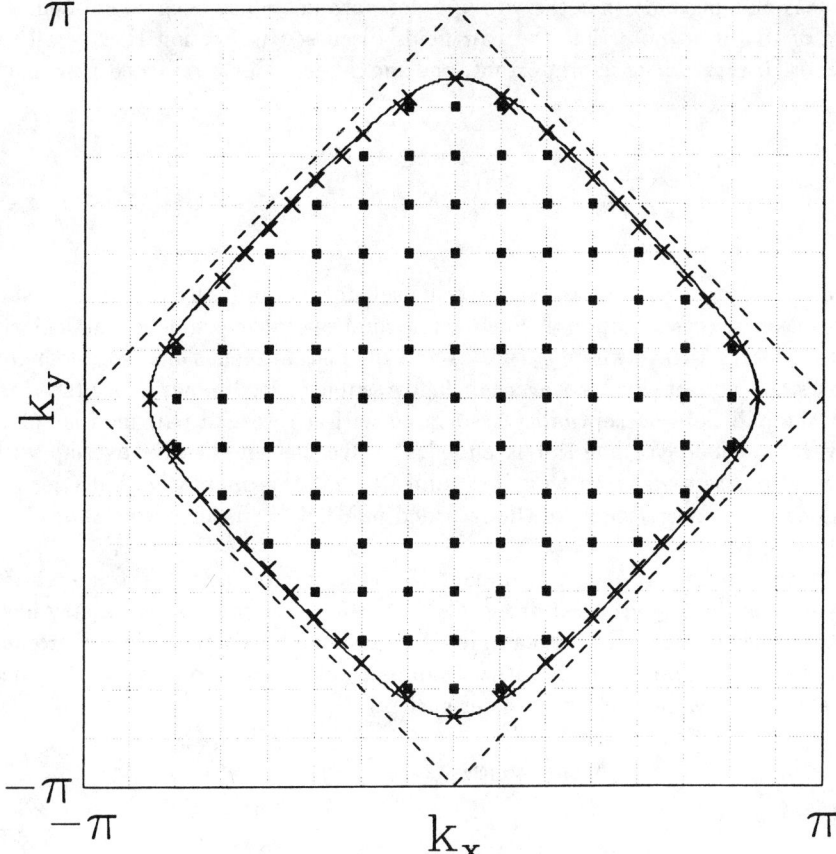

FIGURE 32 This shows the allowed k-points in the Brillouin zone for a 16 × 16 lattice. The solid squares mark k-points at which $\langle n_k \rangle > 0.5$ for a band filling of $\langle n \rangle = 0.87$ with $U/t = 4$ and $\beta = 6/t$. Extrapolating the Monte Carlo data for $\langle n_k \rangle$ on the 16-site lattice to the points in the Brillouin zone where $\langle n_k \rangle$ would equal 0.5 gives the crosses (×). The solid line is the non-interacting ($U = 0$) Fermi surface for $\langle n \rangle = 0.87$ and the dashed line is the non-interacting Fermi surface for the half-filled band $\langle n \rangle = 1.0$. (From Ref. 7.)

It is also possible that the wrong order parameter has been calculated. A mild form of this is to note that the pair fields discussed in Section II were all instantaneous. If retardation is important, one should consider a retarded pair field such as

$$\Delta_\alpha^{\rm ret}(\tau) = \sum_p g_\alpha(p) T \left[c_{-p\downarrow}(\tau) c_{p\uparrow}(\tau - \tau_0) - c_{-p\uparrow}(\tau) c_{p\downarrow}(\tau - \tau_0) \right]. \tag{63}$$

Here τ_0^{-1} corresponds to an energy scale set, for example, by the peak in the spin fluctuation spectral response. Such a retarded operator could be particularly important for an s-wave where $g_\alpha(p) = 1$ and the instantaneous onsite U strongly suppresses the nonretarded s-wave pair-field response. Preliminary results[38] indicate that the pair-field susceptibility associated with a retarded pair field is enhanced. However, we believe that the extended s^*-wave has an effective overlap with the retarded onsite s-wave, so that it is unlikely that the qualitative behavior seen in the long-range correlations of the retarded pair field will differ from that discussed in Section II.

A stronger form of the argument that the wrong order parameter has been measured is the anyon-based proposals[39–41] that time-reversal symmetry has been spontaneously broken. In this case, long-range-flux order signals the occurrence of a superconducting phase. Then rather than pair-field correlations one should measure correlations of a flux-field parameter such as

$$P_{1234} = \phi_{12}\phi_{23}\phi_{34}\phi_{41} - \phi_{14}\phi_{43}\phi_{32}\phi_{21} \tag{64}$$

with

$$\phi_{ij} = \sum_\sigma c_{i\sigma}^\dagger c_{j\sigma}. \tag{65}$$

Here the sites form a plaquette. Initial Monte Carlo studies[42] of the flux-flux correlation functions for the doped Hubbard model with $U = 4$ find that such correlations are only short range. Again, it may be that a J/t parameter range which is greater than can be achieved in the one-band Hubbard model is required for a long-range-flux phase. However, for large J/t where the doped Hubbard model can be mapped onto a frustrated Heisenberg model,[43] recent Lanczos studies[44] find that the susceptibility associated with the order parameter $\langle (\bar{s}_i \cdot \bar{s}_j) \times s_k \rangle$ fails to show interesting structure as a function of frustration (doping). Such structure in this susceptibility would have signaled the possibility that the ground state of the infinite lattice breaks parity and time-reversal symmetry. From the work of Ref. 44, it appears that the gap between the subspaces of states which are even and odd under reflections (parity) is much larger than other gaps in the theory.

Finally, there is the nagging possibility, brought on by the failure of these simulations to see a clear indication of superconductivity, that the Hubbard model rather than missing some small additional feature which tips the balance, *does not contain the basic mechanism for high-temperature superconductivity.* Could electronic charge distortions[45] or even lattice distortions mixed with onsite Coulomb effects play a critical role? Perhaps the antiferromagnetic spin fluctuations so characteristic of the Hubbard model actually act to suppress pairing which arises from some essential terms not included in the Hubbard model. Simulations of the three-band Hubbard model and various Peierls-Hubbard models are being carried out to explore further aspects of the pairing correlations near a metal-insulator transition.

ACKNOWLEDGMENTS

Simulations provide a unique opportunity for theorists to do their own experiments. The work reported here represents the efforts of a number of faculty, postdoctoral, and graduate student colleagues whom I have been privileged to interact with and learn from: R. Blankenbecler, E. Dagotto, J. Gubernatis, J. Hirsch, M. Imada, E. Loh, A. Moreo, R. Scalettar, R. Sugar, D. Toussaint, and S.R. White. I would also like to thank V. Emery, A. Millis, and J.R. Schrieffer for useful discussions about the Hubbard model.

This work was supported in part by the Department of Energy under grant DE-FG03-88ER45197 and by the National Science Foundation under grant DMR86-15454. The numerical calculations reported in this paper were performed at the San Diego Supercomputer Center and at NMFECC. I thank both centers for their support. I would also like to thank the IBM Research Division and the Almaden Research Center for their hospitality and support.

REFERENCES

1. G. Sugiyama and S. E. Koonin, Ann. Phys., **168**, 1, 1986.
2. S. Sorella, S. Baroni, R. Car, and M. Parrinello, Europhys. Lett. **8**, 663, 1989; S. Sorella, E. Tosatti, S. Baroni, R. Car, and M. Parrinello, Int. J. Mod. Phys., **B1**, 993, 1988.
3. S. R. White, D. J. Scalapino, R. L. Sugar, E. Y. Loh, Jr., J.E. Gubernatis, and R.T. Scalettar, Phys. Rev. B **40**, 506, 1989.
4. S. R. White, D. J. Scalapino, R. L. Sugar, and N. E. Bickers, Phys. Rev. Lett. **63**, 1523, 1989.
5. E. Y. Loh, Jr., J. E. Gubernatis, R. T. Scalettar, S. R. White, D.J. Scalapino, and R.L. Sugar, preprint UCSBTH-89-21, submitted for publication.
6. J. E. Hirsch and S. Tang, Phys. Rev. Lett. **62**, 591, 1989.
7. A. Moreo, D. J. Scalapino, R. L. Sugar, S. R. White, and N. E. Bickers, preprint UCSBTH-89-28, to be published in Phys. Rev. B.
8. M. Imada and Y. Hatsugai, J. Phys. Soc. Japan, **58**, No. 10, 1989.
9. S. R. White, D. J. Scalapino, R. L. Sugar, N. E. Bickers, and R. T. Scalettar, Phys. Rev. B **39**, 839, 1989.
10. J. E. Hirsch and H. Q. Lin, Phys. Rev. B **37**, 5070, 1988; J. E. Hirsch, E. Loh, D. J. Scalapino, and S. Tang, Physica, C **153–155**, 549, 1988. In these papers,

the neglect of negative signs in the calculation of the low-temperature pair-field susceptibility leads to a spurious decrease in the d-wave pair-field susceptibility at low temperatures. Nevertheless, as we will discuss, the conclusion that there is no evidence from simulations for long-range superconducting order remains.

11. R. L. Stratonovitch, Dokl. Akad. Nauk SSSR **115**, 1097, 1957; J. Hubbard, Phys. Rev. Lett. **3**, 77, 1959.
12. S. Q. Wang, W. E. Evanson, and J. R. Schrieffer, Phys. Rev. Lett. **23**, 92, 1969.
13. T. Moriya and Y. Takahashi, J. Phys. Soc. Japan **45**, 397, 1978.
14. *Electron Correlations and Magnetism in Narrow-Band Systems*, Springer Series in Solid State Physics **29**, T. Moriya, ed. (Springer-Verlag, Berlin, 1981).
15. R. Blankenbecler, D. J. Scalapino, and R. L. Sugar, Phys.Rev. D **24**, 2278, 1981.
16. J. Hubbard, Proc. R. Soc. London, Ser. A, **276**, 238, 1963; **281**, 401, 1964.
17. H. F. Trotter, Proc. Amer. Math. Soc. **10**, 545, 1959; M. Suzuki, Phys. Lett. A **113**, 199, 1985; R. Fye, Phys. Rev. B **33**, 6271, 1986.
18. J. E. Hirsch, Phys. Rev. B **28**, 4059, 1983.
19. N. Metropolis, A. Rosenbluth, M. Rosenbluth, A. Teller, and E. Teller, J. Chem. Phys.**21**, 1087, 1953.
20. A. Parola, S. Sorella, S. Baroni, R. Car, M. Parinello, and E. Tosatti, preprint, ICTP, Trieste.
21. For other reviews see, for example, D. J. Scalapino, R. L. Sugar, S. R. White, N. E. Bickers, and R. T. Scalettar, Physica Scripta **T27**, 101, 1989; M. Imada, *Proc. Yamada Conf. Strongly Coupled Plasma Physics* (Elsevier, Amsterdam, 1989). In addition, see J. E. Hirsch, Phys. Rev. B31, 4408 (1985) and Journ. Stat. Phys. **43**, 841 (1986).
22. M. C. Gutzwiller, Phys. Rev. **137**, A1726, 1965; for a review see D. Vollhardt, Rev. Mod. Phys. **56**, 99, 1984.
23. J. D. Reger and A. P. Young, Phys. Rev. B **37**, 5978, 1988.
24. D. Baeriswyl, C. Gross, and T.M. Rice, Phys. Rev. B **35**, 8391, 1986.
25. D. A. Huse, Phys. Rev. B **37**, 2380, 1988.
26. H. J. Schulz, to be published.
27. N. E. Bickers, D. J. Scalapino, and S. R. White, Phys. Rev. Lett. **62**, 961, 1989.
28. J. R. Schrieffer, X. G. Wen, and S. C. Zhang, Phys. Rev. B **39**, 11663, 1989.
29. R. T. Scalettar, E. Y. Loh, J. E. Gubernatis, A. Moreo, S. R. White, D. J. Scalapino, R. L. Sugar, and E. Dagotto, **62**, 1407, 1989.
30. S. R. White, to be published.
31. E. Dagotto, A. Moreo, R.L. Sugar, and D. Toussaint, preprint NSF–ITP–89–137; to be published in Phys. Rev. B.
32. M. S. Hybertsen, E. B. Stechel, M. Schluter, and D.R. Jennison, preprint.
33. J. E. Hirsch and D. J. Scalapino, Phys. Rev. Lett. **56**, 2732, 1986.
34. K. Saitoh and S. Takada, J. Phys. Soc. Japan **58**, 783, 1989.
35. R. R. dos Santos, Phys. Rev. B **39**, 7259, 1989.
36. C. M. Varma, S. Schmitt-Rink, and E. Abrahams, Solid St. Comm. **62**, 681, 1987.
37. V. J. Emery, Phys. Rev. Lett. **58**, 3759, 1987; V. J. Emery and G. Reiter, Phys. Rev. B **38**, 4547, 1988.
38. R. Noack, private communication.

39. V. Kalmeyer and R. B. Laughlin, Phys. Rev. Lett. **59**, 2095, 1987.
40. W. G. Wen, F. Wilczek, and A. Zee, Phys. Rev. B **39**, 11413, 1989.
41. P. Lederer, D. Poilblanc, and T. M. Rice, preprint.
42. M. Imada, J. Phys. Soc. Japan **58**, 2650, 1989.
43. M. Inui, S. Doniach, and M. Sabay, Phys. Rev. B **38**, 6631, 1988.
44. E. Dagotto and A. Moreo, Phys. Rev. Lett. **63**, 2148, 1989.
45. J. E. Hirsch, Physica C **158**, 326, 1989.

DISCUSSION

S. Chakravarty: I would like to make a point here. You've pointed out that at (π,π) this doesn't go to zero, and could be a finite temperature effect. If you really think about it a little bit, [what] you are really using is a Feynman-type approximation, to saturate the sum-rule and this doesn't have damping. However, at finite temperature it should have damping, damping due to thermal excitations, and because you are not allowing that, by the sum-rule, this is actually producing a gap in the spectrum. That's the reason why it's not going down to zero. You are right that it is a finite temperature effect, but I think that if you go to zero temperature the q equal to (π,π) spin waves are very well defined, because of the negative curvature of the magnons – that it will go down to zero, and it's completely correct.

The second question I have for you is – I'm sure you've thought about it a lot more than I have – about the sign problem. And sometimes I wonder whether it's not telling us something. In other words, if I have to run the Monte Carlo 20 billion times, or 200 billion times, before the sign settles down is there a physical reason – is there some understanding of that? For example, there are places where the sign problem is not there due to symmetry, or other reasons.

D. J. Scalapino: We've been thinking about the "sign" problem and wondered whether there might be some physical information contained in it. However, we don't understand how to associate a physical observable with it. One of the difficulties is that if you make a different choice for the Hubbard-Stratonovich decoupling, leaving the physics of course the same, the nature of the "sign" problem can change. Recently Scalettar has used a break-up for the Hubbard model in which a Hubbard-Stratonovich field Δ is coupled to $c_{i\uparrow}^\dagger c_{i\downarrow}^\dagger$. In this case there are sign problems even at half-filling where with the usual break-up, which I've discussed this morning, there is no sign problem at half-filling. So, evidently one can find Hubbard-Stratonovich decouplings that have worse sign problems than what we're using. Maybe we can find a better one. Perhaps it's like picking a gauge. Naturally it would be very interesting if the sign problem could be related to the existence of flux phases, chirality, or some quantity like frustration. However, at present I have no idea how to do this.

S. Doniach: Could I make a comment on the sign problem. If you look at the Heisenberg model, with J_2, and then you add the second neighbor...in the original problem Monte Carlo has no sign problem. And, you put that across the diagonal you start to get signs coming in, and in fact we know it's very destabilizing, and the reason for the signs is simple: you get these triangular orbits that are very similar to what Tony Zee was talking about yesterday, when you promote this splitting of the Fermi points, essentially, and you are stabilizing some kind of a flux.

D. J. Scalapino: What Seb is saying is certainly true. For example in the Hubbard model, when you put on a t' next-nearest-neighbor hopping, you break the particle-hole symmetry that we've been using to get rid of the sign problem, and you have a fierce sign problem.

S. Doniach: That may be telling you something about flux-type phases coming in.

E. Abrahams: You said that $\omega_p^{*2} \sim 1-n$. Is that true for all U, and in particular, what happens if U=0?

D. J. Scalapino: Yes, that's what I said, and it isn't fair, right? It's not fair in the following sense: I turned U on to four. When I have a U of four, and I have it half-filled, unfortunately ω_p^2 hasn't gone to zero, because it's just dying like t/U. In fact, it's much worse; its only gone down 20%. Then what we saw was the resurrection of it, the coming back of it. So that what Elihu is saying is right; I'm extrapolating to calculations we haven't done, to large U, where we could drive ω_p decently small, and then turning it on, arguing it would still rise in that way. And we don't know that, but I believe it.

A. Millis: Just briefly on this subject. First of all, there's a clever argument, shown to me by Dr. Susan Coppersmith, which will let you get the coefficient...at half-filling, for example, it falls off like 1/U in the large-U limit, and you can actually get the coefficient exactly in terms of the ground state energy of the Heisenberg model.

D. J. Scalapino: Yes. At large U/t values the optical spectral weight falls like t/U, and one can find the coefficient from $\langle \vec{s}_i \cdot \vec{s}_j \rangle$. In a paper, Moreo et al., we used the numerical results for $\langle \vec{s}_i \cdot \vec{s}_j \rangle$ obtained by Young and Reger. This gives the solid curve on the figure I showed of t_{eff}. The dashed line is the weak coupling perturbation result. You can see that the Monte Carlo provides a good interpolation between these two limiting results.

A. Millis: I wish to comment on the calculation of the compressibility. Very simply, if I take an insulator at half-filling, for example, the spin-density-wave insulator that I would get from the Hubbard model, you know the density of states looks something like this, and it has a gap. So if I imagine now computing the number of particles as a function of the chemical potential...it's clear that if the chemical potential is below the gap and I increase it, the number of particles will go up until my chemical potential gets to the edge of the gap. It will then stay constant for a finite range of chemical potential, and then it will start to increase again. So if I make a plot of n versus μ, what I get is two regions of finite slope connected by a flat region where n is independent of μ.

C. M. Varma: Excuse me, I don't already understand what you're saying. You're taking a fixed density of states, altering the number of particles...

A. Millis: The chemical potential.

C. M. Varma: ...but we know, that for a fixed U, that density of states is only true for a certain chemical potential or a certain filling. So you can't just...

A. Millis: I'm claiming that you can calculate this very straightforwardly for the Hubbard model in the small-U limit, and the same thing must apply in the large-U limit. So that what you would expect for n versus μ is the curve I previously described to get the compressibility, $dn/d\mu$, I'm supposed to differentiate the curve. Now if you look at what Doug has calculated for large U, I would claim that the region where $dn/d\mu = 0$ is the region where the chemical potential is in the gap, and the interesting question is the behavior as μ approaches the gap edge from below.

D. J. Scalapino: You had told me that before, and we just haven't done that. I hear what you're saying, and I have to think about it. But the problem I have with it somehow is, I had this impression, that somehow the half-filled system in the large-U limit would have zero compressibility.

T. M. Rice: It is true, it's flat, $dn/d\mu$ is flat.

D. J. Scalapino: Let me draw on this picture, because I think that the point that Andy has raised is probably true. At very low temperature, it is a question of what $\langle n \rangle$ versus μ looks like at the ends of the flat "gap" portion of this figure. If it comes in with zero slope there, then the compressibility vanishes. However, as you've said, it may come in with a finite slope. To study this we need to go to larger systems and lower temperatures. However, the point I would like to leave with you is the qualitative change U makes in $\langle n \rangle$ versus μ. When $U = 0$, the slope of $\langle n \rangle$ versus μ near $\mu = 0$ goes to infinity, while with $U \neq 0$ (even $U = 1$ or 2) one sees that the slope goes to zero as μ goes to zero.

A. Millis: No, but there's two things. One is the compressibility at n < 1, and the other one is, the region where n is independent of μ, where μ is in the gap. And all I'm saying is that it's necessary to distinguish clearly between those two regimes.

C. M. Varma: I think the question is not well-defined as matters stand, because as you go away from half-filling, you are going to change the state of matter at some n, and so you cannot just take a curve like that and say, you're going to talk about the compressibility at this point, or that point. It's not a well-defined question.

T. M. Rice: The $dn/d\mu$ is the compressibility, but I don't think there's a problem.

J. R. Schrieffer: I agree with Chandra that the density of states is a function of the filling, although for each filling there is a unique value of $dn/d\mu$.

C. M. Varma: How are you going to define at half-filling on that plot?

R. M. Martin: This is just a definition of a gap, and where the Fermi level is in the gap. If there truly exists a gap, then it has nothing to do with filling, it's just that you can move the Fermi energy in the gap and you don't change the filling. And then you start changing it, and in a two-dimensional system it should come in with a finite slope, since the density of states should be a step.

A. Millis: But you don't know whether interactions will do something exotic to that...

R. M. Martin: Well, but that's the simple explanation, which one should test.

A. Millis: Yes.

P. Wiegmann: Let me ask a question which may relate to this discussion. One would expect for considerable doping...considerable deviation from half-filled bands, considerable large U...one could expect some commensurate-incommensurate phenomena, I mean, of some commensurate state with a given lattice doping, something happening which is different from incommensurate doping, so I can expect some kind of Devil's staircase, maybe in that picture. ..I'd like to stress that maybe it is not a gap in the spectrum, but just a lot of singularities like a Devil's staircase.

D. J. Scalapino: About a year ago I had the pleasure of going to the Soviet Union and hearing Paul talk there about this notion of Mott insulating behavior at various fillings. The difficulty we have is the following. If I had known $\langle n \rangle$ versus μ on a larger μ scale, you would see a staircase. However, this is for a 4 × 4 lattice, and what one sees on this scale is complicated by the fact that there is a finite-level structure. We have to try to go to larger lattices and low temperatures. Perhaps we could look near one-third filling where the sign problem isn't so severe as just off of half-filling. We again haven't done that carefully enough, or at large enough U. And part of the other problem is, the simulations—I didn't speak of the limitations. Once we get up to a U that is of order the bandwidth, we find that we get stuck, the way we're running now, we get stuck in some parts of the phase space. And so, again, it may be that in just the interesting regimes you're speaking of, we're blind.

C. M. Varma: May I ask a question about a different thing. Have you calculated things like $\Sigma(k,\omega)$, or some properties of the spectral function, in the attractive Hubbard model?

D. J. Scalapino: We've only studied $\Sigma(k, i\omega_n)$ for the repulsive Hubbard model. We should do it for the attractive one as well.

C. M. Varma: I would like to suggest that the attractive-U Hubbard model in two dimension is not a garden-variety Fermi liquid, in the normal state above the superconducting transition temperature.

J. R. Schrieffer: I just wanted to return to these beautiful calculations of the pairing susceptibility. Doug and I have talked over some period of time why one might not see the pairing correlations in the $c^\dagger c^\dagger$ function, as opposed to a quasiparticle operator-quasiparticle operator function. The fact that when you replace the bare c's or the bare G's by the dressed G's, one does see the curve, instead of going down when you increase U, it actually goes up, and this curve is just the pair susceptibility.

Let me now turn to the question of what is the nature of the quasiparticles in the reasonably heavily-doped material, like lanthanum strontium copper oxide. I personally believe that, while the Fermi liquid theory is broken down, it's broken down in a way which is not simply a renormalization in the usual sense, with small singular terms. It might be on some new quasiparticles there are further singular terms along the line that Chandra is talking about, and I very much subscribe to that. If you take a more primitive point of view, and...for example, it's the one that I guess we've recently been looking at with Arno Kampf...and try to see how the spin fluctuations get rid of the usual quasiparticle peak, which is based on a $c^\dagger$, but actually produced new quasiparticles out of the one usual quasiparticle, and one spin wave (which is really $c^\dagger c^\dagger$ type term). And it seems to me that that is a quasiparticle which is generated out of operators which are not orthogonal, but have relatively weak overlap with the usual renormalization that we know and love from Abrikosov, Gorkov and Dsyalozhiski. If that is in fact the case, it may well be possible to calculate some new amplitudes, some new pairing susceptibilities, with guesses we have for these quasiparticles which would have large overlap on some model operators, compared to very small overlaps, and hence reduce the pair susceptibility below the Monte Carlo noise limit, and perhaps not see it.

An easy thing, I think, to do is not to look directly at the pair susceptibility, but first start out with the quasiparticles in the normal phase, and see if these guesses about what the operators, when put into Green's functions, give you reasonable overlaps on your Monte Carlo states. Only after that should one go on and try to pair these things. But first you should see what are the large-weight combinations of operators, such that these things really do look like quasiparticles in the normal state.

D. J. Scalapino: Bob, that is certainly something that can be done in principle by Monte Carlo. However, as one puts together a pair-field from quasi-particle operators that are themselves composite objects which contain products of the bare creation and destruction operators, the computation becomes quite involved. In addition, our experience is that the more Green's functions that you try to average, the noisier the calculation becomes. It's what's made it difficult, for example, to study chiral correlations.

D. Pines: Doug, we've been discussing at this meeting the extent to which charge and spin are decoupled, as one dopes and goes away from the half-filled phase. I wondered to what extent you have been able to look at two related questions: can one use your program to calculate the extent to which the spins are staying on a site, or moving away from it – that is, the degree of delocalization as one dopes. In some sense, this is connected with the formation of some kind of Fermi Liquid. And the second question, which might be easier, is: can you look at the long-wavelength susceptibility as you start to dope, and see whether it begins to go down, with decreased doping..that is, as we know in O_7, as the susceptibility is larger than it is in $O_{6.7}$, $O_{6.5}$ and so forth.

D. J. Scalapino: We can calculate spatial spin-spin correlations as well as (m_{zi}^2) on a given site. However to address delocalization, it would probably be better to show you data from calculations on a CuO_2 lattice. There we find that the moments tend to stay on the Cu sites as the system is doped. With respect to calculating the $q = 0$ magnetic susceptibility, I did show results for χ for the half-filled band. However, as one goes away from half-filling, the behavior of χ at low temperatures is sensitive to whether one has an even or odd number of electrons. On finite lattices, say

4 × 4 and 6 × 6, we found a low-temperature Curie Law behavior for an odd number of electrons, while χ went to zero for an even number. We found that to determine χ for different dopings we need to carry out a finite-sized scaling analysis, and we have not done this. Clearly there are a number of computations that we'd like to do, such as determine $\chi(q, i\omega_n)$ on 4 × 4, 8 × 8, and perhaps 12 × 12 lattices to compare with conserving approximations and RPA-like phenomenological forms. I believe that this is an important use of Monte Carlo, that is, to check various approximate schemes on finite lattices, and then if the approximation is in reasonable agreement, use it to go to the infinite lattice limit.

V. Emery: Doug, you probably answered this question a number of different ways on the way, but after listening to all of the discussion of the NMR data, and listening to your discussion, there is an obvious question to ask. Are you going to calculate the imaginary part of $\chi(q,\omega)$, and give us what the $1/T_1$ should be in the Hubbard model as a function of doping at the different sites?

D. J. Scalapino: The answer is yes, that's being run. But because we can only go up to, say, a 12 × 12 lattice, it's going to be difficult to extract the low-frequency response needed to determine T_1^{-1}. What we believe is possible is to compare the Monte Carlo results with RPA and conserving approximation calculations on 12 × 12 lattices to get a better feeling of how far we can trust the approximate expressions, which of course can be extended to larger lattices.

Spectral Functions from Quantum Monte Carlo
R. N. Silver
MS B262, Los Alamos National Laboratory, Los Alamos,
NM 87545

In his review, D. Scalapino identified two serious limitations on the application of Quantum Monte Carlo (QMC) methods to the models of interest in High T_c Superconductivity (HTS). One is the "sign problem." The other is the "analytic continuation problem," which is how to extract electron spectral functions from QMC calculations of the imaginary time Green's functions. Throughout this Symposium on HTS, the spectral functions have been the focus for the discussion of normal state properties including the applicability of band theory, Fermi liquid theory, marginal Fermi liquids, and novel non-perturbative states.

The imaginary time Green's function, $G(\tau)$, is related to the spectral function, $A(\omega)$, by

$$G(\tau) = \int_{-\infty}^{\infty} d\omega A(\omega) \frac{e^{-(\omega-\mu)\tau}}{1+e^{-\beta(\omega-\mu)}}$$

where τ extends from 0 to $\beta \equiv 1/k_B T$, and μ is the chemical potential. QMC produces incomplete data about $G(\tau)$ at a discrete set of $\{\tau_k\}$ with statistical error. The goal of analytic continuation is to extract $A(\omega)$ from such data. This is an extremely ill-posed problem similar to the inversion of a Laplace transform from noisy and incomplete data. That is, an infinity of $A(\omega)$ will fit the data according to a χ^2 measure, and small changes in $G(\tau)$ can lead to large changes in $A(\omega)$. Scalapino presented results for the Hubbard model in which analytic continuation was performed by a least squares fitting procedure[1,2] modified to enforce the positivity and smoothness of $A(\omega)$.

In this contribution, we identify analytic continuation as an image reconstruction problem.[3,4] The *image* sought is $A(\omega)$. To get this image, we apply the Maximum Entropy Method (ME), which is to choose the "best" $A(\omega)$ according to probability theory arguments, to provide estimates of the statistical reliability of the results, and to enforce prior knowledge about $A(\omega)$ such as sum rules, symmetry, high frequency behavior, and positivity.

ME is based on Bayes' Theorem on conditional probabilities, which in our case is

$$P[A(\omega)|Data] \propto P[Data|A(\omega)] \times P[A(\omega)]$$

Here, $P[Data|A(\omega)]$ is the conditional probability of the $G(\tau_k)$ data for a given $A(\omega)$ termed the *Likelihood* function, $P[A(\omega)]$ is the probability of $A(\omega)$ before the simulation termed the *Prior*, and $P[A(\omega)|Data]$ is the conditional probability of an $A(\omega)$ given the $G(\tau_k)$ data termed the *Posterior*. The image is obtained by maximizing the Posterior probability, and the statistical errors are obtained from the fluctuations of the Posterior about this maximum. Bayes' theorem quantifies common sense, i.e. the amount of information learned in an experiment depends on both the data and the prior knowledge. For Gaussian independent statistical errors, the Likelihood is given by $exp(-\chi^2/2)$. If the prior probability were ignored, then maximizing the Posterior is equivalent to the familiar fitting procedure of minimizing χ^2. However, we do have prior information, e.g. $A(\omega)$ is positive, so that $P[A(\omega)]$ should be zero for negative $A(\omega)$. More generally, a variety of consistent statistical inference arguments lead to the conclusion that the Prior should be $exp(\alpha S)$, where α is a self-consistent parameter and S is the Shannon-Jaynes information theory entropy relative to a default model

$$S \equiv \sum_i \Delta\omega \left[A_i - m_i - A_i \ln\left(\frac{A_i}{m_i}\right) \right].$$

Here A_i is the value of $A(\omega_i)$, m_i is the corresponding value of the default model, and $\Delta\omega$ is the spacing between ω_i. Maximizing the Posterior probability is equivalent to maximizing the entropy, S, subject to a constraint provided by the data, χ^2, with Langrange multiplier $1/\alpha$. In practice, this procedure is implemented using a third generation commercial code for ME image processing.

The figure shows a simulation of analytic continuation using ME. We see that the central peak is extremely accurately reproduced, that the sidepeaks are broadened but with correct integated intensities, and that one can also estimate errors on gaps in the spectrum. In general, via the Likelihood function one can show that the data contain the most information about features in $A(\omega)$ at small ω, and that the information content of the $G(\tau_k)$ data decreases with increasing ω.

Applying ME to real QMC data introduces important complications which have not been addressed in any of the prior approaches[1,2] to analytic continuation. Statistical control of the QMC calculation requires that the central limit theorem applies, so that the calculated $G(\tau)$ approach a normal distribution. The data should be binned to minimize correlations in Monte Carlo time due to the limited updates of the Hubbard-Stratonovich fields in each sweep. Also the QMC data are not statistically independent in τ-space: one must calculate the covariance matrix, C, of the statistical errors, and generalize the Likelihood function to $\chi^2 \to \delta\vec{G}^T \cdot \overleftrightarrow{C}^{-1} \delta\vec{G}$, where $(\delta\vec{G})_k$ is the difference between the calculated $G(\tau_k)$ data and the prediction for a given $A(\omega)$. In practice, this is implemented by a singular value decomposition of C which provides a unitary transformation to a data space in which the statistical errors are independent.

We have applied this approach to a QMC calculation of the symmetric single impurity Anderson model. The result is in excellent agreement with a calculation[5] of the spectral function by self-consistent second order perturbation theory at parameters where this should be valid. For parameters where this theory should break down, we see deviations from this prediction for the spectral function in the low ω region toward the universal Kondo behavior expected.

Work performed in collaboration with D. S. Sivia, J. E. Gubernatis, and M. Jarrell. Supported by the Department of Energy and the Ohio Supercomputing Center.

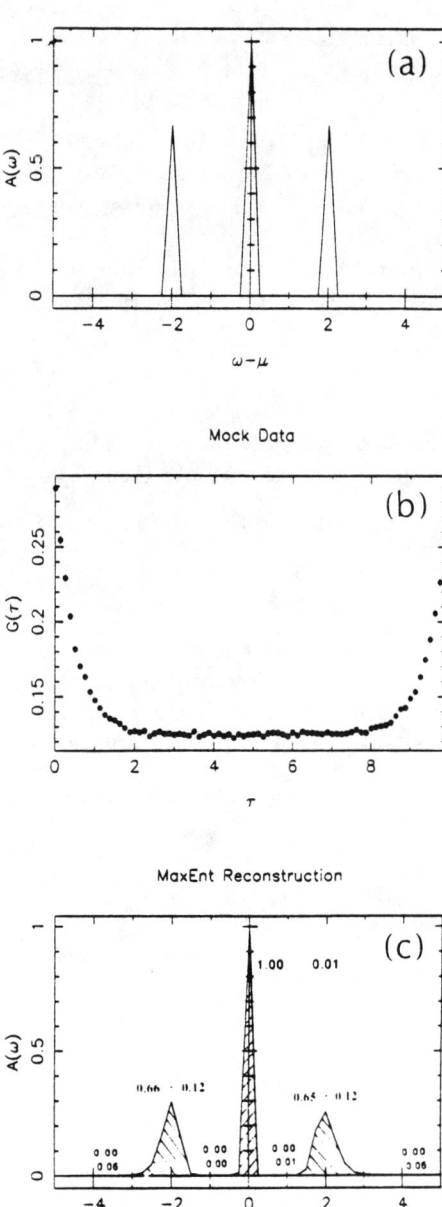

FIGURE 1 An example of the use of the Maximum Entropy Method (ME) for the analytic continuation of Quantum Monte Carlo data. Fig. (a) is a mock spectral function three δ - function peaks. Fig. (b) is the corresponding $G(\tau)$ data to which 1% Gaussian random noise has been added. Fig. (c) is the ME reconstruction of the spectral function from this data. The numbers are the integrated intensities and estimated statistical errors over the corresponding hatched regions.

REFERENCES

1. S. R. White, D. J. Scalapino, R. L. Sugar, N. E. Bickers, Phys. Rev. Lett. **63**, 1523 (1989).
2. See also, M. Jarrell, O. Biham, Phys. Rev. Lett. **63**, 2504 (1989).
3. R. N. Silver, D. S. Sivia, J. E. Gubernatis, Proceedings of *Quantum Simulations of Condensed Matter Phenomena*, Los Alamos, August, 1989, to be published by World Scientific Press.
4. R. N. Silver, D. S. Sivia, J. E. Gubernatis, Physical Review B1 (Feb, 1990).
5. B. Horvatic, V. Zlatic, Phys. Stat. Sol. (b) **99**, 251 (1980).

DISCUSSION

S. Trugman: Can you just enforce the symmetry in the maximum entropy procedure itself, instead of generating new Monte Carlo?

R. Silver: Of course, but naturally they didn't in the Monte Carlo simulation that he did.

S. Trugman: I know, but in principle you shouldn't need that.

R. Silver: No, actually it turns out that you reduce the covariance tremendously by doing that. Because if you have inherently asymmetric data, you have a strong component of that covariance matrix which is asymmetric. Whereas now, with the new data, which I've just taken a look at, it's absolutely zero asymmetry.

From Quantum-Chemical to Strong Coupling Models: Support for a Single-Band Model

E. B. Stechel
Division 1151, Sandia National Laboratories, Albuquerque, NM 87185

Several simplified, strong-coupling Hamiltonians have been suggested for the low-energy electronic structure of CuO_2 planes in Cu-O based superconductors.[1-3] This has generated a large literature concerning the properties of these models, typically emphasizing the phenomenology as parameters are varied. We have adopted[3] an opposite approach, deriving parameters which explicitly reflect the properties of real Cu-O materials, at the outset. Local density (LDA) results accurately predict the cohesive properties of these materials[4] and the LDA energy bands certainly show the central role of CuO_2-$pd\sigma$ derived bands near the Fermi level.[5] This strongly motivates a three-band extended Hubbard Hamiltonian to model the low energy spectrum.

The derivation of the three-band Hubbard model using a constrained LDA has been fully described.[6] The Coulomb interaction parameters follow from the calculated energy for local charge fluctuations and include screening appropriate for the solid. The bare one-electron parameters are derived from the LDA bands by mapping to Hartree-Fock (mean-field) bands of the three-band Hubbard model. This is not an exact procedure due to the local exchange-correlation potential in the LDA. However, this is well-known to give results intermediate between a Hartree and a Hartree-Fock potential. Hence, the derived bare onsite energy difference is a *lower bound*. Finally, the parameter set derived for the three-band Hubbard model is internally consistent and agrees with experimental data on a high energy scale.[7]

The three-band Hubbard, while already a simplified model for the CuO_2 sheets, still has a very large Hilbert space and thus can be solved exactly only on very small clusters. In addition, its Hilbert space spans a large energy range. Thus it is both desirable and conceivable to further reduce it by focusing on low energy states. For example, the Cu_5O_{16} cluster with six holes has ~1.8 million states; the low energy region (lowest 2 eV) has 30.[7] We seek a model Hamiltonian to describe, just those 30 states. Our procedure is to map from results of exact diagonalization on clusters to effective one band models. We stress that the required interaction parameters are *local*, as is well-known, for the Heisenberg J. Size consistency reflects this local nature: J computed from clusters with N Cu sites (Cu_N, N=2-5) agree to a few percent.[7,8] In addition, for Cu_N(N>2) J is overdetermined; i.e., Cu_4 is small enough for exact three-band Hubbard calculations and large enough that the range of interactions is less than the cluster size. Thus this cluster (and similarly Cu_5) also tests the form of the model Hamiltonian. It is non-trivial that this effective Hamiltonian, derived *from first principles, for these materials*,[7,8] yields a Heisenberg model.

To renormalize the Hamiltonian for the metallic state we map the energy spectrum for Cu_N clusters with N±1 holes to strong-coupling models such as t-t'-J-J' (generic t-J) and t-t'-U (generic single-band Hubbard) in order to derive interaction energies (e.g., t, t' and U). Note this is not an extrapolation from the insulating regime. The arguments given for the Heisenberg case likewise apply for the determination of the local parameters, t,t', and U. Furthermore, for Cu_5 the mapping non-trivially tests the couplings in the one-band model plus the wavefunctions and the projection of the subspace. We now summarize the results from Ref. [7].

For the insulator (N holes), calculated low energy spectra map onto a Heisenberg model separated from the charge-transfer excitation spectra by >2.5 eV. The J (nearest-neighbor, nn) and J' (next-nearest-neighbor, nnn) equal 128±5 and 3±1 meV, respectively (error bars indicating cluster-to-cluster consistency). J measured from polarized light scattering is 128±6 meV.[9] This level of agreement is fortuitous considering the accuracy of the derived parameters implies J is determined to 50%. The calculated low energy spectra for N±1 holes (hole and electron doped) exhibit the same number of lines per symmetry per spin state as for the single band model projected to minimum number of double occupancies (in hole language) as in a t-J model. This low energy spectrum in the clusters is separated from the next set of excitations by ~2 eV. The mapping onto the t-J model held J and J' fixed at the non-doped values. For example, the best fit for hole doped Cu_5 has t=440±5 and t'=-60±20 meV (error bars again reflecting cluster-to-cluster consistency) in general agreement with other estimates.[10] The signs refer to hole notation. The mapping is not perfect, but in the absence of a rigorous criterion to determine how good is sufficient, we have identified the t-J Hilbert space within the three-band space by forming a singlet in a CuO_4 unit, described by two fully-correlated holes (~5 variational parameters). This is to be distinguished from the simplified Zhang-Rice treatment.[2] Projecting from the lowest 30 three-band eigenstates to this t-J Hilbert space, we find $TrP_{tJ}/Tr1 \approx 0.96$ (where Tr, P_{tJ} and 1 are trace, projection and identity operators, respectively). With the best t-J fit and using the (identifiable) one to one correspondence between the t-J variational and the exact eigenstates then $|<\Psi_i^{exact}|\Psi_i^{t,J}>|^2 > 0.90$ for all states except two which are degenerate by symmetry and whose projections are 0.80. These two statements together seem to indicate that the t-J-like Hilbert space is an adequate description of the low energy dynamics.

Somewhat surprisingly the effective low energy carrier dynamics exhibit a notable symmetry between electrons and holes, despite the obvious underlying chemical differences (e.g. added electrons are about 75-80% Cu while added holes are about 75-80% O). We find for N-1 holes that, reminiscent of N+1 holes, t and t' equal 410±5 and -70±10 meV, respectively (signs in the hole notation). In fact a single band Hubbard model with U ≈ 5.4 eV, t = 430 meV (U/t ≈ 12.5) and t' = - 70 meV adequately describes all clusters calculated with N-1, N or N+1 holes.

In summary, the important points are: (1) the parameters for the three-band extended Hubbard model have been derived from quantum-chemical calculations specific to La_2CuO_4 [6] and they basically agree with complementary ab-initio parameters[11] and purely semi-empirical parameters.[10] (2) With one hole per CuO_2 unit the three-band Hubbard model with the calculated parameters maps very well onto the Heisenberg model with nn J in agreement with experiment and only a very small nnn J' (J'/J ≈ .02). *This is non-trivial.* In fact, parameters such that the charge-transfer gap is approximately constant, but ε small and U_{pd} (Cu-O Coulomb) large will not give a Heisenberg spectrum. (3) In the metallic state we observe (i) remarkable symmetry for hole and electron doped materials, despite significant underlying chemical asymmetry. (ii) A singlet construction, related to that proposed by Zhang and Rice[2] forms an excellent Hilbert space to describe the low energy states. (iii) The nnn hopping, t', is non-negligible in the generic t-J model and should strongly affect the coherent quasi-particle peak.[12] The present results give strong evidence for the validity of an effective one-band description of low-energy carrier dynamics in the Cu-O materials at modest doping levels. Additional interactions within this description, though small, may be non-trivial. Furthermore, the lowest energy degrees of freedom renormalized away (e.g., intrasite Cu d or Cu-O charge transfer excitations) limit the range of validity in energy of the effective one-band models. Ultimately, determining this validity requires calculation of additional response functions for comparison to experiment.

ACKNOWLEDGEMENTS

This work was done in collaboration with M. S. Hybertsen, M. Schluter and D. R. Jennison. Work at Sandia National Labs is supported by the U.S.DOE under Contract #DC-AC04-76DP00789.

REFERENCES

1. P. W. Anderson, Science **235**, 1196 (1987); V. J. Emery, Phys. Rev. Lett. **58**, 3759 (1987); C. M. Varma, S. Schmitt-Rink and E. Abrahams, Solid State Commun. **62**, 681 (1987); P. A. Lee, Phys. Rev. Lett. **63**, 680 (1989).
2. F. C. Zhang and T. M. Rice, Phys. Rev. B **38**, 4632 (1988).
3. E. B. Stechel and D. R. Jennison, Phys. Rev. B **38**, 4632 (1988).
4. R. E. Cohen, W. E. Pickett and H. Krakauer, Phys. Rev. Lett. **62**, 831 (1989).
5. L. F. Mattheiss, Phys. Rev. Lett. **58**, 1028 (1987); J. Yu, A. J. Freeman and J.-H. Xu, Phys. Rev. Lett. **58**, 1035 (1987).
6. M. S. Hybertsen, M. Schluter and N. E. Christensen, Phys. Rev. B **39**, 9028 (1989).
7. M. S. Hybertsen, E. B. Stechel, M. Schluter and D. R. Jennison, preprint.
8. E. B. Stechel and D. R. Jennison, Phys. Rev. B **40**, 6919 (1989).
9. R. R. P. Singh, P. A. Fleury, K. B. Lyons and P. E. Sulewsky, Phys. Rev. Lett. **62**, 2736 (1989).
10. H. Eskes, G. A. Sawatzky and L. F. Feiner, Physica C **160**, 424 (1989).
11. A. K. McMahan, R. M. Martin and S. Satpathy, Phys. Rev. B **38**, 6650 (1989).
12. W. Stephan, K. J. von Szczepanski, M. Ziegler and P. Horsch, preprint.

DISCUSSION

S. Doniach: I know that Richard Martin here has done work on clusters using real chemistry, CI interactions and all that. Have you tried to compare any of your clusters with real chemistry...I mean, you've always started from these band model parameters...

E. Stechel: Well the LDA bands reflect real chemistry...

S. Doniach: ...Yeah, but that's LDA, and LDA is up for grabs. It's interesting to know how does it map in with chemistry.

E. Stechel: Well Rich Martin and I have spent a lot of time talking, but we haven't fully resolved this question. The problem with the quantum chemistry cluster calculations is that they can't possibly include all polarization effects, so they can't possibly get U correctly. But they can do a better job of truly doing Hartree-Fock, where the Local Density calculation cannot. So each method suffers from different kinds of problems. But maybe it would be better to let Rich L. Martin answer your question, because he's thought more about it.

R. L. Martin: Doug Scalapino this morning talked about uncontrolled approximations, and we have plenty of those as well. We have a Madelung point charge background; in order to treat all correlations involving the d's and the f-functions, the calculations get very big. But with what most people would consider reasonable approximations, we see a wave function for two holes in a small cluster that looks different from what's being discussed here. It looks...rather than being two holes highly correlated on oxygen and copper, for example, it looks like finally we see what chemists would call a "molecular orbital" picture. In other words, the two holes finally enter something that looks like a band structure, a highly covalent one. In the undoped situation, it becomes very ionic-looking to us. So in terms of parameters – we're still working on trying to extract parameters; because of the uniqueness problem we don't frankly know how to do it. In terms of the parameters, the major point is that we have a larger dopant at this point than anyone else. Our best estimates for the energy difference is of the order of six...And so, the type of thing that we see is...when you put in two holes, we get two configurations that are very close in energy. One has a hole in the bonding orbital, and the other has the hole in the anti-bonding orbital, and those two interact very strongly. So I guess the point is that our wave functions look different.

G. Sawatsky: I just want to make a very brief comment. As you know we've also done these types of calculations on somewhat smaller clusters, and we've also done experiments to try to find out what the parameters are, and those parameters fitting photoemission spectra of model systems, like CuO and so on and so forth, agree very well with the parameters of Hybertsen and Shluter, except for perhaps U_d, which is a little bit smaller. It's 8.5, rather than 10.5. I don't think that's terribly serious. But there's one thing we've done, one thing that comes out a little bit different out of our cluster calculations of Cu_2O_7 and Cu_2O_8, similar calculations to what you've done, except smaller clusters, but then taking all of the d-orbitals into account. Now, we also map very beautifully onto t, and t', and J, which are very close to the values you have. But there's one difference and that is, namely, at what would be the equivalent $k\pi/a$ point, we get quite a bit of d_z^2 mixed into the extra-hole state, the state where you put an extra hole into your system; this happens at $k = \left(\frac{\pi}{a}, 0\right)$ not at Γ where there is no mixing. And that might be rather important for explaining these X-ray absorption results, if the holes primarily go somewhere between Γ and X polarization dependent. This could also be important for explaining, perhaps, some influence of the apex oxygen, that some people like to talk about.

E. Stechel: Can I make a comment? We're also now putting in all the d-levels, and investigating that question of the d_z^2 and, in the density functional calculations they also find significant amount of the d_z^2, so we are also, with Cu_2O_{11} including all Cu-d and O-p orbitals and calculating photoemission spectra. We're also going to calculate the optical conductivity and the Raman spectrum to try to get a experimental handle on t and t'.

R. M. Martin: I have one comment on the other Rich Martin's discussion. We have a smaller energy difference $\epsilon_p - \epsilon_d$ because we're trying to do the straight LDA, whereas they're doing Hartree-Fock. We all know truth is in between those two. We've been doing some estimates on that and basically I think we've pushed both answers in the right direction.

Could I comment about your U in your effective part. It seems to me that that's such a big number that surely that's not very accurate, out of fitting just the lowest-energy part of the spectrum.

E. Stechel: What do you mean it's not accurate? The fit is a least squares fit.

R. M. Martin: But if you only fit things over an energy range of the order of a few times t, and you're extracting the parameter that's...

T. M. Rice: The energy scale is of order t^2/U.

E. Stechel: That's right.

T. M. Rice: ..You're getting excellent agreement.

R. M. Martin: Right, exactly. So I'm proposing that that would change as you change clusters, and a lot of other things.

E. Stechel: No, it doesn't change that much. One can calculate U from a CuO_4 cluster or from Cu_2O_7, Cu_4O_{12} or Cu_5O_{16}, etc. We have calculated U from those 4 and other clusters and incredible size consistency persists.

C. M. Varma: I've spent lots of time thinking about this three-band versus one-band model, and I'd like to make a comment. First thing of course is that, in view of the computer-experimental results presented this morning, I think one should start thinking about where one has gone wrong, in thinking about the problem as a one-band model. The comment I would like to make is that any three-band model which does not reduce for low-energy excitations to an effective one-band Hubbard model at half-filling is a priori nonsense. It has to.

E. Stechel: Yes. We take that as given. We agree.

C. M. Varma: And every model that I think of with a three-band nature near half-filling must give me an insulator, must give me the right value of J, and ...

E. Stechel: No, not necessarily.

C. M. Varma: It does in the one that I believe in. Now, the question is, can you trust what you learned from your understanding of the half-filled situation – that effective model – should you trust that when you're in the metallic phase? And I think this is a logical trap, to think that it is the same model. Because in the insulator, you have this gap for charge excitations, and in the metal you do not have that gap for charge excitations. You have new low-energy excitations in the metallic state, and therefore you can, in principle, have new interactions between those new low-energy excitations. The somewhat vague, or approximate way of seeing that, is to take a three-band model, and do a calculation. And rather elaborate calculations of three different kinds have been done, now; one is some kind of generalized RPA, one is a Gutzwiller calculation, and the third is going on, with the slave-boson method, including corrections of the order $1/N$. And in each case you find that near half-filling, for reasonable parameters which are within a factor of 1.5 of the parameters that you gave, and that's the degree to which those parameters should be trusted. You get near half-filling a perfectly nice insulator, and an antiferromagnet, but if you go into the metallic state you generate new instabilities, you generate new interactions. And so, I completely agree with your results near half-filling, when you are looking for the excitation spectra of one-electron, one-hole, or you are looking for low-energy spin excitations. But to conclude from that about the metallic state – and I don't think you are concluding anything about that for the metallic state – but the common assumption in this field has been, if you know the effective Hamiltonian for the half-filled case, you know it also for the metallic case, and I'm saying that is false.

E. Stechel: Yes, but we never said that. These calculations are not done only in the insulating regime. They are also done with an additional hole, $(N+1)$ holes per N Cu's within the three-band Hubbard model.

C. M. Varma: I wasn't commenting or criticizing anything you said, I was making a general remark about the way this field had developed in the last two years.

Comments on the Nature of the Normal Metallic Phase of the CuO Planar Materials

Richard M. Martin

Department of Physics and Science and Technology Center for Superconductivity, University of Illinois, 1110 W. Green Street, Urbana, IL 61801

A key issue in the theoretical work addressing the character of the electronic excitations in the metallic materials containing CuO planes is whether the excitations can be described by quasiparticles in a normal Fermi liquid theory or have quantum numbers which are different from normal quasiparticles in an essential way. This issue has been raised particularly by Anderson[1] who has proposed essential differences in a spin liquid state with separation of spin and charge degrees of freedom and by Laughlin[2] and others who have proposed excitations with fractional statistics in analogy with the fractional Quantum Hall Effect. A phenomenological form for the self energy as a "marginal Fermi liquid" which is different in form from the usual quasiparticle form has been proposed by Varma *et al.*, based upon experimental evidence.[3]

Two characteristics of a normal Fermi liquid in a periodic system are the existence of Fermi surface in k space and excitations which can be described as quasiparticles having the same symmetries as particles in an appropriate non-interacting system. The first criterion is the existence of a Fermi surface which obeys the "Luttinger Theorem," i.e., whose volume is the same as that contained by the Fermi surface of a non-interacting electron system of the same symmetry.[4-6] [It should be pointed that Luttinger himself[4] took great care to state his conclusions in a careful way depending upon the convergence of perturbation theory starting from the non-interacting state. It was later workers[6] who identified the conditions on the Fermi surface as a "theorem" which could apply in the presence of strong interactions and whose validity depended only upon the analytic nature of the self energy of the quasiparticles.] Thus in the metallic state, unless there is some hidden symmetry, the theorem states that the volume enclosed by the Fermi surface includes *all* the electrons or holes. In the case of the CuO planes this means that the Fermi surface should include $(1+\delta)$ holes, where 1 represents the one hole per Cu present in the insulating antiferromagnetic state at half filling, and δ represents the additional holes which are required to stabilize the metallic state. In contrast to this, many models[7] have proposed Fermi surfaces which contain only δ holes, implicitly invoking some remnant of antiferromagnetic order to evade the theorem.

The properties of electrons in one dimension clarify the issues related to the "Luttinger Theorem." The exact solution of the one-dimensional Hubbard model by Lieb and Wu[8] show that it is an insulator only at half filling and a metal away from half filling. A recent analysis of Ogata and Shiba[9] has clarified that away from half filling, the wavevectors at which electrons and holes can be added are exactly the same as in the non-interacting problem, i.e., they obey the Luttinger theorem in its usual form. Thus there is no rigorous distinction between an "upper" and a "lower" Hubbard band and the electron states indeed form a continuum with a well-defined Fermi surface obeying the Luttinger Theorem. In two dimensions there are no exact solutions; nevertheless, simulations[10,11] of the Hubbard model give strong evidence that the Fermi surface exists and is also consistent with the Luttinger theorem. The highest occupied and lowest unoccupied states are remarkably close to the simple square Fermi surface known for the noninteracting case.[11]

Even though the Fermi surface is in agreement with Fermi liquid theory, in one dimension the nature of the quasiparticles is *not* the same as in a normal Fermi liquid. In particular, a signature is the form of the occupation n(k) at the Fermi surface; instead of the step required in the usual Fermi liquid theory there is a power law singularity[7] with $n(k) \sim (k - k_F)^{1/8}$. In the usual Fermi liquid theory, the height of the step in n(k) is 1 in a non-interacting system and a value $Z<1$ in an interacting system. Thus the 1d case represents a qualitative change from $Z \neq 0$ to $Z=0$. In two dimensions, even though simulations suggest a strong variation of n(k) near k_F, this is not sufficient since the 1d example shows that n(k) can vary extremely rapidly near k_F yet have no step. Since it is very difficult to distinguish a form such as the 1/8 power law from a step in numerical simulations, current simulations cannot answer the question of the detailed form of Z near the Fermi energy. In three dimensions, there is abundant evidence that normal Fermi liquids exist with quasiparticle excitations. For the CuO materials it thus becomes crucial whether 1d-like behavior also occurs in an isolated 2d plane, and further whether such behavior can persist in the presence of coupling between planes in the full 3d problem.

Now let us turn to experiments and what they have to say about the Fermi surface and quasiparticle excitations. As is now well known, angle resolved photoemission (ARPES) on $Bi_2Sr_2CaCu_2O_8$[12] and $YBa_2Cu_2O_7$[13] materials have observed dispersive states whose energy as a function of wavevector k approaches the Fermi energy and whose intensity disappears for k beyond wavevectors which can be identified as the Fermi surface. The Fermi surfaces appear consistent with band calculations[14] which, of course, agree with the Luttinger theorem. Although this is not a detailed test of the Fermi surface at the present time, it is strong evidence for the existence of a Fermi surface containing $\sim 1 \gg \delta$ holes. Since a strong variation in n(k) at such a Fermi surface is predicted by 2d simulations and is known in the exact 1d solutions, this seems convincing evidence that experiments indeed confirm one aspect of the expected behavior of electronic excitations in these materials.

In order to probe the character of the excitations, it is essential to examine shapes of the spectra near the Fermi surface. At this point more detailed analysis of the ARPES to show whether the spectra look like quasiparticle poles approaching the Fermi energy or like a branch cut singularity[1,3] in a continuous spectrum. However, I wish to make one general point based upon the spectra. The intensity of the signal in photoemission is the strength of the one-particle Green function, which is $\sim Z$. The experiments give us a qualitative estimate of this strength relative to other states (mainly nonbonding oxygen and Cu) which we expect to be invariant from the insulating antiferromagnetic to the metallic state. In the insulating state the intensity near the Fermi energy E_F is small as expected and the strong signal starts at energies below E_F. These states appear to be the energies for creating holes in an insulating antiferromagnet, i.e., the "lower Hubbard band." In the doped metallic state, the Fermi edge is clearly visible at energies well above (~ 0.5 eV) supported by recent work of Allen, *et al.*,[15] where the Fermi energy is found to vary smoothly from electron-to hole-doped materials. Therefore it appears that states have "filled" the gap as in a band with intensity which indicates that some average value of Z over this energy range is not greatly reduced from 1.

The above observations suggest that, if there is a great renormalization of Z to 0 as appears likely in theories in which the excitations are qualitatively modified from quasiparticles,[1-3] this occurs only on a small scale of energy near E_F. Then the key questions are: Are the excitations consistent with a Fermi liquid having a low energy scale or are they qualitatively different below the characteristic low energy? What is the origin of the low energy scale? Is there another energy scale which determines a crossover to 3d Fermi liquid behavior?

REFERENCES

1. P. W. Anderson, this proceedings.
2. R. B. Laughlin, Science **242**, 525 (1988).
3. C. M. Varma, P. B. Littlewood, S. Schmitt-Rink, E. Abrahams, and A. E. Ruckenstein, Phys. Lett. **63**, 1996 (1989).
4. J. M. Luttinger, Phys. Rev. **119**, 1153 (1960).
5. A. A. Abrikosov, L. P. Gorkov, and E. I. Dzyaloshinski, *Methods of Quantum Field Theory in Statistical Physics* (Printice Hall, 1980).
6. A paper on the "Luttinger theorem" with many earlier references is R. M. Martin, Phys. Rev. Lett. **48**, 362 (1982). In that paper it is pointed out that the Friedel sum rule is the same as the Luttinger condition of the Fermi surface applied to an impurity problem. Since there are exact solutions for impurity problems, this paper concludes that the sum rules are obeyed and the usual Fermi liquuid behavior occurs in strongly interacting impurity problems like the Anderson and Kondo problems. Furthermore it speculates that the Fermi liquid theory also applies to heavy Fermion crystals and gives evidence that this is observed.
7. For example, a specific Fermi surface inconsistent with the Luttinger theorem is given by Y. Guo, J-M Langlois, and W. A. Goddard, Science, **239**, 896 (1988); G. Chen and W. A. Goddard, *ibid*, 899 (1988).
8. E. H. Lieb and F. Y. Wu, Phys. Rev. Lett. **20**, 1445 (1968).
9. M. Ogata and H. Shiba, preprint.
10. S. Sorella, S. Baroni, R. Car, and M. Parrinello, Europhys. Lett. **8**, 663 (1989).
11. A. Moreo, D. J. Scalapino, R. L. Sugar, S. R. White and N. E. Bickers, preprint. An extensive discussion of recent work is given by D. J. Scalapino, this proceedings.
12. T. Takahashi, *et al.*, Phys. Rev B**39**, 6636 (1989); C. G. Olson, *et al.*, Science, **245**, 731 (1989); R. Manske, *et al.*, Europhys, Lett. **9**, 477 (1989).
13. A. J. Arko, *et al.*, Phys. Rev. B **40**, 2268 (1989).
14. A review of the band calculations is given by W. E. Pickett, Rev. Mod Phys. **61**, 433 (1989).
15. J. W. Allen, *et al.*, preprint.

DISCUSSION

A. Millis: Since you don't have absolute intensities or a theory of the photoemission, how do you really know what Z is from the experiment?

R. M. Martin: You don't. Absolutely. And that is certainly not something that one could come up with a number for. However, we have something sitting there. We have a large blob of states that, I think, are basically one-electron, or as one-electron as any old oxide that exists.

A. Millis: But that's all the spaghetti of all of these other bands below the Fermi level.

R. M. Martin: Exactly, but see they form a marker. There's some states here, particularly up near the edge, that are not the ones we love so dearly, and they're the marker to which you refer these other states. But they're there, and we know how many electrons are in them, and all that.

T. M. Rice: But isn't the issue of Z how much weight is in the coherent peak and how much is in the background? How do you distinguish them?

R. M. Martin: Well, now, that is certainly the more detailed point. I do have one other figure that I just decided not to show, on this much more detailed scale...Then you're absolutely right. Now we go down to the next energy scale, which is like a tenth of a volt or below where we start asking in the questions and ...clearly there's a lot of weight here. I think that we can assert; so that the total amount is a rather large weight in this low-energy range, and the next steps I do not pretend to have the solution, but at least one possibility is a band-like calculation with spins added in, with the fluctuations added in, that give rise to the essentially large incoherent part. I certainly wouldn't dispute that.

S. Chakravarty: I didn't understand the answer to the question. Sorry, I mean, I'm asking the same question again. Maybe you understand it, and can explain what the answer was...

R. M. Martin: Okay, I'll attempt to rephrase. We have suggestions for what might be the possibilities: a branch cut, a pole (which is broadened, in the way analyzed by Scott List)...we simply have to weight those things, and analyze it in more detail, the zeroth-order point, which we just simply haven't brought out at this conference, is that there's a lot of weight in this region.

G. Sawatzky: There has to be a lot of weight in that region, because we would all expect the incoherent part of the quasiparticle to extend over only about one eV, from the low-energy-type spin excitations...

J. R. Schrieffer: I would say more on the scale of 0.2 eV. There is spectral weight at higher energy but this comes from inelastic processes which are dominated by the recoil kinetic energy of the hole.

R. M. Martin: Okay, I can see...

C. M. Varma: One thing these measurements do say is that there is a Fermi surface with a Luttinger theorem property, period. You can't say anything about Z. This says there's a Fermi-Dirac function, if anybody imagined there wasn't a Fermi-Dirac function, that's all.

Phase Separation of Holes in Antiferromagnets
V. J. Emery
Brookhaven National Laboratory, Upton, NY 11973

ABSTRACT

It is shown that dilute holes in an antiferromagnet do not have a uniform density but separate into a hole-rich phase and a phase with no holes. The argument is presented for the t-J model but is expected to be more general. When the exchange interaction J exceeds a critical value J_c, one phase consists of all holes, the other all electrons. It is proposed that, for J slightly less than J_c, the hole rich phase is a low density superfluid of electron pairs.

Much of the theoretical work on high-temperature superconductors has been concerned with the behavior of doped holes in antiferromagnets. Typically these studies have assumed that the density of holes is uniform and have investigated the possibility of various states with long-range order, such as antiferromagnetism, spiral states, flux phases or superconductivity. However, we have shown,[1] together with H. Q. Lin and S. Kivelson, that there are good physical reasons to expect that phases of non-uniform density should occur if the concentration of holes is below a critical value x_m (which depends on the parameters of the model). The system separates into two phases, one of which is the undoped antiferromagnet and the other contains all of the holes at concentration x_m. The argument was presented specifically for the t-J model, but the possibility of phase separation should be taken into account for any model of dilute holes in an antiferromagnet.

The Hamiltonian of the t-J model is given by:

$$H = J \sum_{ij} (\vec{S}_i \cdot \vec{S}_j - \frac{1}{4} n_i n_j) - t \sum_{i,j,s} [c^{\dagger}_{i,s} c_{j,s} + H.c.] \tag{1}$$

where $c^{\dagger}_{i,s}$ creates an electron of spin s on site i, and $n_i, \vec{S}_i$ are the electron number and spin operators respectively. The Hamiltonian is supplemented by the constraint that there be no doubly occupied sites. It is straightforward to show that complete phase separation occurs for large J. The point is that the cost to remove an electron from the antiferromagnetic phase is $1.168J$ (two bonds) and this outweighs the gain in kinetic energy which is $-4t$. The value of J/t for which it first becomes profitable to transfer electrons to the hole-rich phase depends on whether the electrons form n-particle bound states. It has been shown[1] that, as J is decreased, the first instability is for the transfer of pairs, and this occurs when $J = J_2 = 3.828t$. If the transition is continuous, the hole-rich phase will consist of a dilute gas of pairs when J is slightly less than J_2, and since the pairs are bosons, they should form a superfluid.[1] For $J \ll t$, arguments similar to these of Visscher[2] and Ioffe and Larkin[3] show that a very low density of holes is unstable to phase separation, with the hole-rich phase ferromagnetic. If, as has been suggested, the lowest state for $J \ll t$ is not ferromagnetic, the argument is only strengthened.

The physical reason for phase separation in the large J limit is that the gain in exchange energy by maximizing the number of antiferromagnetic bonds outweighs the cost in kinetic energy. In the small J limit, phase separation occurs because the motion of a hole in an antiferromagnet is frustrated and it is better to put all of the holes into a region in which they have a lower kinetic energy.

Since there is phase separation for $J \ll t$ and $J \gg t$, it is reasonable to expect that the same qualitiative behavior should be found for all values of J/t and that x_m increases as J/t increases. We have tested this idea by exact diagonalization of the Hamiltonian on a 4 × 4 lattice, interpreting the energy of N_h holes as the energy of a hole concentration $x = N_h/16$, which is reasonable when the number of *electrons* is not too small. Figure 1 is a plot of the concentration x_m below which the uniform state is unstable as a function of J/t.

It may be objected that holes in high T_c superconductors are charged and that long-range Coulomb interactions will prevent condensation. However, holes are often donated by oxygen atoms which are quite mobile at temperature below the consolute temperature of the holes. Hence the background O^{2-} ions may also phase separate and compensate for any local charge imbalance. Precisely such phase separation occurs[4] in oxygen-doped La_2CuO_4. Evidently a complete picture would include the energies of the oxygen defects, but it is clear that the holes themselves may be a force for phase separation and that charge imbalance will not necessarily prevent it.

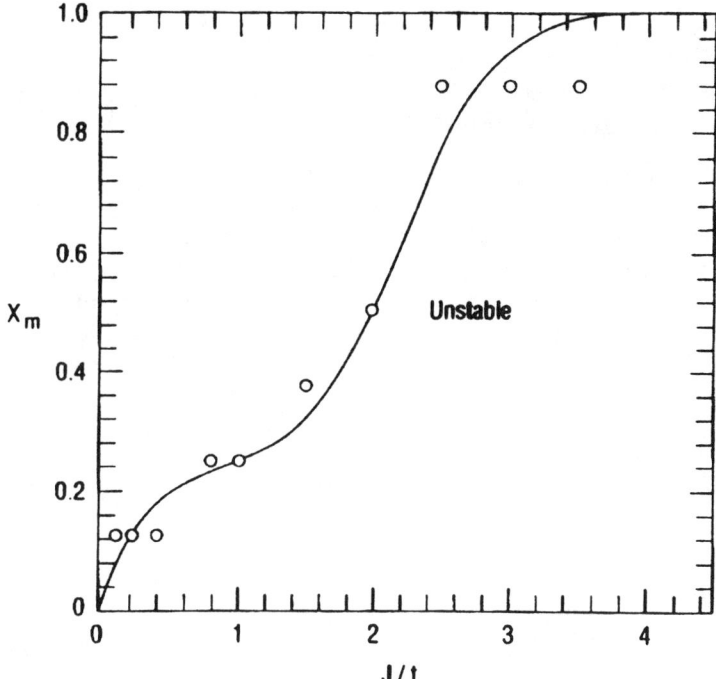

FIGURE 1 x_m, the concentration below which the uniform state is unstable is shown as a function of J/t. The data points are for a 4 × 4 lattice and the curve illustrates the expected behavior for an infinite system.

Phase separation may be a quite general phenomenon for holes in an antiferromagnet. Recently it has been shown[5] to occur in a large-N generalization of the t-J model and there are reasons to believe[1] that it must be taken into account for the single-band Hubbard model and for certain parameter ranges in three band models of the copper oxide planes of high-temperature superconductors.

The work reported here was carried out in collaboration with H. Q. Lin and S. Kivelson. It was supported in part by NSF grant DMR 87-06250 and PHY 82-17853 and supplemented by funds from NASA. It was also supported by the Division of Material Sciences, Basic Energy Sciences, U.S. Department of Energy under Contract No. DE-AC02-76CH00016.

REFERENCES

1. V. J. Emery, S. Kivelson and H. Q. Lin, submitted to Phys. Rev. Lett.
2. P. Visscher, Phys. Rev. B**10**, 943 (1974).
3. L. B. Ioffe and A. I. Larkin, Phys. Rev. B**37**, 5730 (1988).
4. J. D. Jorgensen *et al.* Phys. Rev. B**38**, 11337 (1988).
5. M. Marder, N. Papanicolaou and G. C. Psaltakis, preprint.

DISCUSSION

J. R. Schrieffer: You mentioned long-range forces. How about elastic relaxation: will that give energies comparable to the J's?

V. Emery: I don't know exactly whether or not they're comparable to that. There are a lot of energies in the problem that can move the minimum around. You can see that the sort of concentration that the t-J model wants, at reasonable values for the parameters, is not too far from the concentration that you would want for the 38K materials. But obviously there are other energies, and what I would like to know is where the equivalent evidence is in the 1,2,3 materials. It's suspicious that you have the 60 Kelvin plateau, but I don't think there's any evidence of phase separation in that region. It's quite clear that because the elastic Coulomb energies, etc. are different in different materials, that you may well find that the magic concentration may be different from one material to another. But I think that once you have a situation where you have mobile oxygens, you do at least have the possibility that the background can go around with the holes.

S. Trugman: It's rather difficult to quantitatively calculate the energy of a ferromagnetic polaron if you take into account that the boundaries may not be sharp. The spin correlations may not change from strongly ferromagnetic to strongly antiferromagnetic in a single lattice spacing.

V. Emery: There is no doubt that we need to do a better job of that. I think we can, because I think you can do that with lower bounds. Because it's a very specific question for a single hole in an antiferromagnet where you're looking for that polaron. But that is certainly a way in which we would like to prop up the argument a little better. If you make any reasonable estimate of the energy of the ferromagnetic polaron, there's no doubt that the uniform phase – at very low J for these systems at finite concentration of holes beats it, once you go to low enough J. There's a $\sqrt{2\pi}$ factor that is not precise but it helps you.

S. Trugman: The numerical factor could arise from a poor variational choice for the ferromagnetic polaron, and the relative stability of the two states could reverse with a better variational choice.

V. Emery: There's no doubt that one needs to do that better, and that's why I say I think it's possible to prove a lower bound on that number, which then would make the argument essentially rigorous, I think.

S. Trugman: You haven't yet done a rigorous calculation.

V. Emery: No.

Antiferromagnetic Paramagnons in the Cuprate Oxide Superconductors: A Novel Fermi Liquid

David Pines

Physics Department, University of Ilinois at Urbana-Champaign, 1110 West Green Street, Urbana, IL 61801

The phenomenological model of nuclear magnetic relaxation which Millis, Monien, and I have developed[1] provides, at present, the only quantitative description of NMR, NQR, and Knight shift experiments on the yttrium, planar copper, and planar oxygen sites in $YBa_2Cu_3O_7$. It is a one-component description: one spin degree of freedom, $\underset{\sim}{S}_n$, per CuO_2 unit cell, centered on the Cu-sites, is assumed to be responsible for the spin-lattice relaxation and Knight shifts for the ^{63}Cu, ^{17}O, and ^{89}Y nuclei; its coupling to these nuclei is described by a "local spin" Hamiltonian,

$$H = ^{63}\vec{I}_n \cdot \mathbf{A} \cdot \vec{S}_n + B \sum_\delta {}^{63}\vec{I}_n \cdot \vec{S}_{n+\delta} + C \sum_{\delta\prime} {}^{17}\vec{I}_n \cdot \vec{S}_{n+\delta\prime} + D \sum_{\delta\prime\prime} {}^{89}\vec{I}_n \cdot \vec{S}_{n+\delta\prime\prime} \quad (1)$$

where, $\delta, \delta\prime, \delta\prime\prime$ label the unit cells adjacent to the nuclear spin in question, $\mathbf{A}$ is the direct hyperfine tensor, and B,C, and D, assumed to be isotropic, are transferred hyperfine couplings. The spin dynamics are assumed to be described by mean field theory, with

$$\chi''(q,\omega) = \frac{\pi \chi_0 \omega}{\Gamma}\left[1 + \frac{\beta(\xi/a)^4}{(1+\xi^2q^2)^2 + \frac{\pi^2\omega^2}{\Gamma^2}\beta\left(\frac{\xi}{a}\right)^4}\right] \quad (2)$$

where $\chi''(q,\omega)$ is the imaginary part of the spin susceptibility, χ_0 is the uniform static susceptibility, Γ is a typical spin fluctuation energy (of order the Fermi energy, E_F in a Fermi liquid picture), ξ is the spin correlation length, a the lattice constant, β is a dimensionless parameter (~ 10), and q is measured from the zone corner, $(\pi/a, \pi/a)$. From their fits to experiment, MMP find that:

- Both the direct hyperfine tensor $\mathbf{A}$, and the transferred hyperfine coupling, B, are changed remarkably little from the values these possess in the antiferromagnetic insulator, $YBa_2Cu_3O_6$.

- The temperature-dependent spin correlation length follows a mean-field dependence,

$$[\xi(T)/a]^2 = [\xi^2(T=0)/a^2]\frac{T_x}{T+T_x}, (T_x \sim T_c) \quad (3)$$

above T_c, with $[\xi^2(T)/a^2] \equiv [\xi^2(T_c)/a^2] \cong 8$ below T_c.

- The characteristic energy which sets the range over which antiferromagnetic correlations are significant is, with $\Gamma \sim 0.4$ eV,

$$\hbar\omega_{SF} \equiv \frac{\Gamma}{\beta^{1/2}\pi}\left[\frac{a}{\xi(T_c)}\right]^2 \cong 60K \qquad (4)$$

An additional measure of the success of the model comes from the preliminary fit[2] which it provides to the very recent NMR experiments on the 60K superconductor, $YBa_2Cu_3O_{6.63}$,[3] which show that A,B, and C are essentially unchanged from their values in $YBa_2Cu_3O_7$. When the temperature dependence of χ_0 (between 300K and 60K) revealed by these experiments is taken into account, a quantitative fit to the ^{63}Cu and ^{17}O relaxation rates is obtained with $\Gamma \sim 0.5$ eV, $[\xi^2(T_c)/a^2] \cong 25, \hbar\omega_{SF} \cong 24K$, and $T_x < T_c$.

The MMP model strongly suggests to me that the planar excitations in the cuprate oxides form a quite new kind of Fermi liquid, in which the dominant low frequency excitations are antiferromagnetic paramagnons at wavevectors near the zone corner whose characteristic energies are $\lesssim \hbar\omega_{SF}$. [In the MMP model, the closest approach to antiferromagnetic behavior came at the zone corner, however preliminary calculations[4] suggest that its basic results are little changed if one has a closest approach at incommensurate wave-vectors which are ~10% of $(\pi/a, \pi/a)$]. Its properties are quite different from those of liquid ^3He, where the dominant low frequency excitations are "ferromagnetic" paramagnons at wavevectors near $q = 0$, or the usual Fermi liquid found for electrons in metals. In this liquid, the spins are nearly localized (witness the success of the local spin Hamiltonian description)--but definitely are itinerant, since when the system goes superconducting, the Knight shift and spin lattice relaxation rate behave as though the spins pair in a strong coupling version of BCS theory.[5] Because of the strong anti-ferromagnetic correlations, the liquid differs from the "marginal" Fermi liquid recently proposed by Varma *et al.*;[6] indeed, I would expect that the renormalization constant Z which measures the momentum discontinuity at the Fermi surface will turn out to be finite. On the other hand, it differs from the conventional description of a "failed" Heisenberg antiferromagnetic, for which the dominant excitations at wavevectors $q\xi \gtrsim 1$ are expected to be spin waves with a velocity s and an energy $qs \sim J \gg \omega$.

Both the magnetic and transport properties of this nearly antiferromagnetic Fermi liquid may be expected to differ markedly from those of normal Fermi liquids. First, a natural explanation for the behavior of $\chi_0(T)$ in $YBa_2Cu_3O_{7-x}$ is that the antiferromagnetic spin correlations responsible for the enhancement of χ_q for $q \sim Q_{AF} = (\pi/a, \pi/a)$, lead to a decrease in the uniform susceptibility, χ_0, which for $x = 0$ is nearly independent of temperature, but which, with the substantially larger values of ξ (and $\chi_{Q_{AF}}$) found for $x = 0.37$, lead to an additional, and temperature-dependent, suppression of χ_0. Second there exists a characteristic low energy excitation, $\hbar\omega_{SF}$, which is significantly less than E_F, and is even small compared to kT_c.

The presence of antiferromagnetic paramagnons with a characteristic energy $\lesssim T_c$ means that the normal state transport properties will not be of those of a "normal" Fermi liquid, and might explain the measured very short quasiparticle lifetime, $(\hbar/\tau) \sim T$. It is tempting to conclude that, just as is the case with heavy electron systems and organic superconductors, antiferromagnetic spin fluctuations are the physical origin of the transition to the superconducting state.[7] Since, however, the normal state quasiparticles are so poorly defined, the development of a microscopic theory which demonstrates this poses a major challenge to us all.

REFERENCES

1. A. Millis, H. Monien, and D. Pines, preprint; A. Millis, these proceedings.
2. H. Monien, D. Pines, and D. Thelen, private communication.
3. M. Takigawa, these proceedings; R. Walstedt, these proceedings.
4. H. Monien, private communication.
5. S. Barrett *et al.*, preprint; H. Monien and D. Pines, preprint.
6. C. Varma *et al.*, Phys. Rev. Lett. **63**, 1996 (1989).
7. D. Pines, Physica **B**, in the press.

DISCUSSION

D. J. Scalapino: I agree with what David says regarding the existence of antiferromagnetic fluctuations in the CuO_2 superconductors. However, as I discussed, Monte Carlo results suggest that they lead to short-range d-wave pairing correlations and that over the temperature region we have been able to run the simulations, these correlations don't appear to grow with lattice size. Basically the Monte Carlo results are discouraging for superconductivity, particularly for a nodeless superconducting state.

Thus it seems to me that there is a possibility that what we're seeing in these beautiful antiferromagnetic systems is that superconductivity is being suppressed by the antiferromagnetic fluctuations. Previously Lee and Read raised the question, why is T_c so low, and Chandra has discussed the role of dynamic pair breaking on T_1^{-1} near T_c which is then suppressed as the superconducting gap opens. I believe that there is something in this system, that at least I don't understand, that causes s-wave pairing, and that in addition there are antiferromagnetic fluctuations that suppress T_c. Then below T_c, as the superconducting gap opens, these fluctuations are suppressed, giving rise to a large $2\Delta_0/kT_c$ ratio. If this is the case, those of us who are working on the Hubbard model don't have in the phonon, exciton, charge transfer, or "something else" that we need for superconductivity, and we're being fooled by our ability to get agreement with the magnetic properties.

D. Pines: Doug, if I understand you, you are suggesting that the antiferromagnetic correlations are not the origin of superconductivity...that something else would cause T_c to be very much larger, and therefore they are screwing it up in some sense. And then there's a correlation between T_c and ω_{SF}: "the better the antiferromagnetic correlation, the more they destroy the ability of the system to be, say, a room-temperature superconductor."

D. J. Scalapino: Yes. We clearly need to understand the pair-breaking spectral weight, but I could believe that the correlation between T_c and the spin fluctuation frequency is reflecting the suppression of T_c by the spin fluctuations.

J. R. Schrieffer: I would like to go on record as saying I totally disagree that the primary role of the antiferromagnetic fluctuations is to suppress T_c. I believe that these fluctuations help bring about superconductivity and probably dominate the physics of the pairing in these systems. I agree with Doug, however, that just as thermally excited phonons can depress the transition temperature in conventional materials such as lead, the spin fluctuations can act similarly here.

D. Pines: Bob to clarify things, you totally disagree with Doug and totally agree with me.

J. R. Schrieffer: Okay, I agree.

C. M. Varma: I want to make a comment on this thing of heavy fermions, and also maybe liquid ^{3}He. In heavy fermions there certainly is superconductivity promoted by antiferromagnetic fluctuations, and now we have detailed phase diagrams in which we understand the relationship between the two. And that is a problem which, in fact, maps on at low temperatures to – after you have done the Kondo problem, and the interaction between two Kondo impurities, and so on – maps on indeed to, essentially, a t-J model with J always of order t. And in that case, we end up with an effective t let's say on the scale of 100 degrees, and an effective T_c on the scale of one degree. So you again end up with T_c/E_F in the t-J model, with J of order t, on the scale of 10^{-2}.

Let's take another problem: liquid ^{3}He, which Rice, and Vollhardt, and others have argued is really very well represented by the Hubbard model. Now here we have a problem in which the coupling constant as measured by the effective mass is considerably larger, at least a factor of two larger than the copper oxide materials that we are dealing with; m*/m is six, the compressibility is 50 – you know, it's very enhanced there – and we have there a problem in which it's completely clear, starting from the original ideas of Berk and Schrieffer, that paramagnons, are driving superconductivity and in that case because they are ferromagnetic, it's triplet, but the physics is essentially the same – and we're ending up experimentally with a T_c/E_F of the order of 10^{-3}. And this is another reason that I think it's most unlikely that the copper-oxides have similar physics reason that I think it's most.

D. Pines: I'd like to respond to Chandra. I've also thought about ^{3}He, and there one is dealing with paramagnons, as you pointed out; the characteristic temperature you want to be looking at is not E_F, but it's $(1+F_o^a) E_F$ where $(1+F_o^a)$ is a nontrivial correction.

C. M. Varma: I was comparing these parameters.

D. Pines: No, let me continue. And that the other point is that there, because you're dealing with ferromagnetic, paramagnons, the effect goes away as you go to larger q, and that is what is preventing you from getting something like T_c of the order of the characteristic spin fluctuation energy. On the other hand, in heavy electrons, you're seeing that, and that's what we seem to be seeing in this case. So I think it's good to look at these systems. What we're dealing with, if Andy and Hartmut and I are right, is a theory of antiferromagnetic paramagnons. I wish there were a catchy term for them – antiferromagnons, whatever. But that's really what our theory is. It's a phenomenological theory in the same spirit as the paramagnon theory.

J. R. Schrieffer: I'm sorry to disagree. I think there is a real confusion in people who are not deeply involved in this whole issue that antiparamagnons, like phonons, like paramagnons, when exchanged, create an attraction. Doug Scalapino proved this is definitively not true, at least in an s-wave context; you can get something weak in d-wave, and that weak d-wave thing will not explain high T_c, and I totally agree with Chandra. What I believe, and I think is really deeply true, is that the exchange is not an antiparamagnon at all, but is – I don't know what you want to call it, it is this bag, or what have you – it's not an oscillating thing in space, but it's a depression of a gap, which really causes an attraction which is very, very different. It's not one exchange of a boson, but it's two, etc. So I want to make sure that no one comes away thinking that I believe, at least, that this is the exchange of an antiparamagnon, which I am convinced makes a repulsion.

Neutron Scattering Studies of Spin Fluctuations in Metallic $YBa_2Cu_3O_{6+x}$

J. M. Tranquada

Physics Department, Brookhaven National Laboratory, Upton, NY 11973

In principle neutron scattering measurements on cuprate superconductors are capable of providing a direct measure of the imaginary part of the dynamic susceptibility $\chi''(Q,\omega)$, a function of considerable interest to theorists studying the mechanisms of superconductivity in these compounds. In practice the measurements are limited by the availability of large single crystals of suitable quality. As part of a continuing project at Brookhaven, three large single crystals (~ 1 cm^3 each) of orthorhombic $YBa_2Cu_3O_{6+x}$ have been studied. These crystals, corresponding to $x = 0.4$, 0.45, and 0.5 (determined from the lattice parameters), exhibit superconducting transitions at $T_c = 25$, 45, and 50 K, respectively, in ac susceptibility measurements. Detailed descriptions of the neutron scattering results have been presented elsewhere.[1-3] (Related measurements on a crystal with $x = 0.45$ and $T_c = 35$ K have been reported by Rossat-Mignod et al.[4]) The discussion here will be limited to a brief overview of the results and their significance.

Because of the current neutron drought at Brookhaven, the initial scattering measurements were performed at Risø National Laboratory in Denmark.[1] Scans were performed at fixed energy transfer in (h,h,l) zone, scanning the momentum component h across the 2D antiferromagnetic Bragg rod at $Q = (\frac{1}{2}, \frac{1}{2}, -1.8)$ (in reciprocal lattice units). (Scans were performed at $l = -1.8$ because that point corresponds to a peak in the inelastic structure factor which modulates the scattering due to intra-bilayer correlations.) In antiferromagnetic, insulating crystals, spin waves disperse from the 2D rod with a high velocity due to the strong superexchange in the CuO_2 planes.[5,6] Measurements at $\Delta E = 3$ meV and $T = 11$ K showed a peak on the rod which was reasonably intense for the $x = 0.4$ and 0.45 crystals, but dramatically reduced for $x = 0.5$; in each case the peak width was greater than the spectrometer resolution. The cross section decreased significantly with increasing temperature for $x = 0.4$ and 0.45.

In order to extend the measurements to higher excitation energies, the experiments were continued at Chalk River Nuclear Laboratories in Ontario, Canada.[3] As shown in the figure, the cross section on the 2D AF rod observed at $T = 12$ K decreases with increasing excitation energy for the $x = 0.45$, whereas it rises for $x = 0.5$. The opposite behaviors for the two samples can be explained by an order of magnitude difference in the 2D spin-spin correlation lengths, the correlation length being smaller for the $x = 0.5$ crystal. The longer correlation length in the $x = 0.4$ and 0.45 crystals probably indicates the presence of magnetic and electronic inhomogeneities in these samples, which would be consistent with the results of a recent μSR study.[7] The short correlation length in the $x = 0.5$ crystal, together with the strength of the scattering, suggest that in this sample the CuO_2 planes are more homogeneously doped with holes.

The cross section at 15 meV observed for the $x = 0.5$ crystal at 12 K is comparable to that observed for an insulating sample. The fact that the spin fluctuations survive in metallic $YBa_2Cu_3O_{6+x}$ and remain peaked about the antiferromagnetic

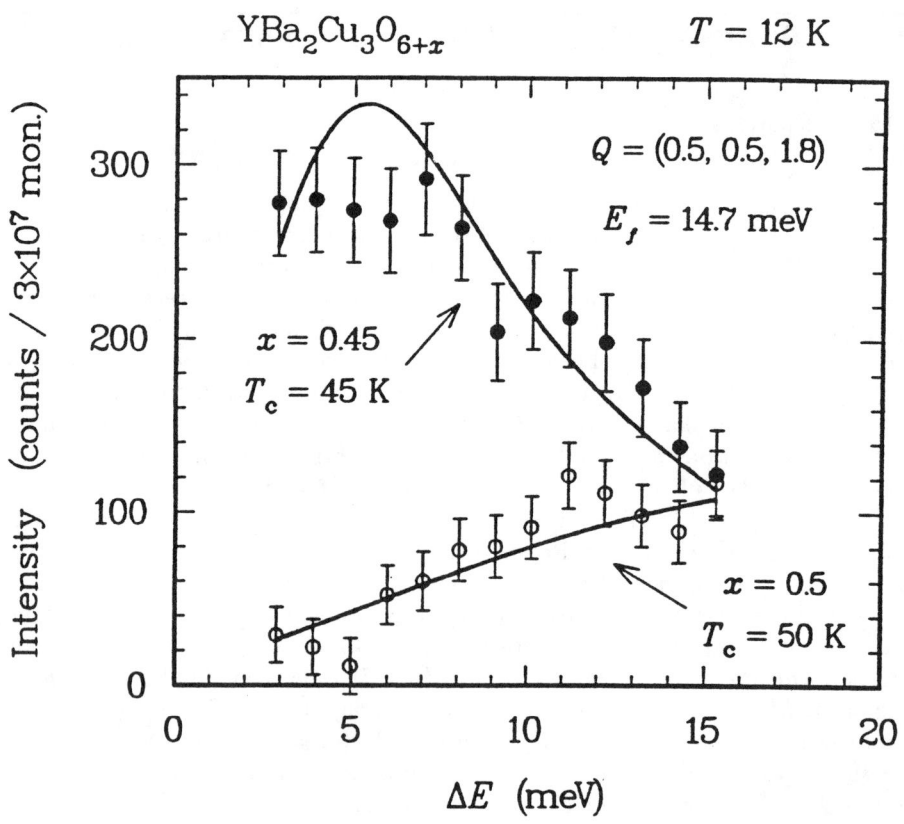

FIGURE 1 Constant-Q scans measured at $Q = (\frac{1}{2}, \frac{1}{2}, 1.8)$ and $T = 12$ K on the $x = 0.45$ (solid circles) and $x = 0.5$ (open circles) crystals. A constant background value has been subtracted from each set of data. The solid lines were calculated using a paramagnetic scattering function as discussed in Ref. 3.

rod is consistent with the enhancement of the nuclear relaxation rate for Cu relative to that for O as observed in $x = 1$ powders.[8] The strength of the inelastic AF cross section at a temperature well below T_c suggests that these spin fluctuations are not significantly suppressed in the superconducting state. This result needs to be reconciled with the observed scaling of the Cu and O relaxation rates below T_c.

The work discussed here is the result of collaborations with J. Als-Nielsen, W. J. L. Buyers, H. Chou, T. E. Mason, M. Nielsen, M. Sato, S. Shamoto, and G. Shirane. Research at Brookhaven National Laboratory is supported by the U.S. Department of Energy, Division of Materials Sciences under contract DE-AC02-76CH00016.

REFERENCES

1. G. Shirane, J. Als-Nielsen, M. Nielsen, J. M. Tranquada, H. Chou, S. Shamoto, and M. Sato, Phys. Rev. B (to be published).
2. J. M. Tranquada, W. J. L. Buyers, H. Chou, T. E. Mason, M. Sato, S. Shamoto, and G. Shirane (submitted to Phys. Rev. Lett.).
3. J. M. Tranquada and G. Shirane, in *Dynamics of Magnetic Fluctuations in High Temperature Superconductors*, edited by G. Reiter (Plenum, to be published).
4. J. Rossat-Mignod, L. P. Regnault, M. J. Jurgens, C. Vettier, P. Burlet, J. Y. Henry, and G. Lapertot, in *Proc. of the Santa Fe Conference on the Physics of Highly Correlated Electron Systems*, Physica B (to be published).
5. J. M. Tranquada, G. Shirane, B. Keimer, S. Shamoto, and M. Sato, Phys. Rev. B **40**, 4503 (1989).
6. C. Vettier, P. Burlet, J. Y. Henry, M. J. Jurgens, G. Lapertot, L. P. Regnault, and J. Rossat-Mignod, Physica Scripta **T29**, 110 (1989).
7. R. F. Kiefl *et al.*, Phys. Rev. Lett. **63**, 2136 (1989).
8. P. C. Hammel, M. Takigawa, R. H. Heffner, Z. Fisk, and K. C. Ott, Phys. Rev. Lett. **63**, 1992 (1989).

Some Comments on Nuclear Spin Relaxation Below T_c
D. J. Scalapino
Department of Physics, University of California, Santa Barbara, CA 93106

It is possible to obtain a remarkably good fit of the T_1^{-1} relaxation times of the planar ^{63}Cu and ^{17}O in YBa$_2$Cu$_3$O$_7$ using an RPA form for $\chi(q,\omega)$.[1] Here $\chi(q,\omega) = \chi_o(q,\omega)/(1 - U\chi_o(q,\omega))$ with $\chi_o(q,\omega)$ the non-interacting band-electron susceptibility

$$\chi_o(q,\omega) = \sum_P \frac{f(\epsilon_{p+q}) - f(\epsilon_p)}{\omega - (\epsilon_{p+q} - \epsilon_p) + i\delta}$$

with $\epsilon_p = -2t(\cos p_x + \cos p_y)$. The hyperfine form factors act to filter the spectral weight

$$Im\chi(q,\omega)/\omega|_0 = \frac{Im\chi_o(q,\omega)/\omega|_o}{(1 - U\chi_o(q,0))^2}$$

so that the strong antiferromagnetic contribution is picked up by the ^{63}Cu nuclei and suppressed at the ^{17}O nuclei. In fitting the relaxation data with a single-band Hubbard model, we used both U and the filling $\langle n \rangle = \langle n_{i\uparrow} + n_{i\downarrow}\rangle$ as parameters. Basically we adjusted $\langle n \rangle$ by means of a chemical potential μ so that the system remained paramagnetic at zero temperature, but had a small low-energy scale T^* set by $\mu/4$. Below T^*, the antiferromagnetic fluctuations saturate and T_1^{-1} on the copper follows a linear Korringa behavior. Above T^*, the antiferromagnetic fluctuations are reduced as the temperature is increased, leading to the observed bending over of the Cu T_1^{-1} response. The oxygen hyperfine coupling $\cos^2(q_x/2)$ or $\cos^2(q_y/2)$ makes it relatively insensitive to the $q = (\pi,\pi)$ antiferromagnetic fluctuations, and we find that $(T_1^{-1})_{oxygen}$ has a nearly Korringa-like behavior. These ideas, originally developed within the RPA,[1] are closely related to the work Millis discussed.[2]

Recently Bulut, Bickers and I have extended this approach to the superconducting state by evaluating $\chi_o(q,\omega)$ using the usual G and F propagators characterized by s- ($\Delta_p = \Delta_0$) or d-wave ($\Delta_p = \Delta_0(\cos p_x - \cos p_y)/2$) gaps.[3] In order to understand the response of the system below T_c, I've shown in Figure 1 $\chi_o(q,0)$ versus q and $D_0(q) = Im\chi_o(q,\omega)/\omega|_0$ versus T/T_c for q near zero and (π,π) for s- and d-wave gaps. For both the s- and d-wave gaps, $\chi_o(q,0)$ is of course suppressed at $q = 0$. This is just the effect of singlet spin pairing seen in the Knight shift. For the s-wave gap, this suppression is exponential at low temperatures, while it varies linearly with T for the d-wave gap. The s-wave result appears to be favored by the ^{17}O Knight shift measurements although there remains some uncertainty with respect to the ^{63}Cu results.[4,5] For the s-wave gap, $\chi_o(q,0)$ is also suppressed near (π,π), since $2\Delta_0$ is larger than T^*. This has the effect of reducing T_1^{-1} on the Cu site and can, for a rapidly opening gap ($2\Delta_0/kT_c \sim 8$), suppress the usual Hebel-Slichter[6] peak for Cu. In the case of a d-wave gap, the contribution of the

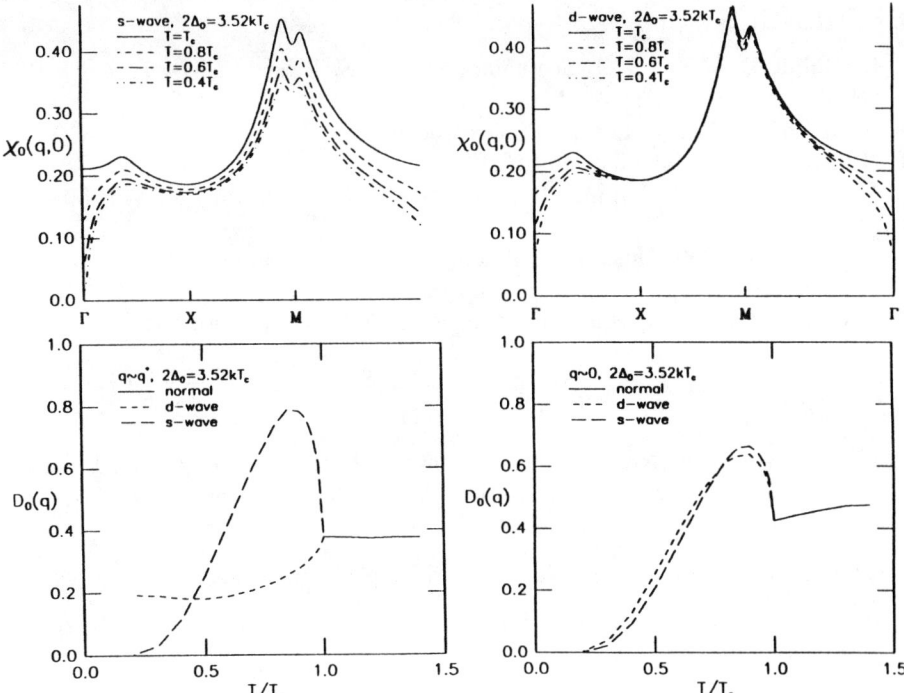

FIGURE 1

The temperature dependence of $D(q) = \mathrm{Im}\chi_o(q,\omega)/\omega|_0$ versus T is shown in the lower part of the viewgraph. For $q \sim 0$, both the s- and d-wave states show a Hebel-Slichter peak. However, for $q = q^*$, the q-value where $\chi(q,0)$ peaks, which is near (π,π), the spectral weight for the d-wave is suppressed below T_c. This is because $\Delta_{p+(\pi,\pi)} = -\Delta_p$ for the d-wave gap, and the resulting coherence factor vanishes.

Putting together the enhancement factor $(1 - U\chi_o(q,0))^{-2}$ and the spectral weight $\mathrm{Im}\chi_o(q,\omega)/\omega|_0$, along with the hyerfine form factors, we have evaluated T_1^{-1} for both the copper and the oxygen. We find that while it is possible to suppress the Hebel-Slichter peak on the Cu using either an s- or d-wave gap, it is difficult to suppress it for the oxygen using an s-wave gap. One needs a strong broadening and a rapidly opening s-wave gap to suppress the Hebel-Slichter behavior on the oxygen. It will be interesting to examine in detail the temperature dependence of the ratio of the relaxation rates on the Cu and O for $T < T_c$.

REFERENCES

1. N. Bulut, D. Hone, D. J. Scalapino, and N. E. Bickers, Phys. Rev. **41** 1797 (1990).
2. A. Millis, H. Monien, and D. Pines, preprint.
3. N. Bulut, D. J. Scalapino, and N. E. Bickers, preprint.
4. M. Takigawa, P. C. Hammel, R. H. Heffner, and Z. Fisk, Phys. Rev. B **39** 7371, (1989).
5. S. E. Barrett, D. J. Durand, C. H. Pennington, C. P. Slichter, T. A. Friedman, J. P. Rice, and D. M. Ginzberg, preprint.
6. L. C. Hebel and C. P. Slichter, Phys. Rev. **113** 1504, (1959).

DISCUSSION

J. M. Tranquada: There's one more point on the neutron scattering. From the data that's available now, I think it's indicating an absence of a gap in the spin fluctuation spectrum around (π, π).

D. J. Scalapino: That's going to pose serious problems for traditional theories.

A. Millis: Sure. It's difficult to reconcile with NMR, also, which seems to show a big gap opening up.

S. Chakravarty: Can I ask you a question about this calculation? The RPA calculation has a divergence, strictly speaking, at finite temperature.

D. J. Scalapino: No, it doesn't if the filling is moved away from half-filling and U is less than a critical value, $1 - U\chi_o(q,0)$ doesn't vanish.

S. Chakravarty: No, no...you tune the calculation to be sufficiently close to the divergence, but not quite at the divergence...

D. J.Scalapino: Correct.

S. Chakravarty: So, there shouldn't be a divergence in the first place...

D. J. Scalapino: And we don't have one.

S. Chakravarty: Wait. If you take the calculation on face value, that $\chi_o(q,\omega)/(1-U\chi_o(q,\omega))$, it has a divergence at a finite temperature, if you don't tune it...

D. J. Scalapino: If I don't dope it with just the right magnitude. Correct.

S. Chakravarty: Now, you tune it sufficiently close, in fact incredibly close between .85 and .86, it's .01 that you're talking about. So you tune it sufficiently close so that the correlation length becomes very large, so that you can pick up that effect.

D. J. Scalapino: Not super large, but large.

S. Chakravarty: Yes, so that you can pick up that effect. So you are relying on an effect which is presumably not there in the first place.

D. J. Scalapino: I think I understand what you are concerned about. We're taking an RPA form, $\chi_o/(1 - U\chi_o)$, and adjusting U and the filling $\langle n \rangle$ to drive the response close to criticality. And you're asking me why I do something like this, when I know RPA is not correct.

The first reason we did it, was to simply see whether a one-component form for $\chi(q,\omega)$, filtered by the hyperfine form factors, could qualitatively reproduce the experimental data. The second reason is that we knew from the Monte Carlo results as well as the conserving approximation calculations of Bickers', that there were antiferromagnetic fluctuations. What we believed was happening as you summed diagrams beyond the RPA or carried out Monte Carlo simulations, was that the antiferromagnetic correlations were being renormalized, and if χ_o was calculated with the dressed propagators, $U\chi_o$ was relatively close to 1. Thus our view is that we are using an RPA form for $\chi(q,\omega)$, but U and $\langle n \rangle$ are not simply related to the bare Coulomb interaction and the doping but are rather quantities which allow us to parameterize $\chi(q,\omega)$.

S. Chakravarty: Yes, you need a fine tuning. That's what I'm objecting to.

D. J. Scalapino: It is a fine tuning. There's no question, we are introducing a low-energy scale $T^* \sim \mu/4$ which is of order 100 to 200 K.

V. Emery: Doug, doesn't that curve already disagree with Takigawa's data?

D. J. Scalapino: No, I don't believe it does.

V. Emery: I thought he showed a beautiful straight line when he plotted the data.

D. J. Scalapino: You see, the straight line they find is for a plot of $(T_1^{-1})_{Cu}$ versus $(T_1^{-1})_O$, but it's on a log-log scale. The relaxation rates decrease by a factor of 10^{-4}, so that on that plot it's difficult to see small variations. It's easier to see what's happening when one plots $R = (T_1^{-1})_{Cu}/(T_1^{-1})_O$ versus T. Then one can see that this ratio varies from a low of 17 to a high of perhaps 21.

V. Emery: So if they plotted that ratio, it wouldn't be a constant, as it appears.

D. J. Scalapino: When you look at a plot of $(T_1^{-1})_{Cu}$ versus $(T_1^{-1})_O$, you see that this ratio remains near its value at T_c. But as I've said, if you look at R versus T/T_c, you do see some structure. I believe that this structure contains interesting information because the hyperfine from factors for Cu allow the (π,π) fluctuations to enter T_1^{-1} while for O the form factor filters them out.

7. PHENOMENOLOGY

Chair — A. P. Malozemoff

S. Doniach
Department of Applied Physics
Stanford University, Stanford, CA 94305

The phenomenology of flux motion in high temperature Superconductors

1. INTRODUCTION

Since the early days of high Tc it has been clear that flux motion is an important part of the phenomenology of these materials in regions of the H-T plane well below the $H_{c2}(T)$ line, in contrast to conventional superconductors. As measurements on single crystals and new compounds such as $Bi_2Sr_2CaCu_2O_8$ (Bi2212, or "BSCCO"), became available, an improved understanding of the nature of these differences is becoming possible.

In this report, we will argue that two primary physical differences distinguish the cuprate-based materials from conventional superconductivity: (1) the short Ginzburg-Landau coherence length, on the scale of 12-15 Å(a few unit cell spacings), and (2) the weak inter-layer coupling of these layered compounds. Note that the discussion will be confined throughout to single crystal properties, leaving aside all consideration of ceramics which are clearly in the "granular superconductivity universality class" and hence behave quite differently from single crystals.

The first thing an experiencd superconductivity person notices when looking at the behaviour of the resistivity in a magnetic field is the similarity to what happens in conventional two- dimensional superconductors, or thin films[1] [Fig. 1]. It appears clear that Abrikosov vortex lines inserted in these strongly type-2 materials above H_{C1} are able to migrate, thus leading to dissipation. This immediately leads to the

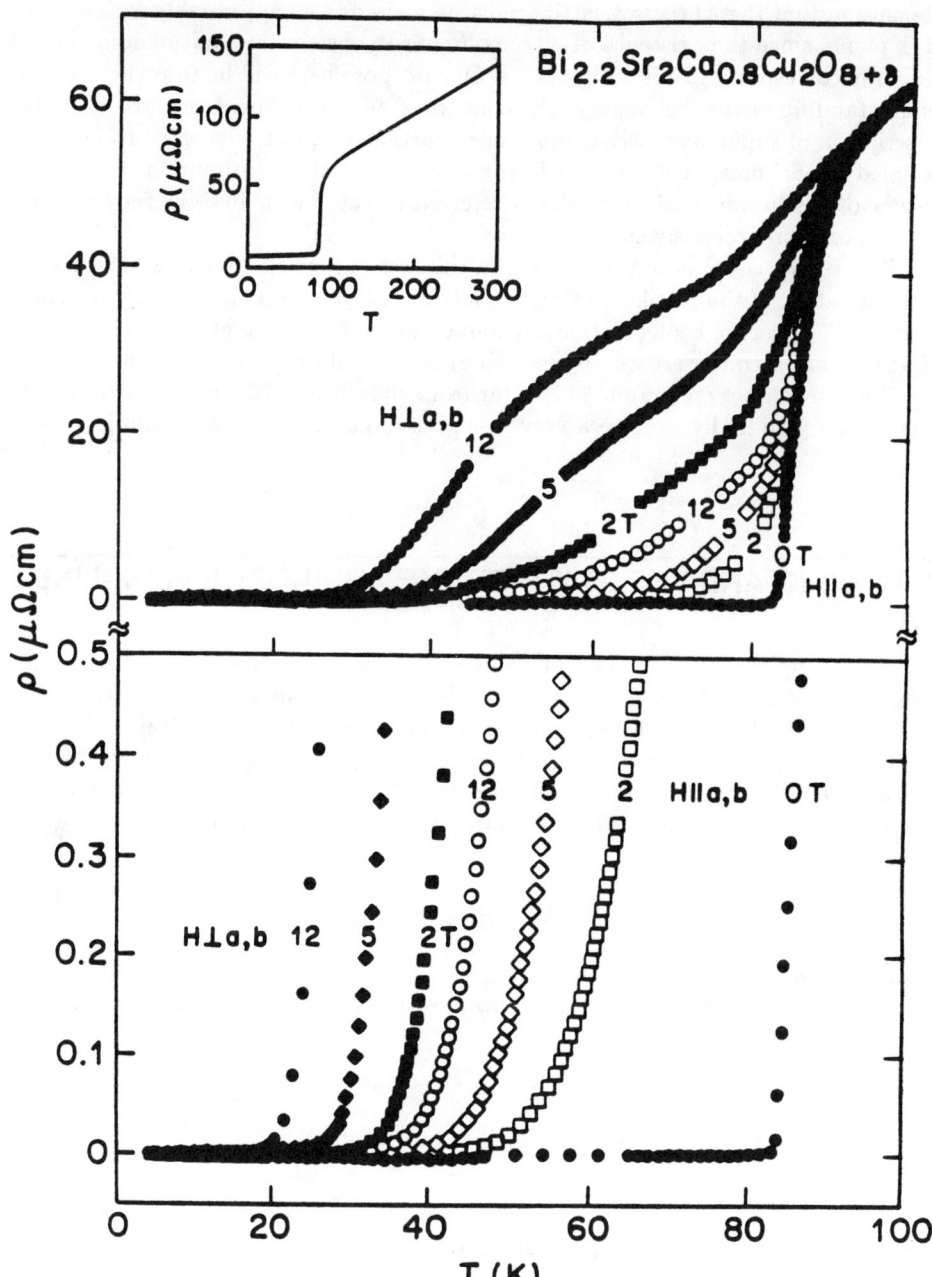

FIGURE 1 Temperature dependence of the electrical resistivity of BSCCO in four selected magnetic fields, oriented $\parallel$ to and $\perp$ to the CuO_2 planes. The lower part of the Fig. is a magnification of about a factor 100. (From Palstra et al.[1])

conclusion that the Abrikosov lattice must be melted in an appreciable region of the H-T plane, since the existence of a true ohmic response in the limit of zero current can only correspond to a fluid-like state. Our purpose here will be to review current understanding of this behavior in the context of an anisotropic Ginzburg - Landau description of superconductivity, and more particularly, in terms of a "Josephson - coupled layers" picture of the high Tc materials. We make no attempt at a complete survey of the literature. The reader is referred to the review by Malozemoff[2] for a much more comprehensive summary.

The report is organized as follows. In the first part of the report we discuss the general nature of the Abrikosov flux line lattice in layered materials: basic effects of pinning(§2), vortex configurations, reconnection and melting of the vortex lattice (§3-5). In the second part we discuss experimental data on vortex kinetics: close to equilibrium (ohmic regime) in §6 and far from equilibrium (Bean critical state and flux creep) §7. Finally we give a very short discussion of pinning mechanisms,§8.

2. PINNING AND THE STATE OF THE ABRIKOSOV FLUX-LINE LATTICE

It was pointed out by Larkin[3] almost 20 years ago that the nature of the long range order of the Abrikosov lattice is severely modified by the existence of randomly distributed pinning centers. The argument is very general and would apply to any periodic order parameter with random pinning (e.g. charge or spin density waves). Larkin writes the free energy of the Abrikosov lattice in terms of a slowly changing displacement $\vec{u}_i(r)$ away from the periodic positions of the flux lines

$$\vec{R}_i = \vec{R}_i^0 + \vec{u}_i(r) \qquad (1)$$

where $\vec{R}_i^0$ are the periodic positions.

Then using elasticity theory and to lowest order in $\vec{u}$, the Ginzburg Landau free energy contains a term

$$\delta F = \frac{1}{2}\int \{(C_{11} - C_{66})(\nabla_\alpha u_\alpha)^2 + C_{66}(\nabla_\alpha u_\beta)^2 + C_{44}(\frac{\partial u_\alpha}{\partial z})^2\}d^3x \qquad (2)$$

where α,β refer to x,y components (implied summation) and the z axis is along the magnetic field. Introducing coupling to a pinning potential, δF may be written in Fourier space as

$$\delta F_{\text{tot}} = \sum_k \{\frac{C_{11} - C_{66}}{2}(k_\alpha u_{k,\alpha})^2 + \frac{C_{66}k_\alpha^2 + C_{44}k_z^2}{2}u_{k,\beta}^2 + u_{k,\alpha}f_{k,\alpha}\}, \qquad (3)$$

where f_k is the Fourier transform of an effective random pinning potential (assumed isotropic in the xy plane). Defining the elastic propagator as

$$\mathbf{G}_{\alpha\beta} = [\frac{P_T}{C_{66}k_\perp^2 + C_{44}k_z^2} + \frac{P_L}{C_{11}k_\perp^2 + C_{44}k_z^2}] \tag{4}$$

where $P_T = (\delta_{\alpha\beta} - k_\alpha k_\beta/k_\perp^2)$ is the transverse projector and $P_L = k_\alpha k_\beta/k_\perp^2$ is the longitudinal projector (with $k_\perp^2 = k_x^2 + k_y^2$), the pair correlation function of the lattice displacements then becomes

$$\overline{[u(R)-u(0)]^2} = \sum_k (1 - \cos \vec{k}\cdot\vec{R}) \bar{f}_k^2 \{Tr[\mathbf{G}(\mathbf{k})^2]\}$$

$$= \frac{\bar{f}_k^2}{4\pi}\{\sqrt{(R_\perp^2 C_{44} + R_\parallel^2 C_{11})/(C_{44}C_{11})^2} + \sqrt{(R_\perp^2 C_{44} + R_\parallel^2 C_{66})/(C_{44}C_{11})^2}\} \tag{5}$$

That is, the pair correlation function grows linearly with $|R|$ so that the lattice always loses its long range order. As pointed out by Larkin, because the defects are fixed in space they interact directly with the deformation of the lattice and not with its gradient (as would be the case for a conventional crystal with impurities). Thus even a weak random pinning potential will always be a relevant operator for destroying the long range periodic order of the flux lattice.

At this point a crucial question arises: even though the long range periodicity is destroyed, what is the energy needed to induce plastic flow, or shear, in the vortex lattice? The vortices after all are still coupled by fairly long range forces (falling off exponentially above the London penetration depth). One way of looking at a liquid is as a system in which dislocation loops grow without bound above a certain nucleation size, so that an infinitesimal shear stress can always induce a corresponding flow. Clearly this *cannot* happen at $T = 0$. The question of whether it can happen at any finite T or whether it can only happen above a critical glass melting temperature, T_g, is still open, though Matthew Fisher[4] has given a rather plausible argument for a second order phase transition at a finite $T_g(H)$. Intuitively it seems clear that a sufficiently large T will be needed to induce the entropic explosion of dislocation loops. Thus we expect a vortex-density, i.e. field-dependent "glass transition line" in the H-T plane below which ohmic resistivity will be *strictly zero*, while above which the system will exhibit a finite ohmic resistance, albeit thermally activated as a result of the existence of the pinning.

Recent experiments by Koch et al.[5] have identified a glass transition temperature $T_g(H)$ by seeing where the I-V curves for expitaxial thin films of YBCO cross over from exponential, i.e. $V \sim \exp(-I_o/I)$, $(T < T_g)$ to ohmic through a power law regime (i.e., $\log V \propto n(\log I)$ where $n = (z+1)/(d-1)$ and z is a dynamic critical exponent. As T_g is approached from above, they argue that the ohmic (linear) component should scale to zero as a power law of $(T/T_g - 1)$,

$$\rho_L \sim (T/T_g - 1)^{\nu(z+2-d)} \tag{6}$$

with ν the critical exponent for the coherence length of the glass transition. Fitting the observed power law of the I-V curve at T_g and the observed temperature dependence of the ohmic component at $T > T_g$, and of a cross-over current j_0^-, they find a consistent set of values of $\nu \sim 1.5 - 2.0$ and $z \cong 4.8$. On the other hand, we note here that a straightforward analysis of ohmic data (see §6 below) suggests an Arrhenius-type activated behavior in apparent conflict with the analysis of Koch et al. It's also possible that the glass transition is First order. This is a pretty new topic and a consistent resolution of these points of view is not yet in.

3. EFFECTS OF WEAK INTERLAYER COUPLING ON VORTEX CONFIGURATIONS

The first thing to be said is that we expect a strong anisotropy in properties such as field-dependent resistivity below $H_{C2}(T)$ as a result of the very large anisotropy of electronic coupling between CuO_2 planes in the high Tc materials. This should happen for compounds like $YBa_2Cu_3O_7$ (123) which is fairly anisotropic, but even more markedly for BSCCO which has a very large anistropy. In fact this is already clear from single crystal measurements (Fig. 1) where the resistivity falls much more rapidly with T for H parallel to the ab-planes than for $H \perp ab$. This may even suggest the existence of intrinsic pinning in which vortex lines become "trapped" between CuO_2 planes for $H \parallel ab$. (see §8 below for further discussion).

The discussion which follows will be focused on the case of $H \perp ab$. If pinning is neglected for a moment, then as discussed by Nelson,[6] the entropic "wiggling" of the vortex lines will be exacerbated by the short coherence lengths ξ_{ab}, ξ_c (c-axis normal to the CuO_2 planes) since the energy to grow additional lengths of vortex core scales as $\xi_{ab}^2 \xi_c$. Now it's clear that for the highly anisotropic cuprate materials, ξ_c will scale down for $T < T_c(H)$ until it rapidly becomes of order the inter-plane spacing, d. Below this line in the H-T plane, the anisotropic Ginzburg-Landau model used by Nelson breaks down and it is more appropriate to think of the vortices in the context of the Lawrence-Doniach[7] model as strings of 2-dimensional vortices connected by weak interplanar cores (Fig. 2).

To make this more quantitative, we write the LD model as

$$F_{LD} = \sum_i T_c^2 N(\varepsilon_F) \int d^2x \{(\xi_{ab}^0)^2 |\nabla \psi_i|^2 + (T/T_c)^{-1} \psi_i^2 + b\psi_i^4 - g[\psi_i * \psi_{i+1} + h.c.]\} \quad (7)$$

where $\psi(\vec{x})$ is a dimensionless order parameter for plane i, $N(\varepsilon_F)$ is the two dimensional electronic density of states at the Fermi level and g is a dimensionless interplanar coupling constant. (Note that the energy scale is often expressed in terms of the thermodynamic critical field, H_c, via $T_c^2 N(\varepsilon_F) \propto H_c^2/8\pi$.)

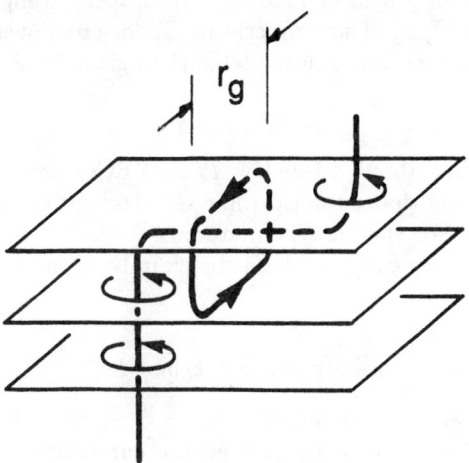

FIGURE 2 An Abrikosov vortex seen as a series of 2D vortices connected by interplanar strings of width r_g.

The effective Ginzburg-Landau coherence length at temperature T is then given by

$$[\xi^0_{ab}/\xi_{ab}(T)]^2 = (1 - T/Tc), \tag{8}$$

and in the continuum limit, $\xi_c(T)$ is given by

$$\xi_c(T)/d = \sqrt{g/(1 - T/T_c)}. \tag{9}$$

The anisotropic coherence lengths are often expressed in terms of the Ginzburg - Landau "effective masses" M_{ab}, M_c for pair motion in the ab planes and normal to them. The correspondence of the ratio M_c/M_{ab} to the coupling constant g is then given by

$$(\xi^0_c/\xi^0_{ab})^2 = \frac{M_{ab}}{M_c} = g(d/\xi^0_{ab})^2. \tag{10}$$

Putting in $\xi^0_{ab} \approx 12\text{Å}$ for the CuO_2 planes, one then has $(d/\xi^0_{ab} \cong 1)$, leading to $(M_{ab}/M_c) \cong g$.

Measurement of the anisotropy ratio M_{ab}/M_c may be made in a number of different ways: the most reliable appears to be via torque magnetometry:- using this technique, Farrel et al.[8] find

$$\sqrt{g} \cong \gamma \equiv (M_{ab}/M_c)^{\frac{1}{2}} \cong 1/55, \tag{11}$$

for BSSCO at $77.5°K$ in a field of 1 Tesla. This is to be compared with a value of $\gamma \cong 1/5$ for $Y_1Ba_2Cu_3O_{1-\delta}$. The temperature T_x for cross over from the anistropic GL model to the Lawrence Doniach model is then given by $\xi_c| \sim d$ leading to

$$(1 - T_x/T_c) \cong g. \tag{12}$$

This results in T_x lying about 4% below T_c in YBCO and within .03% of T_c in BSCCO! So we conclude that an anisotropic GL description really breaks down in most of the H-T plane for the high Tc materials.

In the limit $\xi_c \ll d$, we can consider a vortex in terms of two dimensional vortex currents

$$\frac{\partial}{\partial \theta}\psi(r - R_i) = 1/|r - R_i| \tag{13}$$

for a vortex centered at R_i in plane i.

If R_i is displaced relative to R_{i+1}, then the currents will tunnel between the planes via the coupling $g|\psi|^2 \cos(\phi_i - \phi_{i+1})$ where we write $\psi(r) = |\psi|e^{i\phi(r)}$ and treat $|\psi|$ as constant. For a given pair of planes minimization of the LD free energy leads to a two dimensional Sine-Gordon equation

$$-|\xi_{ab}^0 \nabla \phi|^2 + g\sin(\phi) = 0 \tag{14}$$

where $\phi(x) \equiv \phi_{i+1}(x) - \phi_i(x)$.

The simplest non-trivial solution is a sine-Gordon kink, leading to a "string" or vortex with core running between the planes, in which the relative phase goes from zero to 2π over a length scale determined by

$$r_g = \xi_{ab}^0/\sqrt{g}. \tag{15}$$

As the interplane coupling g goes to zero, the scale of this 'vortex core' diverges. This is analagous to what happens to dislocations in smectic liquid crystals[9]. When the two vortices in adjacent planes lie on top of one another, the phase difference of the order parameter will be zero everywhere and the total energy will be that of the condensate which we can take as our zero of energy. As the vortices are separated beyond $2r_g$, an interplanar core, or "string" will start to form with energy $\approx g^{1/2}$ per unit length. So the interplanar coupling between two two-dimensional vortices may be written

$$\begin{aligned}V(R_i - R_{i+1}) &\cong 0 \text{ for } |R_i - R_{i+1}| < 2r_g, \\ &\cong Ag^{1/2}(|R_i - R_{i+1}| - 2r_g) \text{ for } |R_i - R_{i+1}| > 2r_g.\end{aligned} \tag{16}$$

A 3-dimensional Arbrikosov vortex may then be viewed as a series of 2D vortices connected by interplanar strings (Fig. 2).

To get an estimate of the actual magnitude of r_g, take $\xi_{ab}^0 \approx 12\text{Å}$ in equation (15) leading to

$$r_g \approx 60\text{Å for YBCO and } r_g \approx 660\text{Å for BSCCO}. \tag{17}$$

4. INTERACTIONS BETWEEN ABRIKOSOV VORTICES IN LAYERED SUPERCONDUCTORS

What happens when more than one 3D vortex is injected into a layered material? In the continuum limit there is a large barrier to overlap of two vortex cores which scales as ξ_c^2/ξ_{ab}^3, (treating the interaction between two elementary lengths of vortex as $\vec{dl_1} \cdot \vec{dl_2}/r_{12}^3$). However in the limit of weak interlayer coupling this is considerably reduced since a pair of vortices in each of two adjacent planes will start to "fuse", i.e., will no longer have distinguishable interplanar cores, when their relative distances become of order $2r_g$. The interaction energy for 2D vortices may be written as[10]

$$V_{2D}(R_i - R_i') \cong +q^2 \ln\left|\frac{\xi_{ab}}{R_i - R_i'}\right|, \qquad (17)$$

where $q^2 = (\phi_0/4\pi)^2 d_0/\lambda^2$, with λ the penetration depth and d_0 the layer thickness, measures an effective charge for the vortices. (A positive background term needs to be subtracted to take account of the magnetic energy.) The barrier for fusion of two Abrikosov vortices may then be measured relative to the repulsion at the mean spacing $R_{av} = \sqrt{\phi_0/B}$, with ϕ_0 the unit flux quantum $= hc/2e$ as,

$$V_{max} \cong q^2 \ln(R_{av}\sqrt{g}/\xi_{ab}). \qquad (18)$$

This will be of order $K_B T$ for temperatures approaching the effective Kosterlitz-Thouless transition temperature for an isolated plane. So we see that the physical picture of a melted Abrikosov lattice in a layered material is very different from that in a truly 3D material: rather than "tangling," vortices will have a strong tendency to re-connect (see Fig. 3).

We can now distinguish two different regimes for the vortex lattice: if the density is such that the average vortex spacing $R_{av} \gg r_g$, then the vortices will still behave as independent 3 dimensional objects, with a finite probability of reconnection. On the other hand for $R_{av} \lesssim r_g$, the fluxoids in a given 2D plane will lose the identity of their 'partner' fluxoids in adjacent planes, and the statistical mechanics will be that of a set of 2D vortex gases moving in a random potential due to the adjacent planes. In terms of field this cross over will occur at $B \approx \phi_0 r_g^2 \simeq 4.5$ kGauss for BSCCO but would not occur below H_{C2} for YBCO.

5. MELTING OF THE ABRIKOSOV LATTICE IN LAYERED SUPERCONDUCTORS

The simplest semiquantitative estimate of the melting temperature (at given H) of the Abrikosov lattice is obtained by applying the Lindeman criterion which

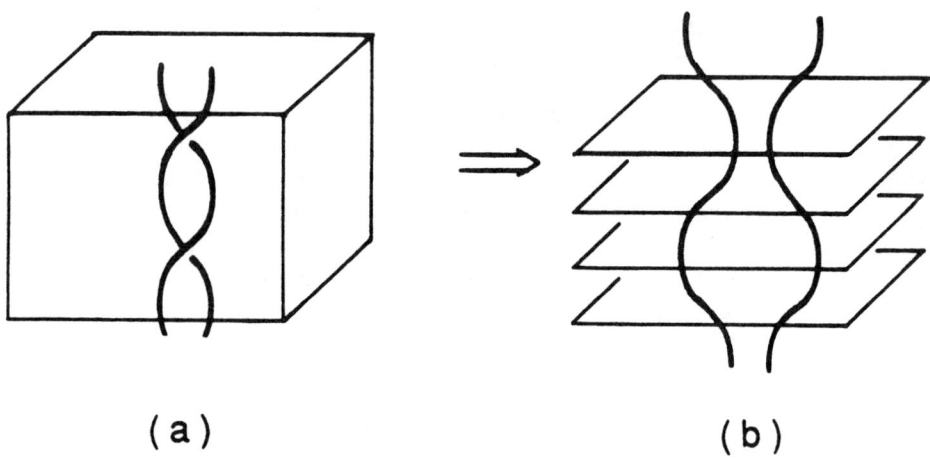

FIGURE 3 "Tangling" of Abrikosov vortices (a), versus reconnection (b).

asserts that a lattice will melt when the mean square displacement $<u^2>$ (with u defined in Eq. (1) becomes of order cR_{av}, where R_{av} is the lattice parameter of the vortex lattice. ($\sqrt{3}R_{av}/2 = \phi_0/B$ for a triangular lattice) and c is an empirical constant of order 0.1. $<u^2>$ may be estimated using the elastic propagator defined in §2 as

$$<u^2> = K_B T \sum_k Tr\mathbf{G}(\vec{k})$$
$$= K_B T \sum_k [\frac{1}{C_{66}k_\perp^2 + C_{44}k_z^2} + \frac{1}{C_{11}k_\perp^2 + C_{44}k_z^2}]. \quad (19)$$

To evaluate this for the vortex lattice, one needs to take account of the non-local character of the vortex lattice: for

$$k \gtrsim k_\lambda = \frac{2\pi}{\lambda}\sqrt{1 - H/H_{C2}}, \quad (20)$$

where λ is the London penetration depth, the vortex lattice is much softer than in the long wavelength limit (Brandt[11], Houghton, et al.[12]), i.e. the elastic constants are strongly k-dependent.

Houghton et al. have estimated $<u^2>$ for the cuprate superconductors principally within the anisotropic Ginzburg-Landau description and find it to be dramatically enhanced relative to conventional superconductors, as a result of a factor (ξ_{ab}/ξ_c) which is large as a result of the weak interplanar coupling. Overall, Houghton et al. find a melting criterion

$$\frac{T/T_{\text{BCS}}}{\sqrt{1-T/T_{\text{BCS}}}} \cong \alpha \frac{1-B/H_{C2}}{\sqrt{1-B/H_{C2}}} \qquad (21)$$

where α is estimated to be of order 0.1 for the cuprate superconductors (using $\xi_{ab}/\xi_c \cong 100, H_{C2} \cong 100\text{T}, T_{\text{BCS}} \cong 100°K$) which suggests that the vortex lattice will be melted over a large region of the H-T plane.

5.1 STATISTICAL MECHANICS OF MELTING

The Lindeman criterion, however, only gives a rule of thumb for estimating the melting temperature. Nelson[13] has given a more detailed discussion of the melting criterion based on a comparison of the free energy of the flux line liquid with that of the crystal. (c.f. an earlier discussion of melting of smectics by Deutsch and Doniach[14]), still within the anisotropic GL model.

Nelson points out that the expression for the free energy of a 3- dimensional Abrikosov lattice at finite temperature T may be mapped on to that of a two dimensional quantum gas of bosons at zero temperature. In this mapping vortices are represented by the "world lines" of the bosons. These exhibit fluctuations as a result of zero-point motion, corresponding to the motion of the vortices in the third dimension which undergo an entropic wandering at finite temperature. (Note however that the vortices are strictly classical objects so that the symmetrization which is an essential feature of the boson model should not be taken literally in this mapping). In the anisotropic Ginzburg - Landau model, the core energy per unit length of a vortex may be written as

$$\tilde{\epsilon}_1 = (\xi_c/\xi_{ab})^2 \epsilon_1 \qquad (22)$$

where ϵ_1 would be the core energy $\sim \xi^2 T_{\text{BCS}}^2 N(\varepsilon_F)$ in an isotropic superconductor. In the boson mapping $K_B T$ becomes an effective "Planck's constant" for the 2D quantum boson gas, and $\tilde{\epsilon}_1$ plays the role of the boson mass. Since $\tilde{\varepsilon} \to 0$ as $T \to T_c$ for low fields, Nelson argues that the melting temperature will drop rapidly at low flux line density (low field) leading to a kind of "re-entrant" phase diagram (Fig. 4).

5.2 EFFECTS OF RECONNECTION ON MELTING

The effects of reconnection on the melting of the Abrikosov lattice seem likely to lower the melting temperature relative to that which would be predicted for models in which the vortices can only tangle but cannot reconnect. This lowering effect may be argued on a number of grounds. As discussed in §4, at sufficiently high vortex density (i.e. applied field), such that the mean vortex spacing R_{av} is $< r_g$, the system may be viewed as a set of two dimensional vortex lattices, each one in a random field due to the influence of the vortices in adjacent planes. As orginally shown by Huberman and Doniach,[15] and further discussed by D. S. Fisher,[16] the

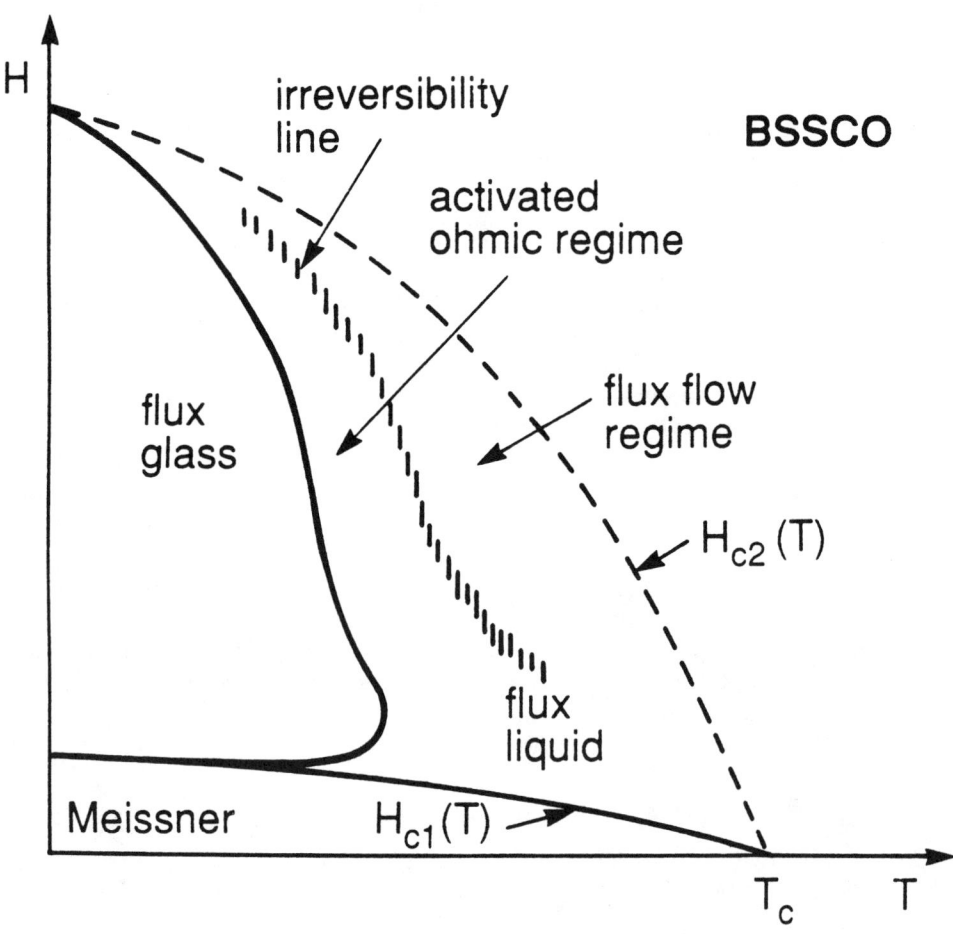

FIGURE 4 Schematic phase diagrams for 3D single crystals of BSCCO for magnetic field ⊥ to the CuO_2 planes. Adapted from Nelson.[13]

2D vortex lattice melting temperature is very much smaller than the Kosterlitz-Thouless transition temperature, with lower limit $T_M^{2D}/T_{KT} = 1/4\pi\sqrt{3}$, which in turn is lower than the BCS transition temperature. The effects of the random fields due to adjacent planes are hard to predict, but seem unlikely to alter the above result drastically. Thus the quasi-2-D character resulting from weak interplanar coupling seems likely to lead to a lower melting temperature.

At lower densities of the vortex lattice, the effect of reconnection is to relieve the constraints imposed by entanglement i.e. to increase the entropy of the system at a given temperature. This in turn seems likely to drive down the melting temperature.

Currently Seungoh Ryu, A. Kapitulnik and S. Doniach[17] are running a numerical Monte Carlo simulation to try and assess these effects more quantitatively.

5.3 RECONNECTION AND THE BOSON MODEL

How could the effects of reconnection be estimated in the context of the boson mapping? The effect of reconnection cannot be represented directly within the boson model, since it is a kinetic (i.e. time dependent) phenomenon. From the point of view of therodynamic equilibrium, the kinetics is irrelevant, and the effect of reconnection is to alter the entropy of the system, i.e. the weighting of reconnected versus tangled configurations.

One possible way to extend the boson model which might simulate this effect is to introduce the possibility of anti-particles into the model. Then, in addition to the world lines of the bosons representing the vortices injected by the external field, there will also be a spontaneous creation of boson-antiboson pairs representing vortex loops. Since the bosons and antibosons will attract each other, this will alter the weighting of the tangled trajectories and may be a way of simulating the effects of reconnection (Fig. 5).

A boson model which includes antibosons will be likely to have a lower melting "temperature" than one with only bosons, since the spontaneous creation of pairs will be an additional source of entropy, driving the system towards the melted state.

Of course, none of the above discussion includes the effects of pinning on the physics of the melting process. However, as discussed in §2, the vortex-vortex interactions are likely to play a crucial role in determining whether the system is in a "glassy" but essentially solid state, or whether "free" (but activated) motion of the vortices can happen at a given temperature.

6. EXPERIMENTAL STUDIES OF VORTEX MOTION: (A) -THE RESISTIVE REGIME

The observations of flux motion which are simplest to interpret at the conceptual level are measurements of electrical resistivity in an applied magnetic field in the ohmic regime. In practice, however, this requires assuming that the sample has come to equilibrium in the field which is hard to achieve (see discussion §7) and also requires measurement of extremely small voltages so that the current applied is well below critical. The latter is also hard to do for temperatures well below Tc,

FIGURE 5 creation and anihilation of boson pairs will reduce barrier for vortex crossing, hence may simulate effects of reconnection.

so that this restricts the region of the H-T plane which can be explored, to lie fairly close to H_{c2}.

The theory of thermally activated flux motion, or creep as it is often called, in the <u>absence</u> of an applied current, i.e. in the linear response regime, involves a number of quite strong assumptions. First of all, as emphasized in §2, there is the implicit assumption of the existence of "free" dislocations, i.e. that the vortex lattice is in a liquid state. Secondly it is assumed that the motion of a true 3-dimensional vortex can proceed by motion of a small element of the vortex, or possibly of a bundle of vortices, corresponding to an elementary "moveable volume" in Tinkham's language.[18] For layered compounds in the $\xi_c < d$ regime, this may amount to the motion of two dimensional fluxoids followed by subsequent reconnection (see Fig. 6).

Such considerations suggest that the elementary moveable volume may be quite small, on the scale of $R_{av}^2 d$ where R_{av} is the vortex lattice spacing $R_{av} = \sqrt{\phi_0/B}$. This is not inconsistent with results of Griessen et al.[19] (see §7 for further discussion). One next needs to make some assumption about the pinning potential. There are, broadly, two possibilities: (a) strong pinning - each fluxoid sits in its own pinning center; (b) weak pinning - the vortex lattice as a whole is pinned at a number of centers whose density is considerably lower than the vortex density. In

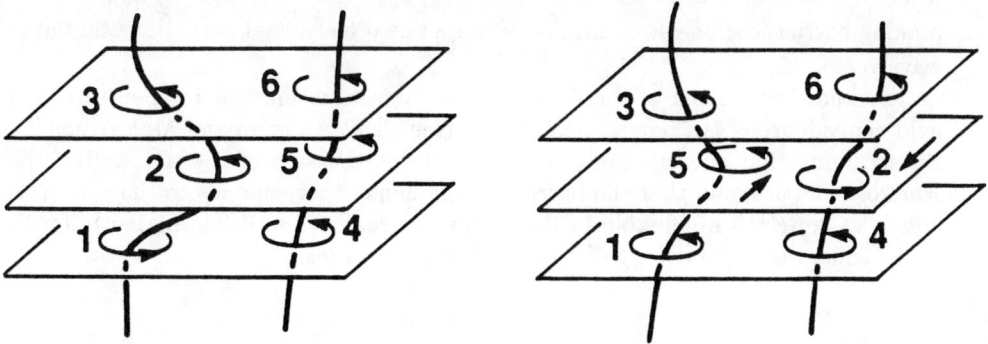

FIGURE 6 Motion of fluxoid 2 followed by that of fluxoid 5 leads to reconnection resulting in net motion of the right hand vortex.

the second case, the pinning potential for a given vortex is effectively provided by the adjacent fluxoids, i.e. is quite density (field) dependent.

A microscopic theory of pinning is quite complex (see discussion in §8). However, the most elementary pinning site in the cuprate superconductors is an oxygen vacancy, from which a rather low pinning energy of a few $°K$ may be deduced (see Woerdenweber and Kes[39]). Thus at temperatures of 10's of $°K$, these materials are probably best discussed in a "collective pinning" regime [Larkin and Ovchinnikov[37]].

For strong pinning one has a simple activated hopping rate

$$1/\tau_{hop} = \nu_0 e^{-U/K_B T} \tag{23}$$

where ν_0 is an attempt frequency.

For weak, collective pinning, one may think of the fluxoid moving in a periodic potential resulting from interaction with the other vortices. (An alternative discussion based on the idea of a network of Josephson junctions has been given by Tinkham.[18] The results are basically the same). The simple picture of the kinetics is an overdamped Brownian kinetics which maybe derived from a Langevin equation in which a random driving force results from thermal motion of the normal electrons in the vortex core. These also provide damping via the Bardeen-Stephen mechanism.[40] The solution of the resulting Fokker-Planck equation for diffusion in a periodic potential has been given by Ambegaokar and Halperin[20] and leads to

$$1/\tau_{hop} = \nu_0 / [I_0(U/2K_B T)]^2 \tag{24}$$

where I_0 is a modified Bessel function ($I_0^{-2}(x/2) \sim xe^{-x}$ for largel x) and U is the pinning barrier/coefficient of $\cos(\vec{K}_0 \cdot \vec{r})$ for a one dimensional periodic potential of wave vector $\vec{K}_0$.

Tinkham has given a simple argument for estimating the temperature and field dependence of the barrier height U in (see also Yeshusrun and Malozemoff[21]). By assuming that the moveable volume scales as the area of the unit cell of the Abrikosov lattice, and that the barrier energy density corresponds to the energy to create an "interstitial" fluxoid in the vortex lattice, i.e., can be estimated in terms of the square-triangle energy difference, Tinkham writes

$$U \propto H_c^2 \xi \phi_0 / B. \tag{25}$$

Since H_c scales as $(1 - T/T_c)$ near T_c and ξ scales as $(1 - T/T_c)^{-1/2}$, U scales as $(1 - T/T_c)^{3/2}$. Rewriting in terms of the thermodynamic critical current $j_c(0)$ at $T = 0, H = 0$, this reduces to

$$U/K_B T \propto (j_c(0)/BT_c) g(T/T_c) \tag{26}$$

where

$$g(t) \approx (1-t)^{3/2}. \tag{27}$$

Thus in this picture the onset of the exponentially activated creep regime is determined by $U/K_B T = \mathcal{O}(1)$ leading to

$$(1 - T_{\text{irrev}}(B)/T_c) \sim B^{2/3} \tag{28}$$

for the "irreversibility line" below which a hysteresis loop opens up in the magnetization M(H) curves (see §7 below).

Inui et al.[22] point out that eq. (22a) should be modified at temperatures $K_B T \sim U$, since the density of fluxoids close to the top of the barrier is then appreciable. This amounts to replacing U by a <u>free energy</u>

$$U_{\text{eff}}(T) = U - T S_{\text{eff}} \tag{29}$$

which considerably alters one's estimate of the pre-factor ν_0 in equation 22a. (This kind of argument is very familiar from general reaction state theory in chemical Physics).

Using the simple Bardeen-Stephen[40] argument that the flux-flow resistance in the high temperature, flux-flow regime ($K_B T \sim U_{\text{eff}}$) is caused by the normal electrons in the vortex cores, and hence proportional to the vortex density, one may then set the scale of the resistivity in the flow regime and write

$$\rho/\rho_n = (B/H_{C2}) \frac{A}{[I_0(U_{\text{eff}}(T)/K_B T)]^2}. \tag{30}$$

where A is a constant.

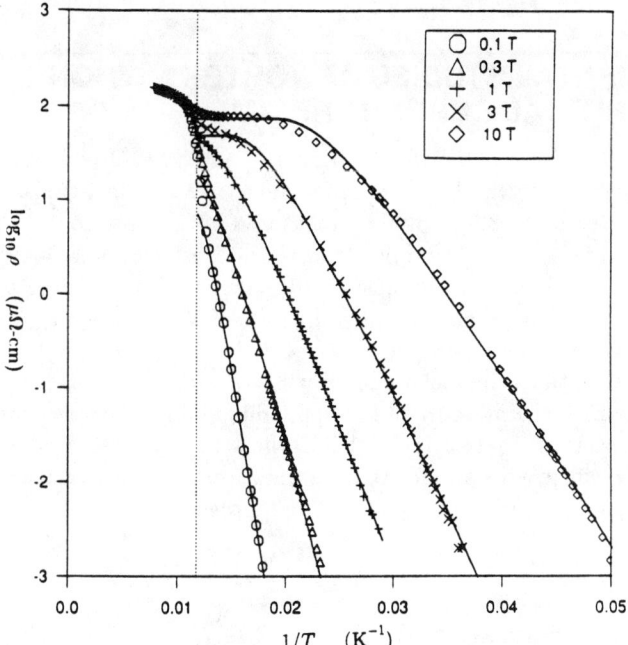

FIGURE 7 Fit of Eq. (30) to resistivity data of Palstra et al.[1] for BSCCO vs inverse temperature. (From Inui et al.[22])

Inui et al.[22] have fit this expression to the data of Palstra et al.,[23] using $U_{eff} = 0$ for temperatures above the 'depinning temperature' $T_{depin} = U/S_{eff}$ estimated from, where the resistivity is in the flux flow regime (see Fig. 7).

From the fit S_{eff} is found to be large, of order 8∼9 which renormalizes the prefactor by a large number, of order 1000.

The barrier height, extrapolated to zero temperature, $U(H)$ is found by Inu et al.[22] to be fairly field dependent, varying roughly as

$$U(H) \sim U_0 + \sigma \ln(H_0/H) \qquad (31)$$

at low fields (below about 2T), with $U_o \simeq 500°K$, $H_0 \simeq 2.5T$ and $\sigma \simeq 200°K$.

This approach is somewhat more detailed than the simple approach of Tinkham and no longer leads to the $B^{2/3}$ law seen in the magnetization data.

It's clear from the observed field dependence that a detailed understanding of the microscopic origin of the flux flow resistance requires a deeper knowledge of the nature of the vortex-vortex correlation function in the pinned vortex liquid.

As suggested above (see Fig. 6), the quasi 2D nature of the vortex array may also influence the microscopic details of the fluxoid diffusion process. This seems to be a promising area for future research.

7. EXPERIMENTAL STUDIES OF VORTEX MOTION: (B) - THE CRITICAL CURRENT REGIME

A class of experiments for which thermally activated flux motion is very important are those in which the time dependence of the relaxation of the diamagnetism of the superconducting sample is measured in response to a changing applied field. As originally observed by Müller et al.[24] and Yeshurun and Malozemoff,[21] the magnetization as a function of applied field relaxes rapidly, leading to a reversible magnetization curve above an "irreversibility line," $H \sim (1 - T/Tc(H))^{2/3}$, while it becomes irreversible below this line.

Experiments of this type may be readily interpreted in terms of the Bean critical state model in which the pressure of the applied field forces vortices into the sample leading to an initial magnetization distribution $M(x)$.

Vortices are subject to an effective Lorentz force proportional to the local supercurrent density $j(x)$ which in a simple one dimensional model is given by

$$j_y(x) = -\frac{dM_z}{dx}. \tag{32}$$

In the simple Bean model* the vortices then relax rapidly until j is reduced to infinitesimally below the critical current density j_c at which point the system stabilizes in the critical state with

$$\left|\frac{dM}{dx}\right| = j_c. \tag{33}$$

Subsequent to the establishment of the critical state, thermal fluctuations will lead to creep-like relaxation of the system towards equilibrium. The characteristic "log of the time" law was explained early on by Anderson and Kim[25] and studied further by Beasley et al.[26] The basic idea is very simple: The effective Lorentz force pressure, $\vec{F}$, on a pinned vortex due to the local current density may be represented (in a simple 1 dimensional model) by writing the effective potential on a fluxoid "moveable volume," δV, to linear order in $\vec{F}$ as

$$U(x, F)\delta V \cong U_0(x)\delta V - |F|\delta V x. \tag{34}$$

The term $U_{\text{pin}} = U_0(x)\delta V$ is then what would be considered the relevant equilibrium pinning potential as applied in the resistivity discussion of §6.

---

*Feigel'man et al.[41] have given a more general discussion of flux creep in which the pinning potential acquires a power law dependence on the driving current j. Also, in a recent preprint, Wolfus et al.[27] have generalized the Bean model by postulating a generalized scaling relation between the critical current and the local magnetic field $J_c = J_{c1}(H/H*)^{-n}$. They argue that this allows a one-parameter scaling of the observed dependence of M on H throughout the irreversible regime.

The creep process which results is considered to take place via a series of discrete "flux-hop" events by an activated process:

$$\frac{1}{\tau_{\text{hop}}} = \nu_0 e^{-\delta U_{\text{hop}}/K_B T} \tag{35}$$

where δU_{hop} represents the average, current dependent barrier per hop and ν_0 is an attempt rate:

$$\delta U_{\text{hop}} \cong U_{\text{pin}}(x_{\text{hop}}) - |\vec{F}| V \delta x_{\text{hop}}. \tag{36}$$

x_{hop} is the average distance between the bottom of the pinning well and the relevant saddle point for escape. Writing

$$\vec{F} = \frac{1}{c} \vec{j} \times \vec{H} \tag{37}$$

we can then derive a simple kinetic equation for the decay of the local current density

$$\frac{dj}{dt} = -\frac{j}{\tau_{\text{hop}}} = -\nu_0 j e^{-U_{\text{pin}}/K_B T} e^{+U'j/K_B T} \tag{38}$$

where $U' = H V \delta x_{\text{hop}}/c$.

Solution of for $(1 - j(t)/j(0)) \ll 1$ yields a characteristic $\ln(t)$ dependence of the form

$$y(t) = A - b \ln(t/\tau_0) \quad \text{for} \quad t \geq \tau_0, \tag{39}$$

where $j(t) = j(t)/j(0)$. Substitution in (38) and the assumption that the system starts from the critical state, i.e. $j(0) \equiv j_c$, gives

$$\frac{j(t)}{j_c} = \frac{U_{\text{pin}}}{U' j_c} - \frac{K_B T}{U' j_c} \ln(t/\tau_0) \tag{40}$$

with the time scale set by

$$\begin{aligned}\frac{1}{\tau_0} &= \nu_0 (U' j_c / K_B T) \\ &= \nu_0 (\frac{H j_c}{c})(V \delta x_{\text{hop}}/K_B T)\end{aligned} \tag{41}$$

In an actual experiment, the barrier height is a function of the local field and $j(x)$ becomes a function of position (following the Bean model) so that one needs to extend the above argument to include spatial dependence of j (Beasley et al.[26]).

In practice (40) is a reasonable zero order approximation, and the characteristic rate of decay of magnetization with $\ln(t)$ may be measured as

$$\frac{d(M(t)/M(0))}{d\ln t} = \frac{K_B T}{U' j_c}$$
$$= \frac{K_B T}{(j_c H/c) V \delta x_{\text{hop}}} \quad (42)$$

where j_c is the critical current density.

Combining and one may also write

$$\frac{d\ln M(t)}{d\ln t} = \left[\frac{U_{\text{pin}}}{K_B T} - \ln(t/\tau_0)\right]^{-1} \quad (43)$$

from which the pinning barrier $(U_{\text{pin}}/K_B T)$ may be extracted.

Experimentally the $\ln(t)$ law works quite nicely but the predicted temperature dependence fails in quite a striking manner (Hagen and Griessen[28], Sun et al.[29]).

For temperatures above about $20-40°K$ McHenry et al.[30] find a reasonable Arrhenius behaviour for Tl(2223) $(Tl_2Ca_2Ba_2Cu_3O_x)$ with strongly field-dependent pinning barriers on the scale of a few 1/10ths of an eV at 1 kgauss dropping to .05 eV above 3 kgauss, with similar behavior for YBCO. However at lower temperatures the Arrhenius law again appears to fail quite strikingly.

7.1 ESTIMATES OF THE "MOVEABLE VOLUME" BASED ON CRITICAL CURRENT MEASUREMENTS

Measurements of the order of magnitude of the pinning barrier via decay of the magnetization coupled with the observed critical current densities allow us to give an order of magnitude estimate of the relevant "moveable volume" for activated flux motion in the critical current regime. Combining the force-dependent activation energy with the Lorentz relation allows us to write

$$j_c = \delta U_{\text{hop}}/\delta U'_{\text{hop}}$$
$$= U_{\text{pin}}/HV\delta x_{\text{hop}} \quad (44)$$

From the observed values of j_c and U_{pin} at a given field H one can then estimate the product of the moveable volume $\times \delta x_{\text{hop}}$.

For $U_{\text{pin}} \approx 25$ meV at H=1 Tesla, $j_c \cong 2 \cdot 10^7 A/cm^2$, Griessen et al.[19] deduce $V\delta x_{\text{hop}} \cong 2 \cdot 10^{-24} cm^4$.

Considering δx_{hop} must be of order R_{av}, the inter vortex spacing, given by $\sqrt{\phi_0/H} \cong 450 \text{Å}$, at 1 Tesla, one then finds $V \approx 4.5 \cdot 10^{-19} cm^3$. This is consistent with an expression

$$V = p R_{\text{av}}^2 R_z \quad (45)$$

where p is a fraction of pinning centers, giving

$$p R_z \approx 2.2 \text{Å}. \quad (46)$$

Thus the observed thickness of the moveable volume normal to the CuO_2 planes is certainly consistent with a scale smaller than the interlayer spacing, i.e. with the fluxoid picture presented in §3.

From this qualitative discussion one sees that the somewhat surprising observation of relatively high critical curents in the high Tc superconductors may be understood despite the relative smallness of the observed activation energies $U_{pin} \approx 25$ meV. They are a consequence of the very small coherence lengths, and specifically of the rather weak interplanar coupling leading to the concept that the "moveable elements" of the vortices are the 2D fluxoids.

7.2 NATURE OF THE CREEP PROCESS AT LOW TEMPERATURE

As mentioned above, the observed temperature dependence of $d\ln M(t)/d\ln t$ at temperatures below about $20°K$ does not appear to show the kind of activated behavior expected from the simple thermally activated creep picture of equation . Hagen and Griessen[28] suggest that this may be interpreted in terms of a distribution of pinning energies. In order to make progress in fitting this model to experiment, they were obliged to make a number of fairly strong assumptions, not all of which appear quite consistent. Nevertheless, the general idea (familiar from many other fields in which hierarchies of barriers occur) seems quite plausible.

The net result is a distribution of barrier heights which has the approximate form of a "log-normal" distribution

$$p(E) = p_m \exp[-\gamma(\ln(E/E_m))^2] \qquad (47)$$

The peak of the distribution, E_m is found to be on the scale of the pinning energies discussed above, ~ 67 meV for a single crystal of YBCO, with $\gamma \sim 5.5$.

The qualitatively remarkable aspect of this distribution is the non negligeable weight at quite low energies, leading to appreciable creep at quite low temperatures.

In fact in other classes of superconductors, A.C. Mota[31] has observed appreciable creep even at temperatures as low as 10's of milliKelvin! This is an area of investigation which seems quite challenging on very general grounds.

7.3 CREEP AND POWER LAW I-V CURVES

J. Sun and collaborators[29] observe that a $\ln(t)$ relaxation behavior may also be deduced on general grounds for a system in which non-ohmic I-V curves of a power low type are observed:

$$V \propto I^n. \qquad (48)$$

Since the observed voltage is due to vortex motion, the above relation may be directly applied to equation (38) to give

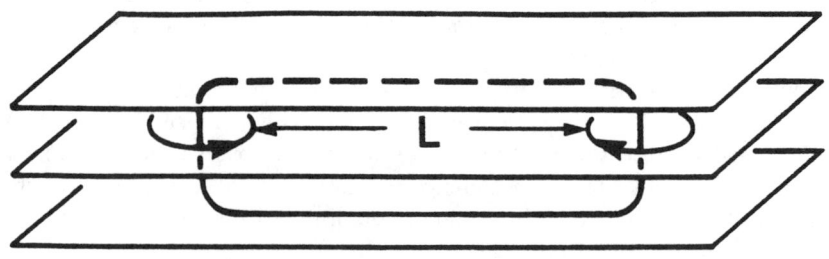

FIGURE 8 A two-dimensional vortex loop of size L.

$$\frac{dj}{dt} \propto j^n, \qquad (49)$$

from which one may deduce

$$\frac{M(t)}{M(0)} \approx \left[1 - \frac{1}{n-1}\ln(t/\tau_1 + 1)\right] \qquad (50)$$

for $(1 - M(t)/M(0)) \ll 1$. $dM/d\ln t$ is now directly given in terms of the power - law exponent n, which turns out to be very large, on the scale of 30-50 at low temperatures.

This picture still does not escape from the requirement that all flux motion must be thermally activated at low enough temperatures.

A simple picture of power law behaviour arises quite generally in defect - dominated systems (such as plastic flow due to dislocation loops etc.). In the two-dimensional limit, which is not all that unreasonable given the discussion in §(7.1), one may envision the current-driven vortex motion as arising from elementary process in which a vortex loop is "blown up" by the Lorentz-induced driving force.

Extending an old argument due to Huberman et al.,[32] a vortex loop of size L (see Fig. 8) in the presence of a current will have an energy of the form

$$E_{\text{loop}} = E_0 + 2E_{\text{string}}L + q^2 \ln\left(\frac{L}{\zeta_{ab}}\right) - 2FVL, \qquad (51)$$

where E_{string} is the line tension.

Using equation one then has, for j greater than a lower critical value

$$j_{c1} = (E_{\text{string}}/HV) \qquad (52)$$

a barrier energy for formulation of unstable loops

$$E_{\text{barrier}} = E_0 + q^2 \left[\ln\left(\frac{q^2}{(j-j_1)HV\xi_{ab}}\right)\right]. \tag{53}$$

Putting this into an activated decay process then gives

$$\frac{dj}{dt} \approx \nu_0 e^{-E_{\text{barrier}}/K_B T} \propto e^{+\frac{q^2}{K_B T}\ln(j-j_1)}. \tag{54}$$

Thus power law behavior occurs quite naturally for generation of large scale defects, with large exponents

$$n \approx q^2/K_B T. \tag{55}$$

However, there is no escaping the thermally activated character predicted for the resulting flux motion. So, to explain the apparent lack of temperature dependence of $d\ln M/d\ln(t)$ at low temperatures, some kind of barrier distribution (as discussed in §7.2) is required.

7.4 IS THE CRITICAL STATE A SELF-ORGANIZED FRACTAL?

Bak et al.[33] have introduced the general notion of self- organized criticality in dynamical systems. The Bean model may be thought of as a mean-field version of an archetypal critical state in which the local current density is treated as an average quantity which is just below the critical current everywhere.

The treatment discussed in §7 above of the subsequent thermally activated decay is one in which individual flux-jumps are treated as uncorrelated events moving in an average or "mean field" current density $j(r)$. Bak et al.'s model is one in which the coupling between jump events (or sand particles in the sand pile model) is a crucial feature. The system is inherently that of a system of many non-linearly coupled degrees of freedom. In fact, their model is spontaneously chaotic, presumably due to the discrete nature of the non-linear map which describes the evolution of their dynamical system.

The analogy with the superconducting critical state is highly suggestive: when a fluxoid moves (presumably due to a thermal fluctuation), it will perturb the local currents over an extended region of space due to the long ranged nature of vortex-vortex coupling. This perturbation may well push other vortices to experience a local Lorentz force which exceeds their local j_c. (In fact, since j_c is likely to vary spatially as a result of the local fields on a given vortex being the summation over many other vortices, this may well enhance the tendency to spontaneously chaotic organization.) Thus one thermally activated event may well trigger an avalanche of related flux jumps which could extend over a large region of space and also occur over an extended time. This could then simulate the effect of having a broad distribution of barrier heights which as we have seen are needed in the independent jump model of Hagen and Griessen[28] (§7.2).

It would be of considerable general interest to measure the noise correlation function of the decay of the critical state. It seems most likely that power law correlations of "$1/f$" character should be seen.†

Note that the idea of thermally activated decay from a critical state is an extension of Bak et al.'s model. In their case the perturbations away from the critical state are introduced by non- thermal agents (e.g. stock-market price swings, etc.).

The influence of quenched randomness (i.e. pinning centers) on the critical state is also an important open question for the superconductors. From the point of view of the Bak model, quenched randomness does not seem to have a major influence on their simulation. It is entirely conceivable that weak pinning is only important to anchor the flux lattice, and that the distribution of barrier heights needed in an independent jump model is in fact a result of a self-induced fractal arrangement of the flux-line lattice in the critical state: in other words it is possible that the Bean-type formation of a critical sate spontaneously introduces a fractal-type power law for the spatial correlation function $< u(r)u(0) >$ in the Abrikosov lattice.

It's intriguing that direct observations of the nature of the flux line lattice by the Bitter - pattern technique have lead Dolan et al.[35] to the conclusion that a "self - pinning" mechanism could provide one possible explanation of the apparent existence of a "dense, uniform pinning source" even in the "best crystals."

The whole question of "dynamically-induced" fractal character in a system which probably has a static glassy character (see §2) is most intriguing and appears completely open at the present time.

8. MICROSCOPIC NATURE OF PINNING IN HIGH Tc SUPERCONDUCTORS

The origin of inhomogenities in the order parameter leading to pinning is a pretty complex subject, involving point defects, and large scale defects such as grain boundaries, twin boundaries, dislocations, precipitates etc.

One of the intriguing questions for the high Tc materials is that of intrinsic pinning. Because of the weak interlayer coupling, the energy of a vortex core parallel to the CuO_2 planes with $H \parallel c$ is much less than that of a core in the planes with $H \perp c$.

One may therefore expect highly anisotropic forces on vortices lying sandwiched between planes: for motion parallel to the planes there is no intrinsic barrier, while for motion normal to the planes, there will be a considerable barrier involved in re-forming the "sine-Gordon" current loop from one pair of planes to the next. Tachiki

---

†This type of "dynamically induced" $1/f$ noise should not be confused with equilibrium $1/f$ noise seen in thin films at very low fields (less than 1 gauss).[34]

and Takahashi[36] have given a discussion of the pinning energy involved using an anisotropic Ginzburg-Landau picture in which the order parameter is treated in a 3-dimensional continuum with correlation lengths,ξ_{ab}, ξ_c. In this model, the variation of order parameter around a vortex centered along z_0 is given by

$$\Delta(r) = \Delta_0(z) \tanh \sqrt{(\frac{x}{\xi_{ab}})^2 + (\frac{z-z_0}{\xi_c})^2}. \tag{56}$$

Assuming a simple form for the z-dependence of the order parameter in the absence of the vortex,

$$\Delta(z) = \Delta_1 + \Delta_2 \cos(2\pi/d) \tag{57}$$

with d the interlayer spacing, a simple estimate of the force on the vortex with core displaced to the position of maximum energy then gives a maximum pinning force of

$$f_{pM} = \frac{H_c^2}{8\pi} 2\pi d (\frac{\xi_{ab}}{\xi_c}) \eta_M \tag{58}$$

where η_M is a dimensionless quantity which depends on Δ_2/Δ_1 and ξ_c/d. For $\Delta_2/\Delta_1 = 0.8$ and $\xi_c/d \cong 0.5$ they find $\eta_M \cong 0.4$.

Putting in the vortex density n_{vortex} at given field, the critical current density may be written

$$\begin{aligned} j_c &= \frac{f_{pM} n_{\text{vortex}}}{B/c} \\ &= c \frac{H_c}{8\pi d}(\frac{\xi_{ab}}{\xi_c}) \eta_M \frac{H_c}{B_z^0}(1 - B/H_{c2}) \end{aligned} \tag{59}$$

where $B_z^0 = \phi_0/2\pi d^2$. Putting in reasonable values of the parameters, Tachigi and Takahashi give an estimate of the critical current density for YBCO at B=0 and $77°K$, which comes out to be $j_c \cong 3 \cdot 10^7 A/cm^2$.

In practice this is likely to be an overestimate, since the vortex can lower the barrier energy for crossing a plane by creating a "kink": see Fig. 9.

A general theory of vortex pinning requires discussion of the vortex-vortex correlation function, nature of the vortex lattice (i.e. "glass" or "liquid") and would require a whole article on its own - see Larkin and Ovchinnikov,[37] Woerdenweber and Kes.[38]

ACKNOWLEDGMENTS

The author would like to thank Masahiko Inui, Jonathan Sun, Ted Geballe, Peter Kes, and, particularly, Aharon Kapitulnik for helpful conversations.

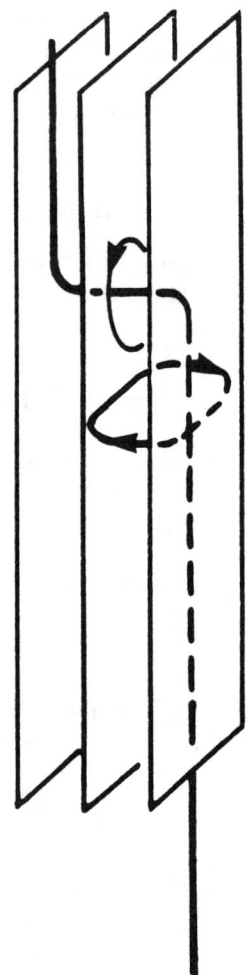

FIGURE 9 Kink motion of a vortex with core parallel to the planes.

REFERENCES

1. T. T. M. Palstra, B. Batlogg, R.B. Van Dover, L.F. Schneemeyer and J.V. Waszczak, Phys. Rev. Lett. **61**, 1662 (1988) and Phys Rev B (1989) to be published.
2. A. P. Malozemoff, in "Physical Properties of High Temperature Superconductors - I," ed D.M. Ginsberg, World Scientific Singapore (1989) pp 71-150.
3. A. I. Larkin, Sov. Phys. JETP **31**, 784 (1970).
4. M. P. A. Fisher, Phys. Rev. Letts. **62**, 1415 (1989).
5. R. H. Koch, V. Foglietti, W. J. Gallagher, G. Koren, A. Gupta and M. P. A. Fisher, Phys. Rev. Lett. (1989).
6. D. R. Nelson, Phys. Rev. Letts. **60**, 1973 (1988).
7. W. E. Lawrence and S. Doniach in Proc 12th Int. Conf. on Low Temperature Physics, Kyoto (1970) ed. E. Kanda (Keigaku, Tokoyo 1971) p 361.
8. D. E. Farrell, S. Bonham, J. Foster, Y. C. Chang, P. Z. Jiang, K. G. Vardervoost, D. J. Lam and V. G. Kogan, Phys. Rev. Letts. (1989).
9. S. Doniach in "Ordering in two Dimensions" ed. S.K. Sinha, (North-Holland, Amsterdam 1980), p 67.
10. S. Doniach and B. A. Huberman, Phys. Rev. Letts. **42**, 1169 (1979).
11. E. H. Brandt, J. Low Temp. Phys. **26** 709 (1977), **28**, 263 and 291 (1977), **42**, 557 (1981).
12. A. Houghton, R. A. Pelcovits and A. Sudbo, preprint (1989).
13. D. R. Nelson, J. Statistical Physics (to be published).
14. C. Deutsch and S. Doniach, Phys. Rev. B **29** 2724 (1984).
15. B. A.Huberman and S. Doniach Phys. Rev. Letts. **43**, 950 (79).
16. D. S. Fisher, Phys. Rev. **B22**, 1190 (1980).
17. Seungoh Ryu, A. Kapitulnik and S. Doniach, work in progress.
18. M. Tinkham, Phys. Rev. Letts. **61**, 1658 (1988).
19. R. Griessen, G. F. J. Flipse, C. W. Hagen, J. Lensink, B. Dam and G. M. Stollman, J. Less-Common Metals (1989) to be published.
20. V. Ambegaokar and D. I. Halperin, Phys. Rev. Letts. **22** 1364 (1969).
21. Y. Yeshuron and A. P. Malozemoff, Phys. Rev. Letts. **60**, 2202 (1988).
22. M. Inui, P. B. Littlewood and S . N. Coppersmith, preprint 1989.
23. T. T. M. Palstra, B. Batlogg, L. F. Schneemeyer and J. V. Waszczak, Phys. Rev. Letts. **61**, 1662 (1988).
24. K. A. Müller, M. Takashige and J. G. Bednorz, Phys. Rev. Letts. **58**, 1143 (1987).
25. P. W. Anderson and B. Kim, Rev. Mod. Phys. **36**, 39 (1964).
26. M. R. Beasley, R. Labusch and W. W. Webb, Phys. Rev. **181**, 682 (1969).
27. Y. Wolfus, Y. Yeshurun, I. Febner and H. Sompolinsky, Preprint (1989.).
28. C. W. Hagen and R. Griessen, Phys. Rev. Letts. **62**, 2857 (1989).
29. J. Z. Sun, C. B. Eom, B. Lairson, J. Bravman and T. H. Geballe, preprint (1989).

30. M. E. McHenry, M. P. Maley, E. L. Venturini and D. L. Girley, Phys. Rev. B (Rapid Communs.) **39**, 4784 (1989).
31. A. C. Mota, A. Pollini, P. Visani, K. A. Muller and J. G. Bednorz Phys. Rev. **B 36**, 4011 (1987).
32. B. A. Huberman, R. J. Meyerson and S. Doniach, Phys. Rev. Letts. **40**, 780 (1978).
33. P. Bak, C. Tang and K. Wiesenfeld, Phys. Rev. **A 38** 364.
34. M. J. Ferrari, M. Johnson, F. C. Wellstood, J. Clarke, P. A. Rosenthal, R. H. Hammond and M. R. Beasley, Appl. Phys. Letts. **52**, 695 (1988).
35. G. J. Dolan, G. V. Chandrashekhar, T. R. Dinger, C. Field and F. Holtzberg, Phys. Rev. Letts. **62**, 827 (1989).
36. M. Tachiki and S. Takahashi, Solid State Communs., **70** 291 (1989).
37. A. I. Larkin and Y. N. Ovchinnikov, J. Low Temp Phys **34** 409 (1979).
38. P. Kes Physica C **153-155** (1988), 1121.
39. R. Woerdenweber and P. H. Kes, Cryogenics **29**, p. 321-327.
40. J. Bardeen and M. J. Stephen, Phys. Rev. **140**, A1197 (1965).
41. M. V. Feigel'man, V. B. Geshkeubein, A. I. Larkin and V. M. Vinokur, Phys. Rev. Lett. **63**, 2303 (1989).

DISCUSSION

Kes: I would like to make a comment on the behavior in conventional superconductors. You have to be very careful about saying that the Lindemann criterion gives rise to melting. In case pinning is negligible this may be correct. Let me show you what happens if the pinning is important. I have a viewgraph which shows the behavior of the pinning force, J_c x B versus B for a very thin film, amorphous Niobium Germanium, a weak pinning material. You see that it follows very well the so-called two-dimensional collective pinning theory, in which the correlated regions are straight, cilinders in the direction of the field. But at this field you see a kink the pinning force starts to deviate from that behavior. It turns out – you can compute it – that you Lindemann's criterion at that "kink" field, but not due to thermal fluctuations, but due to the disorder caused by *static* random pinning forces!! It doesn't give rise to a decrease of the pinning, but it gives rise to an increase of the pinning. Now, what I described here was a situation where R_c, the correlated region, is much larger than a_o. Then you can compute the melting temperature in the right way, but even then you find instead of melting an increase of the pinning force. My point is – and I don't know whether I'm right, that when you start with the condition R_c is of the order a_o, which is the case in strong-pinning superconductors, that you may find a quite different melting criteria.

Doniach: What happens is that the current circulates between the planes. Since the coupling is weak, the current has to spread out over a very big area. So the radius of the string is really a measure of how strong the current is. You've still got one quantum of current circulating and it has to spread out because the tunnelling matrix elements connecting the planes are much smaller than those acting in the planes.

Trugman: Is this a graph at zero temperature that you are showing?

Kes: No, no, this is at any temperature below T_c, and T_c is about 4K. So you don't see melting; melting is predicted at that field by the theory of dislocation mediated melting (Daniel Fisher), but you don't see it. You instead see this onset of a peak in the pinning force.

Trugman: It makes good sense that the pinning force should increase monotonically with the strength of all the pins, right?

Kes: You have to work out collective pinning theory...it makes some sense.

Pines: Consider this in the nature of the light humor, in terms of the beautiful serious talk Seb has given. But the existence of intrinsic pinning in high temperature superfluids, accompanied by avalanches, has been documented for about a decade now. It's just the high temperature superfluids in question the pinned neutron superfluid in the crust of a neutron star, where the only pinning you've got going is intrinsic, you're pinning to the crustal nuclei – and there one sees very clearly both vortex creep and avalanches in the form of glitches. These are seen, these are a characteristic feature in the life of a spinning-down pulsar. And of course it was

Phil Anderson who started us all on that, because he knew about flux pinning in Type II superconductors.

So it's plausible, Seb, that you're going to find that the experimentalists will find avalanche effects associated with a crystal.

Doniach: Seems obvious, but I'm amazed that the terrestrial ones haven't been looked for.

Maley: In your model of the two-dimensional vortices joined together between layers by strings, is the string, say, in a vortex ??? part of the vortex translated with respect to the other, is that supported by currents, now, that flow through the C axis, basically...

Doniach: What happens...the current goes along the plane. Since the coupling is so weak, it has to spread out over a very big area. So this ??? connection is really a measure of how...you've still got a quantum of current going on, so one flux quantum...and it has to spread out because the conductivity, or the tunneling matrix elements were smaller??, that's a measure of how big it is.

Malozemoff: I had one question. You expressed your doubt about the existence of an entanglement regime of the type David Nelson has talked about. Do you think it's possible that there could be some different liquid regimes, some hexatic regime, anything of this sort? For example, a transition from hexatic to an orientationally disordered state?

Doniach: I don't have much of an opinion about it. I think the main point I was trying to make is that it's a probabilistic question of whether you'll really get entanglement or not, and at very low densities, the probability could be high, but about the order, I haven't thought about it.

Campbell: You emphasized that we were essentially in a classical regime. One of the features of classical regimes is that things are easy to calculate, and presumably if we had a decent equation of motion of the flux lines some of these questions you ask about whether they move bit by bit, or in bundles, or what not – these local questions could be answered by direct calculation. Where is the equation of motion of the vortex lines?

Doniach: Well I think that in these systems...It depends. It depends a lot on the damping, and if you don't have any normal core...then the hydrodynamic equations are the right ones. But my feeling is that these things are overdamped, so it's actually Langevin, Brownian motion, basically. It is an extremely simple equation now, you just have a Langevin driving force and then coupling with the potentials.

Campbell: But that doesn't make it uncalculable.

Donaich: No! It's very easy.

Kes: I have another question about the vortex string. I really agree with you that two-dimensionality is very important in this, especially to the thallium compound.

But why do you think that a vortex string should have a kind of continuous order parameter zero, because if the layers are really weakly coupled, very weakly coupled, even that string may be decoupled. I'm not so sure that this vortex string is really a continuous thread or something like that.

Doniach: Well, you have to think in terms of the Ginzburg-Landau equations with at least two coupled planes – and in each plane, you have the phases coming out like this, and then down here you've got them coming out like that. This happens over an extended area so they conflict with each other. So you're forced, by just minimizing the free energy, that if you didn't bend it down, that this would be out of phase with that. That wouldn't matter in the small area, but it's extensive over the whole plane. So it could never coexist without the string. It's simply continuity.

Millis: But if I have a finite density, can't I just attach it differently? If I have a finite density of vortices in each plane can't I just interchange which one goes with which one.

Doniach: Sure, oh absolutely. Yeah, that was my whole point.

Millis: So then why talk about strings I guess is the question.

Doniach: Well you see once they're further apart than R_c, then they do remember. The string will be there, and only when they get close enough will they forget.

Millis: So the strings are a low-density concept.

Doniach: Right.

Dissipation in very Anisotropic Superconductors
P. H. Kes
Kamerlingh Onnes Laboratory, Leiden University, The Netherlands

In the high-temperature superconductors (HTS) the resistive transition in a magnetic field becomes broader with increasing anisotropy, expressed by $\Gamma = m_Z/m$, the effective mass ratio for quasiparticle tunneling between and conduction in the CuO planes. It has been assumed that the broadening is caused by an unusually large flux-creep effect[1] brought about by the driving force on the flux-line lattice (FLL) given by

$$\vec{F}_d = \vec{j} \times \vec{B} \qquad (1)$$

This explanation has been questioned by experiments on films of Bi:2212 in which the field is rotated with respect to the current direction which was fixed along the ab-planes.[2] If at constant T and H the field rotates from the c-direction to the ab-plane, the resistivity depends on the angle θ (with respect to the c-axis) as $\rho = \rho_\perp \cos\theta$ in accord with (1) (the subindex denotes field $\perp$ or $\parallel$ ab planes). However, if the field is rotated in the ab-plane, the resistivity remains constant. We offer an explanation for this important discrepancy.[3]

Farrell et al.[4] obtained Γ from torque measurements. For the Bi:2212 $\Gamma = 3000$. This large Γ expresses the fact that the Bi compound should be considered as a stacking of two-dimensional Josephson junctions instead of an anisotropic 3D superconductor.[5] The dimensional crossover takes place at $t_{co} = T_{co}/T_o$ defined by

$$\Gamma(1 - t_{co}) = 2(\xi_{ab}(0)/s)^2 \qquad (2)$$

with s the distance between the CuO planes and ξ_{ab} the coherence length in the planes. For Bi:2212 (s = 1.2 nm, $\xi_{ab}(0) \approx 2.4$ nm) T_c - $T_{co} \approx 0.3$ K, so practically the 2D behavior prevails. This is exemplified by a cusp of $H_{c2}(\theta)$ at $\theta = \pi/2$, as well as the nature of the superconducting fluctuations in the normal state. In addition, the anisotropic 3D Ginzberg-Landau (GL) theory gives rise to an unphysical Abrikosov lattice for large Γ, since the size of the FLL unit cell in the c-direction $a_{o,c} \approx (\Phi_o/B_\parallel)^{1/2} \Gamma^{-1/4}$ cannot be smaller than s, which would happen at $B_\parallel = 25$ T. This is a relatively small field in comparison with the slope of B_{c2} at T_c of ≈ 45 T/K. The large slope actually reflects the 2D behavior as well, since for 2D $H_c \propto (1-t)^{1/2}$.

We propose that in a Josephson-coupled layered-superconductor for arbitrary field orientation a FLL is formed which has the characteristics of Josephson vortices in a 2D Josephson junction for $H_\parallel$ and the properties reminiscent to Abrikosov vortices for $H_\perp$. A similar idea has recently been put forward by Tachiki and Takahashi.[6] The order parameter is non-zero in the CuO planes and hardly different from zero in between. The pinning of the vortex segment parallel to the planes is therefore negligible in the direction along the planes, and only the 2D vortex "disks" in the planes can be pinned. In this model the resistivity is solely determined by the thermal depinning of the 2D vortices and thus by $H_\perp$ and $F_d = H_\perp j$. When the field is rotated around the c-axis both $H_\perp$ and F_d remain constant. Consequently, the resistance should be independent of the angle of rotation. In Iye's experiment[2] the field could be applied ∥ to the substrate, but presumably not ∥ to the CuO planes because of the granularity of the film. Thus, his results can be explained by our model[3] assuming a slight misalignment of the field with the CuO planes.

For a 2D Josephson junction H_{cl} and M are very small, e.g. $<5 \times 10^5$ T at t = 0.9. This means that the material is almost "transparent" for $H_\parallel$ and the magnetic moment is determined by $M_\perp(H_\perp)$, where $H_\perp = H \cos\theta$ and

$$M_\perp = \frac{H_{c2\perp} \ln \kappa}{\kappa^2} \ln\left(\frac{\eta\, H_{c2\perp}}{H_\perp}\right) \qquad (3)$$

for $H_{cl\perp} \ll H_\perp \ll H_{cl\perp}$, i.e. in case H = 1 T[4] and $H_{cl} \approx 2$ mT for $\theta < 1/2$. The parameter η is a constant ≈ 1. For $\theta \approx \pi/2$ $M_\perp$ should drop linearly as $H_\perp$. Combining this behavior with (3) in a simple interpolation expression, we obtain for the torque[3]

$$\tau(\theta) = f(\theta) H_\perp / (f(\theta) + H_\perp) \qquad (4)$$

where $f(\theta) \propto M_\perp$. With very reasonable values for the parameter η and $H_{c2\perp}$, we got very good agreement with the experimental torque data of Farrell et al.,[4] see Fig. 1.

Implications of practical importance are the large critical field $H_{c\parallel}(0) = 2.5$ kT for a 2D superconducting CuO plane of thickness 0.3 nm and the large j equal to the depairing current. In practice, however, it will be difficult to avoid a $H_\perp$ component, because a/w is very small (w,s the width of some superconducting device) and because the current will generate a field $H_\perp$. Therefore j_c will be limited by $j_c(H_\perp)$ which is mainly determined by the large thermally activated depinning in layered superconductors.

REFERENCES

1. T. T. M. Palstra *et al.*, Phys. Rev. Lett. **61**, 1662 (1988).
2. Y. Iye *et al.*, Physica **159**, 433 (1989).
3. P. H. Kes *et al.*, submitted.
4. D. E. Farrell *et al.*, Phys. Rev. Lett. **63**, 782 (1989).
5. W. E. Lawrence and S. Doniach, Proceedings LT12, Kyato, 1970, p. 361.
6. M. Tachiki and S. Takahashi, Sol. St. Comm., to be published.

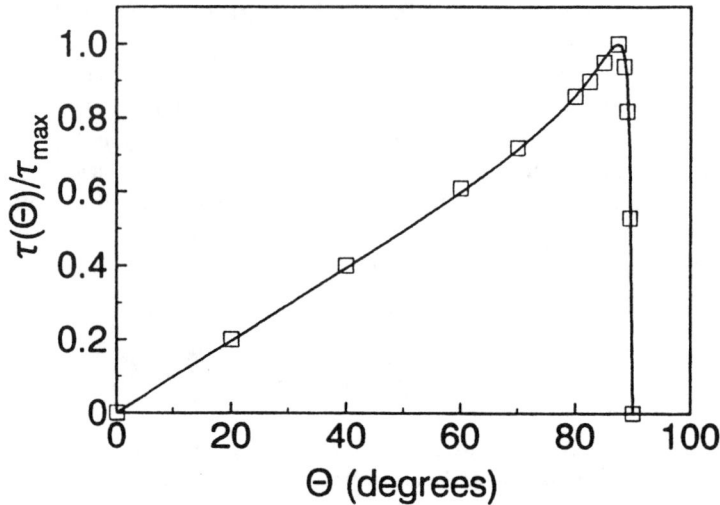

FIGURE 1 Points selected from the normalized torque data of Ref. 4 at Bi:2212 at B - 1T and T = 77.5 K (squares) compared with Eq. (4) (drawn line).

DISCUSSION

Batlogg: Peter, can it therefore be concluded that the extraction of the anisotropy parameters, based on these experiments and their interpretation should not be taken literally in terms of the anisotropy of the mass?

Kes: Yes, I see what you mean. Of course, we considered that too. I think what we have to do is to work out, in this two-dimensional model, what the magnetization would be in the perpendicular field. But the H_{c1} is somewhere hidden in this kind of curve fitting. H_{c1} is usually related to the line energy of a vortex. Now if the vortex consists of discs which are very weakly coupled, then this line energy is not a continuous energy. You have to split it up in parts, and I think the coupling parameter between the layers comes in the prediction for H_{c1}. That has never been considered, as far as I know, so H_{c1} for a layered superconductor has not been computed. And I don't think it's the same as for a continuous superconductor.

Batlogg: What I find interesting experimentally is the anisotropy in H_{c1} which corresponds to the anisotropy in λ, and the anisotropy of H_{c2}, which corresponds to the anisotropy of the coherence length, is the same. Would that follow automatically from these 2-D models?

Kes: I don't know. I think, actually, in 2-D, if it's a real 2-D system, and you see that from the huge slope of H_{c2}, at T_c, that you shouldn't compare it in terms of the 3-D Ginzburg-Landau theory. So one would be proportional to the square root of 1-t, and the other one should be linear with 1-t. So all these estimates are not really more than an estimate, and there's clear evidence that these materials behave as a 2-D superconductor, you see that from the cusp.

Batlogg: On the other hand, as you know...I'm just talking about the data that we have for the 1,2,3 compound, which is the only solid data set that is available, here the anisotropy at T_c, of H'_{c_2} according to George Crabtree and collaborators, can be modeled by the anisotropic Ginzburg-Landau theory.

Kes: True. But YBaCuO is a 3-D superconductor. It has a rounded curve, if you look at the H_{c2} verses angle θ. It behaves as a 3-D superconductor; that's why it has such a nice, and large, critical current density at 77 K.

Maley: Your model seems to fit the torque measurements quite well, but then so does the anisotropic Ginzburg-Landau.

Kes: Sure.

Maley: Is there a difference, though, perhaps in the way in which those curves changes with the magnetic field. Because yours is heavily dependent on the value, at which angle you cross H_{c2}. I'm wondering if you might see some kind of difference.

Kes: Yes, our prediction is actually very simple. You have to divide the torque by $\sin\theta$, and you would find the magnetization. But for these measurements they always give a reduced value, or normalized value. So you never know what the magnetization actually is, and that would be very important to measure, as a function of field and so on. But I think they have problems of fitting their data. If you look very carefully there's a fudge parameter η, which should be a constant – but's it's not a constant.

Chiral spin state and gauge field in the extended $t - J$ Model
Hidetoshi Fukuyama
Institute for Solid State Physics, University of Tokyo, 7-22-1 Roppongi, Minato-ku, Tokyo 106, Japan

One of the most interesting proposals as regards the high Tc superconductivity is based on fractional statistical particles (anyons), especially 1/2-statistics (semions).[1-4] Above all Wen, Wilczek and Zee[4] have investigated the undoped case described by the Heisenberg spin model on the square lattice with exchange interactions not only between the nearest neighbors (J) but also between the next nearest neighbors (J'), and pointed out the possibility of a state violating the time-reversal symmetry and parity, which they called a chiral spin state. In this chiral state the gauge fields representing the fluctuations around the mean field solutions turn out to have the Chern-Simon term which transforms a test particle into a semion. Based on this finding they have developed a phenomenology for the semion superconductivity. Further detailed studies on anyon superconductivity have been made by Fetter et al.[5] and Chen et al.[6] The explicit derivation of anyons, however, in the doped case, where high Tc is actually observed, has not yet been reported.

In this short communication we report the result of our recent study on the doped systems based on the extended t-J model, which has both transfer integral, t', and the exchange interaction, J', between the next nearest neighbors together with those between nearest neighbors, t and J.

Our model is given by

$$H = J \sum_{n.n.} S_i \cdot S_j + J' \sum_{n.n.n.} S_i \cdot S_j$$
$$- t \sum_{n.n.} c_{i\sigma}^\dagger (1 - n_{i,-\sigma})(1 - n_{j,-\sigma}) c_{j\sigma}$$
$$- t' \sum_{n.n.n.} c_{i\sigma}^\dagger (1 - n_{i,-\sigma})(1 - n_{j,-\sigma}) c_{j\sigma} - \mu \sum_{i,\sigma} c_{i\sigma}^\dagger c_{i\sigma} \quad (1)$$

Representing the electron field operator in terms of the slave bosons, $b_i^\dagger$ and b_i, fermion fields, $f_{i\sigma}^\dagger$ and $f_{i\sigma}$, as $c_{i\sigma} = f_{i\sigma} b_i^\dagger$ etc, we transform eq. (1) into the following mean field Hamilitonian

$$H = H_F + H_B , \quad (2.a)$$

$$H_F = -\sum_{ij}(\phi_{ij}^* \chi_{ij} + h.c.) - \lambda_F \sum c_{i\sigma}^\dagger c_{i\sigma} , \quad (2.b)$$

$$H_B = -\sum (\beta_{ij}^* b_i^\dagger b_j + h.c.) - \lambda_B \sum b_i^\dagger b_i , \quad (2.c)$$

where $\phi_{ij} = t_{ij} <b_i^\dagger b_i> +(J_{ij}/2)<\chi_{ij}>, \beta_{ij} = t_{ij}<\chi_{ij}>$ are mean field bond order parameters, $\lambda_F = \mu + \lambda + (J+J')(1-\delta), \lambda_B = \lambda$ with λ and $\delta \equiv <b_i^\dagger b_i> = 1-\sum_\sigma <f_{i\sigma}^\dagger f_{i\sigma}>$ being the spatial average of λ_i and the doping rate, respectively. Here $\chi_{ij} = \sum_\sigma <f_{i\sigma}^\dagger f_{j\sigma}>$.

For the t-J model ($t' = 0$, $J' = 0$), we investigated three different kinds of states; (a) Peierls (b) staggered flux and (c) uniform. Each of these states is characterized as follows by bond order parameters between nearest neighbors; (a) $|<\chi_{ij}>|$ alternating every other bond in one direction (b) $<\chi_{ij}> = fe^{i\Phi/4}$ with uniform $f = |<\chi_{ij}>|$ (c) $<\chi_{ij}> = f, f$ being a positive number. The result is essentially the same as that by Marston and Affleck,[7] and Kotliar,[8] who examined a similar but different model or the same model by a different aproximation.

In the presence of t' and J' we are particularly interested in the possibility of the chiral spin state, which naturally emerges from the staggered flux state. We have found that there exist critical lines in the $t'-J'$ plane for the onset of the chiral spin state and that this state has a lower energy than the Peierls state.

The effects of fluctuations around the mean field solutions for the chiral spin state have been investigated in terms of the gauge fields $a_\mu (\mu = 0,1,2)$ introduced by the local constraint by $\lambda_i = \lambda + a_0(i)$ and those of the order parameters by $\phi_{ij} \Rightarrow \exp[i \int_i^j \vec{a} d\vec{l}]\phi_{ij}$ and $\beta_{ij} \Rightarrow \exp[i \int_i^j \vec{a} d\vec{l}]\beta_{ij}$. After integrating over fermion fields we have obtained not only the Chern-Simons term but also a term proportional to $a_\mu K^{\mu\nu} a_\nu$ due to the finite density of state at the Fermi energy. This result indicates that anyons, in its rigorous sense, do not exist in the present state.

More detailed description of the present result will appear in J. Phys. Soc. Jpn.

This work has been performed in collaboration with O. Narikiyo and K. Kuboki, and is financially supported by Grant-in-Aid for Scientific Research on Priority Areas, Mechanisms of Superconductivity (01631005) and New Functionality Materials-Design, Preparation and Control (01604015) from the Ministry of Education, Science and Culture.

REFERENCES

1. V. Kalmeyer and R. B. Laughlin, Phys. Rev. Lett. **59** (1987) 2095; Phys. Rev. B **39** (1989) 11879.
2. R. B. Laughlin, Phys. Rev. Lett. **60** (1988) 2677.
3. S. A. Kivelson and D. S. Rokhsar, Phys. Rev. Lett. **61** (1988) 2630.
4. X. G. Wen, F. Wilczek and A. Zee, Phys. Rev. B**39** (1989) 11413.
5. A. L. Fetter, C. B. Hanna and R. B. Laughlin, Phys. Rev. B **39** (1989) 9679.

6. Y.-H. Chen. G. Wilczek, E. Witten and B. I. Halperin, Int. J. Mod. Phys. B3 (1989) 1001.
7. J. B. Marston and I. Affleck, Phys. Rev. **B 39** (1989) 11538.
8. G. Kotliar, preprint.

DISCUSSION

A. Zee: I wanted to ask you about your generation of the A^2 term, as we discussed. Since your Lagrangian is explicitly gauge invariant, and you generate this gauge-invariant A^2 term, so some gauge non-invariance must have come into your calculations, perhaps in the diagonalization.

H. Fukuyama: Yes...

Z. Zee: Presumably you used a gauge non-invariant regulator.

H. Fukuyama: Yes, I think you can put...only spins component has a finite mass, which reflects the finite density of the fermion field, just at the Fermi energy. Because of the doping, chemical potential shifts downward to the continuum. That leads to the finite density of states, that leads to the finite value for the density-density correlation function. That's only for the time-component. And for the space part, that's k^2.

Comment on the Transport Properties of YBaCuO Superconductors

A. P. Malozemoff
IBM Research, Yorktown Heights, NY 10598-0218

I wanted to briefly mention in this meeting some results of my colleagues at IBM Yorktown, Roger Koch and his collaborators studying high current density YBCO thin films,[1] and Tom Worthington and his collaborators studying low current density YBCO crystals.[2] Both groups have made some of the most detailed studies to date on the current-voltage or E-J (electric-field vs. current density) characteristics of these materials. My main point is to emphasize the contrast in the thin film and single crystal results: many detailed features are different. Theorists should be careful, therefore, in formulating explanations, to consider both types of behavior.

In films, Koch and collaborators found a very beautiful scaling behavior in the E-J characteristics which is evident in a log-log plot.[1] At high temperatures, for fixed applied field in the Tesla range, the slope is 1, corresponding to ohmic behavior. With decreasing temperature, a higher slope develops, first at high currents, but eventually this higher slope, typically about 3, characterizes the entire measured range of current. They call the temperature at which this happens a glass transition temperature T_g, following the ideas of Fisher et al.[3,4] Below T_g, the data begin to show a downward curvature at the lowest electric field levels, with an almost vertical slope becoming more pronounced with decreasing temperature. This region of vertical slope indicates the appearance of a true critical current density, the hallmark of a true superconducting state with phase coherence or "off-diagonal long-range order."

The crossovers between the ohmic behavior and slope-3 behavior above T_g and between the slope-3 behavior and the vertical slope below T_g delimit a roughly symmetrical region in the J-T plane described by a crossover exponent $J_{cross} \propto |T_g - T|^{2\nu}$. The E-J data can be scaled in both regions using this crossover critical current density J_{cross}.

This beautiful scaling behavior reinforces the interpretation of Fisher et al.[3,4] in terms of a kind of glassy phase transition at T_g, analogous to the spin glass phase transition, with the frustration coming from the randomness in the location of pinning centers for the vortices.

By contrast, the recent results of Worthington et al.[2] on an YBaCuO crystal show no such simple power law behavior as was used by Koch et al. to define the glass transition temperature in the thin film case. Instead Worthington et al. see a rapid appearance of a steep downward slope with decreasing temperature in a log-log plot, which breaks away from simple ohmic behavior. Worthington has pointed out that a linear rather than log-log plot is more illuminating in this case. Here, all E-J curves are revealed to be linear, except that below some temperature (which is dependent on the magnetic field), the curves begin to shift from the origin in a way which can be interpreted in terms of the onset of a finite critical current density. The temperature for this "critical current transition" trends downward in a systematic way with increasing applied field. The results appear to fit into the classical picture of the onset of flux flow beyond a field-dependent current-density threshold.

In the context of the glass transition model,[4] these results imply such a weak pinning that the scaling region is suppressed to immeasurably small current densities. Some other model, perhaps a classic picture of crossover between flux creep and flux flow,[5] seems a better starting point to explain the data.

Another interesting aspect of these data are the appearance of another feature, a "bump," in the resistivity-versus-temperature, in the ohmic regime above the critical-current transition. If the ohmic region corresponds to flux flow, then there must be two viscous states. Various suggestions have been made to interpret such an effect, including the possibility of a "hexatic" or "entangled" phase of the vortex liquid.[6]

In summary, so far, any one of the many competing models appears to be inadequate to explain the full range of observed data.

REFERENCES

1. R. H. Koch, V. Foglietti, W. J. Gallagher, G. Koren, A. Gupta and M. P. A. Fisher, Phys. Rev. Lett. **63**, 1511 (1989).
2. T. K. Worthington, F. H. Holtzberg and C. A. Feild, Phys. Rev. Lett., to be published.
3. M. P. A. Fisher, Phys. Rev. Lett. **62**, 1415 (1989).
4. D. S. Fisher, M. P. A. Fisher and D. A. Huse, to be published.
5. A. P. Malozemoff, T. K. Worthington, E. Zeldov, N.-C. Yeh, M. W. McElfresh and F. Holtzberg, in Springer Series in Solid State Sciences, Vol. 89, ed. H. Fukuyama, S. Maekawa and A. P. Malozemoff (Springer-Verlag, Berlin, 1989), p. 349.
6. D. R. Nelson, Phys. Rev. Lett. **60**, 1973 (1988); M. C. Marchetti and D. R. Nelson, Phys. Rev. B, to be published.

DISCUSSION

L. Campbell: There's a preprint from IBM by Zeldov, *et al.* that does not mention any downturn.

A. Malozemoff: The Zeldov data are very similar to the data of Koch *et al.*, except for the detailed systematics of the downturn in log E vs log J at the lowest temperatures. On a log-log plot Zeldov's data tend to be linear, though with increasing slopes at lower temperature. The difference is in a regime where they are pushing the experimental limits, and perhaps Koch has pushed a bit further to see the downturn which Zeldov has missed.

L. Campbell: So, the preprint may be withdrawn?

A. Malozemoff: I think it is still an interesting preprint, showing how closely the data can be described by a flux creep model, except of course for the downturn.

P. Kes: Alex, did you ever consider the possibility that this first bump may be related to guided motion?. It might be possible that, especially in YBCO flux lines are along the twin boundaries... I'm just suggesting that below this bump, it might be guided motion, which then turns into hopping over the barriers for fields above the bump.

A. Malozemoff: It is an interesting suggestion. One has to make sure that whatever explanation is proposed, it shouldn't involve any kind of thermal activation barrier, because then one would get a nonlinear I-V curve. You would only see linearity at very low voltages, which is the limit you were considering in your TAFF (thermally activated flux flow) model. You should always be able to tell if you're in this regime because if you drive higher, you should see nonlinearity, and that is not seen in the region you are referring to.

J. Schirber: Is that bump always at the same resistance?

A. Malozemoff: The bump drifts down a little bit with decreasing field. If you look back at the earlier literature on YBCO crystals, this bump is pretty universal. Maybe not in all cases is it this well defined, but we found as the crystals got better and better, and the contacts got better and better, this bump appears more clearly, as does the second "critical current" transition to nonlinear I-V behavior at lower temperature.

A. Millis: In the previous transparency how close is that slope really to three, and how sample-dependent is it?

A. Malozemoff: 2.9 plus or minus 0.2, I believe. So far I have not heard of numbers significantly different from three in other samples, but this is an ongoing project, and not so many samples have been studied yet.

M. Maley: Let's see, that was a film?

A. Malozemoff: Yes, whereas the other data I was talking about is for crystals. The second transition, the "bump" I described in the crystals, is never seen in the films to my knowledge.

B. Batlogg: Which might be taken as an indication about sample quality.

A. Malozemoff: Maybe, or it could be that because of pinning, the bump is hidden, that is, that the critical current transition is moved up in temperature because of the increased pinning in films as compared to crystals. But an interesting point is that Roger Koch's glass temperatures actually lie below the bump temperatures in the crystals. So there are still some mysteries there.

8. SUMMARY

Chair — J. L. Smith & A. R. Bishop

Z. Fisk
Center for Materials Science
Los Alamos National Laboratory, Los Alamos 87545

Experimental Summary

These remarks are not a real summary, but rather comments derivative from the beautiful overview given by Bertram Batlogg. My purpose is to emphasize certain of Bertram's points in light of other presentations given at this meeting.

Bertram's first point was that there are serious materials problems with almost all of the high T_c materials. The reason that so much of the best work has been done on $YBa_2Cu_3O_{7-\delta}$ is because the materials problems are best understood and best under control in this system. But experiments must be done on all high T_c materials, and I want to discuss some materials aspects of the other high T_c superconductors.

To those like myself with limited acquaintance with oxides prior to the Bednorz-Mueller discovery, the conducting oxides seem very strange. It was amazing to me that nature would want to make some of these many layer materials, with huge repeat distances, and whose unit cells are built up from such heterogeneous building blocks. Riding along with the structural complication of the materials were questions as to how can one properly characterize the specimens they want to make physical measurements on. The whole history of high T_c experiment is shot through with superb experiments on mystery materials.

One point to emphasize is that of the stoichiometry and disorder found in high T_c materials I do this with data from our own work. The first material that I will talk about is $La_2CuO_{4+\delta}$, the Bednorz-Mueller prototype compound. You probably remember that it took more than one year for people to realize that δ was not a negative number, and that the compound does not easily (or at all) go below 4.0 in oxygen stoichiometry.

From the start there were reports that pure La_2CuO_4 itself was superconducting. It was Jim Schirber from Sandia who helped us understand what was going on. He loaded La_2CuO_4 with excess oxygen in a 3-4 kbar bomb at 600°C. This extra oxygen, at the $\delta \approx .04$ level, goes into the out-of-plane oxygen sites found in the T'-phase of heavy rare earth R_2CuO_4. This was determined independently by Chaillout[1] and Jorgenson.[2] Jorgenson also found that there is a phase separation in the oxygen loaded $La_2CuO_{4+\delta}$ near room temperature. We have some data on this for single crystals.[3] The data of Hundley shown in Fig. 1 for the in-plane and c-direction electrical resistivities shows a clear transition near 260 K between two different linear in T in-plane resistivities and a superconducting $T_c \simeq 37$ K. The c-axis resistivity does virtually nothing near 260 K, and the c-axis superconducting resistive transition is not complete, at least in this particular sample.

What is going on with the oxygen at this transition? Most oxide chemists will tell you that oxygen would not be expected to move easily in these materials at 260 K. The resistivity data looks as though there is some kind of stacking of oxygen rich and oxygen poor (4.0) materials developing at 260 K in this case. How might this happen?

A possibility is that the phase separation is magnetically driven. La_2CuO_4, after all, orders antiferromagnetically slightly above 300 K, and this magnetic phase transition might drive the phase separation. The fact that the separation in the oxygen loaded La_2CuO_4 happens near 260 K could indicate that the magnetic order parameter in the oxygen poor phase already has a finite value at the phase separation temperature. The phase separation is only very weakly, if at all, visible in the magnetic susceptibility data.

This idea receives support from La NMR in the oxygen loaded crystals done by Hammel (Fig. 2). The sharp, split line is from magnetically ordered La_2CuO_4 - the origin of the splitting is not understood. If you plot the average field shift of the split peak as a function of T (Fig. 3) there results a smoothly decreasing value of this shift with increasing T until ~ 260 K, where there is an abrupt loss of the signal at a point where the shift is still finite, indicating a weakly first order phase transition. The broad feature in Fig. 2 underlying the sharp resonance is the signal from the oxygen rich phase, and it is closely similar to the signal found in Sr-doped La_2CuO_4. So it is plausible that the phase separation in $La_2O_{4+\delta}$ is driven by the magnetic phase transition. This is a good example of the squirrelly behavior you find in oxide materials.

Now let me turn to Sr-doped La_2CuO_4, and remark on an earlier observation of Bertram which he did not mention in his overview. This is the observation that there is something peculiar about the superconductivity of $La_{2-x}Sr_x$ CuO_4 as a function of x. T_c peaks at x = .15, but both the shielding and Meissner fraction peak there also. The simplest interpretation of this is that there is really only one superconducting composition. You have to ask yourself here whether there is some hidden variable at play which has not yet been discovered. And remember too that work by Axe and others[5] at Brookhaven on Ba-substituted La_2CuO_4

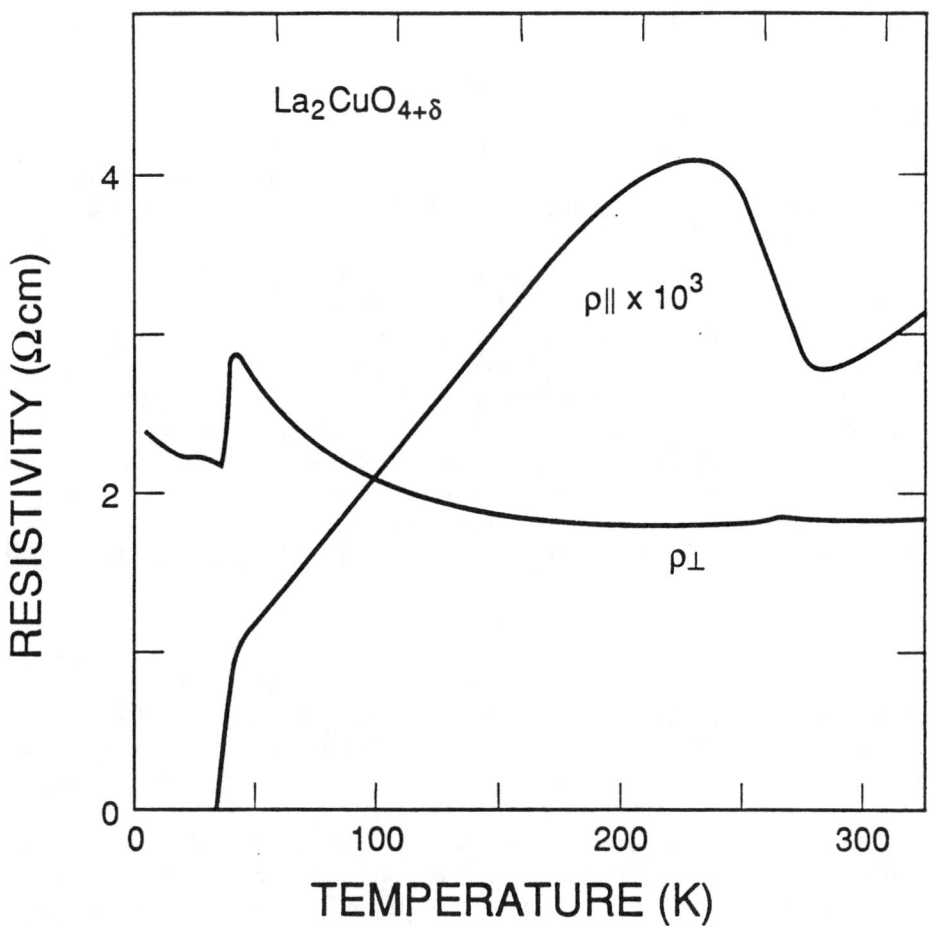

FIGURE 1 Temperature dependent electrical resistivity of single crystal $La_2CuO_{4+\delta}$ (from ref. 3.)

does find subtle phase transitions as a function of x. So it is important to keep in mind that there can be variables connected, for example, with oxygen positions which may be extremely hard to uncover. A probe such as neutron diffraction will not necessarily find this variable unless the experiment is designed to look for it.

The third materials point I want to make concerns the n-doped T′ phases, such as $Nd_{1-x}Ce_xCuO_4$. In the T′ phase there is no apical oxygen above each Cu. Rather, the Cu's are in square planar co-ordination, and the out of plane O's are at (0, 1/2,

Experimental Summary 451

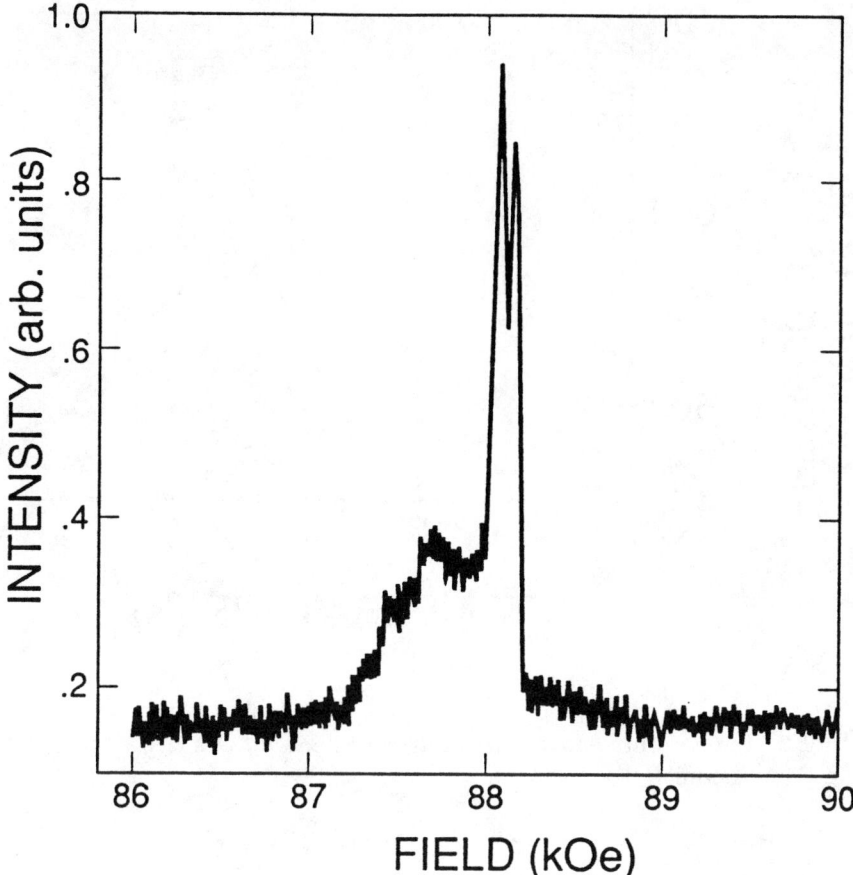

FIGURE 2 ^{139}La NMR in single crystals La$_2$CuO$_{4+\delta}$ (from P. C. Hammel).

1/4) and related positions. These materials become superconducting in a narrow range of x near .15 after a careful reduction procedure. In most polycrystalline samples, the temperature dependence of the electrical resistivity is more or less flat, with T$_c$ increasing as the resistance approaches more and more a metallic characteristic. What I wish to emphasize here is that there is something intrinsically difficult about making a decent superconductor out of these n-doped materials and that these materials are almost always, in some sense, polyphase. This shows up in nuclear resonance experiments. Fig. 4 is again some of Hammel's data. This is a nuclear quadrupole resonance spectrum on superconducting polycrystalline

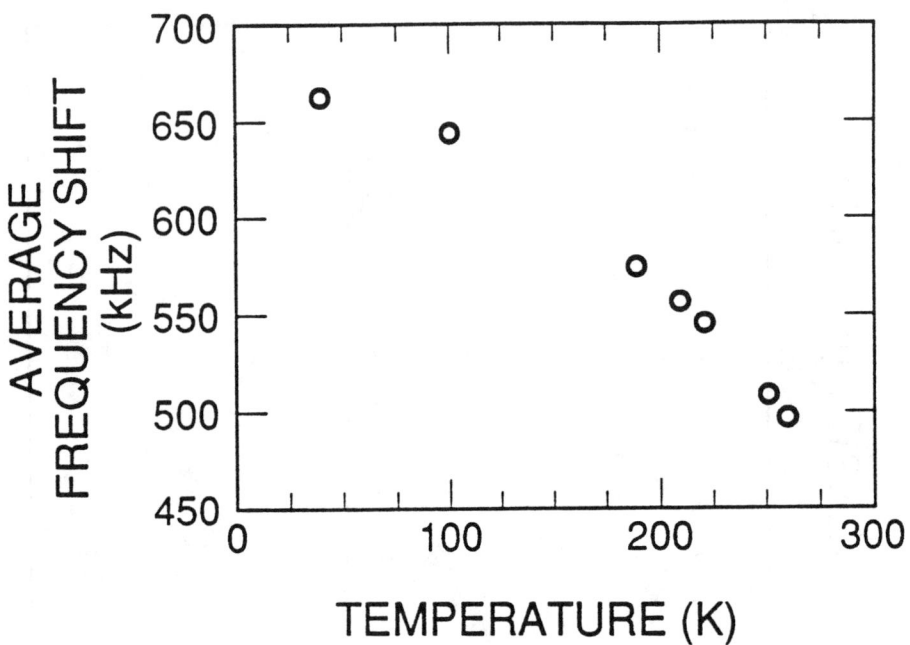

FIGURE 3 Temperature dependence of average line shift of sharp feature in Fig. 2 (from P. C. Hammel).

$Nd_{1.85}Ce_{.15}CuO_4$ material. This feature is essentially the same as one found in pure Nd_2CuO_4. There is another feature seen in NMR of Cu (Fig. 5) which is not found in pure Nd_2CuO_4. Yasuoka[7] has shown that this sharp line belongs to the superconducting phase and claims that they have succeeded in preparing samples which have only this narrow NMR feature and none of the broad quadrupole feature.

This sharp feature in NMR is the only diagnostic that I know of that tells you how much of the material is the right (i.e. superconducting) stuff. The process of reducing the as prepared $Nd_{1.85}Ce_{.15}CuO_4$ material to maximize the amount of superconducting material appears not to be at all understood in detail. Typical screening and Meissner fractions of 20% are found, and I believe only 20% of such samples is superconducting and the rest is this second phase exhibiting the quadrupole resonance feature.

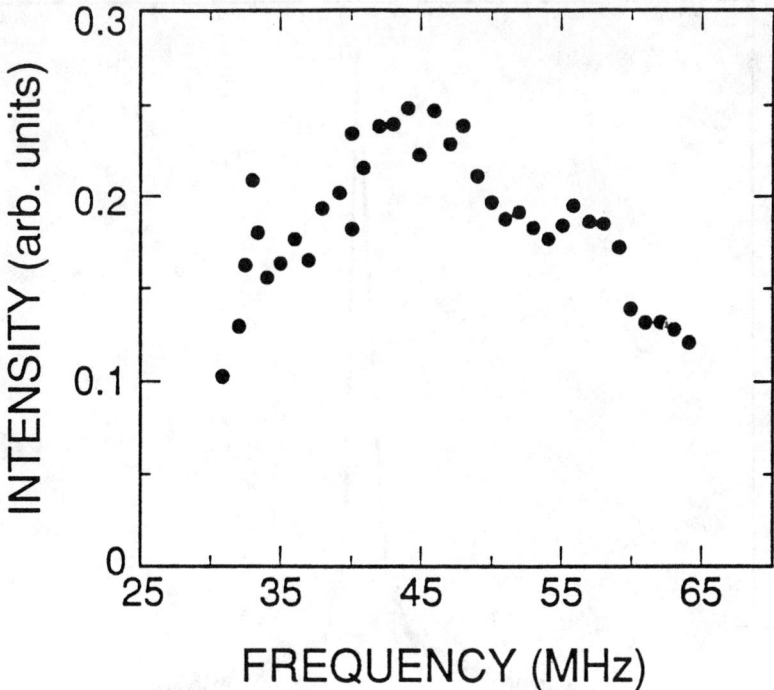

FIGURE 4 Cu NQR in polycrystalline $Ncl_{1.85}Ce_{.15}CuO_{4-y}$ (from P. C. Hammel).

E. Abrahams: Zach, does that mean that you think the superconducting composition is different than .15?

Z. Fisk: No, I think it's a problem of getting the oxygen right. I think it's very hard to equilibrate the oxygen properly in this material.

C. M. Varma: Do we know the electron concentration of the superconducting phase?

Z. Fisk: I don't know. Yasuoka claims to have samples with only the sharp NMR feature. My point is that the only data we should believe for these n-doped materials is material which passes the NMR test.

T. M. Rice: Zach, do you have any idea of the structural differences between these two?

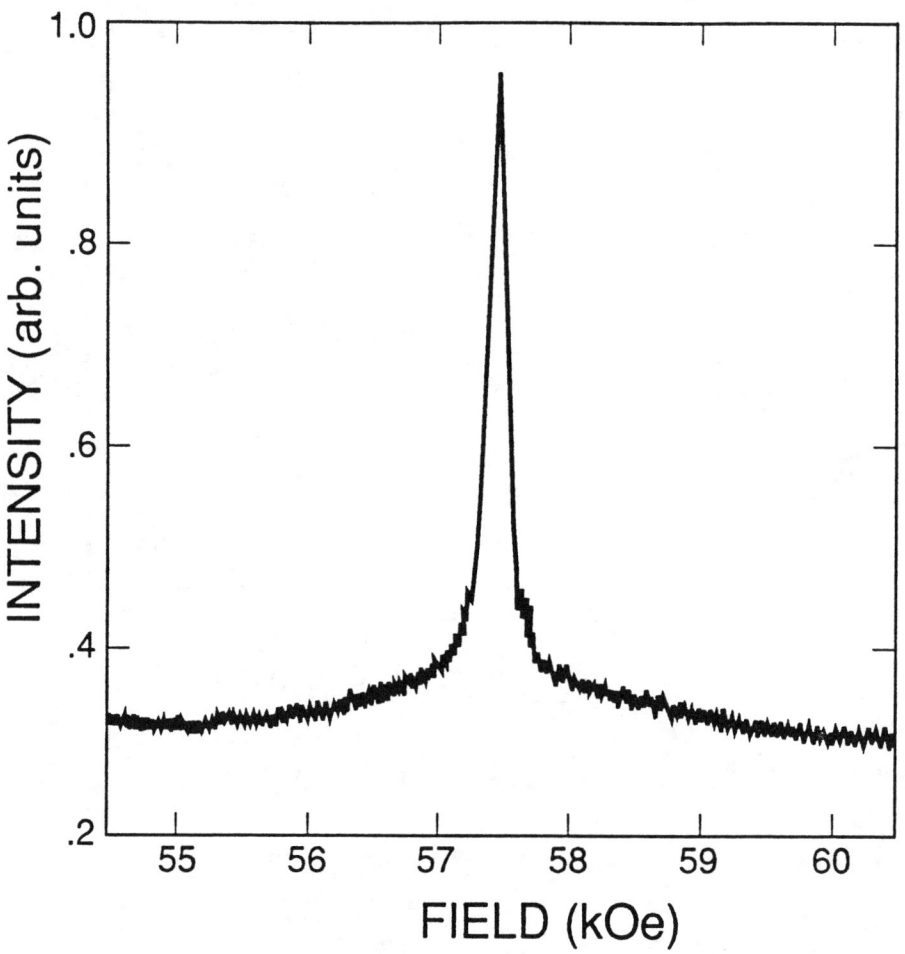

FIGURE 5 Cu NMR in same sample as Fig. 4.

Z. Fisk: No. I believe it is probably a problem of getting the out-of-plane oxygens in the right place and getting the amount of this oxygen right too.

F. Mueller: Why wouldn't you accept the notion that this unusual condition is simply impossible to get as a clean phase and only occurs in these mixed situations?

Z. Fisk: This would be a shame if it were so, but Yasuoka tells us that he can get the superconducting phase by itself.

R. M. Martin: Is it only 20% Messiner effect?

Z. Fisk: No. Not for Yasuoka's samples. What I'm saying is that most of the data one sees reported is for material that is only partially superconducting and that it is possible to do much better.

Now I turn to normal-state properties and consider first aspects of the electrical resistivity and its temperature dependence. Much has been made of the linear-in-T temperature dependence of the electrical resistivity, and I can give an example of how such a resistivity develops as a function of, for example, oxygen processing pressure. The data I show in Fig. 6 is for the electrical resistivity of T*-phase $La_{.85}Sm_{1.0}Sr_{.15}CuO_4$ oxygenated at ambient, 150 bar and 3 kbar oxygen pressure.[8] This polar phase forms with a roughly equal mixture of large and small rare earth ions and is a kind of interpolation between the T- and T'-phases.

We see how the resistance of these polycrystalline samples becomes more and more metallic as the oxygen processing pressure increases, being nearly linear for the highest pressure oxygenation done by Jim Schirber. But these samples still have only 60% shielding, and this makes me wonder if a better job of oxygenation would lead to a truly linear resistivity.

Next consider the resistance anisotropy of n-doped $Nd_{1.85}Ce_{.15}CuO_4$. Figure 7 is single crystal data from Hidaka and Susuki.[9] We don't know how much of the sample is superconducting in these single crystals. The temperature dependence of the resistivity in two different directions is shown, parallel to the planes, and perpendicular. The same metallic characteristic is seen in both directions, although they don't give an absolute number for the c-axis data. This resistivity makes this material different than the p-type materials, where you have a non-metallic characteristic perpendicular to the planes. It is my guess that these resistivities, if they really follow the same temperature dependence in the two directions, probably have absolute values within an order of magnitude of each other. It's hard for me to imagine -- again, this is just a speculation -- a solid that has an identical temperature dependence in these two directions with an extreme anisotropy. This means that there is something different in what you might call the electronics of these n-doped materials. And this would make the anisotropy irrelevant to the high T_c problem. It is very much worthwhile to find out in what ways these T- and T'-type materials really are similar and how their properties are the same or different from other multilayer cuprate-type materials.

The Hall effect has been an important parameter in high T_c problems. Its temperature dependence remains one of the stumbling blocks for theories of the electronic structure. And, more generally, much discussion has developed around the question of whether or not we are dealing with a Fermi liquid in the high T_c compounds. Any experiments which get at what might be a Fermi surface, and photoemission is a good candidate, should tell us alot about this question. It is also possible that de Haas-van Alphen experiments might be possible on sufficiently low T_c cuprates, such as T_c=7K bismuth strontium cuprate.

S. Chakravarty: What was not discussed in this context at all was positron annihilation studies. Can you comment on that?

Z. Fisk: I am aware of some position annihilation experiments reported at the Interlaken Conference. This is clearly a possible technique, although photoemission appears simpler.

C. M. Varma: It is a question of the best momentum resolution - in positron annihilation experiments this is at best a tenth of the Brillouin Zone and these [photoemission] people are doing about a thirtieth of the Brillouin Zone.

F. Mueller: The other mitigating factor is the possibility of annihilating defects, which also mask the structure, and so on.

Z. Fisk: Yes, they are more difficult experiments, but I don't think they should be thrown out.

T. M. Rice: The structure seen [by positron annihilation] in the metallic phase, which they felt had something to do with the Fermi surface, they also found [a similar structure] in the insulating phase.

Many experiments on normal state properties involve controlled doping, such as light-doping of La_2CuO_4. Quite apart from making sure such materials are good, there's always a question as to what these tell us about the metallic state. We don't know the answer to this, but it's very much worthwhile doing these experiments. Photoinduced absorption is a particular clean way to investigate certain of the very low doping properties.

Now about the magnetism. One of the beautiful things that has come out of the high T_c work is the lovely work of Chakravarty, Halperin, Nelson[10] on the 2D spin-1/2 problem and its realization in La_2CuO_4. This material is one of the best-understood magnetic materials. Gabe Aeppli and his collaborators[11] have followed the spin wave spectrum all the way out to the magnon edge near 250 meV, and it follows classical theory perfectly, except that the parameters are renormalized exactly as predicted in the CHN theory. There remains the question as to what the magnetism has to do with the superconductivity. An analogy that I like to make, when people ask me if I think high T_c could possibly be from magnetic mechanism, is to tell them that they first have to understand how you can get an attractive interaction out of something that's essentially repulsive. You can understand how that works if you understand the supermarket. The way the supermarket works is that because of the extremely low prices they lose a little on each transaction. However, they do so many transactions that they end up making a profit.

NMR seems to be one of the finest probes in these materials. It probes where you want to be. This has been emphasized by Andy Millis, Doug Scalapino and others. A lot of information can be obtained from these experiments as to what the wave vector dependent magnetic susceptibility is, and now neutron studies are starting to get some handle on whether the χ that you need to explain the NMR results is actually there. It's clear that as x (meaning the oxygen x in 6+x 123 materials) in these experiments gets up to $\sim.9$ some definitive statements can be made about whether we're really understanding what's going on. This is a development we can look forward to.

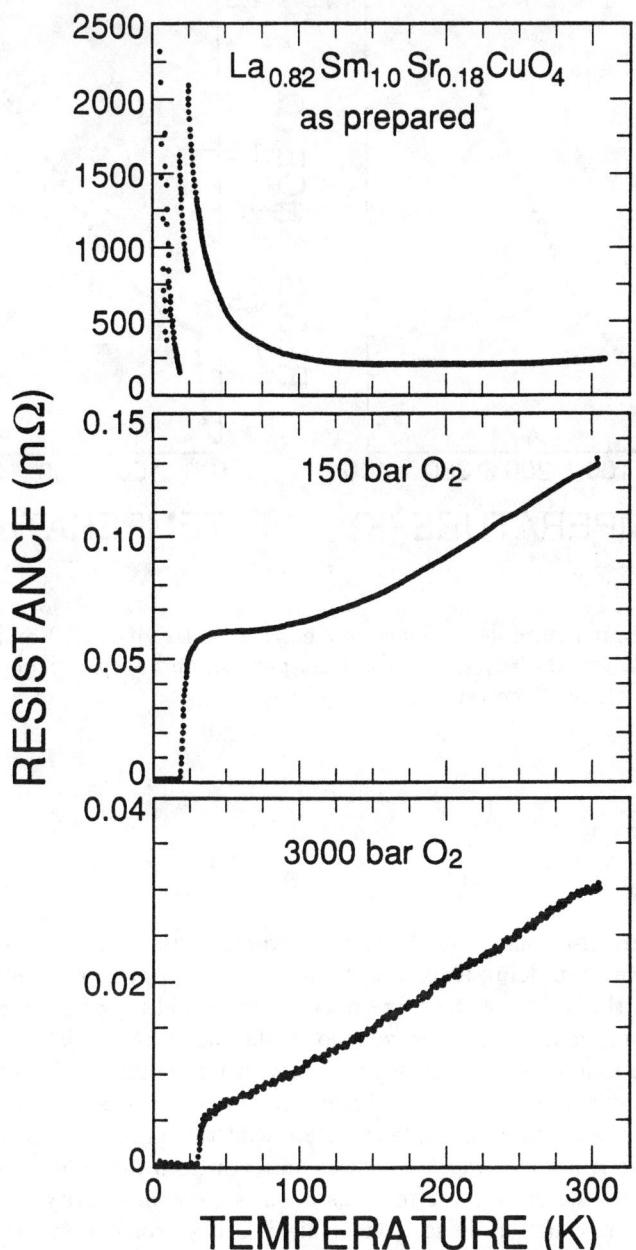

FIGURE 6 Temperature dependence of resistance of $La_{.85}Sm_{1.0}Sr_{.15}CuO_{4+x}$ for different oxygenation treatments (from ref. 3).

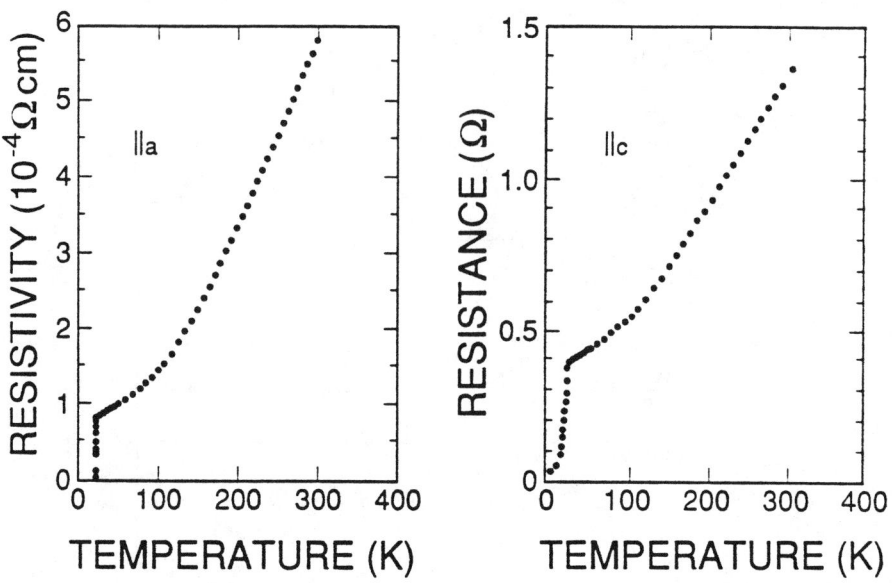

FIGURE 7 Temperature dependence of electrical resistivity (11a) and resistance (11c) of single crystal $Nd_{1.84}Ce_{.16}CuO_{4-y}$ parallel and perpendicular to the copper oxide plane (from ref. 9).

There are also some poor man's type experiments regarding magnetic properties that we've been doing involving doping into the copper oxide planes. As most people know, the purity of the plane must be protected in order to get good superconducting properties. Anything you do in the plane has a disastrous effect, and putting some kind of stable valence impurity in for copper, such as magnesium or zinc, creates magnetic moments. It turns out that you can see these moments with EPR. People are trying to analyze in detail what this kind of impurity should look like, and there is perhaps much that we can learn from that type of experiment.

Finally let me make some remarks about superconductivity. This of course is what makes these materials so wonderful. The big problem and the big interest is related to the nature of the superconducting gap. There have been a lot of experiments now on $2\Delta/kT_c$, and all the values seem to be larger than the weak coupling value. One should first say that any experimental problem, such as surface stoichiometry or oxygen stoichiometry will tend to make Δ small. So one's guess is that in measuring $2\Delta/kT_c$ you're probably getting a lower limit to what the value really is. Optical probes and tunneling probes are surface sensitive, and so we could be getting a low value from these. A reasonable operating hypothesis is that the superconductivity is in the strong coupling regime. I and many others have wondered about temperature effects on T_c: when T_c is around 100 K, there are real excitations such as phonons, magnetic and electronic excitations which can break pairs. These will affect the temperature development of the order parameter. Much can be learned from good studies of the temperature dependence of the gap. It will be important to have some way of distinguishing the real temperature effects from the intrinsic gap temperature dependence. And it may be that certain gap ratios might be independent of these real temperature effects, such as comparing the gap at different sites, as for example in NMR experiments. There is also the suggestion that you could have an order parameter which has nodes which are hidden in some directions in k-space where there isn't any Fermi Surface. Fermi surface studies can address this.

I've covered what I wanted to say about the NMR probes before. Again, for the superconducting properties it is one of the good ways to get at the microscopics. Another topic that didn't come up in Bertram's talk is whether barium potassium bismuth oxide superconductivity is the same as that in the cuprates. Most people, with perhaps one exception, think that phonons are a real mechanism for some superconductivity. In heavy fermions there's pretty good evidence for a magnetic mechanism. As soon as you get away from one mechanism, as soon as you have two real mechanisms, there's less problem with more than two mechanisms. And its not just possible, there probably are more than two. I don't see any a priori reason that the bismuthate and cuprate superconductors have to be the same thing. The feature of the cuprates that distinguishes them from the bismuth compounds is the copper oxide plane. This structural entity in the cuprates can't be fooled with without doing something terrible to superconductivity. There's no such object in these BiO_3-type materials. You can substitute for the bismuth, or you can substitute on the other site, and T_c goes up. The point that Chandra (Varma) would make about this is that as soon as you fool around with the copper oxide plane you make magnetic impurities which have a long healing length, and this is disastrous for the superconductivity. Maybe the "purity" of this structural element is just an artifact that's not really relevant to the comparison. There isn't at the moment any compelling argument, saying that cuprates and bismuthates are either the same or different. But they are both perovskite-based materials.

Not much discussion has been given at this conference to exactly what it is that varies in the physical properties as you go up from one layer, to two layer, to three layer materials. Such investigations might be a very fruitful way to actually pin down more about the mechanism as you go from $T_c = 40$ K to 90 K to 120 K. There's a lot to be learned from going back to some of the lower T_c materials, just to see how the properties compare in detail with the high T_c materials. I leave with this pitch for good measurements in 10 K oxide superconductors.

A. Malozemov: I'd like to address your comments about the homogeneity of the lanthanum strontium cuprates and the point you made about the .15 composition being perhaps the only really homogeneous composition. I think this is a very central point that goes back to the kinds of issues that Patrick Lee was raising on Wednesday, because a large part of that argument hinges on the concentration-dependent data, most of it precisely from the lanthanum-strontium-cuprate system. And so, it's a very important issue whether in fact we're to believe that, for instance, these key experiments of Suzuki studying the films of those materials and finding that the spectral weight is linear in the concentration of strontium. Are we to believe that or not? Because without that, I think that this key discrepancy that Patrick raised may to some extend disappear, and I guess that I'd just like to put on record that I still have some hope for that concentration dependence. I really don't trust these arguments based on Meissner Effect. We've done a lot of studies of the Meissner Effect and it really is very tricky to distinguish what is pinning and what is the real superconducting volume fraction. And especially when you're in a region where the T_c is sloping as a function of concentration. Assume that it is, and you're much more likely to be sensitive to slight local variations, and therefore to have increased pinning, which would reduce your Meissner Effect. And my judgment in the experimental data is that there is reasonable arguments for at least a moderately homogeneous material. I think maybe if you anneal the heck out of these materials, eventually they may phase separate, but the way they're prepared, particularly in some of these thin film experiments that were done in Japan, I think that there is a reasonable hope that there is close to atomistic homogeneity, and that that type of data can be trusted, and should not be disregarded.

Z. Fisk: There's no definite answer to your question. I agree with what you said about the Meissner effect, but let's just look at screening. You make Yttrium 123 fully oxygenated: any high school student will make you a sample that has 100% shielding. That's not true with these lanthanum samples. You really have to work to get 100% shielding. And what strikes me is that if you don't get 100% shielding, there's something wrong. And it is a fact that the Meissner fraction does the same thing as the shielding. You go off from the particular composition ($\sim$.15) and your shielding falls right off. I don't think a pinning argument is going to explain that. I'm suggesting that one has to keep this lanthanum cuprate problem in the back of their mind.

J. M. Tranquada: On the same point—the Meissner fraction seems to peak in the strontium .15, but it seems that microscopically they have to be inhomogeneous because you're putting a small amount of strontium into lanthanum; if that goes in randomly, you don't have a homogeneous system. Now maybe when you get the strontium .15 you get the largest Meissner fraction and somehow more of the sample is superconducting. This doesn't mean that the sample itself is actually homogeneous. I think that there may be just some limitations in these systems and things may peak at that point, but it doesn't necessarily mean that's an incredibly magic point. So some of the same problems which you have at lower doping you still have at strontium .15. So it's just a complicated system.

Z. Fisk: Yes, but if you're going to make that argument, then you have to compare a compositional fluctuation distance with a coherence length, or something like that. And this distance should just go smoothly through .15. As you go to higher and higher strontium concentrations you should have a better and better superconductor based on that argument.

J. M. Tranquada: Sure but that brings up the question why is the superconductivity going away. There may be two parts of the problem, one on the rising side and one on the falling side.

Z. Fisk: Yes, possibly. There are probably more than two parts of the problem.

C. M. Varma: A comment about this bismute-ate and cuprate. If and when we know the normal-state properties of the bismuth compounds and the materials are good enough that we really can trust them, and if and when the normal-state peculiarities turn out to be the same as the copper oxide peculiarities, we are forced to at least say that whatever may be the microscopic reason, it may be different, but finally the excitation spectrum in those materials, in cuprates and the bismuth-ates in the metallic phase is the same, qualitatively the same...

Z. Fisk: This is a prediction

C. M. Varma: .. that will be... yes.

Z. Fisk: You're putting your money there?

C. M. Varma: Yes.

Z. Fisk: Do I have any takers?

E. Abrahams: I wanted to raise a question about the T'-phase which is connected to what people are saying. What I'm wondering is, what's your opinion as to whether or not those who claim that the superconducting composition pops out of the antiferromagnetic phase, so they almost coexist. I've seen other data where the superconducting part of the phase diagram is well separated from the antiferromagnetic part of the phase diagram. You say that you have similar temperature dependencies in the resistivity and a relatively low anisotropy, and then there's other data which says the anisotropy is a thousand.

Z. Fisk: Let me go back to the anisotropy. There have been occasional reports that in $YBa_2Cu_3O_7$ you can get a metallic characteristic perpendicular to the planes, and you might speculate that in these 2- and 3-layer materials it might be very easy to have an occasion fault which gives you an insulting layer that you have to tunnel through. But these anisotropies appear so reproducible that it's hard not to accept them. For the n-doped materials we are in a more primitive state in understanding how to make really good samples, and the problem is apparently that the methods used for reducing the material are giving you this mixture. But I think that this will not fake the resistivity, because improperly-reduced material probably has an extremely large resistivity, some kind of Mott-Hubbard material. Then all you're seeing is the metallic component, these portions of the sample shorting out the rest.

It's not unreasonable to think that the similar temperature dependence along both directions is real.

E. Abrahams: But then what is one to make of data on the T'-phase in the superconducting compositions where resistivity has an enormous anisotropy?

Z. Fisk: The question is not settled. I think the data favors low anisotropy.

S. Doniach: Can I come back to your pure La_2CuO_4 with a smidgen of oxygen? It seems to me that this is very remarkable stuff because even if it phase separates, the oxygen-rich regions is a not very high concentration, apparently, so any doping argument would give a rather low hole concentration.

Z. Fisk: About 30% of the sample is believed to be the oxygen-rich material, and so you can get up to about perhaps .09. Then you have to decide, for this extra oxygen, how many charges does it pull out?

S. Doniach: I can phrase it as a question. What do you think the hole concentration is in those superconducting patterns?

Z. Fisk: I'd be amazed if it wasn't similar to that in $La_{1.85}Sr_{.15}CuO_4$.

C. M. Varma: I think that Jorgensen from the bond lengths deduces it to be about 10%.

Z. Fisk: There's another point that I should make about this phase separation. The diffraction lines from the two phases are sharp. This means that the domain size of the separated phases is many hundred Angstroms. This says that you are probably pushing oxygens over substantial distances, although there are other ways you could think about it - maybe there's already some built-in structure and it's a more subtle kind of condensation, perhaps the type that Vic Emery had in mind.

S. Doniach: You said you saw the oxygen in the T' position?

Z. Fisk: Yes.

S. Doniach: So is it simultaneously both T' and T?

Z. Fisk: Yes, there's an occasional cell that has both, and within this cell both positions then it relax.

M. Maley: A question about the phase separation in the lanthanum-strontium system, the doped system: have there ever been any diffraction lines seen that indicate that the phase separation scale is very very small?

Z. Fisk: I don't know of any evidence that there is a phase separation in lanthanum-strontium samples.

F. Mueller: Couldn't it be that those things are simply mixed up so that you couldn't see them in fact in any diffraction experiment that one could envisage, plaquettes, and rotators, and so forth?

Z. Fisk: In lanthanum-strontium?

F. Mueller: Yes.

Z. Fisk: Yes, and as Chandra just mentioned there is something peculiar in the μSR in $La_{1.85}Sr_{.15}CuO_4$.

F. Mueller: It may be there.

Z. Fisk: Yes.

REFERENCES

1. C. Chaillout et al., Physica C**158**, 183 (1989).
2. J. D. Jorgensen et al., Phys. Rev. B**38**, 11337 (1988).
3. M. F. Hundley et al., submitted to Phys. Rev. B.
4. P. C. Hammel, private communication.
5. J. D. Axe et al., Phys. Rev. Lett. **62**, 2751 (1989).
6. P. C. Hammel, private communication.
7. H. Yasuoka, private communication.
8. M. F. Hundley et al., Phys. Rev. B **40**, 5251 (1989).
9. Y. Hidaka and M. Suzuki, Nature **338**, 6351 (1989).
10. S. Chakravarty et al., Phys. Rev. Lett. **60**, 1057 (1988).
11. G. Aeppli, private communication.

Elihu Abrahams
Serin Physics Laboratory, Rutgers University, Piscataway, NJ 08855

Summary Remarks on the Theory of High-T_c Superconductivity

We have had a conference with comprehensive review talks on all the major issues in high-temperature superconductivity. A real theory summary is somewhat out of place, as the meeting itself has brought us up to date. Therefore, I will restrain myself from reporting what you have already heard and I will offer some personal reflections on some of the main questions which have survived this meeting.

We have had three years of high-T_c superconductivity and it is natural to ask, "What progress have we made in theoretical understanding during the last eighteen months?" It seems to me that we have not made much new progress in developing a real understanding of the basic mechanism of high-T_c superconductivity. There are some people who disagree, but I think that the consensus is, and here I represent the consensus, that we have not substantially improved our comprehension of what is going on during the last eighteen months.

On the other hand, a lot of things have happened. The whole problem of high-T_c superconductivity and the enthusiasm with which it is being attacked has made an enormous difference to the practice of experimental condensed matter physics and also has had an important influence on several areas of theoretical physics which are attached in one way or another to high-T_c research. There has also been a remarkable stimulation of the science and techniques of materials preparation. It has already been emphasized at this conference that after an initial period of dismaying inconsistency in experimental results, now different laboratories and research groups are getting essential agreement on what the measured transport, thermodynamic and spectroscopic properties are. This happy development is a consequence of improved materials preparation and is also due to the development of more precise and elegant experimental methods.

I want to mention two of these in particular, because both of them have received a good deal of attention at this conference. They are two of the many areas of investigation in the high-T_c field in which we have the close collaboration between theorists and experimentalists which has always characterized condensed matter physics. One is nuclear magnetic resonance, which, it seems to me, began to disappear from the mainstream of condensed matter physics in the early sixties. Aside from a few laboratories using NMR to study liquid ^{3}He, one would have to go to a chemistry department to see a state of the art NMR rig. In the early sixties, NMR was a classy way of getting information on condensed matter problems and now again, we have wonderful results from talented people (and their gifted students) who stuck with NMR, or who returned to it. Everyone is now paying an enormous amount of attention to these experiments which involve one of the very few, and perhaps the most precise, local probes that we have to study the high-T_c materials. We have unusual and challenging data which may turn out to be of central importance in the development of our understanding of high-T_c superconductivity. It is interesting that we have phenomenological theories for the NMR data which contain very much more information than the experiments themselves, which is a rather unusual situation.

The other experiments which I want to mention is angle-resolved photoemission (ARPES) spectroscopy. It has generally been the case that many-body theorists have not paid much attention to photoemission experiments. The reason for this is that the energy scales which were probed were rather large, whereas, as everyone knows, one's real interest is in the low energy excitations above the ground state. People who were doing photoemission were very popular with band structure theorists, but such experiments were, in the main, ignored by the other condensed matter theorists. However, now we see that photoemission has become a glamor experiment. Fantastic improvements in energy and angular resolution in just the last year have made ARPES the experiment to watch for, the one that is going to give us the answers to some of the really basic questions of both the normal state (is it a Fermi liquid?) and the superconducting state (what are the characteristics of the gap?). It seems that the hopes we once pinned on numerical computation techniques to understand the Hubbard model have now been transferred to ARPES.

There is also a set of theoretical developments which, while not revealing the truth of high-T_c have illuminated a variety of problems, some of which involve new quantum states which we never before considered. Thus, just in the last two years, we have seen extraordinary advances and new insights in the following areas, most of which we have heard about at this conference and all of which are representatives of the class of problems called "strongly-interacting electrons:"

One-dimensional Physics
Quantum Antiferromagnetism
Hubbard Model

New States: Quantum Spin Liquid, Flux Phases, Chiral Phases and Anyons
Numerical Methods

The reason for these developments in strong-correlation physics is, of course, that the high-T_c compounds have some connection, it is believed, with various parts of the list. On the other hand, despite the intense activity, progress and excitement about these problems, all of the developments still have not answered the question of what the basic underlying high-T_c mechanism is. The message here is that in the last three years there have been great advances in a broad range of the experimental and theoretical physics connected in one way or another with high-T_c, but still that is no reason to expect that we should know "the answer" only three years after the discovery of the materials.

A hot topic has been the question of the character of the normal state of high-T_c superconductors. Is it a Fermi liquid, not a Fermi liquid, not a "simple" Fermi liquid, a "marginal" Fermi liquid, and so forth? P. W. Anderson[1] has repeatedly stressed that the normal state properties of the cuprate superconductors are unlike those of Fermi liquids and this means that the quasiparticle picture fails, i.e. the quasiparticle weight Z vanishes in the neighborhood of the Fermi surface. This idea is central to his approach[2] which is based on the separation of the energy scales of charge and spin excitations. It is evident that whatever the ground state of high-T_c superconducting materials, no matter how exotic - quantum spin liquid, RVB, anyons, flux phase..., experimental probes measure electron correlation functions. Examples of these are the density-density or spin density-spin density correlation functions:

$$\langle \rho(r,t)\rho(r',t') \rangle$$
$$\langle \sigma(r,t)\sigma(r',t') \rangle$$

which are particle-hole correlation functions, or two-particle Green's functions. These quantities are measured in a variety of transport and thermodynamic measurements. There are experiments, tunneling and ARPES, for example, which measure directly the one-particle Green's function $G(k,\omega)$. The behavior of the Green's functions determines the excitation spectra. One can read in these proceedings[3] just what the three conditions are that most people believe a Fermi liquid needs to satisfy. Here, we need just to say that in the conventional picture, the Fermi liquid has rather well-defined single particle excitations which are quasiparticles and that they are long-lived for low excitation energy. It is convenient to describe this by the behavior of $\mathcal{A}(k,\omega)$, the imaginary part of the single-particle Green's function $G(k,\omega)$. It is called the spectral function as it describes the distribution of energies ω in the single particle excitation of wave vector k. The single particle (tunneling) density of states and the momentum distribution are given in terms of $\mathcal{A}$ by

$$\mathcal{N}(\omega) = \sum_k \mathcal{A}(k,\omega) \text{ and } n(k) = \int d\omega \mathcal{A}(k,\omega) .$$

What is the behavior of $\mathcal{A}(k,\omega)$ which describes the physics of quasiparticles?[4,5] At a given k, there is, as a function of ω, a rather symmetric peak centered around the quasiparticle energy E_k (which can often be written $k^2/2m^*$). The spectral weight in the peak is called Z_k, the quasiparticle weight. The width of the peak is the quasiparticle's inverse lifetime and in a Fermi liquid, as you move the momentum k to the Fermi surface, the quasiparticle width disappears as the second (or larger) power of E_k. This decay rate of the quasiparticle is gotten from the imaginary part of the single particle self-energy which enters the Green's function. The weight in the quasiparticle peak is not unity because in an interacting system, the addition of a single particle can create multiple particle-hole pairs or other more complicated excitations. Therefore, there is a weight $1 - Z_k$ in what is usually taken to be a broad featureless background called the incoherent part. It is just this spectral function which can be measured in ARPES so that with sufficiently high resolution and the ability to subtract extrinsic contributions to the spectrum, ARPES can then inform us as to the precise nature of the spectral function and consequently answer the Fermi-liquid-or-not question. From such Green's functions, one can go further and discuss two-particle Green's functions which are related to the collective modes. But for our purposes it is just necessary to establish a dictionary of what people are talking about when they refer to Z_k, the quasiparticle width, the coherent part, and the incoherent part.

In what follows, I would like to return to the question of the phenomenology of the normal state properties of high-T_c superconductivity which was discussed by several people at the beginning of the conference. The terminology discussed above is used in what has been called the paper of the Five Friends[6] which discusses properties of the normal state of the high-T_c superconductors. Actually, the phrase was the "five erstwhile friends," but since no one knows what erstwhile means, and if you do, you cannot tell whether it is an intrinsic or extrinsic property, we'll leave it as the Five Friends. In that paper, an attempt is made to understand the various unusual normal state properties (resistivity, Raman scattering, NMR relaxation rate, tunneling, ARPES, thermal conductivity) on the basis of a single simple assumption. The assumption is that the imaginary part of a susceptibility $\chi(q,\omega)$ has, for a range of q, the following simple ω dependence:

$$Im\chi(q,\omega) \propto \begin{cases} \omega/T & \text{for } \omega < T \\ const. & \text{for } T < \omega < \omega_c \end{cases} \qquad (1)$$

If you want to take a simple view of the matter, you can say, "Well, you just look at the experimental low-frequency Raman spectrum and this [Eq. (1)] is what you see."[7] That is to say, you can explain the Raman spectrum with this assumption. Of course in the Raman spectrum, the momentum transfer is zero, but the Five Friends made an assumption that this frequency behavior is essentially independent of q for some appreciable range of q.

From the simple assumption of Eq. (1), which, as I will discuss in a minute, is rather unusual - but then the data is rather unusual - you can derive the single-particle self energy $\Sigma(k,\omega)$ due to the interaction of an electron with whatever the excitation is that produces the strange susceptibility, Eq. (1). Then, it is only a few steps to get $G(k,\omega)$ and the spectral function $\mathcal{A}(k,\omega)$. The results are:[6]

$$\Sigma(k,\omega) = \lambda(\omega \ln \max(|\omega|,T) - i\pi\max(|\omega|,T)/2], \qquad (2)$$

$$G(k,\omega) = [\omega - \varepsilon_k - \Sigma(k,\omega)]^{-1},$$

$$\mathcal{A}(k,\omega) = -ImG(k,\omega) = \frac{\lambda\pi\max(|\omega|,T)/2}{[(\omega - \varepsilon_k - \lambda\omega \ln \max(|\omega|,T))^2 + (\lambda\pi\max(|\omega|,T)/2)^2]}. \qquad (3)$$

What is interesting about the result for the self energy is that its imaginary part is linear in ω (or T), whereas in a Fermi liquid it goes at least as fast as ω^2. So it appears that the quasiparticles are not going to be well defined, as their decay rate ($Im\Sigma$) will be too big. Incidentally, this is precisely the form of the self energy which you get in perturbation theory in the one-dimensional Hubbard model. The arguments here have nothing to do with dimension, however.

The spectral function, for k near to k_F, rises from zero at $\omega = 0$ (the Fermi energy) has a sharp peak near the kinetic energy E_k and then falls off slowly as $1/\omega$ as you go well away. For $k < (>)k_F$, the peak is below (above) the Fermi energy, but there remains a substantial tail on the other side. If the momentum is precisely on the Fermi surface, then the spectral function is one symmetric peak, essentially $1/|\omega|$.

Now you can go to calculate Z_k, the quasiparticle weight. It is simply given[4] by $[1 - \partial\Sigma/\partial\omega]^{-1}$ at $\omega = E_k$. As E_k goes to the Fermi surface, it is easy to see from Eq. (2) that Z_k goes to zero [as $1/\ln(\omega_c/E_k)$]. So there is no quasiparticle at all when $k = k_F$ in spite of the fact that the spectral function has this peak $1/|\omega|$; it is entirely incoherent - the Fermi liquid picture fails. As noted above, the $Z = 0$ behavior was already suggested by Phil Anderson some time ago.[1] However, it is reached here by a different route. The logarithmic behavior of Z^{-1} is the origin of the term "marginal" Fermi liquid.[6] The Green's function, then, is made up of something much more complicated than just a single one-electron excitation. This is an example where the incoherent part of the spectral function is not just a broad featureless background.

What would you measure, if this is really what is happening, in ARPES? If you take the spectral function $\mathcal{A}(k,\omega)$ and multiply it by a Fermi function $f(\omega/T)$, then you get precisely what is measured at momentum k and energy ω measured from the Fermi energy. You can see by comparing the experimental spectra shown by List[8] at this meeting and the theoretical spectral function in the paper[6] of the Five Friends that they look pretty much the same. So it would be interesting to try to fit the spectral function of the marginal Fermi liquid to the data rather than using the conventional fit to a Lorentzian which is only sensible when there is a conventional Fermi liquid. If this marginal Fermi liquid picture is correct, it suggests that the broad tails seen in the photoemission spectra at energies below the peak are intrinsic, a consequence of the broad incoherent tail of the spectral function. It is also the case that when $k > k_F$ there will still be a lot of weight below the Fermi energy which will remain after the peak marches through the Fermi surface as you change the angle...and of course all this is just what is seen in the experiments!

An important question, then, and one which was discussed several times in this conference is,...how do we understand the origin of the behavior of the quasiparticle renormalization constant Z? In the one-dimensional Hubbard model, we have an example of $Z = 0$; the reason for this is has to do with the charge and spin separation which occurs in that model. So if you believe that the anomalous normal-state properties mean that we are not dealing with a simple Fermi liquid, then you might be attracted by the example of one-dimension. A key problem is how do you get from one dimension, about which we know much more now than we did three years ago,[9] to two dimensions. In connection with that, we have the question, "What does $Z = 0$ really mean?" That is to say, if one has charge-spin separation (as in one dimension) then one can believe that one gets $Z = 0$, but there may be other ways. So an interesting question is "What is the nature of an excitation spectrum which admits $Z = 0$?" By $Z = 0$, I mean $Z = 0$ in the kind of marginal way described above, not $Z = 0$ because a gap has formed at the Fermi surface.

I want to emphasize that the way the Five Friends reached $Z = 0$ was to derive it from the assumption that $Im\chi$ goes like ω/T for small ω. You are almost forced to do that because there is an experiment, the Raman scattering,[7] which says that is what $Im\chi$ does. As people have pointed out, that is rather unusual because the energy scale here is the temperature T and not some electronic energy like the Fermi energy, or the exchange energy. So this is a rather peculiar behavior and if you can understand it, you can understand at least one scenario for $Z = 0$.

Some results reported by John Tranquada here[10] incite me to address a comment to those who are depending, in their thinking about high-T_c, on the existence of a reasonable range of antiferromagnetic order in the superconducting compounds. It seems to me that the information we have so far from neutron scattering is that as you make the superconductor better and better, that is more and more single phase and higher and higher T_c, that the antiferromagnetic correlation length is going down.[10] If this trend continues and one finds magnetic correlation lengths as small as the order of a lattice spacing in a $T_c = 92K$ 1-2-3 material, then some theoretical approaches will have to be reexamined.

I conclude with the remark that there are a whole class of other two-dimensional materials, organic and inorganic and some of them superconducting, which are good laboratories to study. Many of them, for example, show linear resistivities. There are people in the world who can make these materials and tune the parameters so you can get a spectrum of Hubbard models in two dimensions. This area can be an important adjunct to high-T_c research which may help us to finally discover the answers.

REFERENCES

1. P. W. Anderson, in "Frontiers and Borderlines in Many-Particle Physics," edited by J. R. Schrieffer and R. A. Broglia (North Holland, Amsterdam, 1988); P. W. Anderson, in "Strong Correlation and Superconductivity," edited by H. Fukuyama, S. Maekawa and A. Malozemoff (Springer Verlag, Berlin, 1989).
2. P. W. Anderson, these proceedings.
3. P. A. Lee, these proceedings.
4. Philippe Nozières, "Theory of Interacting Fermi Systems," Benjamin, New York, 1964.
5. A. A. Abrikosov, L. P. Gor'kov, I. E. Dzyaloshinskii, "Methods of Quantum Field Theory in Statistical Physics," Prentice-Hall, Englewood Cliffs, NJ, 1963.
6. C. M. Varma, P. B. Littlewood, S. Schmitt-Rink, E. Abrahams, and A. E. Ruckenstein, Phys. Rev. Lett. **63**, 1996 (1989).
7. M. V. Klein, S. L. Cooper, F. Slakey, J. P. Rice, E. D. Bukowski, and D. M. Ginsberg, in "Strong Correlation and Superconductivity," edited by H. Fukuyama, S. Maekawa and A. Malozemoff (Springer Verlag, Berlin, 1989).
8. R. S. List, these proceedings and references therein.
9. F. D. M. Haldane, these proceedings.
10. J. M. Tranquada, these proceedings.

List of Attendees

Professor Elihu Abrahams
Rutgers University
Physics and Astronomy Dept.
Piscataway, NJ 08844

Dr. Robert Albers
Group T-11/B262
Los Alamos National Laboratory
Los Alamos, NM 87545

Professor Philip W. Anderson
Princeton University
Dept. of Physics
Jadwin Hall, P.O. Box 708
Princeton, NJ 08544

Dr. Ivo Batistic
Group T-11/B262
Los Alamos National Laboratory
Los Alamos, NM 87545

Dr. Bertram Batlogg
AT&T Bell Laboratories, ID-369
Physics & Academic Affairs Division
600 Mountain Avenue
Murray Hill, NJ 07974

Dr. Laurence Campbell
Group T-11/B262
Los Alamos National Laboratory
Los Alamos, NM 87545

Dr. Sudip Chakravarty
Department of Physics
University of California/Los Angeles
Los Angeles, CA 90024

Dr. Albert M. Clogston
CMS/K765
Los Alamos National Laboratory
Los Alamos, NM 87545

Dr. Kevin Bedell
Group T-11/K765
Los Alamos National Laboratory
Los Alamos, NM 87545

Dr. Alan Bishop
Group T-11/B262
Los Alamos National Laboratory
Los Alamos, NM 87545

Dr. A. Michael Boring
CMS/K765
Los Alamos National Laboratory
Los Alamos, NM 87545

Dr. Baird Brandow
Group T-11/B262
Los Alamos National Laboratory
Los Alamos, NM 87545

Dr. David Campbell
CNLS/B258
Los Alamos National Laboratory
Los Alamos, NM 87545

Dr. Victor Emery
Brookhaven National Laboratory
Physics Dept., 510A
20 Pennsylvania Avenue
Upton, NY 11973

Dr. Zachary Fisk
CMS/K765
Los Alamos National Laboratory
Los Alamos, NM 87545

Professor Hidetoshi Fukuyama
Institute for Solid State Physics
University of Tokyo
7-22-1, Roppongi, Minato-ku
Tokyo 106, JAPAN

Dr. Dermot Coffey
CMS/K765
Los Alamos National Laboratory
Los Alamos, NM 87545

Professor Seb Doniach
Department of Applied Physics
Stanford University
Stanford, CA 94305-4090

Dr. P. Chris Hammel
Group P-10/K764
Los Alamos National Laboratory
Los Alamos, NM 87545

Dr. Jeff Hay
ERDC/K763
Los Alamos National Laboratory
Los Alamos, NM 87545

Dr. Arno Kampf
CMS/K765
Los Alamos National Laboratory
Los Alamos, NM 87545

Dr. Robert Heffner
Group P-10/K764
Los Alamos National Laboratory
Los Alamos, NM 87545

Dr. Peter H. Kes
Kamerlingh Onnes Laboratory
Nieuwsteeg 18
2311 SB Leiden
THE NETHERLANDS

Dr. Martin Maley
ERDC/K763
Los Alamos National Laboratory
Los Alamos, NM 87545

Dr. James Gubernatis
Group T-11/B262
Los Alamos National Laboratory
Los Alamos, NM 87545

Professor F. Duncane M. Haldane
University of California/San Diego
Physics Department
La Jolla, CA 92093

Professor Gabriel Kotliar
Rutgers University
Serin Physics Lab
P.O. Box 849
Piscataway, NY 08855

Dr. George Kwei
Group CLS-2/H805
Los Alamos National Laboratory
Los Alamos, NM 87545

Professor Anatolii I. Larkin
Department of Physics
University of California/Los Angeles
Los Angeles, CA 90024

Professor Patrick Lee
Department of Physics
Massachusetts Institute of Technology
77 Massachusetts Avenue
Cambridge, MA 02139

Dr. Scott List
P-10/K764
Los Alamos National Laboratory
Los Alamos, NM 87545

Dr. Albert Migliori
Group P-10/K764
Los Alamos National Laboratory
Los Alamos, NM 87545

Dr. Alex P. Malozemoff
IBM
Thomas J. Watson Research Center
Yorktown Heights, NY 10598

Dr. M. Brian Maple
Mail Code B-019
University of California/San Diego
La Jolla, CA 92093

Professor Richard M. Martin
Department of Physics
University of Illinois at Urbana
1110 W. Green Street
Urbana, IL 61801

Dr. Richard L. Martin
Group T-12/J569
Los Alamos National Laboratory
Los Alamos, NM 87545

Dr. Don Parkin
CMS/K765
Los Alamos National Laboratory
Los Alamos, NM 87545

Professor David Pines
University of Illinois
Physics Department
1110 West Green Street
Urbana, IL 61801

Dr. Roger Pynn
LANSCE/H805
Los Alamos National Laboratory
Los Alamos, NM 87545

Professor T. Maurice Rice
Physics Department
ETH-Honggerberg
8093 Zurich
SWITZERLAND

Professor Andrei Ruckenstein
Department of Physics
Rutgers University
Piscataway, NJ 08854

Dr. Andrew Millis
AT&T Bell Laboratories
600 Mountain Avenue
Murray Hill, NJ 07974

Dr. Fred Mueller
CMS/K765
Los Alamos National Laboratory
Los Alamos, NM 87545

Professor Nai-Phuan Ong
Princeton University
Dept. of Physics
Jadwin Hall, P.O. Box 708
Princeton, NJ 08544

Professor Dr. Hans Ott
Laboratorium fum Festkorperphysik
ETH-Honggerberg
CH-8093 Zurich
SWITZERLAND

Dr. George A. Sawatzky
Laboratory of Applied Physics
University of Groningen
Nijenborg 18
9747 AG Groningen
THE NETHERLANDS

Professor Douglas Scalapino
Department of Physics
University of California/Santa Barbara
Brodia Hall
Santa Barbara, CA 93106

Dr. James Schirber
Sandia National Laboratory
Dept. 1090
Albuquerque, NM 87815

Professor J. Robert Schrieffer
Inst. for Theoretical Physics
University of California/Santa Barbara
Santa Barbara, CA 93105

Dr. Richard Silver
T-11/B262
Los Alamos National Laboratory
Los Alamos, NM 87545

Dr. Darryl Smith
Group MEE-11/D429
Los Alamos National Laboratory
Los Alamos, NM 87545

Dr. James L. Smith
ERDC/K763
Los Alamos National Laboratory
Los Alamos, NM 87545

Dr. Ellen Stechel
Sandia National Laboratory
Org. 1151
Albuquerque, NM 87185

Dr. Masashi Takigawa
P-10/K764
Los Alamos National Laboratory
Los Alamos, NM 87545

Dr. Zlatko Tesanovic
T-11/B262
Los Alamos National Laboratory
Los Alamos, NM 87545

Professor Paul Wiegmann
Landau Institute for Theoretical Physics
Vorobjevskoe Shosse 2
Moscow, USSR

Professor Anthony Zee
University of California/Santa Barbara
ITP
Santa Barbara, CA 91306

Dr. John M. Tranquada
Physics Department
Brookhaven National Laboratory
Upton, NY 11973

Dr. Stuart Trugman
T-11/B262
Los Alamos National Laboratory
Los Alamos, NM 87545

Dr. Joe Thompson
Group P-10/K764
Los Alamos National Laboratory
Los Alamos, NM 87545

Dr. Chandra M. Varma
AT&T Bell Labs
1D-365
Murray Hill, NJ 07974

Dr. Russell Walstedt
AT&T Bell Labs
Rm. 1D362
Murray Hill, NJ 07974

Dr. Jeffrey Willis
P-10/K764
Los Alamos National Laboratory
Los Alamos, NM 87545

The Addison-Wesley **Advanced Book Program** would like to offer you the opportunity to learn about our new physics and scientific computing titles in advance. To be placed on our mailing list and receive pre-publication notices and special offers, just **fill out this card completely** and return to us, postage paid. Thank you.

Title and Author of this book: _____ **Date purchased:** _____

Name _____
Title _____
School/Company _____
Department _____
Street Address _____
City _____ State _____ Zip _____
Telephone/s(___) _____ (___) _____

Where did you buy/obtain this book?
☐ Bookstore   ☐ Mail Order   ☐ School (Required for Class)
☐ Campus Bookstore   ☐ Toll Free # to Publisher   ☐ Professional Meeting
☐ Other _____   ☐ Publisher's Representative

What professional scientific and engineering associations are you an active member of?
☐ AAPT (Amer Assoc of Physics Teachers)   ☐ APS (Amer Physical Society)   ☐ SPS (Society of Physics Students)
☐ AIP (Amer Institute of Physics)   ☐ Sigma Pi Sigma   ☐ AAAS (Amer Assoc for the Advancement of Science)
☐ Other _____

Check your areas of interest.

⑩ ✔ **Physics**
11 ☐ Quantum Mechanics   18 ☐ Materials Science   25 ☐ Geophysics
12 ☐ Particle/Astro Physics   19 ☐ Biological Physics   26 ☐ Medical Physics
13 ☐ Condensed Matter   20 ☐ High Polymer Physics 27 ☐ Optics
14 ☐ Mathematical Physics   21 ☐ Chemical Physics   28 ☐ Vacuum Physics
15 ☐ Nuclear Physics   22 ☐ Fluid Dynamics
16 ☐ Electron/Atomic Physics   23 ☐ History of Physics
17 ☐ Plasma/Fusion Physics   24 ☐ Statistical Physics
29 ☐ Other _____

Are you more interested in: ☐ theory ☐ experimentation?

Are you currently writing, or planning to write a textbook, research monograph, reference work, or create software in any of the above areas?
☐ Yes   ☐ No
Area: _____

(If Yes) **Are you interested in discussing your project with us?**
☐ Yes   ☐ No

Physics

fold and staple

No Postage
Necessary
if Mailed in the
United States

BUSINESS REPLY MAIL
FIRST CLASS PERMIT NO. 828   REDWOOD CITY, CA   94065

Postage will be paid by Addressee:

ADDISON-WESLEY
PUBLISHING COMPANY, INC.®
Advanced Book Program
390 Bridge Parkway, Suite 202
Redwood City, CA 94065-1522